QUANTUM NANOCHEMISTRY

(A Five-Volume Set)

**Volume IV:
Quantum Solids and Orderability**

QUANTUM NANOCHEMISTRY

(A Five-Volume Set)

Volume IV:
Quantum Solids and Orderability

Mihai V. Putz

Assoc. Prof. Dr. Dr.-Habil. Acad. Math. Chem.
West University of Timişoara,
Laboratory of Structural and Computational Physical-Chemistry
for Nanosciences and QSAR, Department of Biology-Chemistry,
Faculty of Chemistry, Biology, Geography,
Str. Pestalozzi, No. 16, RO-300115, Timişoara, ROMANIA
Tel: +40-256-592638; Fax: +40-256-592620

&

Principal Investigator of First Rank, PI1/CS1
Institute of Research-Development for Electrochemistry
and Condensed Matter (INCEMC) Timisoara,
Str. Aurel Paunescu Podeanu No. 144 ,
RO-300569 Timişoara, ROMANIA
Tel: +40-256-222-119; Fax: +40-256-201-382

E-mail: mv_putz@yahoo.com
URL: www.mvputz.iqstorm.ro

APPLE
ACADEMIC
PRESS

Apple Academic Press Inc.	Apple Academic Press Inc.
3333 Mistwell Crescent	9 Spinnaker Way
Oakville, ON L6L 0A2 Canada	Waretown, NJ 08758 USA

©2016 by Apple Academic Press, Inc.

First issued in paperback 2021

Exclusive worldwide distribution by CRC Press, a member of Taylor & Francis Group
No claim to original U.S. Government works

ISBN 13: 978-1-77463-102-7 (pbk)
ISBN 13: 978-1-77188-136-4 (hbk)

Library and Archives Canada Cataloguing in Publication

Putz, Mihai V., author
Quantum nanochemistry / Mihai V. Putz (Assoc. Prof. Dr. Dr. Habil. Acad. Math. Chem.) West
University of Timişoara, Laboratory of Structural and Computational Physical-Chemistry for
Nanosciences and QSAR, Department of Biology-Chemistry, Faculty of Chemistry, Biology,
Geography, Str. Pestalozzi, No. 16, RO-300115, Timişoara, ROMANIA, Tel: +40-256-592638;
Fax: +40-256-592620, & Institute of Research-Development for Electrochemistry and Condensed
Matter (INCEMC) Timişoara, Str. Aurel Paunescu Podeanu No. 144, RO-300569 Timişoara,
ROMANIA Tel: +40-256-222-119; Fax: +40-256-201-382, E-mail: mv_putz@yahoo.com, URL:
www.mvputz.iqstorm.ro.

Includes bibliographical references and index.
Contents: Volume I: Quantum theory and observability -- Volume II: Quantum atoms and
periodicity -- Volume III: Quantum molecules and reactivity -- Volume IV: Quantum solids and
orderability -- Volume V: Quantum structure–activity relationships (Qu-SAR).
Issued in print and electronic formats.
ISBN 978-1-77188-133-3 (volume 1 : hardcover).--ISBN 978-1-77188-134-0 (volume 2:
hardcover).--ISBN 978-1-77188-135-7 (volume 3 : hardcover).-- ISBN 978-1-77188-136-4
(volume 4 : hardcover).--ISBN 978-1-77188-137-1 (volume 5 : hardcover).--ISBN 978-1-4987-
2953-6 (volume 1 : pdf).--ISBN 978-1-4987-2954-3 (volume 2 : pdf).--ISBN 978-1-4987-2955-0
(volume 3 : pdf).--ISBN 978-1-4987-2956-7 (volume 4 : pdf).--ISBN 978-1-4987-2957-4 (volume 5 : pdf) 1. Quantum chemistry. 2. Nanochemistry. I. Title.

| QD462.P88 2016 | 541'.28 | C2015-908030-4 | C2015-908031-2 |

Library of Congress Cataloging-in-Publication Data

Names: Putz, Mihai V., author.
Title: Quantum nanochemistry / Mihai V. Putz.
Description: Oakville, ON, Canada ; Waretown, NJ, USA : Apple Academic Press,
[2015-2016] | "2015 | Includes bibliographical references and indexes.
Identifiers: LCCN 2015047099| ISBN 9781771881388 (set) | ISBN 1771881380
(set) | ISBN 9781498729536 (set ; eBook) | ISBN 1498729533 (set ; eBook) | ISBN 9781771881333
(v. 1 ; hardcover) | ISBN 177188133X (v. 1 ; hardcover) | ISBN 9781498729536 (v. 1 ; eBook) |
ISBN 1498729533 (v. 1 ; eBook) | ISBN 9781771881340 (v. 2 ; hardcover) | ISBN 1771881348
(v. 2 ; hardcover) | ISBN 9781498729543 (v. 2 ; eBook) | ISBN 1498729541 (v. 2 ; eBook) | ISBN
9781771881357 (v. 3 ; hardcover) | ISBN 1771881356 (v. 3 ; hardcover) | ISBN 9781498729550
(v. 3 ; eBook) | ISBN 149872955X (v. 3 ; eBook) | ISBN 9781771881364 (v. 4 ; hardcover) |
ISBN 1771881364 (v. 4 ; hardcover) | ISBN 9781498729567 (v. 4 ; eBook) | ISBN 1498729568
(v. 4 ; eBook) | ISBN 9781771881371 (v. 5 ; hardcover) | ISBN 1771881372 (v. 5 ; hardcover) |
ISBN 9781498729574 (v. 5 ; eBook) | ISBN 1498729576 (v. 5 ; eBook) Subjects: LCSH: Quan-
tum chemistry. | Chemistry, Physical and theoretical. | Nanochemistry. | Quantum theory. | QSAR
(Biochemistry)
Classification: LCC QD462 .P89 2016 | DDC 541/.28--dc23
LC record available at http://lccn.loc.gov/2015047099

"If someone would have a huge computer, probable would be able to solve the Schrödinger problem for each (solid) and thus obtained interesting physical values... which most likely will be in concordance with the experimental measures, but nothing vast and new could be obtained by such a procedure. Instead, it is preferable a lively picture of the wave functions' behavior as basis for the description of the essence of the (structural) factors... and for understanding the origin in the variation of properties from a (solid) to another."
(Wigner & Seitz, 1955)

To XXI Scholars

CONTENTS

LIST OF ABBREVIATIONS

AFM	atomic force microscopy
AO	atomic orbitals
BCC	body centered cubic
BZ	brillouin zone
CC	cubic compact type
CCF	compact cubic network
CN	coordination number
CNT	carbon nanotubes
CO	crystalline orbitals
CVD	chemical vapor deposition
DFT	density functional theory
DL	direct lattice
DS	dispersion surface
EA	electronic affinity
FCC	face cubic compact
GNR	graphene nanoribbons
HC	compact hexagonal prism
HREM	high resolution electron microscopy
IP	Ionization potentials
LED	light emitting diode
LJ	Lennard-Jones potential
NP	non-primitive
P	primitive
PC	primitive cubic
RL	reciprocal lattice
SC	structural class
SEB	strengths of electrons in bonding
STM	scanning tunneling microscopy
SW	standing waves

SW	Stone-Wales rotation
SWW	Stone-Wales waves
TEM	transmission electron microscopy
VB	valence band
WSC	Wigner-Seitz unit

PREFACE TO FIVE-VOLUME SET

Dear Scholars (Student, Researcher, Colleague),

I am honored to introduce *Quantum Nanochemistry*, a handbook comprised of the following five volumes:

Volume I: Quantum Theory and Observability
Volume II: Quantum Atoms and Periodicity
Volume III: Quantum Molecules and Reactivity
Volume IV: Quantum Solids and Orderability
Volume V: Quantum Structure–Activity Relationships (Qu-SAR)

This treatise, a compilation of my lecture notes for graduates, post-graduates and doctoral students in physical and chemical sciences as well as my own post-doctoral research, will serve the scientific community seeking information in basic quantum chemistry environments: from the fundamental quantum theories to atoms, molecules, solids and cells (chemical–biological/ligand–substrate/ligand–receptor interactions); and will also creatively explain the quantum level concepts such as observability, periodicity, reactivity, orderability, and activity explicitly.

The book adopts a three-way approach to explain the main principles governing the electronic world:

- firstly, *the introductory principles* of quantumchemistry are stated;
- then, they are analyzed as *primary concepts* employed to understand the microscopic nature of objects;
- finally, they are explained through *basic analytical equations* controlling the observed or measured electronic object.

It explains the first principles of quantum chemistry, which includes quantum mechanics, quantum atom and periodicity, quantum molecule and reactivity, through two levels:

- *fundamental* (or *universal*) character of matter in isolated and interacting states; and
- the primary concepts elaborated for a beginner as well as an advanced researcher in quantum chemistry.

Each volume tells the "story of quantum chemical structures" from different viewpoints offering new insight to some current quantum paradoxes.

- The **first volume** covers the concepts of nuclear, atomic, molecular and solids on the basis of quantum principles—from Planck, Bohr, Einstein, Schrödinger, Hartree–Fock, up to Feynman Path Integral approaches;
- The **second volume** details an atom's quantum structure, its diverse analytical predictions through reviews and an in-depth analysis of atomic periodicities, atomic radii, ionization potential, electron affinity, electronegativity and chemical hardness. Additionally, it also discusses the assessment of electrophilicity and chemical action as the prime global reactivity indices while judging chemical reactivity through associated principles;
- The **third volume** highlights chemical reactivity through molecular structure, chemical bonding (introducing bondons as the quantum bosonic particles of the chemical field), localization from Hückel to Density Functional expositions, especially how chemical principles of electronegativity and chemical hardness decide the global chemical reactivity and interaction;
- The **fourth volume** addresses the electronic order problems in the solid state viewed as a huge molecule in special quantum states; and
- The **fifth volume** reveals the quantum implication to bio-organic and bio-inorganic systems, enzyme kinetics and to pharmacophore binding sites of chemical–biological interaction of molecules through cell membranes in targeting specific bindings modeled by celebrated QSARs (Quantitative Structure–Activity Relationships) renamed here as Qu–SAR (Quantum Structure–Activity Relationships).

Thus, the five-volume set attempts, for the first time ever, to unify the introductory principles, the primary concepts and the basic analytical equations against a background of quantum chemical bonds and interactions (short, medium and long), structures of matter and their properties: periodicity of atoms, reactivity of molecules, orderability of solids, and activity of cells (through an advanced multi-layered quantum structure–activity unifying concepts and algorithms), and observability measured throughout all the introduced and computed quantities (Figure 0.0).

It provides a fresh perspective to the "quantum story" of electronic matter, collecting and collating both research and theoretical exposition the "gold" knowledge of the quantum chemistry principles.

The book serves as an excellent reference to undergraduate, graduate (Masters and PhDs) and post-doctoral students of physical and chemical sciences; for it not only provides basics and essentials of applied quantum theory, but also leads to unexplored areas of quantum science for future research and development. Yet another novelty of the book set is the intelligent unification of the quantum principles of atoms, molecules, solids and cells through the qualitative–quantitative principles underlying the observed quantum phenomena. This is achieved through unitary analytical

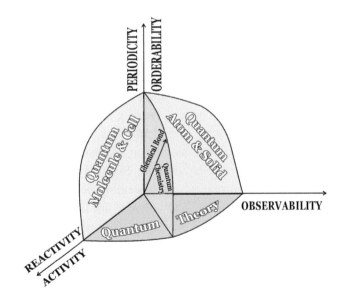

FIGURE 0.0 The featured concepts of the "First Principles of Quantum Chemistry" five-volume handbook as placed in the paradigmatic chemical orthogonal space of atoms and molecules.

exposition of the quantum principles ranging from quanta's nature (either as ondulatory and corpuscular manifestations) to wave function, path integral and electron density tools.

The modern quantum theories are reviewed mindful of their implications to quantum chemistry. Atomic, molecular, solid-state structures along cell/biological activity are analytically characterized. Major quantum aspects of the atomic, molecular, solid and cellular structure, properties/activity features, conceptual and quantitative correlations are unitarily reviewed at basic and advanced physical-chemistry levels of comprehension.

Unlike other available textbooks that are written as monographs displaying the chapters as themes of interests, this book narrates the "story of quantum chemistry" as *an extended review paper*, where theoretical and instructional concepts are appropriately combined with the relevant schemes of quantization of electronic structures, through path integrals, Bohmian, or chemical reactivity indices. The writing style is direct, concise and appealing; wherever appropriate physical, chemical and even philosophical insights are provided to explain quantum chemistry at large.

The author uses his rich university teaching experience of 15 years in physical chemistry at West University of Timisoara, Romania, along with his research expertise in treating chemical bond and bonding through conceptual and analytical quantum mechanical methods to explain the concepts. He has been a regular contributor to many physical-chemical international journals (*Phys Rev, J Phys Chem, Theor Acc Chem, Int J Quantum Chem, J Comp Chem, J Theor Comp Chem, Int J Mol Sci, Molecules, Struct Bond, Struct Chem, J Math Chem, MATCH*, etc.).

In a nutshell, the book amalgamates an analysis of the earlier works of great professors such as Sommerfeld, Slater, Landau and Feynman in a methodological, informative and epistemological way with practical and computational applications. The volumes are layered such that each can be used either individually or in combination with the other volumes. For instance, each volume reviews quantum chemistry from its level: as quantum formalisms in Volume I, as atomic structure and properties in Volume II, as detailed molecular bonding in Volume III, as crystal/solid state (electronic) in Volume IV, and as pharmacophore activity targeting specific bindings in Volume V.

To the best of my knowledge, such a collection does not exist currently in curricula and may not appear soon as many authors prefer to publish well-specialized monographs in their particular field of expertise. This multiple volumes' work, thus, assists academic and research community as a complete basic reference of conceptual and illustrative value.

I wish to acknowledge, with sincerity, the quantum flaws that myself and many researchers and professors make due to stressed delivery of papers using computational programs and software to report and interpret results based on inter-correlation. I feel, therefore, the need of a new comprehensive quantum chemistry reference approach and the present five-volume set fills the gap:

- *Undergraduate students* may use this work as an *introductory and training textbook* in the quantum structure of matter, for basic course(s) in physics and chemistry at college and university;
- *Graduate (Master and Doctoral) students* may use this work as the *recipe book* for analytical research on quantum assessments of electronic properties of matter in the view of chemical reactivity characterization and prediction;
- *University professors and tutors* may use this work as a *reference textbook* to plan their lectures and seminars in quantum chemistry at undergraduate or graduate level;
- *Research (Academic and Institutes) media* may use this work as a *reference monograph* for their results as it contains many tables and original results, published for the first time, on the atomic-molecular quantum energies, atomic radii and reactivity indices (e.g., electronegativity, chemical hardness, ionization and electron affinity results). It also has a collection of original, special and generally recommended literature, integrated results about quantum structure and properties.
- *Industry media* may use this work as a *working tool book* while assessing envisaged theoretical chemical structures or reactions (atoms-in-molecule, atoms-in-nanosystems), including molecular modeling for pharmaceutical purposes, following the presented examples, or simulating the physical–chemical properties before live production;

- *General media* may use this work as an *information book* to get acquainted with the main and actual quantum paradigms of matter's electronic structures and in understanding and predicting the chemical combinations (involving electrons, atoms and molecules) of Nature, because of its educative presentation.

I hope the academia shares the same enthusiasm for my work as the author while writing it and the professionalism and exquisite cooperation of the Apple Academic Press in publishing it.

Yours Sincerely,

Mihai V. Putz,
Assoc. Prof. Dr. Dr.-Habil. Acad. Math. Chem.
West University of Timişoara
& R&D National Institute for Electrochemistry and Condensed Matter Timişoara
(Romania)

ABOUT THE AUTHOR

Mihai V. PUTZ is a laureate in physics (1997), with an MS degree in spectroscopy (1999), and PhD degree in chemistry (2002), with many post-doctorate stages: in chemistry (2002-2003) and in physics (2004, 2010, 2011) at the University of Calabria, Italy, and Free University of Berlin, Germany, respectively. He is currently Associate Professor of theoretical and computational physical chemistry at West University of Timisoara, Romania. He has made valuable contributions in computational, quantum, and physical chemistry through seminal works that appeared in many international journals. He is Editor-in-Chief of the *International Journal of Chemical Modeling* (at NOVA Science Inc.) and the *New Frontiers in Chemistry* (at West University of Timisoara). He is member of many professional societies and has received several national and international awards from the Romanian National Authority of Scientific Research (2008), the German Academic Exchange Service DAAD (2000, 2004, 2011), and the Center of International Cooperation of Free University Berlin (2010). He is the leader of the Laboratory of Computational and Structural Physical Chemistry for Nanosciences and QSAR at Biology-Chemistry Department of West University of Timisoara, Romania, where he conducts research in the fundamental and applicative fields of quantum physical-chemistry and QSAR. In 2010 Mihai V. Putz was declared through a national competition the Best Researcher of Romania, while in 2013 he was recognized among the first Dr.-Habil. in Chemistry in Romania. In 2013 he was appointed Scientific Director of the newly founded Laboratory of Structural and Computational Physical Chemistry for Nanosciences and QSAR in his alma mater of West University of Timisoara, while from 2014, he was recognized by the Romanian Ministry of Research as Principal Investigator of first rank/degree (PI1/CS1) at National Institute for Electrochemistry and Condensed Matter (INCEMC) Timisoara. He is also a full member of International Academy of Mathematical Chemistry.

FOREWORD TO *VOLUME IV: QUANTUM SOLIDS AND ORDERABILITY*

Quantum Solid and Orderability offers a major survey of quantum theory of solids and crystallographic properties based on Prof. Putz's comprehensive expertise on these topics, consolidated through many years of original scientific research and successful teaching experiences at West University of Timisoara – not a marginal aspect for this valuable text book intended for graduate students in Chemistry and Physics and other professionals in various scientific disciplines.

Personally, I was impressed by the way the book deals with crystallography; four out of five chapters (from Chapter 2 on *Geometrical Crystallography* to Chapter 5 on *X-Ray Crystallography*) are related to that subject. Going back 20 years or more, when I was a research student at the Structural Chemistry Department at Parma University, I still remember the *vivid reshuffle* crystallography induced on my fresh knowledge of quantum mechanics, opening the door to the real (observable) world that – undoubtedly – requires much more than the capacity of describing details of the electronic levels in a C_{60}^{+} molecule or Compton-like effects. Solid aggregates of atoms unfold their feeling for geometry by mimicking, somehow surprisingly, all 230 groups of symmetry, which are *mathematically possible* in 3D spaces, creating beautiful natural structures. Silicon and Aluminum sp^3 atoms, for example, fully exploit space symmetries in minerals like Faujasite, a relevant representative of the Zeolite family belonging to the Tectosilicates group treated in Chapter 4 on *Chemical Crystallography*, built with 192 symmetry operators – the maximum number one can apply in 3D spaces – to produce a porous silicon mesh consisting of sodalite cages (the truncated octahedron depicted by Leonardo da Vinci under the name of *octocedron abscisus vacuus*) interconnected through hexagonal prisms. Directly from the pages of Luca Pacioli's book

'De Divina Proportione' right in the middle of a framework of covalent chemical bonds used as molecular sieve! Only Science I think is able to produce such connections between different times and different length-scales, a short-circuit that eventually generates new Science; surely, readers will find a source of inspiration in the multiple sections of this book dealing with crystallography, quantum mechanics and other topics.

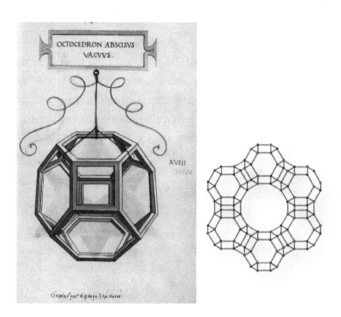

These extraordinary facts are probably hard to explain as a consequence of sole quantum mechanics; nevertheless the author devotes Chapter 3 on *Quantum Roots of Crystals and Solids* to a clear exposition of the quantum mechanical principal elements forming the conceptual ground for the crystalline status and for the diverse behaviors of the electrons in the solids. The task of "unfolding the crystal and solid state quantum paradigm" gets here originally fulfilled.

Finally, let me jump back to Chapter 1 on *Bondons on Graphenic Nanoribbons with Topological Defects* in which novel features of the bosonic particle associated with the chemical bonding field – *the bondon* – recently created by Prof. Putz to fulfill the particle–wave duality (another basal symmetry present in Nature) are disclosed. The existence of these new particles induces various honeycomb defective nanosystems modified by sequel of Stone–Wales rotations, typical phase-transition showing typical bondonic fast critical time and bonding energies. Moreover, Chapter 1 displays the (computational) power of the topological lattice descriptors in characterizing complex systems by capturing the essence of the information stored in the long-range connectivity properties of the atomic networks. Having Prof. Putz studies on those systems, it is a real pleasure to see that Topological Modeling finds such fertile applications in solids, gaining already some space in this modern five-volume set on *Quantum Nanochemistry*.

Every section of the book has interesting examples; I strongly recommend the volume to anyone searching for an updated and complete overview of quantum nanochemistry of solids.

Ottorino ORI
Actinium Chemical Research, Rome
Editor Assistant of *Fullerenes, Nanotubes and Carbon Nanostructures*
(Journal at Taylor & Francis)
November 2015

PREFACE TO *VOLUME IV:*
QUANTUM SOLIDS AND
ORDERABILITY

THE SCIENTIFIC PREMISES

One of the fundamental chemical bonds of matter is the metallic bond, which is specific to the solid-state bodies and occurs by joining of atoms in the compact structures (clusters or crystals).

The metallic bond differs from the bonds between the atoms in a molecule; the former requires a large number of atoms whereas for the latter even two atoms are enough.

What are metallic bonds and how do they compare or relate to with the ionic and the covalent bond? Metallic bond is an unsaturated covalent bond, i.e., the valence electrons are very weakly bound to the atoms from the metallic bond. In other words, the waves associated with these electrons are much diffused (spread) within the bond space and not concentrated between the bound atoms as in a covalent bond. The dispersed waves of the metallic bond account for its increased electronic conductibility.

This can be explained by the Uncertainty Principle (Heisenberg principle), which says that if Δx is the uncertainty in a particle spatial location (i.e., also the electron), it is inversely proportional to the uncertainty in determining the particle momentum (or velocity) of ΔP. Thus, Δx is quite large due to the greatly diffused wave associated with the valence electrons of the metallic bond and results in a decreased ΔP (from the uncertainty principle), i.e., in the kinetic energy (energy of motion) of valence electrons.

The decreased kinetic energy of the electrons in the metallic bonds reduces the possibility of disordered electron movements, facilitating the transport of electrical charges and heat. It is these features of electrical and thermal conductivity that make the crystalline metallic bond differ from the ionic bond in crystals (e.g., structure of NaCl).

As the electrons valence number of an element or a metallic bond increases, their connection with the core or with the core system to which they belong also increases, resulting in more precise localization and enhanced covalent character of the possible or formed bonds.

For instance, transition metals, which have characteristics of both covalent and metallic bond:

$Fe(1s^2 2s^2 2p^6 3s^2 3p^6 3d^6 4s^2)$

$Co(1s^2 2s^2 2p^6 3s^2 3p^6 3d^7 4s^2)$

$Ni(1s^2 2s^2 2p^6 3s^2 3p^6 3d^8 4s^2), ...$

The transition from the covalent bond to the metallic one is most obvious in the fourth group elements: C is completely inclined to make covalent bonds (see the peptide chain, for example), while Si and Ge favor the metallic character of the formed bonds; Sn appears in two crystalline forms – one covalently dominant and the other metallically dominant, while and Pb makes almost completely metallic bond.

But, how is a metallic structure formed?

Let us consider, for example, the potassium atom. Each potassium atom has one electron non-coupled in the orbital 4s on the valence layer.

When two potassium atoms unite, the two non-coupled electrons in the isolated potassium atoms form a bond with two electrons having opposite spin directions (moments) following the Pauli's Exclusion Principle.

However, the principle prohibits the eventuality that a third potassium atom could possibly contribute its valence electron to the bond orbital $(4s^2)$ of the first two atoms bond because it is already complete.

The third electron, coming from the third potassium atom, will occupy (and will also form) the sub-level 4p, energetically very close to the 4s state, thus not coming into conflict with the Pauli's Exclusion Principle.

Successive addition of the potassium atoms results in a potassium crystal, with a centered cubical structure, where each potassium atom is surrounded by other 8(!) potassium atoms. The structure has a network of positive potassium ions immersed in a quasi-bound "sea of electrons" from the formed metallic bond.

Atomic joining in the metallic bond unifies the energy levels of the constituent atoms, generating energy bands (many energy levels, very close as energy, even overlapped, actually united) in the solid form.

Two fundamental energy bands are formed in solid bodies: *valence band* (the band of the electrons quasi-bounded to the atoms from which they come) and *conduction band* (the free electron band).

The energy difference between these energy bands in a solid determines the type of solid to be formed. Thus, a solid is an *insulator* if the energy bands are widely separated, is a *semiconductor* when the bands are separate but close enough to allow the electrons jump from the valence band into the conduction one at thermal excitements over 0K, and a *conductor* when the valence band and the conduction one have an overlapping area (Figure P.1).

Fermi level is an important feature in the analysis of a solid's chemical and physical properties. It corresponds to the energy beyond which the electrons of the valence band cannot pass at 0K. Above the absolute zero temperature, a certain fraction of the valence band electrons can pass (being thermally excited) into the conduction band and populate it based on a probabilistic law called the *Fermi function* or the *Fermi–Dirac* statistics (Figure P.2).

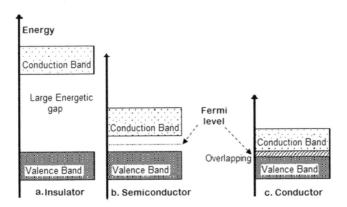

FIGURE P.1　The energetic bands schematization in solids and their classification according to the energetic width between these bands.[1,2]

[1]HyperPhysics (2010). http://hyperphysics.phy-astr.gsu.edu/Hbase/hframe.html.

[2]Putz, M. V. (2006). *The Structure of Quantum Nanosystems*, West University of Timişoara Publishing House, Timişoara.

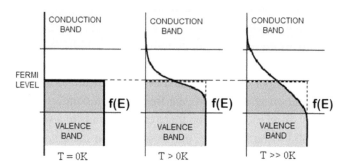

FIGURE P.2 Representation of the Fermi Function $f(E)$, the bold line, along its population with electrons of the valence and conduction bands (the under-curve area) for the temperatures 0K, above 0K and far above 0K, from the left to the right, respectively.[3,4]

In fact, the electronic bands in solid are not regular (Figure. P.2) and that is why the Fermi level is an indicator to separates these bands (Figure P.3 – right).

As can be seen in Figure P.3, fullerenes—the polypeptide chains are the fundamental unit of the modern chemistry because they are based on carbon bonds and they impact the biological structures in combination with metals to generate a real "carbon chemistry".

The discovery of fullerenes has an astronomical root too. Sir Harry W. Kroto in 1985 investigated if the spectrometrically detected (emitted and recorded radiations) carbon chains coming from the red giant stars can be obtained in laboratory on Earth.

Robert Curl, Harold Kroto and Richard Smalley carried out an experiment in autumn 1985 in Houston (USA) and observed that *Carbon can also exist in very stable spherical forms* (Figure P.4). Curl, Kroto and Smalley, together with J. R. Heat and J. C. O'Brien, used graphite to achieve carbon clusters containing mainly 60 or 70 atoms of carbon, which were quite stable.[5–9]

[3]HyperPhysics (2010). http://hyperphysics.phy-astr.gsu.edu/Hbase/hframe.html.

[4]Putz, M. V. (2006). *The Structure of Quantum Nanosystems*, West University of Timişoara Publishing House, Timişoara.

[5]Aldersey-Willliams, H. (1995). The Most Beautiful Molecule: An Adventure in Chemistry, Aurum Press, London.

[6]Baggott, J. (1994). Perfect Symmetry: The Accidental Discovery of Buckminsterfullerene, Oxford University Press.

[7]Hargittai, I. (1995). Discoverers of Buckminsterfullerene. The Chemical Intelligencer 1(3), 6–26.

[8]Kroto, H. W., Allaf, A. W., Balm, S. P. (1991). C60: buckminsterfullerene. Chem. Rev. 91:1213–35.

[9]Smalley, R. E. (1991). Great Balls of Carbon; The story of buckminsterfullerene. The Sciences March/April, 22–28.

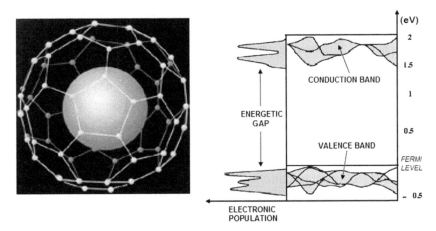

FIGURE P.3   Left: cage cluster C_{60}, "Buckminsterfullerene", containing a metal; right: the density of real electronic population and the energetic levels in the conduction and valence bands for the cluster C_{60}.[10,11]

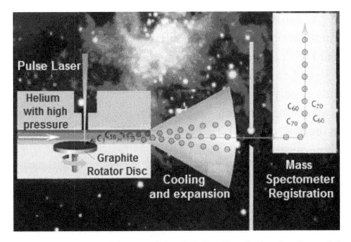

FIGURE P.4   The drawing of the Dr. Richard E. Smalley device that formed the clusters Carbon (C_{60} and C_{70}) by the evaporation of the graphite under the laser pulses and the Helium jet pression.[12]

---

[10]Golden, M., Bernard, D. (2002). *Lectures Notes on Condensed Matter Science*, Master program at University of Amsterdam and Free University of Amsterdam.

[11]Putz, M. V. (2006). *The Structure of Quantum Nanosystems*, West University of Timişoara Publishing House, Timişoara.

[12]The Nobel Prize in Chemistry (1996). Nobelprize.org. Nobel Media AB 2013. Web. 29 Mar 2014. http://www.nobelprize.org/nobel_prizes/chemistry/laureates/1996/index.html

This leads to an interesting question: do carbon atoms exist in forms other than graphite and diamond? The experimental evidence confirms affirmatively. The next question springs up, "What is the constitution of these clusters?"

The discovery of the structure of these new compounds was quite accidental. The three researchers were inspired by the building of the architect *Buckminster Fuller* for the Montreal Expo67 dome projection (Figure P.5 – right), which contained hexagons and a small number of pentagons to represent the curved space.

The C_{60} cluster structure was assumed comprising 12 pentagons and 20 hexagons, similar to a soccer ball (Figure P.5 – left), and it was named "buckminsterfullerene". In 1996, Robert F. Curl, Jr. (Rice University, Houston, USA), Sir Harry W. Kroto FRS (University of Sussex, Brighton, UK), and Richard E. Smalley (Rice University, Houston, USA) received the Nobel Prize in Chemistry for discovering the fullerenes.[13]

The cluster C_{60}, with a density of 1.72 g/cm^3 and a molecular diameter of about 7Å, can easily accept electrons and become a negative ion. It allows introduction of metals or noble gases in the cage formed (Figure P.3 – left) to form new superconducting materials (the conduction band in Figure P.3-right is sufficiently electron populated) or create new organic compounds or polymeric matter.

With alkali metals such as potassium, C_{60} can form a crystalline superconducting material (leads around 19K) built with the triple

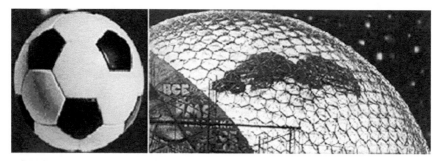

FIGURE P.5   Left: an European soccer ball; right: the dome of the Mondial exhibition of Montreal "Expo67" as projected by the architect Buckminster Fuller.[13]

---

[13]The Nobel Prize in Chemistry (1996). Nobelprize.org. Nobel Media AB 2013. Web. 29 Mar 2014. http://www.nobelprize.org/nobel_prizes/chemistry/laureates/1996/index.html

negative ion C_{60} and with three positive potassium ions: K_3C_{60} (Figure P.6, above).

The ability of C_{60} to accept and donate electrons in a reversible way makes it an ideal catalyst in chemical processes where otherwise precious and/or toxic metals must be used. By modifying the synthesis methods of fullerenes, one can obtain the smallest tubes ever, the nanotubes ($1nm = 10^{-9}m$) in a circular or linear fashion (Figure P.6 – down) with unique electrical and mechanical properties that find use in the electronics industry (high tech).

The last decade saw synthesis of more than a thousand fullerene-based new compounds, with more than one hundred patents approved, but their industrial-scale production was still an expensive affair.

The investigation methods adopted play an important role in the study of metallic bonds and the metal-ligand interactions (as illustrated in the above-mentioned K_3C_{60}).[14]

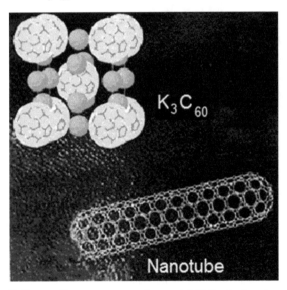

FIGURE P.6 Fullerene C_{60} type compounds: crystal K_3C_{60} (up) and a sample from a nanotube (down).[15]

[14]Russo, N., Salahub, D. R. (Eds.) (2000). *Metal-Ligand Interactions in Chemistry*, Physics and Biology, NATO Science Series, Series C: Mathematical and Physical Sciences, Kluwer Academic Publishers, Dordrecht, Vol. 546.

[15]The Nobel Prize in Chemistry (1996). Nobelprize.org. Nobel Media AB 2013. Web. 29 Mar 2014. http://www.nobelprize.org/nobel_prizes/chemistry/laureates/1996/index.html.

X-ray analysis (using matter radiation) is one of the most widely used methods to investigate the structure and properties of matters. X-rays have very small wavelength (the spatial interval between two locations having the same intensity and amplitude) of the Ångström order ($1Å=10^{-10}$ m, about a thousand million times smaller than a meter), which makes it highly penetrable and a unique tool to investigate the molecular details at the atomic level.

X-rays are also emitted by some specific elements (Cu, Ag, ...) on disintegration/decay of deeper electronic levels called K lines (Figure P.7)

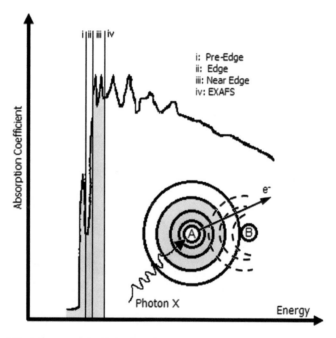

FIGURE P.7　The representation of various regions in the X-ray absorption spectrum; the scheme of EXAFS formation by the interference of the wave associated to the photoelectron emitted by the atom A under the influence of the photon X (the concentric circles in continuous line) with the wave back scattered, coming from the neighboring atoms (atom B for example) collided by the photoelectron there is also rendered. On the abscissa the energy of the incident photons X, is represented in eV.[16]

[16]Abruna, D. H., White, J. H., Albarelli, M. J., Bommarito, G. M., Bedzyk, M. J., McMillan, M. (1988). Is there any beam yet? Uses of synchrotron radiation in the in situ study of electrochemical interfaces. *J. Phys. Chem.* 92(25), 7045–7052.

when excited through energy (by bombarding the material – mounted as anode – with electrons emitted by the cathode incandescence and then accelerated through an appropriate potential difference between cathode and anode).

Advances in synchrotron era (emergence of huge facilities, circular or linear, wherein the electrons can be accelerated at controlled speed to emit electromagnetic radiation of varying types) have widened the applications and possibilities of investigating the matter structure using X-rays. Consequently, the number of experiments that can be done in situ (in its original position) and in electrochemical interface include EXAFS (Extended X-Ray Absorption Fine Structure: the structure determination by the extended, i.e., deep X-ray absorption), XSW (X-ray Standing Waves: standing X-rays, i.e., resonant) besides the increased diffraction on surfaces. The energy of the incident photons X on the abscissa is represented in eV.

EXAFS refers to modulation (the spectral shape) of the X-ray absorbed beyond an absorption edge (a K-line for example) and is manifested about 15% lower than the one near to the edge jump (the exact absorption of the K line, for example) in an energy range covering approx. 1000 eV after the edge (the full absorption spectrum lies between 200 and 3500 eV) (Figure P.7).

Experimentally, EXAFS involves measuring the absorption coefficient of the investigated material or of a parameter correlated with it, depending on the energy of the incident photon X.

The absorption coefficient is a measure of the probability that the incident photon X be absorbed by the atom under investigation and, therefore, depends on the initial and final state of the colliding electron.

The initial electron state is determined by the energy level of its atom. The final state is represented by the ionized electron (expelled, called photoelectron) represented as an emerging wave (which is leaving) from the center of the absorber atom (atom A in Figure P.7). A few free photoelectrons may be scattered near the ionized atom from other neighboring atoms (as atoms B in Figure P.7), but not further than 5Å of the ionized atom because EXAFS are sensitive only to very short distances.

Oscillations called EXAFS interference are the waves associated with the released photoelectrons by the atom A (Figure P.7) with the one of the photoelectrons backscattered by the atom B (Figure P.7).

Structural information about the atom under investigation is collected through frequency and amplitude of the oscillation. The frequency of these oscillations depend on the distance between the atom absorbed by X photons and its closest neighbors, *neighbors of the first order*, while the oscillation amplitude provides information about *the number and the type of these neighbors*.

Remarkably, this type of X-ray analysis, EXAFS, can be applied to any form of matter to provide information on distances, number and the type of constituent atoms.

The previous example involves approximation of single-electron scattering (X-ray irradiation results in a single photoelectron) with the resultant photoelectron having energy high enough to be analytically represented by plane waves (with the oscillation and amplitude ranging from $-\infty$ to $+\infty$).

For this reason, the EXAFS analysis is suitable for atoms with energies higher than 50eV for the absorption edge in the absorption spectrum (Figure P.7). Also other absorption regions in the X-ray absorption spectrum give very important information about the structure (Figure P.7).

For example, the *pre-edge region* records maximum absorption (also called peaks, or local absorption peaks) corresponding to the excitations in bounded states (orbital states allowed in atoms and between the atoms of the irradiated material), which give information on the orbitals' location, the symmetry structure, and the electronic configuration.

The *edge region* provides information on the effective charge of the absorbing atom. The atom position and charge can be correlated to determine the oxidation state of the absorber.

Finally, in the *near edge region* (known as XANES: *X-Ray Absorption Near-Edge Structure*, i.e., the structure determination by X-ray absorption near the absorption edge) the wave associated with the photoelectron has a very weak impulse, which invalidates the single-electronic collision and the approximation in plane waves. Therefore, in this region, the approximation in spherical waves (the waves which decrease once with propagation, from 0 up to infinity) as well as the multiple scattering can be analyzed. This implies that the diffusion of electrons is greater around the emitting atom, thus registering very rich structural information about the investigated material in the absorption spectrum.

X-rays analysis is greatly used in electrochemistry surfaces studies as well. It can be used to analyze deposition of a metallic layer on an electrode from a different material under a potential difference (typically of hundreds of millivolts).

These techniques use differential surface deposition to study the material under examination. It is called Under Potential Deposition technique (UDP) when there is deposition of a layer from fractions to a whole one; it is referred to as Bulk Deposition (structure deposition, i.e., of more layers) technique when there are more overlapping deposited layers, usually under high potential differences.

Examples of such studied systems are Cu/Au(111), Ag/Au(111), Pb/Ag(111), Cu/Pt(111)/I. The fluorescence spectrum of Cu/Pt(111)/I detected in situ by X-ray absorption (Figure P.8) shows a half of copper layer on the face (111) – the face that unites three opposite diagonal corners in a cube, while the monocrystal Pt is used as an electrode within iodine vapor.

Five well-defined oscillations are clearly distinguished in the EXAFS spectrum region (after the A region, Figure P.8). Data analysis (by the Fourier transformation of the EXAFS frequencies) indicates that the Cu–Cu distance was 2.85 Å, very close to the Pt–Pt distance in (111), suggesting that Cu was occupying the gaps with trigonal symmetry (in multiples of 120° at rotation) on the Pt surface. Moreover, one of the oscillations identified as many as six closest neighbors of Cu.

The analysis suggests that the hemi-layer of Cu covering the surface of Pt (111) is rather represented by an ordered Cu atoms cluster (Figure P.8 – B) than any random deposition by Cu atoms, occupying widely separated positions on the surface of Pt (Figure P.8 – C).

Another example of the X-ray spectral information, this time from catalysis, is the hydrogenation reaction of *cinnamaldehyde* (CAL) :

$$\langle\text{benzene ring}\rangle - CH = CH - CHO$$

and its transformation to *cinnamyl alcohol* (COL):

$$\langle\text{benzene ring}\rangle - CH = CH - CH_2OH$$

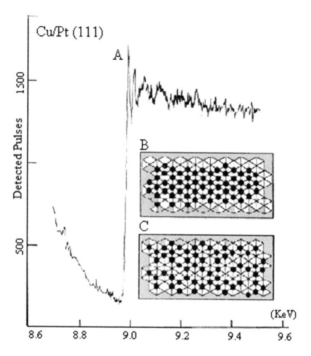

FIGURE P.8   The fluorescence spectrum detected for the in situ X-ray absorption by the atoms of Cu deposited on the face (111) of the Pt electrode, and the structural models investigated (B and C); after there is also rendered. On the abscissa the energy of the incident photons X is represented in eV.[17]

Figure P.9 (left) shows the conditions under which a catalyst, for example Ru/ZrO$_2$, can streamline the reaction CAL $\rightarrow$ COL, the diffracted X-ray (scattered) spectra by this catalyzer in amorphous state (solid but non-periodic, ordered only at short distances) and in crystalline state (solid and periodic, ordered at distance), both reduced by Ru deposition on the surface of ZrO$_2$, at various temperatures.

One notes how the deposited Ru atoms' contribution (grown on the substrate of ZrO$_2$) arranged in the plane (111) appears only in the spectra for the *Ru/ZrO$_2$ crystal* reduced at 973K.

[17]Abruna, D. H., White, J. H., Albarelli, M. J., Bommarito, G. M., Bedzyk, M. J., McMillan, M. (1988). Is there any beam yet? Uses of synchrotron radiation in the in situ study of electrochemical interfaces. *J. Phys. Chem.* 92(25), 7045–7052

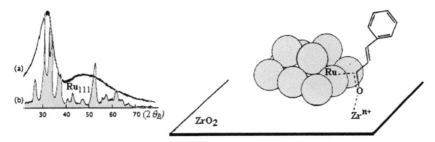

FIGURE P.9 Left: the (scattering) diffraction spectrum at angle $2\theta_B$, with $\theta_{B \text{ the}}$ Bragg angle, of the X-ray by the catalysts (a) – Ru623/ZrO$_2$ (amorphous) and (b) – Ru973/ZrO$_2$ (crystal), right: the cinnamaldehyde hydrogenation by "attaching" the carbonyl group to the Zr^{n+1} locations from the catalytic substrate.[18]

This system is typical for any metal–ligand interaction. The interaction between Ru and ZrO$_2$ can change the epitaxial growth of the metallic particles from the increased rounded edges to the increased planes, which allow the selection of the planes with low crystallographic index, such as the (111), as being those specific to the efficiency of the COL production (Figure P.9. – right).[18]

Turning to the biomolecules register, the metal–ligand interactions are transposed into properties that the biomolecules acquire in the presence of metallic complexes. For instance, the peptide chain may form a ligand for the functional metal groups, transporting them through the cellular membranes, or towards the DNA/RNA, aiming to create new functionality in the protein structures that it generates.

Thus, the metal insertions in biopolymers act as real noncarcinogenic agents or stimulate the molecular recognition (coupling) at the biological level. Significantly, the peptide ligands allow the formation of the metalopeptides by the regular/ordered insertion of metallic complexes (Figure P.10), thus playing an important role in the formation of anti-tumor proteins and the selective recognition (the mixing) of the sequences of DNA and RNA.

Moreover, the presence of metallic centers in biopolymers give them the ability to participate in oxidation–reduction reactions (redox), a property

[18]Coq, B., Kumbhar, P. S., Moreau, C., Moreau, P., Figueras, F. (1994) Zirconia-supported monometallic Ru and bimetallic Ru–Sn, Ru–Fe catalysts: role of metal support interaction in the hydrogenation of cinnamaldehyde. *J. Phys. Chem.* 98(40), 10180–10188

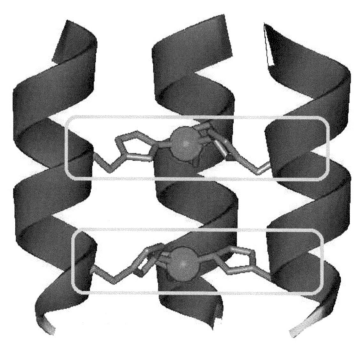

FIGURE P.10   The prototype of the metallic insertion (spherical representation) in polypeptides chains of the proteins structure aiming in forming the metalopeptides.[19]

absent in such structures without metal insertions, which improves the consequences of anti-tumor genetic combinations, selection, and treatment.

Finally, one needs to discuss the determination of protein structures. The human body, for example, contains different (thousands) proteins performing various roles and metabolic functions, all vital for life.

Yet, the functions of these proteins are determined by their structure, i.e., by the molecular composition and the 3D spatial stereochemistry they present. Therefore, the proteins structure determination represents a fundamental step in understanding the metabolism of the living bodies and to design (synthesize) new proteins with a decisive role in regulating or correcting the metabolic cycle.

However, the only complete method that could indicate the entire molecular structure of a protein is the crystallographic one – using the

[19]Purves, W. K., Orians, G. H., Heller, H. C. (2001). *Life: The Science of Biology*, 6th Edition, by Sinauer Associates, Sunderland (MA), and WH Freeman, New York.

data given by X-ray diffraction. To investigate a protein structure with the X-ray method, it should be first crystallized and requires working with the "freezing" bonds and structure.

This is the most difficult step, experimentally, which involves the optimization growth process of the protein crystals. Also so because the crystal should display an internal long-range (at distance) order, i.e., to be periodical, while any disorder generated in the X-ray spectrum due to a mixture of intensities from different structural components cannot be accurately individualized and interpreted in the absence of the structural order. NASA (The National Aeronautics and Space Administration, USA) launched a program in 1985 to grow proteins in space by using the spatial launched satellites developing micro-gravity conditions specific to the cosmic space, thus increasing the precision for the PCG (Protein Crystal Growth) experiments.

After the protein crystal growth, it is exposed to X-ray diffraction where thousands of diffraction spots arranged in a geometry associated to the spatial structure geometry of the analyzed protein is recorded. The interpretation of these spots is performed by applying the specific geometrical

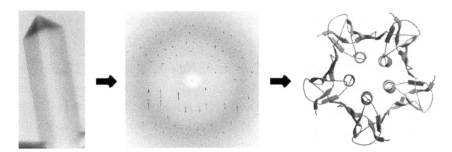

FIGURE P.11 The illustration of the protein structure determination: protein crystallization (left), followed by X-ray experiments with the diffraction figure recording (middle: each spot came from a series of atoms or groups of atoms, identically arranged or in precise crystallographic planes), towards the final interpretation (by geometrical measurements on the diffraction figure) and the determination of the protein basic structure (right) through identifying the structure which obeys the repeatedly crystallized motif, thus provided the type of atoms or of functional groups and their geometrical arrangement in the 3D space.[20]

[20]Purves, W. K., Orians, G. H., Heller, H. C. (2001). *Life: The Science of Biology*, 6th Edition, by Sinauer Associates, Sunderland (MA), and WH Freeman, New York; Advanced Protein Crystallization Facility NASA-APCF (2002). http://www.nasa.gov/centers/marshall/news/background/facts/apcf.html

and mathematical procedures, and the data processing employs numerical calculation programs. The output is the real structure of the protein, specific to the molecular type involved, with the distances and bonding in the 3D spatial form (Figure P.11).

The whole process can take from several months to several years; however, from the results one can understand the function and the role of the respective protein in the metabolic cycle and eventually one can decide the way in which this function and the basic structure should be changed – for example, by a metallic insert – to comply with specific needs of structural (chemical) adjustment and metabolic (biological) action.

VOLUME LAYOUT

The present volume is the *fourth* part of the five-volume work on :

Volume I: Quantum Theory and Observability
Volume II: Quantum Atoms and Periodicity
Volume III: Quantum Molecules and Reactivity
Volume IV: Quantum Solids and Orderability
Volume V: Quantum Structure–Activity Relationships (Qu-SAR)

This book consists of the following chapters:

Chapter 1 (Bondons on Graphenic Nanoribbons with Topological Defects): Recently introduced bosonic quasi-particle "bondon" (see Chapter I of Volume III of the set) accounts for the emergence of long-range interaction in one-dimensional graphenic nanoribbons opening the door for possible phase-transitions effect. Current simulations also benefit from adopting pure topological potential (used as potential energy in the statistical treatment) that greatly simplify, as usual, the computational tasks without sacrificing the physical information stored in the connectivity of the chemical structures. This chapter advances the modeling of bondonic effects on graphenic and honeycomb structures, with two generalizations: (i) by employing the fourth order path integral bondonic formalism through the high-order (second and fourth) derivatives of the given 1D potential for a certain network, here identified with the Wiener topological potential; and (ii) by modeling a class of honeycomb defective structures admitting graphenic as the carbon-based reference case and then generalizing the treatment to Si (Silicene), Ge (Germanene), Sn (Stannene) by using the fermionic two-degenerate statistical states function in terms of electronegativity, a useful parameter easily extendable to related hetero-combinations C–Si, C–Ge, C–Sn, Si–Ge, Si–Sn, Ge–Sn. The honeycomb nanostructures present η-sized Stone–Wales topological defect, the isomeric dislocation dipole originally called by authors as the Stone–Wales wave (SWw). For these defective nanoribbons, the bondonic formalism provides a specific phase-transition whose critical behavior shows typical bondonic fast critical time and bonding energies. The quantum transition of the ideal-to-defect structural transformations is fully described by computing the caloric capacities of nanostructures triggered by the η-sized topological isomerizations;

*Chapter 2 (**Geometrical Crystallography**)*: The crystal (motif + lattice) system is analyzed in its stable ground state by means of geometrical characterization of axes, inversion center and mirroring planes, as the main symmetrical elements driving the allowed operation in such ordered giant molecular structure: they were provided crystallographic systems (7), then extended to Bravais lattice (14), then enlarged to the point groups (32) when all elements of symmetry that intersects on a point are comprehensibly combined, and finally to the spatial groups (230) when the translation comes into play too; these classification schemes correspond with the most resumed way of chemical classification of substances by their crystalline groups (points or spatial) and correspond with the maps obtained by X-ray diffraction or by other optical action (optical activity by laser action or by magnetization) especially through the involved reciprocal space (with the allied theorems of faces represented by normal axes and axes represented by nodes in reciprocal space and of its projective maps) – where the quantum behavior is recovered by means of wave vector, reciprocally related with the acting yet in resonance with electronic wave functions, specific to living electrons in crystals and ordered solids.

*Chapter 3 (**Quantum Roots of Crystals and Solids**)*: The quantum mechanics postulates are shortly reviewed for their application in providing the basic crystal Bloch theorem, further specialization to the quantum modeling of crystals in reciprocal space, on various levels of electronic behavior from free to quasi-free, to quasi-binding, to tight-binding models, as well as for modeling solids by quantum statistical means, especially grounded on Fermi statistics driving the valence-to-conducting levels' displacements so providing the quantum framework for semiconducting electrons and to chemical related moletronics.

Chapter 4 (*Chemical Crystallography*): By summarizing the accumulated experimental results, V.M. Goldschmidt stated *fundamental law of crystal chemistry* that meets certain factors which determine the forming of a certain crystalline structure from a set of particles. Under this law, a crystal structure is determined by the number and size of particles in the elementary cell, by their nature and by the nature of chemical bonds that are established between them. The present chapter is devoted to systematic study of this law, thus developing the crystal chemistry.

Chapter 5 (*X-Ray Crystallography*): The journey toward the depths of the structure of condensed state eventually proceeds through an "*external non-destructive intervention*", with the aid of the X-radiations on the crystals; we should note that X-rays interact with the electrons of the atoms or the groups of atoms from the network and not with their nuclei. Thus, the picture of diffraction (reflection or scattering) of the X-ray will generate an "electronic map" of the bodies investigated, thus characterizing their structure. Moreover, this electronic map is a manifestation of the so-called structure factor of the crystal analyzed – a quantity in direct (Fourier) relationship with the associated electronic density. This way, the union between the experimental methods with X-ray diffraction on crystals and their quantum characterization is retrieved by means of electron densities characteristic to the structures. Practical ways of refining this merging between crystals and X-ray diffraction are the main focus of this chapter.

Thus, the present volume has special features as it:

- Presents the solids and crystals from the quantum nature perspective, in various instances, from bondonic particles on graphenic nanoribbons to geometric and morphologic characterization of crystals, to reciprocal space of electronic behavior, to chemical bonding driving crystals' orderability to quantum observability by X-ray diffraction;
- Continues the bondonic characterization of the chemical systems from Volume III and the path integral quantum methods of Volumes I and II of the present five-volume set;

- Introduces the bondonic phase-transition algorithm to model the topological defects on nanosystems by analytical and quantum simulation;
- Characterizes the solid state as being of chemical nature in a dynamical equilibrium between short- and long-range neighbors alike;
- Reviews the geometrical principles of crystallography and the main crystals' classification in systems/syngonies, symmetry/punctual groups, Bravais lattices, space and magnetic groups alike;
- Formulates the basic stability principles of crystal chemistry on the quantum basis of ionic and dispersive interactions;
- Formalizes the chemical crystal principles under qualitative rules (including Pauling rationalization);
- Predicts the crystal syngonies and bonding, either by geometrical reducing cell method or by authors' X-ray related (Pendellösung) density maps description of electrons in ideal and elastically deformed solids at the unity cell (factor structure) level, respectively.

The author expresses his kind thanks to individuals, universities, institutions, and publishers that inspired and supported the topics included in the present volume; a few of them are mentioned below:

- *Supporting individuals*: Dr. Ottorino Ori (Actinium Chemical Research, Rome); Prof. Hagen Kleinert (Free University of Berlin); Priv. Doz. Dr. Axel Pelster (Free University of Berlin); Prof. Nino Russo (University of Calabria); Prof. Eduardo A. Castro (University La Plata, Buenos Aires); Prof. Dr. Adrian Chiriac (West University of Timişoara)
- *Supporting universities*: West University of Timişoara (Faculty of Chemistry, Biology, Geography/Biology-Chemistry Department/ Laboratory of Computational and Structural Physical Chemistry for Nanosciences and QSAR); Free University of Berlin (Physics Department/Institute for Theoretical Physics/Research Center for Einstein's Physics, Centre for International Cooperation);
- *Supporting institutions & grants*: DAAD (German Service for Academic Exchanges) by Grants: 322 A/17690/2004, 322 A/05356/2011; CNCSIS (Romanian National Council for Scientific

Research in Higher Education) by Grant: AT54/2006–2007; CNCS-UEFISCDI (Romanian National Council for Scientific Research) by Grant: TE16/2010–2013;

- *Supporting publishers*: Elsevier (Amsterdam); Wiley (New York); Nova Science Inc. (New York); Multidisciplinary Digital Publishing Institute – MDPI (Basel, Switzerland); University of Kragujevac (Serbia). Multidisciplinary Digital Publishing Institute – MDPI (Basel); Wiley (Hoboken, NJ), Springer Science (Berlin – Germany, London- UK, and New York – USA).

Additionally, the author takes special moments to thank his family and his lovely little daughters *Katy and Ela* for providing him with necessary energy and creating the work-and-play atmosphere—making the writing a fun. Hopefully, the fun is transmitted to the readers and students too.

Last but not the least, the author would like to make a special mention about the publisher, Apple Academic Press (AAP) team and in particular to Ashish (Ash) Kumar, the AAP President and Publisher, and Sandra (Sandy) Jones Sickels, Vice President, Editorial and Marketing, for professionally supervising the manuscript through production and delivering the five-volume set *Quantum Nanochemistry*.

The vast field of *quantum (physical-)chemical theory of solids and orderability* will be ever expanding and its importance in the years to come in science and technology will only keep enhancing. Thus, any constructive observations, corrections and suggestions are welcome from the readers; such peer contribution is kindly appreciated.

Keep close and think high!

Yours Sincerely,

Mihai V. Putz,
Assoc. Prof. Dr. Dr.-Habil. Acad. Math. Chem.
West University of Timişoara
& R&D National Institute for Electrochemistry and Condensed Matter Timişoara
(Romania)

CHAPTER 1

BONDONS ON GRAPHENIC NANORIBBONS WITH TOPOLOGICAL DEFECTS

CONTENTS

ABSTRACT

Recently introduced bosonic quasi-particle "bondon", see Chapter 1 of the Volume III of the present five-volume set (Putz, 2016a) is shown in this chapter to account for the emergence of long-range interaction in one-dimensional graphenic nanoribbons opening the door to possible phase-transitions effect. Current simulations also benefit from adopting pure topological potential (used as potential energy in the statistical treatment) that greatly simplify, as usual, the computational tasks without sacrificing the physical information stored in the connectivity of the chemical structures. This chapter advances the modeling of bondonic effects on graphenic and honeycomb structures, with an original two-fold generalization: (i) by employing the fourth order path integral bondonic formalism through considering the high order (second and fourth) derivatives of the given 1D potential for a certain network, here identified with the Wiener topological potential; and (ii) by modeling a class of honeycomb defective structures admitting graphenic as the carbon-based reference case and then generalizing the treatment to Si (Silicene), Ge (Germanene), Sn (Stannene) by using the fermionic two-degenerate statistical states function in terms of electronegativity, an useful parameterization easily extendable to related hetero-combinations C-Si, C-Ge, C-Sn, Si-Ge, Si-Sn, Ge-Sn. The honeycomb nanostructures present η-sized Stone-Wales topological defect, the isomeric dislocation dipole originally called by authors Stone-Wales wave or SWw. For these defective nanoribbons the bondonic formalism individuates a specific phase-transition whose critical behavior shows typical bondonic fast critical time and bonding energies. The quantum transition of the ideal-to-defect structural transformations is fully described by computing the caloric capacities for nanostructures triggered by η-sized topological isomerisations.

1.1 INTRODUCTION

Nowadays, graphenic structures represent the recognized new frontier in physics of electronic systems with reduced dimensionality. Quasi-two-dimensional systems are well known for outstanding behaviors in the condensed matter physics as the quantum Hall effect and the high-temperature copper-oxide superconductors. According to literature, what makes graphene "qualitatively new" (Herbut, 2006) is its semi-metallic nature with low-energy quasiparticles behaving as "relativistic" Dirac spinors in the conducting band. Within this context the very recent identification of the "bondon" as quantum particle of the chemical bond, rooting either in Bohm quantum mechanics as well in Dirac relativistic quantum theory (Putz 2010a-b, 2011a, 2012a-d), may provide a comprehensive tool for modeling physical properties of extended nanosystems in general, and those of graphite layer in special, as linking the microscopic behavior of electrons in condensed systems with macroscopic observable quantities described by quantum statistics; the present discussion follows (Putz & Ori, 2014).

With the irresistible rise of graphene, large attention has been posed by the scientific community on the spectacular properties of this carbon monolayer, the "Nobel prized" new carbon allotrope who – after a decade from its discover by Novoselov et al. (2004) – still promises innovative *technological* solutions for many sectors in physics and nanotechnology. Clearly, the real *breakthrough discovery* opening the graphene *golden-age* is still missing (Geim & Novoselov, 2007). This remains an unachieved goal, a severe scientific challenge that pushes experimentalists worldwide to solve the barriers, both technological and cost-wise, which obstacle mass-applications of this *one-atom-thick fabric of carbon* with its "extreme" mechanical and electronic features. The risk for graphene of a frustrating record with no-applicative results, the same fate which (somehow unexpectedly) prevented so far any *practical* C_n fullerene applications, has been recently denied by the most authoritative review on the subject (Novoselov et al., 2012), waiting for a "manufacturing" turning-point; these authors in fact repute that graphene will eventually become attractive for industrial applications providing that "mass-produced graphene will guarantee the same performances as the best samples obtained in research laboratories".

The bondons represents (Putz 2010a, 2012a) the bosonic counterpart of the bonding electrons' wave function carrying the quantized mass related with the bonding length and the energy, written in the ground state as:

$$M_B = \frac{\hbar^2}{2E_{bond} X_{bond}^2} \tag{1.1}$$

with a Heisenberg type interdependency constraint available also for extended bonding systems:

$$E_{bond}[kcal/mol]X_{Bond}[\overset{o}{A}] = a, \quad a = 182019 \tag{1.2}$$

For instance, Eqs. (1.2) prescribes that a nano-system with hundred atoms that comprise an energy about 1,000 kcal/mol may have a spanned maximum bonding for 182.019 Å that it can be detected across the ends of a nanotube of length 18 nm, while a bonding system may support along 100 Å (e.g., along a nanotube of length 10 nm) maximum of 1,820 kcal/mol due to bosonic condensation of bondons; as such, the bonding-fermionic and condensing-bosonic properties are unified by Bohmian/entangled quantum field quantized by the bondon quasi-particle (Putz, 2010a, 2012a) carrying the electronic elementary charge (like a fermion) with almost velocity of light (like a light boson) in the femtoseconds range of observation (Martin et al., 1993) either along bonds (pairing electrons) or networks (connecting many-atoms in nanosystems), respectively.

These general features highly encourages further quest of the bondonic properties at extended systems, since bondons supports no locality spreading at both qualitatively and quantitative descriptions, according based on its basic Bohmian derivation and of extended bonding length-energy relationship of Eq. (1.2), respectively. Furthermore, the specific properties of bondons possibly reverberate in other phenomenon, one of the most interesting being the emergence of phase transitions in one-dimensional (1D) equilibrium systems where such phase transitions are normally excluded by the well known Landau-Lifshitz when only short-range interactions are present (Evans, 2001). In contrast to that, ordered states may occur when the domain walls interact over some finite distance threshold. The present endeavor aims to explore first steps in such fruitful conceptual and physical systems, having graphite fragments as current example.

The chapter aims in supporting the importance of having nanodomains in the graphene lattice, whose 2D nature makes it easy to add, remove or move carbon atoms, to create a variety of structural defects. Recent ab-initio density functional theory (DFT) very detailed investigations (Manna & Pati, 2011; Basheer et al., 2011; Dutta et al., 2009) evidence in fact that hybrid graphene nanoribbons base the fine tuning of their electronic behaviors, from semiconducting to half-metallic or metallic, on the presence of $B_x N_y$ nanodomain with peculiar geometries and concentrations. The theoretical descriptions based on bondon quasiparticles represents a useful tool for nanomaterials engineering, including silicene, the Si-analogue of graphene, a promising nanostructure for hydrogen storage (Jose & Datta, 2011).

1.2 GRAPHENIC TOPOLOGICAL ISOMORPHISM

1.2.1 GRAPHENIC'S TYPE RIBBONS AND THEIR STONE-WALES DEFECTS

From a general perspective, ten years of investigations on graphenic honeycomb lattices point out the scientific relevance of monolayer materials like hexagonal BN, MoS_2 and others, whose 2D crystals present a rich parade of physico-chemical properties that can be furthermore specialized by combining variable stack of heterostructures (often called van der Waals heterostructures due to the presence of van der Waals-like forces gluing the layers together (Butler et al., 2013) as in normal graphite crystals) with applications, for example, in vertical tunneling transistors (Britnell et al., 2012). It is however commonly accepted (Novoselov et al., 2012) that at least for microprocessors, graphene-based logic elements will replace the silicon technology only after, 2025, the main physical limit being so far represented by the reduced value of the induced bandgap in graphene still limited to 360 meV with a reduction of performances of a factor 103 if compared to current silicon devices. This impasse is one of the main reasons moving today research focus in microelectronics from carbon-based toward silicon-based hexagonal systems; the present discussion follows (Putz & Ori, 2014).

The natural candidate for such a class of material is *silicene*, the honeycomb monolayer theoretically introduced as the all-silicon made version of graphene (Takeda & Shiraishi, 1994; Guzman-Verri & Lew Yan Voon, 2007) and recently synthesized by chemically exfoliation of calcium disilicide resulting in silicon 2D nanosheets of 0.37 nm thickness (Nakano et al., 2006) or, more recently, by epitaxial growth on metallic surfaces. With deposition techniques under ultra-high vacuum conditions on silver (110) plane, exceptional silicon 1D Si metallic nanowires (or silicon nanoribbons Si-NR), one-atom thick, have been created (Leandri et al., 2005). Metallic Si-NR are suitable for being promoted to n or p-type semiconductors by chemical doping. Such extended nanoribbons are built by the action of the two-components Si $2p$ quantum levels, corresponding to the electronic contributions coming from the two distinct silicon atomic sites. Subsequent experimental and theoretical investigations by scanning tunneling microscopy (STM) and ab initio calculations based on DFT (Aufray et al., 2010, Sahaf et al., 2007) state that these silicene 1D stripes, grown on the silver substrate with an exceeding 100 nm length and with 1.6 nm "magic" width, posses a genuine honeycomb, graphene-like structure represented in Figure 1.1(a). Silicene NR are moreover able to reach a "self-organized", regular coverage on the substrate surface with a spacing of about 2 nm (Sahaf et al., 2007). These important structural results have been consecrated (Vogt et al., 2012) by the same team of researchers succeeding in the epitaxial formation of silicene 2D sheets on a silver (111) substrate. By mean of STM and angular-resolved photoemission spectroscopy measures, in conjunction with DFT simulations, the study ultimately confirms the silicene *buckled* honeycomb arrangement, see Figure 1.1 (b).

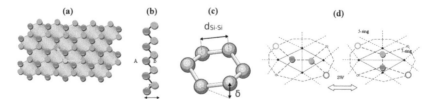

FIGURE 1.1   (a) The honeycomb mesh characterizing the 1D nanoribbons with the two independent atoms and the two unit cell vectors which, by translation, cover the entire structure; (b) side view of the lattice and (c) of the buckling structural parameter δ spacing silicene hexagonal sublattices A and B; the Si-Si bond distance is also depicted; in silicene typical distortion parameter is δ=0.44 Å with d(Si-Si)=2.25 Å; (d) The 5|7|7|5 Stone-Wales (SW) rotation seen in the direct and dual representation in the nanoribbon honeycomb mesh, see text and Putz & Ori (2014).

Topologically, silicene and SiNR graphene share the same kind of hexagonal mesh made of 3-connected nodes (atoms), *structurally* hoverer they show different features, the main distinguishing character being for silicene the *buckling* distortion δ, see Figure 1.1(c) which moves out the system from (graphene) perfect planarity. Such a chair-like puckering of the Si 6-rings corresponds to distortion parameter of δ=0.44 Å and to a bond length of d(Si-Si)=2.25 Å according to Cahangirov et al. (2009). In graphene the approximate inter-atomic distance is d(C-C)=1.42Å. The chemical stability of buckled honeycomb structures is substantially granted by the "puckering induced" dehybridization effect which allow p_z orbitals, oriented normally to the layer, making linear combinations with the s orbitals, forming π bonding and hence π and π* bands, similarly to graphene case. Comparative values for the based on the modified Harrison bond orbital method are reported in Davydov (2010) resulting in an atomic binding energy of E_{atom}=13.5 eV and E_{atom}=7.1 eV for graphene and silicene respectively. It is worth noticing that, from the structural point of view, the two independent atoms that constitute the graphene unit cell generate in silicene the two distinct sublattices A and B, laying in two δ-spaced parallel planes, as in Figure 1.1(b) (Zhang et al., 2002).

A relevant topological mechanism enriching the physico-chemical behavior of honeycomb lattices is represented by the possibility of undertaking isomeric transformations, respecting both the number of atoms and the number of bonds. Such a particular one-bond rotation, the so-called Stone-Wales (SW) rotation or SW topological defect, is known for creating a double pair of 5|7 rings, Figure 1.1(d), inducing changes of the band configurations graphene. It even seems able to alter the planarity of the graphenic layer over regions with many nanometers size, reaching large out-of-plane deformations δ≈1.7 Å, according to recent Monte-Carlo simulations on the subject (Ma et al., 2007), which deals with cases of *buckled SW transformations* as a possible mechanism facilitating fullerene and nanotubes formations. The possibility to induce such a SW defect *solely* depends from the 3-connectivity of the lattice atoms. SW-compatible patterns reflect the properties of the topological adjacency matrix of the system. Typical values for the energy barrier E_b opposing the SW rotations are E_b ≈5 eV and E_b ≈2.8 eV, respectively, in graphene (Ma et al., 2007) and Ag(111)-grown silicene

(Sahin et al., 2013). This large difference reflects the basic structural fact that having silicene an inter-atomic distance larger compared to the graphitic layers, an easier formation of topological defects in silicene may be expected, maintaining however a peculiar stability even at high temperatures. Among the low-dimensional systems, have a really notable richness of electronic properties, one has to consider that SW rotations immediately open a band gap reaching the value of 0.1 eV for silicene fragments with sparse SW defects; the present discussion also follows (Ori-Cataldo-Putz, 2011).

On the other hand, for hexagonal systems like graphene layers, graphene nanoribbons (GNR's) and carbon nanotubes (CNT's), the isolated pentagon–heptagon *single* pair [also called 5|7 pair, 5|7 defect, 5|7 dislocation or the *pearshaped* polygon (Liu & Yakobson, 2010)] and pentagon–heptagon *double* pair 5/7/7/5 arising from the celebrated Stone-Wales transformation [SW transformation or SW rotation (Stone & Wales, 1986)] are important structural defects largely influencing chemical, mechanical, and electronic properties (Terrones et al., 2010). Figure 1.2 represents the general Stone-Wales transformation $SW_{q/r}$ (Figure 1.2a), associated to the most studied variants, the $SW_{6/6}$ in graphene of Figure 1.2(b) often called Stone-Thrower-Wales rotation and the $SW_{5/6}$ in fullerenes, see Figure 1.2(c) the so-called pyracylene rearrangement.

For better understanding of the SW transformations and of their features relating the bondonic behavior on extended lattices, in what follows we like to illustrate the topological properties of SW rotations in hexagonal systems investigating, in particular, the family of isomeric SW transformations able to *generate and propagate* 5|7 defects in graphenic fragments, graphene nanoribbons and carbon nanotubes. We present initially (see next paragraph) an original graphic tool able to modify the hexagonal patterns of carbon atoms under the action of subsequent SW bond rotations generating 5 and 7-membered carbon rings. Our tool operates in the dual space and, more generally, it creates various kinds of defective layers with no limitations on the composition of the modified rings that may have any number of members $m = 3, 4, 5, 6, \ldots$ The topological simulations confirm moreover that SW double pairs 5/7/7/5 possess a peculiar anisotropy, matching, from a pure *topological* point of view similar *ab-initio* results on sp^2-carbon systems recently appeared in literature, see Samsonidze et al. (2002) and related references.

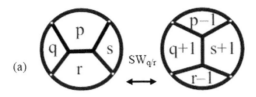

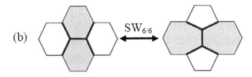

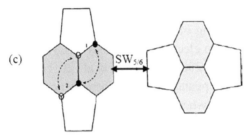

FIGURE 1.2   (a) Local transformation $SW_{q/r}$ changes a group of four proximal faces with p, q, r, s atoms in four new rings with p–1, q+1, r–1, s+1 atoms; (b) On the graphene layer (p=q=r=s=6) $SW_{6/6}$ reversibly flips four hexagons in a 5|7 double pair; (c) $SW_{5/6}$ reversible flip on the fullerene surface (Ori-Cataldo-Putz, 2011).

Another interesting topological effect is also introduced, consisting in the diffusion of a 5|7 pair in the hexagonal network as a consequence of iterated SW rotations; this topology-based mechanism *that produces a linear rearrangement of the hexagonal mesh* is called here the *SW wave*. Whereas mechanically exfoliated monolayer graphene is structurally (almost) perfect in atomic scale (Terrones et al., 2010), graphene layers produced by chemical vapor deposition (CVD) techniques present a parade of defect structures, which are due to the growth on substrates with surface defects and/or other irregularities. New stable carbon allotropes have been therefore proposed (Lusk & Carr, 2008) by considering the presence of periodical arrangements of defective building blocks such as Stone-Wales defects, inverse Stone-Wales defects, vacancy defects, and other structural modifications of the pristine hexagonal plane. The first experimental observation of a particular type of linear topological defects is reported in Lahiri et al. (2010), where extended chains of octagonal and pentagonal sp²-hybridized carbon rings, detected by STM images, function as a quasi-one-dimensional metallic wire and may be the building blocks for new all-carbon electronic devices. This important experimental finding enforces meanwhile the theoretical role of the SW waves that are in principle structurally simpler than the penta-gons-octagons chain reported in Lahiri et al. (2010), as a possible *hexago-nal inter-grain spacing* (see the visualizations given in next sub-section) between graphenic fragments. Molecular mechanics simulations show that in graphene the presence of cylindrical curvature energetically facilitates such a split of the 5/7/7/5 SW dislocation dipole (Samsonidze et al., 2002), assigning to this class wave-like atomic-scale rearrangements a fundamental role in nanoengineering of graphenic lattices. One has however to notice that other transmission electron microscopy (TEM) detailed measurements point out (Meyer et al., 2008, Kotakoski et al., 2011a) that the migration and the separation of the pentagon-heptagon pairs does not happen on planar graphene membranes where the 5–7 defects relax back reconstructing the original graphene lattice. These experiments indicate that extended *disloca-tion dipole*, favored by the presence of structural strain, preferably appear in curved graphitic structures or systems like CNT or fullerene molecules. In epitaxial graphene grown at high temperatures on mechanically-polished SiC(0001), a characteristic 6-fold "flower" defect results from STM mea-sures (Rutter et al., 2007, Meyer et al., 2011). We note that the observed rota-tional grain boundaries is conveniently describable as *radial* type of the *SW wave* suggesting that the wave-like theoretical mechanism presented here, may have a general applicability.

The SW rotations applied in the present studies derive from the gen-eral, see Figure 1.2(a), Stone-Wales local and isomeric transformation $SW_{p/r}$ varying the internal connectivity of four generic carbon rings made of *p, q, r, s* atoms to produce four new adjacent rings with *p–1, q+1, r–1, s+1* atoms without changing the network of the surrounding lattice. $SW_{p/r}$ reversibly rotates the bond shared by the two rings *p* and *r*, preserving both, the total number of carbon atoms

$$v = p + q + r + s - 8 \qquad (1.3)$$

and the total number of carbon-carbon bonds

$$e = v + 3 \qquad (1.4)$$

On the graphene ideal surface, made only of hexagonal faces, the $SW_{6/6}$ rotation transforms four hexagons in two 5|7 adjacent pairs in Figure 1.2(b) symbolized in literature (Samsonidze et al., 2002; Carpio, 2008) as 5/7/7/5 defect and also quoted as the *SW defect* or the *dislocation dipole*. We remember here that the SW rotations play an important role in connecting the isomers of a given C_n fullerene with different symmetries. In the crucial case of the C_{60} fullerene, its 1812 isomers are grouped by the *pyracylene* rearrangements $SW_{5/6}$ as in Figure 1.2(c) in 13 inequivalent sets (the larger one consisting of 1709 cages) connected to the buckminsterfullerene (C_{60}-I_h) through one or more SW transformations (Kumeda & Wales, 2003), leaving 31 isomers unconnected to any of these sets. This limitation has been overcome by the introduction of non-local generalized Stone-Wales transformations (Babic et al., 1995) to generate the whole C_{60} isomeric space starting from just one C_{60} isomer.

Theoretical investigations based on plane-wave density-functional methods (Kumeda & Wales, 2003) set to no less than 6.30 eV the uphill energy barrier dividing the buckminsterfullerene from the SW connected isomer with C_{2v} symmetry; this barrier reaches 9 eV for hexagonal systems like nanotubes or large graphene portions. Using the extended Hückel method, enlarging the relaxation region around the SW defect, it can be found that the formation energy of a SW defect considerably decreases to 6.02 eV for a flat graphene fragment case. This result has been verified by using *ab initio* pseudopotential (Zhoua & Shib, 2003). This result seems to preclude the formation of any SW 5/7/7/5 defect in nature, but as it has been reported (Collins, 2011; Ewels et al., 2002) that this barrier drops rapidly, reducing to 2.29 eV the creation barrier of SW rotations due to the catalyzing action of interstitials defects or ad-atoms present in the hexagonal networks. Pentagon–heptagon pairs have been predicted to be stable defects also in important theoretical articles (Nordlund et al., 1996; Krasheninnikov et al., 2001) showing that energetic particles, as electrons and ions, generate 5|7 pairs in graphite layers or CNT's as a result of knock-on atom displacements. On the experimental side, accurate high-resolution TEM studies made on single-walled carbon nanotubes (Hashimoto et al., 2004) or electron-irradiated pristine graphene (Kotakoski et al., 2011b) document *in situ* formation of SW dislocation dipoles. TEM measures also evidence (Chuvilin, 2009) stable grain boundaries with alternating sequence of pentagons and heptagons that show the relevance of wave-like defects during graphene edge reconstruction.

Extended theoretical investigations (Jeong et al., 2008) by means of first-principles density-functional computations, demonstrate that, on graphene layers, the dislocation dipole 5/7/7/5 defects become particularly stable – in comparison to other possible local defective structures as haeckelite units with three pentagons and three heptagons—when the two 5|7 pairs are separated by lattice vacancies in the number of ten or over. Moreover, recent literature [see the review (Terrones et al., 2010)] on GNR's constructed from haeckelites considers systems with SW defects as new hypothetical nano-architectures with fascinating applications in electronics. Isolated 5|7 pairs could also appear at grain boundary in graphene fragments, changing their edge termination and electronic properties, forming *hybrid* GNR's. These hybrids exhibit half metallicity in the absence of an electric field, and could be used to transport spin-polarized electrons; which could be a step forward in new spintronic devices.

Considering the above experimental and theoretical evidences of the structural stability of hexagonal systems with 5|7 defects, this theoretical note aims to investigate the topological, wave–like mechanisms leading the diffusion (or annihilation) of pentagon–heptagon pairs.

Figure 1.3 shows the fundamental topological operations for the generation and the propagation of SW waves in the graphene lattice. The first rotation $SW_{6/6}$ Figure 1.3(a) of the chemical bond (arrowed) shared by the two hexagons creates the two 5|7 pairs (the SW defect 5/7/7/5). The second operator $SW_{6/7}$ turns the bond between the heptagon and the nearby shaded hexagon and inserts the 6|6 couple of shaded hexagons between the two original 5|7 pairs (this topological defect is also referenced in Samsonidze et al. (2002) as 5/7/6/6/7/5), leading to the overall structural effect of initiating the propagation of the SW wave along the dotted direction Figure 1.3(b). Iterated transformations $SW_{6/7}$ will successively drift the 5|7 pairs in the lattice (along the dotted directions in Figure 1.3(b), producing the topological SW wave (SWW).

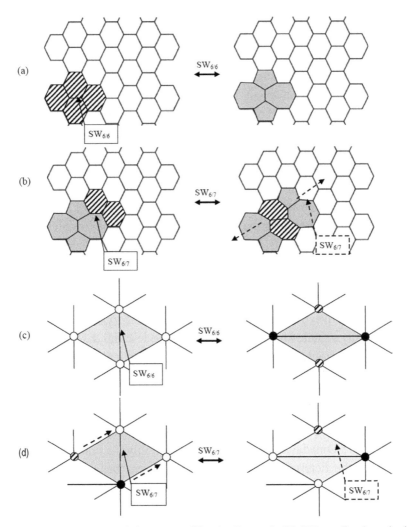

FIGURE 1.3   (a) SW$_{6/6}$ originates two 5|7 pairs (in gray); (b) SW$_{6/7}$ splits the pairs by swapping one of them with two nearby hexagons (shaded). Dotted SW$_{6/7}$ pushes the SW wave in the dashed direction; (c-d) Mechanisms (a,b) in the graphene dual plane. Hexagons, pentagons, heptagons are represented by white, shaded, black circles, respectively (Ori-Cataldo-Putz, 2011).

SWW mechanism provides theoretical support to recent studies on graphenic structures. Some authors (Liu & Yakobson, 2010) emphasize the importance of 5|7 dislocations monopole at the grain boundaries of polycrystalline graphene, stating that these defects cannot be annealed by any local reorganization of the lattice. SW waves allow 5|7 dislocations also to anneal by just involving surrounding 6|6 pairs and moving *backward*, being all transformations in Figure 1.3 completely *reversible*.

Theoretically, SW defects and isolated 5|7 pairs have been extensively investigated in Meyer et al., (2011) where *ab-initio* simulations of the electronic properties are reported; authors conclude that a single hepta-gon–pentagon dislocation is a stable defect whereas the Stone-Wales adja-cent pairs are dynamically unstable. These two conformations may easily find a unified description considering that these lattice defects correspond to different propagation steps of the same SW wave.

Considering the very rich variety and complexity of all possible paths that SW waves may describe on the graphenic surface, involving a vari-able numbers of 5|7 pairs, this article just focuses on the topological prop-erties exhibited by the linear propagation of the basic SW defect, the 5|7 double pair Figure 1.2(b). This choice limits the $SW_{p/r}$ rotations to just to the operators $SW_{6/6}$ and $SW_{6/7}$. In spite of the apparent simplicity of our model, *SW waves present* an evident and *marked topological anisotropy* immediately signaled by the Wiener index (Todeschini & Consonni, 2000) $W(N)$ of the graphenic system under study (graphene fragments, CNT's and GNR's).

It is really important to note that, more and more, various anisotro-pic effects are evidenced in literature (Terrones et al., 2010; Samsonidze et al., 2008; Bhowmick & Waghmare, 2010; Huang et al., 2009; Zeng et al., 2011; Dinadayalane et al., 2007) by applying first-principle tech-niques to the determination of the energy-stress behaviors of different configurations of SW defects on graphene nanotubes and nanoribbons. Similar effects appear in the theoretical distribution of magnetic dipoles in defective carbon metallic nanotubes (Im et al., 2011).

Remarkable, on the basal electronic side, silicene and Si nano-ribbons (SiNR), like graphene, exhibit massless relativistic Dirac fermions arising, for the nanoribbons case (Padova et al., 2010), from the 1D projection of π and π^* Dirac cones. Moreover [see the relevant summary of Kara et al. (2009) and related references], the 1D topology characterizing metallic SiNR structures plays to favor electrons interactions according with the *Luttinger liquid* model which implies the emergence of *bosonic quasi particles effects* coexisting with the Dirac fermionic characters expected for 2D silicene. On top of this, superconducting phenomena may be also expected for 1D silicene stripes matching similar effects measured at 8 K in hexagonal metallic silicon, possibly with an augmented T_c, as argued in Padova et al. (2010); such features further advocates the actual temperature depending bondonic long-range signal modeling on defective honeycomb nanoribbons lattices (see Section 1.5.2).

1.2.2 STONE-WALES REARRANGEMENTS' GENERATION AND PROPAGATION

While modeling the SW wave propagation, it is worth introducing the graphic tool used to generate this kind of defects on the hexagonal structures. The effectiveness of such an algorithm derives from the choice to operate in the *dual topological representation* of the graphenic layers as shown in Figure 1.4. Also the topological modeling will be conducted in the dual space; the present discussion follows (Ori-Cataldo-Putz, 2011).

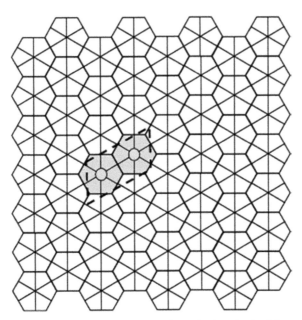

FIGURE 1.4 Dual representation of the graphene lattice obtained by replacing each hexagonal face by the central 6-connected graph node. Graphene plane is then equivalently tiled by hexagons (direct space) or by starred nodes (dual space). The x-periodic (y-periodic) direct graphene nanoribbon has the armchair (zig-zag) orientation. The framed unit cell has been used to build this 4 × 7 graphenic fragment (Ori-Cataldo-Putz, 2011).

The generation of the SW rotations is greatly facilitated by considering the dual representation of the graphene layer by assigning to each hexagonal face the corresponding 6-connected (starred) vertex. Figure 1.4 visually overlaps both direct and dual graphene representations showing their topological equivalency: each pair of adjacent faces in the direct lattice corresponds in fact to a pair of bonded nodes in the dual graph and vice versa. The graphene fragment is taken in its *armchair* orientation along *x* and Figure 1.4 evidences a 4 × 7 dual lattice and its unit cell. In the dual lattice the generic $SW_{p/r}$ *simply rotates* the internal edge between the *p*- and the *r*-connected nodes, making the study of the SW rearrangement very simple and suitable for automatic procedures.

On the graphene dual layer the $SW_{6/6}$ rotation, Figure 1.3(c), then changes four 6-connected nodes (white circles) into two 5-connected (shaded circles) and two 7-connected (black circles) vertices, matching the standard transformation in the direct lattice of four hexagons in two pentagons and two heptagons Figure 1.3(a). Moreover 5|7 pairs may also *migrate* in the graphene lattice, pushed by consecutive Stone-Wales transformations of $SW_{6/7}$ type that rotate, in the dual space, the *vertical* edge between the 6-, and the 7-connected vertices, driving the *diagonal* diffusion of a 5|7 pair in the graphene lattice. Figure 1.3(d) gives more details about the swapping mechanism between the 5|7 and the 6|6 couples. The repeated action of the $SW_{6/7}$ operator originates the topological SW wave in both lattice representations.

Figure 1.5(a) represents the diagonal diffusion of the SW wave (dislocation dipole) after four $SW_{6/7}$ rearrangements, evidencing with the dashed arrows the increasing distance between the two 5|7 pairs of the original $SW_{6/6}$ dislocation. At each step, the pentagon (shaded circle) and the heptagon (black circle) interchange their locations with those of two hexagons (white circles) producing the *diagonal SW wave*, a large dislocation dipole that modifies the landscape of direct and dual lattices, Figure 1.5(a) and Figure 1.5(b). Being η the size of the dislocations (e.g., η equals the number of 6|6 pairs included between the two 5|7 pairs) both examples in Figure 1.5 have size $\eta = 4$, assuming size $\eta = 0$ for the basic $SW_{6/6}$ rotation of Figure 1.2(b). Equivalently, η equals the number of $SW_{6/7}$ rearrangements used to generate the dislocations in both spaces. A SW wave produces, see Figure 1.5(b) a characteristic *hexagonal inter-grain spacing*, isomeric to the pristine graphene layer that represents therefore a good theoretical model for the boundary between graphenic fragments (Ori-Cataldo-Putz, 2011).

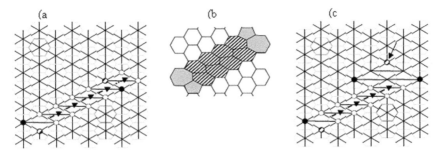

FIGURE 1.5 (a) Diagonal SW wave (dislocation dipole) in the dual graphene layer after the generation and four propagation steps (size $\eta = 4$); at each step (dashed arrows) $SW_{6/7}$ swaps the pair made by one pentagon (dashed circle) and one heptagon (black circle) with two connected hexagons (dotted circles); dotted arrow indicates the next available translation of the 5|7 pair; (b) The topological modification (a) originates, in the direct lattice, a hexagonal inter-grain spacing (dashed rings); (c) After a few more SW rotations, an isolated pentagon (arrowed), forming a small nanocone, is generated (Ori-Cataldo-Putz, 2011).

The dual space represents the natural arena for studying all sorts of SW flips, avoiding the graphical difficulties that one usually encounters in redistributing the carbon atoms and bonds in the direct lattice. One easily generates the *vertical* SW wave by applying in fact our graphical algorithm to the *diagonal* edges of the graphene dual lattice (Figure 1.6). SW transformations produce also very complex rearrangements of the graphenic layer including isolated pentagonal nanocones, as the very little one on top of the diagonal SW wave in Figure 1.5(c), creating fullerenic-like regions in the graphenic plane (Cataldo et al., 2010). The proposed dual space graphical algorithm appears therefore capable to handle complex combinations of general Stone-Wales rotations $SW_{p/r}$ to create novel classes of isomeric rearrangements, with rings made of various numbers of atoms, of fullerene (dimensionality $D = 0$), nanotubes ($D = 1$), graphenic structures ($D = 2$) or crystals ($D = 3$) as schwarzites or zeolites.

The some-how arbitrary definition of *diagonal* or *vertical* direction assigned to the SW waves on closed surfaces of graphenic fragments, nanoribbons, nanotubes considers the *armchair* graphene orientation selected in Figure 1.4. Topologically, the extension of the region interested by the dislocation dipole may be arbitrarily enlarged by applying more and more $SW_{6/7}$ rotations.

Energetically, the situation is more articulated; as explained elsewhere (Ori-Cataldo-Putz, 2011), the lattice shows in fact anisotropic reactions to

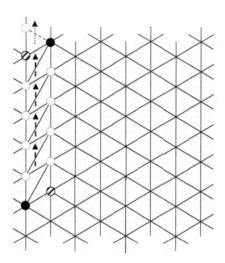

FIGURE 1.6   SW vertical wave in the dual graphene layer after four propagation steps (dashed arrows); $SW_{6/7}$ swaps the pentagon (dashed circle) heptagon (black circle) pair with two hexagons (dotted circles); dotted arrow indicates the next possible translation of the 5|7 pair, induced by a $SW_{6/7}$ rotation of the hexagon-heptagon diagonal dashed bond. The SW wave generates anti-diagonal hexagons-hexagons bonds with respect to the unrotated one (Ori-Cataldo-Putz, 2011).

the propagation of the SW waves along different directions when a closed graphene fragment is considered.

In summary, on the armchair-oriented graphene (Figure 1.4), the simplest propagation mechanisms available for the 5|7 pairs are (Ori-Cataldo-Putz, 2011):

- Diagonal SW wave, Figure 1.5: $SW_{6/7}$ rotates the vertical bond of the graphene dual lattice between the 6-connected node and the 7-connected node of the diffusing 5|7 pair, causing the diagonal drift of the pair and the creation of a new horizontal hexagon-hexagon bond, Figure 1.3(d) gives some more details);
- Vertical SW wave, Figure 1.6: $SW_{6/7}$ rotates the diagonal bond of the graphene dual lattice between the 6-connected node and the 7-connected node of the diffusing 5|7 pair, with the overall effect to vertically shift the pair, generating a new anti-diagonal hexagon-hexagon bond.

Above diffusion processes apply to an isolated 5|7 dislocation monopole as well to the 5|7 double pair arising from a $SW_{6/6}$ rearrangement. In the following we mainly study this latter case, focusing on the mechanisms leading to the creation of diagonal or vertical *extended dislocation dipoles* in the graphene lattice.

It is worth noting that similar topological tools are used in other disciplines like in Biology where wave-like diffusion mechanisms model cells proliferation processes (Pyshnov, 1980).

Finally, we observe that from the pure topological point of view one may consider each lattice configuration illustrated in this work as the result of an *instantaneous* transformation caused by a *single, non-local* SW rotation. This new class of transformations represents a further generalization, potentially infinite, of the non-local rearrangements early proposed (Babic et al., 1995) to generate the entire isomeric space of a given C_n fullerene starting for a limited number of inequivalent cages.

1.3 FOURTH ORDER QUANTUM CONDITIONAL DENSITY AND PARTITION FUNCTION

1.3.1 PATH INTEGRAL SEMICLASSICAL TIME EVOLUTION AMPLITUDE

Turning to analytically modeling the quantum nature of the bondonic propagation on the long-range nanoribbons, path integrals quantum formalism represents the viable integral alternative to the differential orbital approaches of many-electronic systems at nanoscale; accordingly, by employing path integral formalism one actually avoids the cumbersome computation and modeling of the orbital wave-function, see the Volumes I and II of the present five-volume set. Current method is therefore best suited for the path integral approach to the extended systems in which topological and bondonic chemistry (see the Volume III of the present five-volume set) appropriately describe long-range structures and the long-range interactions, respectively; the present discussion follows (Putz & Ori, 2014).

In this path-integral context, going to find the evolution amplitude of a particle with mass M moving in D dimension under the influence of an external potential $V(x)$ its path integral representation within the imaginary-time approach may be assumed (Putz, 2009, 2011b; Putz & Ori, 2014):

$$\left(x_b\tau|x_a0\right)= \int\limits_{x(0)=x_a}^{x(\tau)=x_b}\!\!\!\mathcal{D}\eta\exp\left\{-\frac{A[x]}{\hbar}\right\} \tag{1.5}$$

with the Euclidian action

$$A[x]=\int\limits_0^\tau d\tau_1\left[\frac{M}{2}\overset{\bullet}{x}^2(\tau_1)+V\left(x(\tau_1)\right)\right] \tag{1.6}$$

In order to derive the semiclassical approximation of (1.5) the decomposition of the path $x(\tau)$ into the path average of the end points

$$\bar{x}=\frac{x_a+x_b}{2} \tag{1.7}$$

and the fluctuation factor $\eta(\tau)$,

$$x(\tau)=\bar{x}+\eta(\tau) \tag{1.8}$$

is firstly considered to provide the imaginary time amplitude:

$$\left(x_b\tau|x_a0\right)= \int\limits_{\eta(0)=-\Delta x/2}^{\eta(\tau)=\Delta x/2}\!\!\!\mathcal{D}\eta\exp\left\{-\frac{1}{\hbar}\int\limits_0^\tau d\tau_1\left[\frac{M}{2}\overset{\bullet}{\eta}^2(\tau_1)+V\left(\bar{x}+\eta(\tau_1)\right)\right]\right\} \tag{1.9}$$

with the notation

$$\Delta x(x_b+x_a) \tag{1.10}$$

Note the difference between the fluctuation factor $\eta(\tau)$- as a function – in equation (1.9) and the chemical hardness η – as a functional; since first one is a local (time dependent) quantum (fluctuation) effect the second stands as a global index of an electronic system, so no confusion should arise.

In what follows we evaluate the path integral (1.9) extending the standard procedure, (see Kleinert, 2004), upto the fourth order. As such, for smooth potentials we can expand $V\left(\bar{x}+\eta(\tau)\right)$ into a Taylor expansion of the fluctuation $\eta(\tau)$

$$V\big(\bar{x}+\eta(\tau)\big)=V\big(\bar{x}\big)+\partial_iV\big(\bar{x}\big)\eta_i(\tau)+\frac{1}{2}\partial_i\partial_jV\big(\bar{x}\big)\eta_i(\tau)\eta_j(\tau)$$

$$+\frac{1}{6}\partial_i\partial_j\partial_kV\big(\bar{x}\big)\eta_i(\tau)\eta_j(\tau)\eta_k(\tau)+\frac{1}{24}\partial_i\partial_j\partial_k\partial_lV\big(\bar{x}\big)\eta_i(\tau)\eta_j(\tau)\eta_k(\tau)\eta_l(\tau)+...$$

$$(1.11)$$

and rewrite the above path integral upto the forth order in the fluctuation. However, in order to evaluate the obtained path integral there is useful to make use of the auxiliary harmonic imaginary-time amplitude

$$\big(\Delta x/2,\tau\big|-\Delta x/2,0\big)=\int_{-\Delta x/2}^{\Delta x/2}\mathcal{D}\eta\exp\left\{-\frac{1}{\hbar}\int_0^\tau d\tau_1\frac{M}{2}\dot{\eta}^2(\tau_1)\right\} \qquad (1.12)$$

and of the harmonic expectation values:

$$\big\langle F[\eta]\big\rangle=\frac{1}{\big(\Delta x/2,\tau\big|-\Delta x/2,0\big)}\int_{-\Delta x/2}^{\Delta x/2}\mathcal{D}\eta F[\eta]\exp\left\{-\frac{1}{\hbar}\int_0^\tau d\tau_1\frac{M}{2}\dot{\eta}^2(\tau)\right\} \quad (1.13)$$

With these ingredients the working path integral casts as (Putz, 2009, 2011b; Putz & Ori, 2014):

$$\big(x_b\tau\big|x_a0\big)=e^{-\frac{\tau}{\hbar}V\big(\bar{x}\big)}\big(\Delta x/2,\tau\big|-\Delta x/2,0\big)$$

$$\times\left\{1-\frac{1}{\hbar}\left[\partial_iV\big(\bar{x}\big)\int_0^\tau d\tau_1\big\langle\eta_i(\tau_1)\big\rangle+\frac{1}{2}\partial_i\partial_jV\big(\bar{x}\big)\int_0^\tau d\tau_1\big\langle\eta_i(\tau_1)\eta_j(\tau_1)\big\rangle\right.\right.$$

$$+\frac{1}{6}\partial_i\partial_j\partial_kV\big(\bar{x}\big)\int_0^\tau d\tau_1\big\langle\eta_i(\tau_1)\eta_j(\tau_1)\eta_k(\tau_1)\big\rangle$$

$$+\frac{1}{24}\partial_i\partial_j\partial_k\partial_lV\big(\bar{x}\big)\int_0^\tau d\tau_1\big\langle\eta_i(\tau_1)\eta_j(\tau_1)\eta_k(\tau_1)\eta_l(\tau_1)\big\rangle+...\bigg]$$

$$+\frac{1}{2\hbar^2}\left[\partial_iV\big(\bar{x}\big)\partial_jV\big(\bar{x}\big)\int_0^\tau d\tau_1\int_0^\tau d\tau_2\big\langle\eta_i(\tau_1)\eta_j(\tau_1)\big\rangle\right.$$

$$+\partial_k\partial_iV\big(\bar{x}\big)\partial_jV\big(\bar{x}\big)\int_0^\tau d\tau_1\int_0^\tau d\tau_2\big\langle\eta_k(\tau_1)\eta_i(\tau_1)\eta_j(\tau_2)\big\rangle$$

$$+\frac{1}{3}\partial_l\,\partial_k\,\partial_i V(\bar{x})\partial_j V(\bar{x})\int_0^\tau d\tau_1\int_0^\tau d\tau_2\langle\eta_l(\tau_1)\eta_k(\tau_1)\eta_i(\tau_1)\eta_j(\tau_2)\rangle$$

$$\left.+\frac{1}{4}\partial_k\,\partial_i V(\bar{x})\partial_l\partial_j V(\bar{x})\int_0^\tau d\tau_1\int_0^\tau d\tau_2\langle\eta_k(\tau_1)\eta_i(\tau_1)\eta_j(\tau_2)\eta_l(\tau_2)\rangle+...\right]$$

$$-\frac{1}{6\hbar^3}\left[\partial_i V(\bar{x})\partial_j V(\bar{x})\partial_k V(\bar{x})\int_0^\tau d\tau_1\int_0^\tau d\tau_2\int_0^\tau d\tau_3\langle\eta_i(\tau_1)\eta_j(\tau_2)\eta_k(\tau_3)\rangle\right.$$

$$\left.+\frac{3}{2}\partial_l\partial_i V(\bar{x})\partial_j V(\bar{x})\partial_k V(\bar{x})\int_0^\tau d\tau_1\int_0^\tau d\tau_2\int_0^\tau d\tau_3\langle\eta_l(\tau_1)\eta_i(\tau_1)\eta_j(\tau_2)\eta_k(\tau_3)\rangle+...\right]$$

$$\left.+\frac{1}{24\hbar^4}\partial_i V(\bar{x})\partial_j V(\bar{x})\partial_k V(\bar{x})\partial_l V(\bar{x})\int_0^\tau d\tau_1\int_0^\tau d\tau_2\int_0^\tau d\tau_3\int_0^\tau d\tau_4\langle\eta_i(\tau_1)\eta_j(\tau_2)\eta_k(\tau_3)\eta_l(\tau_4)\rangle+...\right\}$$

$$(1.14)$$

Since the auxiliary harmonic imaginary-time amplitude (1.12) reads explicitly

$$(\Delta x/2,\tau|-\Delta x/2,0)=\left(\frac{M}{2\pi\hbar\tau}\right)^{D/2}\exp\left\{-\frac{M}{2\pi\hbar\tau}(\Delta x)^2\right\}\qquad(1.15)$$

the harmonic expectation values can be calculated from the generating functional

$$(\Delta x/2,\tau|-\Delta x/2,0)[j]=\int_{\eta(0)=-\Delta x/2}^{\eta(\tau)=\Delta x/2}\otimes\eta\exp\left\{-\frac{1}{\hbar}\int_0^\tau d\tau_1\left[\frac{M}{2}\dot{\eta}^2(\tau_1)-j(\tau_1)\eta(\tau_1)\right]\right\}\qquad(1.16)$$

whose explicit solution looks like

$$(\Delta x/2,\tau|-\Delta x/2,0)[j]=\left(\frac{M}{2\pi\hbar\tau}\right)^{D/2}\exp\left\{\begin{array}{l}-\dfrac{M}{2\pi\hbar\tau}(\Delta x)^2+\dfrac{1}{\hbar}\int_0^\tau d\tau_1\eta^{cl}(\tau_1)j(\tau_1)\\[2mm]+\dfrac{1}{2\hbar^2}\int_0^\tau d\tau_1\int_0^\tau d\tau_2 G(\tau_1,\tau_2)j(\tau_1)j(\tau_2)\end{array}\right\}\qquad(1.17)$$

with the abbreviations for the classical (average of the quantum fluctuation) path

$$\eta^{cl}(\tau_1)=\left(\frac{\tau_1}{\tau}-\frac{1}{2}\right)\Delta x\qquad(1.18)$$

and the Green function temporally correlating two quantum events

$$G(\tau_1, \tau_2) = \frac{\hbar}{M} \frac{\theta(\tau_1 - \tau_2)(\tau - \tau_2)\tau_2 + \theta(\tau_2 - \tau_1)(\tau - \tau_2)\tau_1}{\tau} \qquad (1.19)$$

in terms of the Heaviside step-function

$$\Theta(\tau - \tau') = \begin{cases} 1 ... \tau > \tau' \\ \dfrac{1}{2} ... \tau = \tau' \\ 0 ... \tau < \tau' \end{cases} \qquad (1.20)$$

Now, the expectation values of type (1.13) in (1.14) can be evaluated from the functional derivatives (Putz, 2009, 2011b):

$$\langle \eta(\tau_1)..\eta(\tau_n) \rangle = \frac{1}{(\Delta x/2, \tau | - \Delta x/2, 0)} \left(\hbar \frac{\delta}{\delta_j(\delta_1)} \right)..\left(\hbar \frac{\delta}{\delta_j(\delta_n)} \right)$$
$$\times \left. (\Delta x/2, \tau | - \Delta x/2, 0)[j] \right|_{j=0} \qquad (1.21)$$

Taking into account the solution (1.17) we obtain, for the first order path average:

$$\langle \eta_i(\tau_1) \rangle = \eta_i^{cl}(\tau_1) \qquad (1.22a)$$

with the working specialization

$$\langle \eta_i(\tau_1) \rangle = \left(\frac{\tau_1}{\tau} - \frac{1}{2} \right) \Delta x_i \qquad (1.22b)$$

Analogously, for the second order path average we have in general

$$\langle \eta_i(\tau_1)\eta_j(\tau_2) \rangle = \eta_i^{cl}(\tau_1)\eta_j^{cl}(\tau_2) + G(\tau_1, \tau_2)\delta_{ij} \qquad (1.23a)$$

and with the working expressions as for:

- different events-times' moment (correlation)

$$\langle \eta_i(\tau_1)\eta_j(\tau_2)\rangle = \left(\frac{\tau_1}{\tau} - \frac{1}{2}\right)\left(\frac{\tau_2}{\tau} - \frac{1}{2}\right)$$

$$+ \frac{\hbar}{M}\frac{\theta(\tau_1 - \tau_2)(\tau - \tau_1)\tau_2 + \theta(\tau_2 - \tau_1)(\tau - \tau_2)\tau_1}{\tau}$$

$$(1.23b)$$

- as well as for two events-at the same time moment

$$\langle \eta_i(\tau_1)\eta_j(\tau_1)\rangle = \left(\frac{\tau_1}{\tau} - \frac{1}{2}\right)^2 \Delta x_i \Delta x_j + \frac{\hbar}{M}\frac{(\tau - \tau_1)\tau_1}{\tau}\delta_{ij} \qquad (1.23c)$$

In the way the third order path average has the general form

$$\langle \eta_i(\tau_1)\eta_j(\tau_2)\eta_k(\tau_3)\rangle = \eta_i^{cl}(\tau_1)\eta_j^{cl}(\tau_2)\eta_k^{cl}(\tau_3) + G(\tau_1,\tau_2)\delta_{ij}\eta_k^{cl}(\tau_3)$$

$$+ G(\tau_1,\tau_3)\delta_{ik}\eta_j^{cl}(\tau_2) + G(\tau_2,\tau_3)\delta_{jk}\eta_i^{cl}(\tau_1)$$

$$(1.24a)$$

with the specializations as:

- three events-at three different times formulation

$$\langle \eta_i(\tau_1)\eta_j(\tau_2)\eta_k(\tau_3)\rangle = \left(\frac{\tau_1}{\tau} - \frac{1}{2}\right)\left(\frac{\tau_2}{\tau} - \frac{1}{2}\right)\left(\frac{\tau_3}{\tau} - \frac{1}{2}\right)\Delta x_i \Delta x_j \Delta x_k$$

$$+ \frac{\hbar}{M}\frac{\theta(\tau_1 - \tau_2)(\tau - \tau_1)\tau_2 + \theta(\tau_2 - \tau_1)(\tau - \tau_2)\tau_1}{\tau}\left(\frac{\tau_3}{\tau} - \frac{1}{2}\right)\Delta x_k \delta_{ij}$$

$$+ \frac{\hbar}{M}\frac{\theta(\tau_1 - \tau_3)(\tau - \tau_1)\tau_3 + \theta(\tau_3 - \tau_1)(\tau - \tau_3)\tau_1}{\tau}\left(\frac{\tau_2}{\tau} - \frac{1}{2}\right)\Delta x_j \delta_{ik}$$

$$+ \frac{\hbar}{M}\frac{\theta(\tau_2 - \tau_3)(\tau - \tau_2)\tau_3 + \theta(\tau_3 - \tau_2)(\tau - \tau_3)\tau_2}{\tau}\left(\frac{\tau_1}{\tau} - \frac{1}{2}\right)\Delta x_i \delta_{jk} \quad (1.24b)$$

- three events-at two different times formulation

$$\langle \eta_k(\tau_1)\eta_i(\tau_1)\eta_j(\tau_2)\rangle = \left(\frac{\tau_1}{\tau} - \frac{1}{2}\right)^2\left(\frac{\tau_2}{\tau} - \frac{1}{2}\right)\Delta x_i \Delta x_j \Delta x_k$$

$$+ \frac{\hbar}{M}\left(\frac{\tau_1}{\tau} - \frac{1}{2}\right) \frac{\theta(\tau_1 - \tau_2)(\tau - \tau_1)\tau_2 + \theta(\tau_2 - \tau_1)(\tau - \tau_2)\tau_1}{\tau} \left(\Delta x_k \delta_{ij} + \Delta x_i \delta_{kj}\right)$$

$$+ \frac{\hbar}{M}\left(\frac{\tau_2}{\tau} - \frac{1}{2}\right) \frac{(\tau - \tau_1)\tau_1}{\tau} \Delta x_j \delta_{ki}$$

(1.24c)

- and three events-at equal times formulation

$$\left\langle \eta_i(\tau_1)\eta_j(\tau_1)\eta_k(\tau_1) \right\rangle = \left(\frac{\tau_1}{\tau} - \frac{1}{2}\right)^3 \Delta x_i \Delta x_j \Delta x_k$$

$$+ \frac{\hbar}{M} \frac{(\tau - \tau_1)\tau_1}{\tau}\left(\frac{\tau_1}{\tau} - \frac{1}{2}\right)\left(\Delta x_i \delta_{jk} + \Delta x_j \delta_{ik} + \Delta x_k \delta_{ij}\right)$$

(1.24d)

Finally, for the fourth order path average one has the general formulation

$$\left\langle \eta_i(\tau_1)\eta_j(\tau_2)\eta_k(\tau_3)\eta_l(\tau_4) \right\rangle = \eta_i^{cl}(\tau_1)\eta_j^{cl}(\tau_2)\eta_k^{cl}(\tau_3)\eta_l^{cl}(\tau_4) + G(\tau_1,\tau_2)\delta_{ij}\eta_k^{cl}(\tau_3)\eta_l^{cl}(\tau_4)$$

$$+ G(\tau_1,\tau_3)\delta_{ik}\eta_j^{cl}(\tau_2)\eta_l^{cl}(\tau_4) + G(\tau_1,\tau_4)\delta_{il}\eta_j^{cl}(\tau_2)\eta_k^{cl}(\tau_3)$$

$$+ G(\tau_2,\tau_3)\delta_{jk}\eta_i^{cl}(\tau_1)\eta_l^{cl}(\tau_4) + G(\tau_2,\tau_4)\delta_{jl}\eta_i^{cl}(\tau_1)\eta_k^{cl}(\tau_3)$$

$$+ G(\tau_3,\tau_4)\delta_{kl}\eta_i^{cl}(\tau_1)\eta_j^{cl}(\tau_2) + G(\tau_1,\tau_2)G(\tau_3,\tau_4)\delta_{ij}\delta_{kl}$$

$$+ G(\tau_1,\tau_3)G(\tau_2,\tau_4)\delta_{ik}\delta_{jl} + G(\tau_1,\tau_4)G(\tau_2,\tau_3)\delta_{il}\delta_{jk}$$

(1.25a)

with the specializations as:
- four events-at four different times formulation

$$\left\langle \eta_i(\tau_1)\eta_j(\tau_2)\eta_k(\tau_3)\eta_l(\tau_4) \right\rangle = \left(\frac{\tau_1}{\tau} - \frac{1}{2}\right)\left(\frac{\tau_2}{\tau} - \frac{1}{2}\right)\left(\frac{\tau_3}{\tau} - \frac{1}{2}\right)\left(\frac{\tau_4}{\tau} - \frac{1}{2}\right)\Delta x_i \Delta x_j \Delta x_l \Delta x_k$$

$$+ \frac{\hbar}{M}\left(\frac{\tau_1}{\tau} - \frac{1}{2}\right)\left(\frac{\tau_2}{\tau} - \frac{1}{2}\right)\frac{\theta(\tau_3 - \tau_4)(\tau - \tau_3)\tau_4 + \theta(\tau_4 - \tau_3)(\tau - \tau_4)\tau_3}{\tau}\Delta x_i \Delta x_j \delta_{kl}$$

$$+ \frac{\hbar}{M}\left(\frac{\tau_1}{\tau} - \frac{1}{2}\right)\left(\frac{\tau_3}{\tau} - \frac{1}{2}\right)\frac{\theta(\tau_2 - \tau_4)(\tau - \tau_2)\tau_4 + \theta(\tau_4 - \tau_2)(\tau - \tau_4)\tau_2}{\tau}\Delta x_i \Delta x_k \delta_{jl}$$

$$+\frac{\hbar}{M}\left(\frac{\tau_1}{\tau}-\frac{1}{2}\right)\left(\frac{\tau_4}{\tau}-\frac{1}{2}\right)\frac{\theta(\tau_2-\tau_3)(\tau-\tau_2)\tau_3+\theta(\tau_3-\tau_2)(\tau-\tau_3)\tau_2}{\tau}\Delta x_i \Delta x_l \delta_{jk}$$

$$+\frac{\hbar}{M}\left(\frac{\tau_3}{\tau}-\frac{1}{2}\right)\left(\frac{\tau_4}{\tau}-\frac{1}{2}\right)\frac{\theta(\tau_1-\tau_2)(\tau-\tau_1)\tau_2+o\theta(\tau_2-\tau_1)(\tau-\tau_2)\tau_1}{\tau}\Delta x_k \Delta x_l \delta_{ij}$$

$$+\frac{\hbar}{M}\left(\frac{\tau_2}{\tau}-\frac{1}{2}\right)\left(\frac{\tau_4}{\tau}-\frac{1}{2}\right)\frac{\theta(\tau_1-\tau_3)(\tau-\tau_1)\tau_3+\theta(\tau_3-\tau_1)(\tau-\tau_3)\tau_1}{\tau}\Delta x_j \Delta x_l \delta_{ik}$$

$$+\frac{\hbar}{M}\left(\frac{\tau_2}{\tau}-\frac{1}{2}\right)\left(\frac{\tau_3}{\tau}-\frac{1}{2}\right)\frac{\theta(\tau_1-\tau_4)(\tau-\tau_1)\tau_4+\theta(\tau_4-\tau_1)(\tau-\tau_4)\tau_1}{\tau}\Delta x_j \Delta x_k \delta_{il}$$

$$+\frac{\hbar^2}{M^2}\frac{\theta(\tau_1-\tau_2)(\tau-\tau_1)\tau_2+\theta(\tau_3-\tau_4)(\tau-\tau_3)\tau_4}{\tau}\frac{\theta(\tau_3-\tau_4)(\tau-\tau_3)\tau_4+\theta(\tau_4-\tau_3)(\tau-\tau_4)\tau_3}{\tau}\delta_{ij}\delta_{kl}$$

$$+\frac{\hbar^2}{M^2}\frac{\theta(\tau_1-\tau_3)(\tau-\tau_1)\tau_3+\theta(\tau_3-\tau_1)(\tau-\tau_3)\tau_1}{\tau}\frac{\theta(\tau_2-\tau_4)(\tau-\tau_2)\tau_4+\theta(\tau_4-\tau_2)(\tau-\tau_4)\tau_2}{\tau}\delta_{ik}\delta_{jt}$$

$$+\frac{\hbar^2}{M^2}\frac{\theta(\tau_1-\tau_4)(\tau-\tau_1)\tau_4+\theta(\tau_4-\tau_1)(\tau-\tau_4)\tau_1}{\tau}\frac{\theta(\tau_2-\tau_3)(\tau-\tau_2)\tau_3+\theta(\tau_3-\tau_2)(\tau-\tau_3)\tau_2}{\tau}\delta_{il}\delta_{jk}$$

$$(1.25b)$$

- four events-at equal times formulation

$$\langle \eta_i(\tau_1)\eta_j(\tau_1)\eta_k(\tau_1)\eta_l(\tau_1)\rangle = \left(\frac{\tau_1}{\tau}-\frac{1}{2}\right)^4 \Delta x_i \Delta x_j \Delta x_k \Delta x_l$$

$$+\frac{\hbar}{M}\frac{(\tau-\tau_1)\tau_1}{\tau}\left(\frac{\tau_1}{\tau}-\frac{1}{2}\right)^2 \binom{\Delta x_i \Delta x_j \delta_{kl}+\Delta_i\Delta_k\delta_{jl}+\Delta_i\Delta_l\delta_{jk}}{+\Delta x_k \Delta x_l \delta_{ij}+\Delta x_j \Delta x_l \delta_{ik}+\Delta x_j \Delta x_k \delta_{ie}}$$

$$+\frac{\hbar^2}{M^2}\frac{(\tau-\tau_1^2)}{\tau^2}\left(\partial_{ij}\partial_{kl}+\delta_{ik}\delta_{jl}+\delta_{il}\delta_{jk}\right)$$

$$(1.25c)$$

- four events-at two symmetrical different times formulation

$$\langle \eta_k(\tau_1)\eta_i(\tau_1)\eta_j(\tau_2)\eta_i(\tau_2)\rangle = \left(\frac{\tau_1}{\tau}-\frac{1}{2}\right)^2\left(\frac{\tau_2}{\tau}-\frac{1}{2}\right)^2 \Delta x_i \Delta x_j \Delta x_k \Delta x_l$$

$$+\frac{\hbar}{M}\frac{(\tau-\tau_2)\tau_2}{\tau}\left(\frac{\tau_1}{\tau}-\frac{1}{2}\right)^2 \Delta x_k \Delta x_j \delta_{jl}+\frac{\hbar}{M}\frac{(\tau-\tau_1)}{\tau}\left(\frac{\tau_2}{\tau}-\frac{1}{2}\right)^2 \Delta x_j \Delta x_i \delta_{kl}$$

$$+\frac{\hbar^2}{M^2}\frac{(\tau-\tau_1)\tau_1(\tau-\tau_2)\tau_2}{\tau^2}\delta_{ki}\delta_{jl}$$

$$+\frac{\hbar}{M}\left(\frac{\tau_1}{\tau}-\frac{1}{2}\right)\left(\frac{\tau_2}{\tau}-\frac{1}{2}\right)\frac{\theta(\tau_1-\tau_2)(\tau-\tau_1)\tau_2+\theta(\tau_2-\tau_1)(\tau-\tau_2)\tau_1}{\tau}\binom{\Delta x_k \Delta x_j \delta_{il}+\Delta x_k \Delta x_i \delta_{ij}}{+\Delta x_i \Delta x_j \delta_{kj}+\Delta x_i \Delta x_j \delta_{kl}}$$

$$+\frac{\hbar^2}{M^2}\frac{\theta(\tau_1-\tau_2)(\tau-\tau_1)\tau_2+\theta(\tau_2-\tau_1)(\tau-\tau_2)\tau_1}{\tau^2}\left(\delta_{kj}\delta_{il}+\delta_{ki}\delta_{ij}\right)$$

$$(1.25\text{d})$$

- four events-at two asymmetrical different times formulation

$$\langle\eta_i(\tau_1)\eta_k(\tau_1)\eta_l(\tau_1)\eta(\tau_2)\rangle=\left(\frac{\tau_1}{\tau}-\frac{1}{2}\right)^3\left(\frac{\tau_2}{\tau}-\frac{1}{2}\right)\Delta x_i\Delta x_j\Delta x_k\Delta x_l$$

$$+\frac{\hbar}{M}\left(\frac{\tau_1}{\tau}-\frac{1}{2}\right)^2\frac{\theta(\tau_1-\tau_2)(\tau-\tau_1)\tau_2+\theta(\tau_2-\tau_1)(\tau-\tau_2)\tau_1}{\tau}\left(\Delta x_i\Delta x_k\delta_{ij}+\Delta x_i\Delta x_i\delta_{kj}+\Delta x_k\Delta x_i\delta_{ij}\right)$$

$$+\frac{\hbar}{M}\left(\frac{\tau_1}{\tau}-\frac{1}{2}\right)\left(\frac{\tau_2}{\tau}-\frac{1}{2}\right)\frac{(\tau-\tau_1)\tau_1}{\tau}\left(\Delta x_i\Delta x_j\delta_{ki}+\Delta x_i\Delta x_j\delta_{ik}+\Delta x_k\Delta x_j\delta_{ii}\right)$$

$$+\frac{\hbar^2}{M^2}\frac{(\tau-\tau_1)\tau_1}{\tau}\frac{\theta(\tau_1-\tau_2)(\tau-\tau_1)\tau_2+\theta(\tau_2-\tau_1)(\tau-\tau_2)\tau_1}{\tau}\left(\delta_{ik}\delta_{ij}+\delta_{ii}\delta_{kj}+\delta_{ij}\delta_{ki}\right)$$

$$(1.25\text{e})$$

- four events-at three different times formulation

$$\langle\eta_i(\tau_1)\eta_l(\tau_1)\eta_j(\tau_2)\eta_k(\tau_3)\rangle=\left(\frac{\tau_1}{\tau}-\frac{1}{2}\right)^2\left(\frac{\tau_2}{\tau}-\frac{1}{2}\right)\left(\frac{\tau_3}{\tau}-\frac{1}{2}\right)\Delta x_i\Delta x_j\Delta x_k\Delta x_l$$

$$+\frac{\hbar}{M}\frac{(\tau-\tau_1)\tau_1}{\tau}\left(\frac{\tau_2}{\tau}-\frac{1}{2}\right)\left(\frac{\tau_3}{\tau}-\frac{1}{2}\right)\Delta x_j\Delta x_k\Delta x_{li}$$

$$+\frac{\hbar}{M}\left(\frac{\tau_1}{\tau}-\frac{1}{2}\right)^2\frac{\theta(\tau_2-\tau_3)(\tau-\tau_2)\tau_3+\theta(\tau_3-\tau_2)(\tau-\tau_3)\tau_2}{\tau}\Delta x_i\Delta x_i\Delta x_{jk}$$

$$+\frac{\hbar}{M}\left(\frac{\tau_1}{\tau}-\frac{1}{2}\right)\left(\frac{\tau_2}{\tau}-\frac{1}{2}\right)\frac{\theta(\tau_1-\tau_3)(\tau-\tau_1)\tau_3+\theta(\tau_3-\tau_1)(\tau-\tau_3)\tau_1}{\tau}\left(\Delta x_i\Delta x_j\delta_{ik}+\Delta x_i\Delta x_j\delta_{lk}\right)$$

$$+\frac{\hbar}{M}\left(\frac{\tau_1}{\tau}-\frac{1}{2}\right)\left(\frac{\tau_3}{\tau}-\frac{1}{2}\right)\frac{\theta(\tau_1-\tau_2)(\tau-\tau_1)\tau_2+\theta(\tau_2-\tau_1)(\tau-\tau_2)\tau_1}{\tau}\left(\Delta x_i\Delta x_k\delta_{ij}+\Delta x_i\Delta x_k\delta_{ij}\right)$$

$$+\frac{\hbar^2}{M^2}\frac{(\tau-\tau_1)\tau_1}{\tau}\frac{\theta(\tau_2-\tau_3)(\tau-\tau_2)\tau_3+\theta(\tau_3-\tau_2)(\tau-\tau_3)\tau_2}{\tau}\delta_{il}\delta_{jk}$$

$$+\frac{\hbar^2}{M^2}\frac{\theta(\tau_1-\tau_2)(\tau-\tau_1)\tau_2+\theta(\tau_2-\tau_1)(\tau-\tau_2)\tau_1}{\tau}\frac{\theta(\tau_1-\tau_3)(\tau-\tau_1)\tau_3+\theta(\tau_3-\delta_1)(\tau-\tau_3)\tau_1}{\tau}\left(\begin{array}{c}\delta_{ij}\delta_{ik}\\+\delta_{ik}\delta_{ij}\end{array}\right)$$

$$(1.25\text{f})$$

With this recipe some imaginary-time integrals over above expectation values vanish (Putz, 2009, 2011b):

$$\int_0^\tau d\tau_1 \langle \eta_i(\tau_1) \rangle = \int_0^\tau d\tau_1 \langle \eta_i(\tau_1) \eta_j(\tau_1) \eta_k(\tau_1) \rangle = \int_0^\tau d\tau_1 \int_0^\tau d\tau_2 \langle \eta_k(\tau_1) \eta_i(\tau_1) \eta_j(\tau_2) \rangle$$

$$= \int_0^\tau d\tau_1 \int_0^\tau d\tau_2 \int_0^\tau d\tau_3 \langle \eta_i(\tau_1) \eta_j(\tau_2) \eta_k(\tau_3) \rangle = 0$$

$$(1.26)$$

while for the non-vanishing imaginary-time integrals appearing in Eq. (1.14) one yields

$$\int_0^\tau d\tau_1 \langle \eta_i(\tau_1) \eta_j(\tau_1) \rangle = \frac{\tau}{12} \Delta x_i \Delta x_j + \frac{\hbar}{M} \frac{\tau^2}{6} \delta_{ij} \qquad (1.27)$$

$$\int_0^\tau d\tau_1 \langle \eta_i(\tau_1) \eta_j(\tau_1) \eta_k(\tau_1) \eta_l(\tau_1) \rangle = \frac{\tau}{80} \Delta x_i \Delta x_j \Delta x_k \Delta x_l$$

$$+ \frac{\hbar}{M} \frac{\tau^2}{120} \left(\Delta x_i \Delta x_j \delta_{kl} + \Delta x_i \Delta x_k \delta_{jl} + \Delta x_i \Delta x_l \delta_{jk} + \Delta x_k \Delta x_l \delta_{ij} + \Delta x_j \Delta x_l \delta_{ik} + \Delta x_j \Delta x_k \delta_{il} \right)$$

$$+ \frac{\hbar^2}{M^2} \frac{\tau^3}{30} \left(\delta_{ij} \delta_{kl} + \delta_{ik} \delta_{jl} + \delta_{il} \delta_{ik} \right)$$

$$(1.28)$$

$$\int_0^\tau d\tau_1 \int_0^\tau d\tau_2 \langle \eta_i(\tau_1) \eta_j(\tau_2) \rangle = \frac{\hbar}{M} \frac{\tau^3}{12} \delta_{ij} \qquad (1.29)$$

$$\int_0^\tau d\tau_1 \int_0^\tau d\tau_2 \langle \eta_k(\tau_1) \eta_i(\tau_1) \eta_j(\tau_2) \eta_l(\tau_2) \rangle = \frac{\tau^2}{144} \Delta x_i \Delta x_j \Delta x_l \Delta x_k$$

$$+ \frac{\hbar}{M} \frac{\tau^3}{72} \left(\Delta x_k \Delta x_i \delta_{jl} + \Delta x_j \Delta x_l \delta_{ki} \right)$$

$$+ \frac{\hbar}{M} \frac{\tau^3}{720} \left(\Delta x_k \Delta x_j \delta_{il} + \Delta x_k \Delta x_l \delta_{ij} + \Delta x_i \Delta x_l \delta_{kj} + \Delta x_i \Delta x_j \delta_{kl} \right)$$

$$+ \frac{\hbar^2}{M^2} \frac{\tau^4}{36} \delta_{ki} \delta_{jl} + \frac{\hbar^2}{M^2} \frac{\tau^4}{90} \left(\delta_{kj} \delta_{il} + \delta_{kl} \delta_{ij} \right)$$

$$(1.30)$$

$$\int_0^\tau d\tau_1 \int_0^\tau d\tau_2 \left\langle \eta_l(\tau_1)\eta_k(\tau_1)\eta_i(\tau_1)\eta_j(\tau_2)\right\rangle = \frac{\hbar}{M}\frac{\tau^3}{240}\binom{\Delta x_l \Delta x_k \delta_{ij}}{+\Delta x_l \Delta x_i \delta_{kj} + \Delta x_k \Delta x_i \delta_{lj}}$$

$$+\frac{\hbar^2}{M^2}\frac{\tau^4}{60}\left(\delta_{lk}\delta_{ij} + \delta_{li}\delta_{kj} + \delta_{lj}\delta_{ki}\right)$$

$$(1.31)$$

$$\int_0^\tau d\tau_1 \int_0^\tau d\tau_2 \int_0^\tau d\tau_3 \left\langle \eta_l(\tau_1)\eta_i(\tau_1)\eta_j(\tau_2)\eta_k(\tau_3)\right\rangle = \frac{\hbar}{M}\frac{\tau^4}{144}\Delta x_l \Delta x_i \delta_{jk}$$

$$+\frac{\hbar^2}{M^2}\frac{\tau^5}{72}\delta_{li}\delta_{jk}$$

$$+\frac{\hbar^2}{M^2}\frac{\tau^5}{120}\left(\delta_{lj}\delta_{ik} + \delta_{lk}\delta_{ij}\right)$$

$$(1.32)$$

$$\int_0^\tau d\tau_1 \int_0^\tau d\tau_2 \int_0^\tau d\tau_3 \int_0^\tau d\tau_4 \left\langle \eta_i(\tau_1)\eta_j(\tau_2)\eta_k(\tau_3)\eta_l(\tau_4)\right\rangle = \frac{\hbar^2}{M^2}\frac{\tau^6}{144}$$

$$\left(\delta_{ij}\delta_{kl} + \delta_{ik}\delta_{jl} + \delta_{il}\delta_{jk}\right)$$

$$(1.33)$$

Thereby, the result for the semiclassical evolution amplitude upto fourth order in *D-dimension* is obtained (Putz, 2009, 2011b):

$$(x_b,\tau|x_a,0) = \left(\frac{M}{2\pi\hbar\tau}\right)^{D/2}\exp\left\{-\frac{M}{2\pi\hbar\tau}\Delta x^2 - \frac{\tau}{\hbar}V(\bar{x})\right\}$$

$$\times\left\{1 - \frac{1}{\hbar}\left[\frac{1}{2}\partial_i\partial_j V(\bar{x})\left(\Delta x_i \Delta x_j \frac{\tau}{12} + \delta_{ij}\frac{\hbar}{M}\frac{\tau^2}{6}\right) + \frac{1}{24}\partial_i\partial_j\partial_k\partial_l V(\bar{x})\right.\right.$$

$$\times\left\langle \Delta x_i \Delta x_j \Delta x_k \Delta x_l \frac{\tau}{80} + \left(\Delta x_i \Delta x_j \delta_{jl} + \Delta x_i \Delta x_k \delta_{jl} + \Delta x_i \Delta x_l \delta_{jk} + \Delta x_k \Delta x_l \delta_{ij}\right.\right.$$

$$\left.\left. + \Delta x_j \Delta x_i \delta_{ik} + \Delta x_j \Delta x_k \delta_{il}\right)\frac{\hbar}{M}\frac{\tau^2}{120} + \left(\delta_{ij}\delta_{kl} + \delta_{ik}\delta_{jl} + \delta_{il}\delta_{jk}\right)\frac{\hbar^2}{M^2}\frac{\tau^3}{30}\right\rangle\right]$$

$$+\frac{1}{2\hbar^2}\left[\partial_i V(\bar{x})\partial_j V(\bar{x})\delta_{ij}\frac{\hbar}{M}\frac{\tau^3}{12} + \frac{1}{4}\partial_k\partial_i V(\bar{x})\partial_i\partial_j V(\bar{x})\right.$$

$$\times \left\langle \Delta x_i \Delta x_j \Delta x_l \Delta x_k \frac{\tau^2}{144} + \left(\Delta x_k \Delta x_i \delta_{jl} + \Delta x_j \Delta x_l \delta_{ki} \right) \frac{\hbar}{M} \frac{\tau^3}{72} \right.$$

$$+ \left(\Delta x_k \Delta x_j \delta_{il} + \Delta x_k \Delta x_l \delta_{ij} + \Delta x_l \Delta x_j \delta_{kj} + \Delta x_i \Delta x_j \delta_{kl} \right) \frac{\hbar}{M} \frac{\tau^3}{720}$$

$$\left. + \delta_{ki} \delta_{ji} \frac{\hbar^2}{M^2} \frac{\tau^4}{36} + \left(\delta_{kj} \delta_{il} + \delta_{kl} \delta_{ij} \right) \frac{\hbar^2}{M^2} \frac{\tau^4}{90} \right\rangle$$

$$+ \frac{1}{3} \partial_l \partial_k \partial_i V(\overline{x}) \partial_j V(\overline{x}) \left\langle \left(\Delta x_l \Delta x_k \delta_{ij} + \Delta x_l \Delta x_i \delta_{kj} + \Delta x_k \Delta x_i \delta_{lj} \right) \frac{\hbar}{M} \frac{\tau^3}{240} \right.$$

$$\left. + \left(\delta_{lk} \delta_{ij} + \delta_{li} \delta_{kj} + \delta_{lj} \delta_{ki} \right) \frac{\hbar^2}{M^2} \frac{\tau^4}{60} \right\rangle \Bigg]$$

$$- \frac{1}{6\hbar^3} \left[\frac{3}{2} \partial_l \partial_i V(\overline{x}) \partial_j V(\overline{x}) \partial_k V(\overline{x}) \left\langle \Delta x_l \Delta x_i \delta_{jk} \frac{\hbar}{M} \frac{\tau^4}{144} + \delta_{li} \delta_{jk} \frac{\hbar^2}{M^2} \frac{\tau^5}{72} \right.\right.$$

$$\left.\left. + \left(\delta_{lj} \delta_{ik} + \delta_{lk} \delta_{ij} \right) \frac{\hbar^2}{M^2} \frac{\tau^5}{120} \right\rangle \right]$$

$$+ \frac{1}{24\hbar^4} \partial_i V(\overline{x}) \partial_j V(\overline{x}) \partial_k V(\overline{x}) \partial_l V(\overline{x}) \left\langle \delta_{ij} \delta_{kl} + \delta_{ik} \delta_{jl} + \delta_{il} \delta_{jk} \right\rangle \frac{\hbar^2}{M^2} \frac{\tau^6}{144} \Bigg\}$$

$$(1.34)$$

As a note, let's remark that the above time evolution amplitude corresponds to the second order semiclassical expansion of the mass M of the particle as can be clearly seen from unfolded expression (1.14), since $\partial_i[\bullet] \propto \sqrt{M}$. Such behavior is a special reflection of considering quantum corrections to the classical action of a system allowing mass to be coordinate-dependent from the outset (Putz, 2009, 2011b). The usefulness of this result can be visualized in many ways, from describing the atomic systems via semiclassical quantum time evolution amplitude and partition function upto describing interesting physical systems as compound nuclei, where the collective Hamiltonian contains coordinate-dependent collective mass parameters (Putz, 2009, 2011b). However, in the present study we will consider the semiclassical conditional probability (1.34) to the one-dimensional case to unfold the bondonic propagation study on the graphenes with topological (Stone-Wales) defects.

1.3.2 SCHRÖDINGER'S CONDITIONAL PROBABILITY DENSITY

Before proceeding with employment of the conditional probability (1.34) to further developments, giving its complexity, worth checking it for Schrödinger equation fulfillment, however for 1-dimensional case written in the quantum amplitude form:

$$-\hbar\frac{\partial}{\partial\tau_b}\left(x_b,\tau_b|x_a,\tau_a\right) = \left\{-\frac{\hbar^2}{2M}\frac{\partial^2}{\partial x_b^2} + V(x_b)\right\}\left(x_b,\tau_b|x_a,\tau_a\right) \quad (1.35)$$

or equivalently in the time-space evolution conditional probability form

$$-\hbar\frac{\partial}{\partial\Delta\tau}\left(\Delta x,\Delta\tau\right) = \left\{-\frac{\hbar^2}{2M}\frac{\partial^2}{\partial\Delta x^2} + V(x_a + \Delta x)\right\}\left(\Delta x,\Delta\tau\right) \quad (1.36)$$

Equation (1.36) is asked to have the formal (path integral) solution, i.e., factorizing the thermal amplitude, see Volumes I and II of the present five volume set (Putz, 2016b-c)

$$\left(\Delta x,\Delta\tau\right) = \sqrt{\frac{M}{2\pi\hbar\Delta\tau}}\exp\left\{-\frac{M\Delta x^2}{2\hbar\Delta\tau} - \frac{\Delta\tau}{\hbar}V(x_a)\right\}F\left(\Delta x,\Delta\tau\right) \quad (1.37)$$

by the actual fourth order semiclassical/quantum fluctuation correction $F(\Delta x, \Delta\tau)$. Next we are seeking the conditions such correction should fulfill and then checking for our actual forth order expansion. To this aim, one plug the formal solution (1.37) in evolutionary probability density Schrödinger equation (1.36) by its temporal derivative

$$\frac{\partial(\Delta x,\Delta\tau)}{\partial\Delta\tau} = \left\{\left[-\frac{1}{2\Delta\tau} + \frac{M\Delta x^2}{2\hbar\Delta\tau^2} - \frac{1}{\hbar}V(x_a)\right]F(\Delta x,\Delta\tau) + \frac{\partial F(\Delta x,\Delta\tau)}{\partial\Delta\tau}\right\}(\Delta x,\Delta\tau)^{(0)}$$

$$(1.38)$$

with

$$\left(\Delta x,\Delta\tau\right)^{(0)} = \sqrt{\frac{M}{2\pi\hbar\Delta\tau}}\exp\left\{-\frac{M\Delta x^2}{2\hbar\Delta\tau} - \frac{\Delta\tau}{\hbar}V(x_a)\right\} \quad (1.39)$$

Along the spatial derivative obtained by the first derivation

$$\frac{\partial(\Delta x, \Delta \tau)}{\partial \Delta x} = \left[-\frac{M\Delta x}{\hbar \Delta \tau} F(\Delta x, \Delta \tau) + \frac{\partial F(\Delta x, \Delta \tau)}{\partial \Delta x} \right] (\Delta x, \Delta \tau)^{(0)} \qquad (1.40)$$

and then to the following second order

$$\frac{\partial^2 (\Delta x, \Delta \tau)}{\partial (\Delta x)^2} = \left\{ \begin{array}{l} -\dfrac{M}{\hbar \Delta \tau} F(\Delta x, \Delta \tau) + \dfrac{M^2 \Delta x^2}{\hbar^2 \Delta \tau^2} F(\Delta x, \Delta \tau) \\[2mm] -2 \dfrac{M\Delta x}{\hbar \Delta \tau} \dfrac{\partial F(\Delta x, \Delta \tau)}{\partial \Delta x} + \dfrac{\partial^2 F(\Delta x, \Delta \tau)}{\partial \Delta x^2} \end{array} \right\} (\Delta x, \Delta \tau)^{(0)} \qquad (1.41)$$

All together, Eqs. (1.38) and (1.41) in Eq. (1.36) provides the actual Schrödinger equation for the conditional probability quantum correction to be:

$$-\hbar \frac{\partial F(\Delta x, \Delta \tau)}{\partial \Delta \tau}$$
$$= -\frac{\hbar^2}{2M} \frac{\partial^2 F(\Delta x, \Delta \tau)}{\partial \Delta x^2} + \frac{\hbar \Delta x}{\Delta \tau} \frac{\partial F(\Delta x, \Delta \tau)}{\partial \Delta x} + \{V(x_a + \Delta x) - V(x_a)\} F(\Delta x, \Delta \tau)$$

$$(1.42)$$

However, when effectively considering now the fourth order expansion, for the potential contribution

$$V(x_a + \Delta x) - V(x_a) = V'(x_a)\Delta x + \frac{1}{2} V''(x_a)\Delta x^2 + \frac{1}{6} V'''(x_a)\Delta x^3 + \frac{1}{24} V''''(x_a)\Delta x^4 \qquad (1.43)$$

and, respectively, for the quantum fluctuation correction too,

$$F(\Delta x, \Delta \tau) = F_0(\Delta \tau) + F_1(\Delta \tau)\Delta x + F_2(\Delta \tau)\Delta x^2 + F_3(\Delta \tau)\Delta x^3 + F_4(\Delta \tau)\Delta x^4 \qquad (1.44)$$

the actual equation (1.42) becomes:

$$-\hbar \frac{\partial F_0(\Delta \tau)}{\partial \Delta \tau} - \hbar \frac{\partial F_1(\Delta \tau)}{\partial \Delta \tau}\Delta x - \hbar \frac{\partial F_2(\Delta \tau)}{\partial \Delta \tau}\Delta x^2 - \hbar \frac{\partial F_3(\Delta \tau)}{\partial \Delta \tau}\Delta x^3 - \hbar \frac{\partial F_4(\Delta \tau)}{\partial \Delta \tau}\Delta x^4$$
$$= \frac{\hbar \Delta x}{\Delta \tau} \left[F_1(\Delta \tau) + 2F_2(\Delta \tau)\Delta x + 3F_3(\Delta \tau)\Delta x^2 + 4F_4(\Delta \tau)\Delta x^3 \right]$$
$$-\frac{\hbar^2}{2M} \left[2F_2(\Delta \tau) + 6F_3(\Delta \tau)\Delta x + 12F_4(\Delta \tau)\Delta x^2 \right]$$

$$+ F_0(\Delta\tau)V'(x_a)\Delta x + F_1(\Delta\tau)V'(x_a)\Delta x^2 + F_2(\Delta\tau)V'(x_a)\Delta x^3 + F_3(\Delta\tau)V'(x_a)\Delta x^4$$

$$+ F_0(\Delta\tau)\frac{1}{2}V''(x_a)\Delta x^2 + F_1(\Delta\tau)\frac{1}{2}V''(x_a)\Delta x^3 + F_2(\Delta\tau)\frac{1}{2}V''(x_a)\Delta x^4$$

$$+ F_0(\Delta\tau)\frac{1}{6}V'''(x_a)\Delta x^3 + F_1(\Delta\tau)\frac{1}{6}V'''(x_a)\Delta x^4$$

$$+ F_0(\Delta\tau)\frac{1}{24}V''''(x_a)\Delta x^4$$

$$(1.45)$$

Equation (1.45) allows for one-to-one correspondences for the spatially involved orders in quantum evolution. This way the five-fold system is formed by the individual respective equations:

$$(\Delta x)^0 : -\hbar\frac{\partial F_0(\Delta\tau)}{\partial\Delta\tau} = -\frac{\hbar^2}{M}F_2(\Delta\tau) \tag{1.46}$$

$$(\Delta x)^1 : -\hbar\frac{\partial F_1(\Delta\tau)}{\partial\Delta\tau} = \frac{\hbar}{\Delta\tau}F_1(\Delta\tau) - 3\frac{\hbar^2}{M}F_3(\Delta\tau) + F_0(\Delta\tau)V'(x_a) \tag{1.47}$$

$$(\Delta x)^2 : -\hbar\frac{\partial F_2(\Delta\tau)}{\partial\Delta\tau} = 2\frac{\hbar}{\Delta\tau}F_2(\Delta\tau) - 6\frac{\hbar^2}{M}F_4(\Delta\tau) + F_1(\Delta\tau)V'(x_a) + \frac{1}{2}V''(x_a)F_0(\Delta\tau)$$

$$(1.48)$$

$$(\Delta x)^3 : -\hbar\frac{\partial F_3(\Delta\tau)}{\partial\Delta\tau} = 3\frac{\hbar}{\Delta\tau}F_3(\Delta\tau) + F_2(\Delta\tau)V'(x_a) + \frac{1}{2}F_1(\Delta\tau)V''(x_a) + \frac{1}{6}V'''(x_a)F_0(\Delta\tau)$$

$$(1.49)$$

$$(\Delta x)^4 : -\hbar\frac{\partial F_4(\Delta\tau)}{\partial\Delta\tau}$$

$$= 4\frac{\hbar}{\Delta\tau}F_4(\Delta\tau) + F_3(\Delta\tau)V'(x_a) + \frac{1}{2}F_2(\Delta\tau)V''(x_a) + \frac{1}{6}F_1(\Delta\tau)V'''(x_a) + \frac{1}{24}F_0(\Delta\tau)V''''(x_a)$$

$$(1.50)$$

Practically, now, we have to check that the one-dimensional conditional probability (1.34) satisfies the above coupled system of equations. To achieve this goal, we can evenly find the analytic expressions of the coefficient functions of Eq. (1.44) from the expression (1.34). As such, they can spring out by performing in Eq. (1.34) all the relevant combinations of the expansions of the potential $V(x)$ and of its derivatives upto the fourth order:

$$(\Delta x, \Delta \tau) \cong \sqrt{\frac{M}{2\pi\hbar\Delta\tau}} \exp\left\{-\frac{M\Delta x^2}{2\hbar\Delta\tau} - \frac{\Delta\tau}{\hbar}\left[V(x_a) + \frac{1}{2}V'(x_a)\Delta x + \frac{1}{2}V''(x_a)\left(\frac{\Delta x}{2}\right)^2\right.\right.$$

$$\left.\left. + \frac{1}{6}V'''(x_a)\left(\frac{\Delta x}{2}\right)^3 + \frac{1}{24}V''''(x_a)\left(\frac{\Delta x}{2}\right)^4\right]\right\}$$

$$\times\left\{1 - \frac{\Delta\tau^2}{12M}V'' - \frac{\Delta\tau}{\hbar}\frac{1}{24}(\Delta x)^2 V''\left(\frac{x_a+x_b}{2}\right) - \frac{\Delta\tau^2}{48M}\Delta x V'''\left(\frac{x_a+x_b}{2}\right)\right.$$

$$-\frac{\Delta\tau}{24\cdot 80\hbar}\Delta x^4 V''''\left(\frac{x_a+x_b}{2}\right) - \frac{\Delta\tau^2}{24\cdot 20M}(\Delta x)^2 V''''\left(\frac{x_a+x_b}{2}\right)$$

$$-\hbar\frac{\Delta\tau^3}{24\cdot 10M^2}V''''\left(\frac{x_a+x_b}{2}\right) + \frac{1}{24M}\frac{\Delta\tau^3}{\hbar}V'\left(\frac{x_a+x_b}{2}\right)^2$$

$$+\frac{1}{8}V''\left(\frac{x_a+x_b}{2}\right)^2\left(\frac{\Delta\tau^2}{144\hbar^2}\Delta x^4 + \frac{\Delta\tau^3}{30\hbar M}\Delta x^2 + \frac{29}{180M^2}\Delta\tau^4\right)$$

$$+\frac{1}{6}V'''\left(\frac{x_a+x_b}{2}\right)V'\left(\frac{x_a+x_b}{2}\right)\left(\frac{\Delta\tau^3}{80\hbar M}\Delta x^2 + \frac{\Delta\tau^4}{20M^2}\right)$$

$$-\frac{1}{4}V''\left(\frac{x_a+x_b}{2}\right)V'\left(\frac{x_a+x_b}{2}\right)^2\left(\frac{\Delta\tau^4}{144\hbar^2 M}\Delta x^2 + \frac{11\Delta\tau^5}{360\hbar M^2}\right) + \frac{1}{24}V'\left(\frac{x_a+x_b}{2}\right)^4\frac{\Delta\tau^6}{48\hbar^2 M^2}\right\}$$

$$(1.51)$$

Now we take into account that also the potential derivatives can be expanded towards the fourth order contribution, thus providing the respective contributions:

$$V'\left(\frac{x_a+x_b}{2}\right) = V'\left(x_a + \frac{x_b-x_a}{2}\right)$$

$$\cong V'(x_a) + V''(x_a)\frac{\Delta x}{2} + \frac{1}{2}V'''(x_a)\left(\frac{\Delta x}{2}\right)^2 + \frac{1}{6}V''''(x_a)\left(\frac{\Delta x}{2}\right)^3 \qquad (1.52)$$

$$V''\left(\frac{x_a+x_b}{2}\right) = V''\left(x_a' + \frac{x_b-x_a}{2}\right)$$

$$\cong V''(x_a) + V'''(x_a)\frac{\Delta x}{2} + \frac{1}{2}V''''(x_a)\left(\frac{\Delta x}{2}\right)^2 \qquad (1.53)$$

$$V'''\left(\frac{x_a + x_b}{2}\right) = V'''\left(x_a + \frac{x_b - x_a}{2}\right) \cong V'''(x_a) + V''''(x_a)\frac{\Delta x}{2} \qquad (1.54)$$

$$V''''\left(\frac{x_a + x_b}{2}\right) = V''''\left(x_a + \frac{x_b - x_a}{2}\right) \cong V''''(x_a) \qquad (1.55)$$

When Eqs. (1.52)–(1.55) back in Eq. (1.51) one has the detailed expansion:

$$(\Delta x, \Delta \tau) \cong \sqrt{\frac{M}{2\pi\hbar\Delta\tau}} \exp\left\{-\frac{M\Delta x^2}{2\hbar\Delta\tau} - \frac{\Delta\tau}{\hbar}V(x_a)\right\}$$

$$\times \left\{1 - \frac{\Delta\tau^2}{12M}V''(x_a) + \left(\frac{V'(x_a)^2}{24Mh} - \frac{\hbar V''''(x_a)}{240M^2}\right)\Delta\tau^3 + \left(\frac{29}{8\cdot 180}\frac{V''(x_a)^2}{M^2} + \frac{V'(x_a)V'''(x_a)}{120M^2}\right)\Delta\tau^4\right.$$

$$-\frac{11}{4\cdot 360}\frac{V''(x_a)V'(x_a)^2}{\hbar M^2}\Delta\tau^5 + \frac{V'(x_a)^4}{24\cdot 48M^2\hbar^2}\Delta\tau^6$$

$$+\Delta x\left[-\frac{V'(x_a)}{2\hbar}\Delta\tau - \frac{V'''(x_a)}{16M}\Delta\tau^2 + \frac{V'(x_a)V''(x_a)}{12Mh}\Delta\tau^3 - \frac{V'(x_a)^3}{48Mh^2}\Delta\tau^4\right]$$

$$+\Delta x^2\left[-\frac{V''(x_a)}{6\hbar}\Delta\tau + \left(\frac{V'(x_a)^2}{8\hbar^2} - \frac{11}{480}\frac{V''''(x_a)}{M}\right)\Delta\tau^2 + \left(\frac{7}{160}\frac{V'(x_a)V'''(x_a)}{Mh} + \frac{1}{40}\frac{V''(x_a)^2}{\hbar M}\right)\Delta\tau^3\right.$$

$$\left. -\frac{11}{288}\frac{V''(x_a)V'(x_a)^2}{Mh^2}\Delta\tau^4 + \frac{V'(x_a)^4}{192Mh^3}\Delta\tau^5\right]$$

$$+\Delta x^3\left[-\frac{V'''(x_a)}{24\hbar}\Delta\tau + \frac{V'(x_a)V''(x_a)}{16\hbar^2}\Delta\tau^2\right]$$

$$\left. +\Delta x^4\left[-\frac{V''''(x_a)}{120\hbar}\Delta\tau + \left(\frac{V'(x_a)V'''(x_a)}{96\hbar^2} + \frac{V''(x_a)^2}{192\hbar^2}\right)\Delta\tau^2 - \frac{V'(x_a)^2V''(x_a)}{192\hbar^3}\Delta\tau^3\right]\right\}$$

$$(1.56)$$

With these, after straight albeit cumbersome algebra for one-by-one iden-tifying the Eqs. (1.56) with that obtained by considering Eqs. (1.43) and (1.44) in the solution (1.37), the decomposition functions of Eq. (1.44) are obtained with the forms:

$$F_0(\tau) = 1 - \frac{V''(x_a)}{12M}\tau^2 + \left(\frac{V'(x_a)^2}{24Mh} - \frac{V''''(x_a)}{240M^2}\right)\tau^3$$

$$+\left(\frac{V''(x_a)^2}{160M^2} + \frac{V'(x_a)V'''(x_a)}{120M^2}\right)\tau^4 - \frac{11}{1440}\frac{V''(x_a)V'(x_a)^2}{\hbar M^2}\tau^5 + \frac{V'(x_a)^4}{1152M^2\hbar^2}\tau^6$$

$$(1.57)$$

$$F_1(\tau) = -\frac{V'(x_a)}{2\hbar}\tau - \frac{V'''(x_a)}{24M}\tau^2 + \frac{V'(x_a)V''(x_a)}{12M\hbar}\tau^3 - \frac{V'(x_a)^3}{48M\hbar^2}\tau^4 \quad (1.58)$$

$$F_2(\tau) = -\frac{V''(x_a)}{6\hbar}\tau + \left(\frac{V'(x_a)^2}{8\hbar^2} - \frac{V''''(x_a)}{80M}\right)\tau^2$$

$$+ \left(\frac{V'(x_a)V'''(x_a)}{30M\hbar} + \frac{V''(x_a)^2}{40\hbar M}\right)\tau^3 - \frac{11}{288}\frac{V''(x_a)V'(x_a)^2}{M\hbar^2}\tau^4 + \frac{V'(x_a)^4}{192M\hbar^3}\tau^5$$

$$(1.59)$$

$$F_3(\tau) = -\frac{V'''(x_a)}{24\hbar}\tau + \frac{V'(x_a)V''(x_a)}{12\hbar^2}\tau^2 - \frac{V'(x_a)^3}{48\hbar^3}\tau^3 \quad (1.60)$$

$$F_4(\tau) = -\frac{V''''(x_a)}{120\hbar}\tau + \left(\frac{V'(x_a)V'''(x_a)}{48\hbar^2} + \frac{V''(x_a)^2}{72\hbar^2}\right)\tau^2$$

$$- \frac{V'(x_a)^2 V''(x_a)}{48\hbar^3}\tau^3 + \frac{V'(x_a)^4}{384\hbar^4}\tau^4 \quad (1.61)$$

One can easily check that the parametric functions (1.57)–(1.61) solve the system (1.46)–(1.50) in the fourth order truncated expansions. Thus, considered (1.43) and (1.44) in the solution (1.37) it actually provides the fourth order expanded solution of the Schrödinger equation (1.36). Therefore, the same is true also for their originating expression, the semi-classical conditional probability (1.34).

1.3.3 PARTITION FUNCTION QUANTUM EXPANSION

One considers a particle (here the bondon) with mass M moving between the space-points x_a and x_b under the potential $V(\bar{x})$ to be further identified with the molecular net topological potential. The associate quantum evolution may be described by semiclassical propagator obeying the Schrödinger 1D equation, with the path integral solution being found in semiclassical expansion upto fourth (IV) order to look like, as abstracted from Eqs. (1.51), see also (Putz, 2009, 2011b; Putz & Ori, 2014)

$$\left(x_b,\hbar\beta;x_a,0\right)^{(IV)} = \sqrt{\frac{M}{2\pi\hbar^2\beta}}\exp\left\{-\frac{M}{2\hbar^2\beta}(\Delta x)^2 - \beta V(\overline{x})\right\}$$

$$\times\left\{1-\frac{1}{\hbar}\left[\frac{1}{2}V''(\overline{x})\left(\frac{\hbar\beta}{12}\Delta x + \frac{\hbar^3}{M}\frac{\beta^2}{6}\right) + \frac{1}{24}V''''(\overline{x})\left(\frac{\hbar\beta}{80}(\Delta x)^4 + \frac{\hbar^3}{M}\frac{\beta^2}{20}(\Delta x)^2 + \frac{\hbar^5}{M^2}\frac{\beta^3}{10}\right)\right]\right.$$

$$+\frac{1}{2\hbar^2}\left[\frac{\hbar^4}{M}V''(\overline{x})^2\frac{\beta^3}{12} + \frac{1}{4}V''(\overline{x})^2\left(\frac{\hbar^2\beta^2}{144}(\Delta x)^4 + \frac{\hbar^5}{M}\frac{\beta^3}{30}(\Delta x)^2 + \frac{\hbar^6}{M^2}\frac{\beta^4}{20}\right)\right.$$

$$\left.+V'''(\overline{x})V'(\overline{x})\left(\frac{\hbar^4}{M}\frac{\beta^3}{240}(\Delta x)^2 + \frac{\hbar^6}{M^2}\frac{\beta^4}{60}\right)\right] - \frac{1}{6\hbar^3}\left[\frac{3}{2}V''(\overline{x})V'(\overline{x})^2\left(\frac{\hbar^5}{M}\frac{\beta^4}{144}(\Delta x)^2 + \frac{11}{360}\frac{\hbar^7}{M^2}\beta^5\right)\right.$$

$$\left.\left.+\frac{1}{1152\hbar^4}V'(\overline{x})^4\frac{\hbar^8}{M^2}\beta^6\right\}$$

$$(1.62)$$

in terms of the classical path dependence connecting the end-points (1.7) as well as on the path difference (1.10); here and through the whole paper β stays for the inverse of the thermal energy $k_B T$ and $\hbar$ the reduced Planck constant. With Eqs. (1.3), one can form the partition function for the periodical quantum orbits by considering close integration over the classical or average path; the present discussion follows (Putz & Ori, 2014):

$$z^{[IV]}(\beta) = \int\left(x_b,\hbar\beta;x_a,0\right)^{[IV]}_{x_a=x_b=\overline{x}}d\overline{x}$$

$$= \sqrt{\frac{M}{2\pi\hbar^2\beta}}\int\exp\left\{\begin{array}{l}-\beta V(\overline{x}) - \dfrac{\hbar^2}{M}\dfrac{\beta^2}{12}\nabla^2 V(\overline{x}) - \dfrac{\hbar^4}{M^2}\dfrac{\beta^3}{240}\nabla^4 V(\overline{x}) \\[2mm] +\dfrac{\hbar^2}{M}\dfrac{\beta^3}{24}[\nabla V(\overline{x})]^2 + \dfrac{\beta^4}{160}\dfrac{\hbar^4}{M^2}[\nabla^2 V(\overline{x})]^2 + \dfrac{\beta^4}{120}\dfrac{\hbar^4}{M^2}\nabla^3 V(\overline{x})\nabla V(\overline{x}) \\[2mm] -\dfrac{11}{1440}\dfrac{\hbar^4}{M^2}\beta^5\nabla^2 V(\overline{x})[\nabla V(\overline{x})]^2 + \dfrac{1}{1152}\dfrac{\hbar^4}{M^2}\beta^6[\nabla V(\overline{x})]^4\end{array}\right\}d\overline{x}$$

$$(1.63)$$

One may apply the Gauss theorem successively for integrals of gradient of a given long range defined quantity $(\therefore)$, successively as (Putz & Ori, 2014):

$$0 = \int\nabla\left\{\therefore[\nabla\therefore]\right\}d\overline{x} = \int\therefore\left[\nabla^2\therefore\right]d\overline{x} + \int[\nabla\therefore]^2 d\overline{x} \qquad (1.64)$$

$$0 = \int\nabla\left\{\therefore\nabla^3\therefore\right\}d\overline{x} = \int[\nabla\therefore]\left[\nabla^3\therefore\right]d\overline{x} + \int\therefore\left[\nabla^4\therefore\right]d\overline{x} \qquad (1.65)$$

$$0 = \int \nabla \left[\nabla \left\{ \because \left[\nabla^2 \because \right] \right\} \right] d\bar{x} = \int \nabla \left\{ \nabla \because \left[\nabla^2 \because \right] + \because \left[\nabla^3 \because \right] \right\} d\bar{x} + \int \left[\nabla^2 \because \right]^2 d\bar{x}$$

$$= 2 \int \left[\nabla^2 \because \right]^2 d\bar{x} + 2 \int \left[\nabla \because \right] \left[\nabla^3 \because \right] d\bar{x} + \int \because \left[\nabla^4 \because \right] d\bar{x}$$

$$= 2 \int \left[\nabla^2 \because \right]^2 d\bar{x} + \int \left[\nabla \because \right] \left[\nabla^3 \because \right] d\bar{x}$$

$$= 2 \int \left[\nabla^2 \because \right]^2 d\bar{x} - \int \because \left[\nabla^4 \because \right] d\bar{x} \tag{1.66}$$

$$0 = \int \nabla \left\{ \because \left[\nabla \because \right]^3 \right\} d\bar{x} = \int \left[\nabla \because \right]^4 d\bar{x} + \int \because \nabla \left[\nabla \because \right]^3 d\bar{x}$$

$$= \int \left[\nabla \because \right]^4 d\bar{x} + 3 \int \because \left[\nabla \because \right]^2 \left[\nabla^2 \because \right] d\bar{x} \tag{1.67}$$

then specializing them for the attractive (bonding) potential to the working relationships:

$$\int [\nabla V(\bar{x})]^2 \, d\bar{x} = Abs\left(-\frac{1}{\beta} \int \nabla^2 V(\bar{x}) d\bar{x} \right) = \frac{1}{\beta} \int \nabla^2 V(\bar{x}) d\bar{x} \tag{1.68}$$

$$\frac{1}{\beta} \int \nabla^4 V(\bar{x}) d\bar{x} = -\int \nabla^3 V(\bar{x}) \nabla V(\bar{x}) d\bar{x} = 2 \int \left[\nabla^2 V(\bar{x}) \right]^2 d\bar{x} \tag{1.69}$$

$$\int \nabla^2 V(\bar{x}) [\nabla V(\bar{x})]^2 \, d\bar{x} = Abs\left(-\frac{\beta}{3} \int [\nabla V(\bar{x})]^4 \, d\bar{x} \right) = \frac{\beta}{3} \int [\nabla V(\bar{x})]^4 \, d\bar{x} \tag{1.70}$$

with the help of which the Eqs. (1.63) for fourth order partition function is provided under the actual working form (Putz & Ori, 2014):

$$z^{[IV]}(\beta) = \sqrt{\frac{M}{2\pi\hbar^2 \beta}} \int \exp \left\{ \begin{array}{l} -\beta V(\bar{x}) \\ -\dfrac{\hbar^2}{M}\dfrac{\beta^2}{12}\nabla^2 V(\bar{x}) + \dfrac{\hbar^2}{M}\dfrac{\beta^2}{24}\nabla^2 V(\bar{x}) \\ -\dfrac{\hbar^4}{M^2}\dfrac{\beta^3}{240}\nabla^4 V(\bar{x}) + \dfrac{\beta^3}{320}\dfrac{\hbar^4}{M^2}\nabla^4 V(\bar{x}) - \dfrac{\beta^3}{120}\dfrac{\hbar^4}{M^2}\nabla^4 V(\bar{x}) \\ -\dfrac{11}{4320}\dfrac{\hbar^4}{M^2}\beta^6 [\nabla V(\bar{x})]^4 + \dfrac{1}{1152}\dfrac{\hbar^4}{M^2}\beta^6 [\nabla V(\bar{x})]^4 \end{array} \right\} dx$$

$$= \sqrt{\frac{M}{2\pi\hbar^2 \beta}} \int \exp \left\{ \begin{array}{l} -\beta V(\bar{x}) - \dfrac{\hbar^2}{M}\dfrac{\beta^2}{24}\nabla^2 V(\bar{x}) \\ -\dfrac{3\beta^3}{320}\dfrac{\hbar^4}{M^2}\nabla^4 V(\bar{x}) - \dfrac{29\beta^6}{17280}\dfrac{\hbar^4}{M^2}[\nabla V(\bar{x})]^4 \end{array} \right\} d\bar{x}$$

$$\tag{1.71}$$

1.4 TOPOLOGICAL-BONDONIC ALGORITHM ON EXTENDED NANOSTRUCTURES

The second conceptual instrument used in the present analysis regards the physical-to-topological passage; the present discussion follows (Putz & Ori, 2012; 2014):

- The evolution of the nanoribbon defective structure is controlled by a pure topological potentials Ξ expressing the long-range, collective effects of the network on the network stability itself in terms of distance-based topological invariants computed on the nanoribbon chemical graph composed by n nodes.

Equally important, *topological potentials Ξ* are subject to a minimization principle. In spite of this apparent simple statement, appropriate approximation demonstrates in several cases a substantial predictive power when the *topological potentials Ξ* are applied for studying the *isomeric* evolution of *complex systems*, like the *SWw-surfed nanoribbons* under present investigation. An overview, from "fullerene to graphene", of *topological modeling* simulations is provided in Iranmanesh et al. (2012), whereas article by Putz and Ori (2012) presents the first investigation of the influence of the collective topological properties of honeycomb lattices over the collective bosonic behavior of sp^2 electrons.

It is worth to remember here the "basal properties of distance-based topological potentials" making those mathematical object exceptionally suitable for determining delocalized bondonic properties (Putz & Ori, 2012, 2014):

(i) *physically,* the topological potential Ξ considers *by definition* the collective long-range effects produced by the mutual interactions of all atoms pairs of the chemical system;

(ii) *numerically,* Ξ features an easily-manageable *polynomial behavior* in term of the parameter expressing the *size of the system* (that parameter may be n or even η) with the leading coefficient of the respective polynomial only depending *from the dimensionality* D of the system – see the recent review on topological modeling methods and results (Iranmanesh et al., 2012).

Actually, a practical introduction to lattice topological descriptors is provided by looking to the nanoribbon structure in Figure 1.1 as an hexagonal network with n atoms. While indicating with d_{ij} the ij-element of the $n \times n$ distance matrix D of the graph, the first important lattice descriptor is represented by the topological Wiener index W, for example, the semi sum of the n^2 entries of:

$$W = \sum_{i>j} d_{ij} \quad with \ d_{ij} = 0 \tag{1.72}$$

The invariant (1.72) provides a powerful rank of isomeric chemical graphs, privileging the most compact structures (Iranmanesh et al., 2012); for this reason, the Wiener index is a natural choice for the role of chemical potential of the system, here with the involvement of the energetic calibration slope (α)

$$\Xi^W = \alpha W \tag{1.73}$$

Systems like graphene and, to some extent, the related one, including silicene, who are rich in sp^2 electrons, are conveniently described by introducing an explicit term in the electronic potential energy to convey the effects of *conjugation forces* among the occupied states of the unfilled π-bands. As recently demonstrated in De Corato et al. (2013) that electronic *conjugation* term involves the lattice topology, being *directly* proportional to a combination of the Wiener index W (1.72) and the order s corrections:

$$W^{(s)} = \sum_{i>j} d_{ij}^s \quad with \ d_{ii}=0 \tag{1.74}$$

In case of large structures only the first terms $s = 1, 2, 3, 4,\ldots$ significantly contributes to the global energy, and proper scale factors γ_s have to be computed in the relative expression for the topological potential:

$$\Xi^W = \sum_s \gamma_s W^{(s)} \quad with \ s = 1,2,3,4\ldots \tag{1.75}$$

The interested readers may find the formal derivation of Eq. (1.74) contributions and related asymptotic properties in De Corato et al. (2013) original work. For $s=1$ (this is the case of large lattices), Eq. (1.75) reduces to

Eq. (1.72) with γ_1 being the energy scale factor one may interpolate by ab-initio results.

Next, one should fix the energy-length realm of the bondon in the 0th order of the partition function which renders the classical observability by the involved thermal length, here mapped into the topological space and energy so defining the bondonic unitary cell of action; to this aim, one firstly runs the 0th partition function as abstracted from the 0th order of Eq. (1.71) while considering the bondonic quantum information as given by Eqs (1.1) (Putz & Ori, 2012)

$$z_B^{[0]}\left(\beta, \chi_M, \Xi^{[0]}\right) = \frac{1}{2X_{Bond}} \sqrt{\frac{1}{\pi E_{Bond}\beta}} \int_0^{X_{Bond}} \exp\left[-\beta \Xi^{[0]}\right] d\bar{x} = \frac{1}{2}\sqrt{\frac{1}{\pi E_{Bond}\beta}} \exp\left[-\beta \Xi^{[0]}\right]$$

(1.76)

Then, Eqs. (1.76) is used for internal energy computing of the bondon as the average energy condensed in the network responsible for bonding at periodical-range action (Putz & Ori, 2012)

$$\left\langle E_{Bond}\right\rangle [kcal/mol] = u_B^{[0]} = -\frac{\partial \ln z_B^{[0]}\left(\beta, E_{Bond}, \Xi^{[0]}\right)}{\partial \beta}$$

$$= \frac{1}{2\beta} + \Xi^{[0]} = \begin{cases} \Xi^{[0]}, ... \beta \to \infty (T \to 0K) \\ \infty, ... \beta \to 0 (T \to \infty K) \end{cases}$$

(1.77)

It immediately fixes the long-range length of periodic action of bondon by recalling Eq. (1.2) (Putz & Ori, 2012)

$$\left\langle X_{Bond}\right\rangle [\overset{o}{A}] = \frac{a}{\left\langle E_{Bond}\right\rangle [kcal/mol]} = \frac{2a\beta}{1 + 2\beta \Xi^{[0]}} = \begin{cases} \frac{a}{\Xi^{[0]}}, ... \beta \to \infty (T \to 0K) \\ 0, ... \beta \to 0 (T \to \infty K) \end{cases}$$

(1.78)

Remarkable, when the asymptotic limits are considered for both periodic energy and length of bondon, one sees that they naturally appears associated with the topological potential and with the Coulombian interaction for the low-temperature case, while rising and localizing the bonding information (like the delta-Dirac signal) for the high-temperature range, respectively, being the last case an observational manifestation of bondonic chemistry. This feature will be used in a moment below.

This result has the rationale in predicting no bonding (length) for indefinitely heated nanostructure (when all periodicity and bonding structures are destroyed), yet with a finite β-dependency for stable periodic nets with a ground state limit in terms of topological index Ξ as an energetic measure for the bonding in the system, consistent with Eqs. (1.2).

The bondonic length of Eq. (1.76) may be further employed into the full partition expression (see below) for modeling the transition between two phases of a nanosystems, for example, an ideal and a modified (with defects) ones, as follows: let's consider the topological index computed for ideal net system Ξ_0 as well as for the net with (propagating) defects Ξ_D alike, the so called critical temperature $1/\beta_{CRITIC}$ for the phase transition between these two systems may be carried out through the caloric capacity equation (Putz & Ori, 2012, 2014)

$$C_B\left(\beta_{CRITIC},\Xi_0\right)_{IDEAL} = C_B\left(\beta_{CRITIC},\Xi_D\right)_{DEFECTS} \tag{1.79}$$

1.4.1 SECOND ORDER IMPLEMENTATION

Considering a particle with mass M moving between the space-points x_a and x_b under the potential $V(\bar{x})$ with Eq. (1.17) the associate quantum statistical propagator may be expanded upto the second order in Eq. (1.10), as abstracted for instance from Eq. (1.62) (Putz & Ori, 2012),

$$(x_b\hbar\beta;x_a 0) = \sqrt{\frac{M}{2\pi\hbar^2\beta}}\exp\left\{-\frac{M}{2\hbar^2\beta}(\Delta x)^2 - \beta V(\bar{x})\right\}$$

$$\times\left\{1 - \frac{\hbar^2\beta^2}{12M}\nabla^2 V(\bar{x}) - \frac{\beta}{24}(\Delta x\nabla)^2 V(\bar{x}) + \frac{\hbar^2\beta^3}{24M}[\nabla V(\bar{x})]^2\right\} \tag{1.80}$$

with β being the inverse of the thermal energy $k_B T$ and $\hbar$ the reduced Planck constant, as usually. Upon close integration over the classical or average path it provides the immediate partition function (Putz & Ori, 2012):

$$z(\beta) = \int(x_b\hbar\beta;x_a 0)\Big|_{x_a=x_b=\bar{x}}d\bar{x}$$

$$= \sqrt{\frac{M}{2\pi\hbar^2\beta}}\int\exp\left[-\beta V(\bar{x}) - \frac{\hbar^2\beta^2}{24M}\nabla^2 V(\bar{x})\right]d\bar{x} \tag{1.81}$$

So corresponding to the first terms of Eq. (1.71) too. By further specialization for the bondonic evolution and mass (1.1) the working form casts as (Putz & Ori, 2012)

$$z_B\left(\beta, X_{bond}, E_{bond}\right) = \frac{1}{2X_{bond}}\sqrt{\frac{1}{\pi E_{bond}\beta}} \int \exp\left[-\beta V(\bar{x}) - \frac{\beta^2}{12} E_{bond}\bar{x}^2\nabla^2 V(\bar{x})\right]d\bar{x}$$

(1.82)

where, for consistency, X_{bond} was considered as $\bar{x}$ under spatial integration.

The connection with topological properties of the periodic nets is achieved through considering the correspondence of the driving potential with a proper topological descriptor for the periodic cell $V(\bar{x}) \rightarrow \Xi$ as considered in Eq. (1.76) to superior orders

$$\nabla^2 V(\bar{x}) \rightarrow \Xi^{[2]}$$

(1.83)

also holds. Two basal properties of *distance-based* topological descriptors perfectly fit in describing bondons properties (Putz & Ori, 2012, 2014):

(i) topological potentials Ξ behave as long-range interatomic potentials connecting all atoms pairs in the system;

(ii) topological invariants exhibit, also in case of periodic structures, peculiar polynomial behaviors with leading coefficients only depending from the dimensionality D of the system – see the recent review on topological modeling methods and results (Iranmanesh et al., 2012).

Analytically, for a net with N-periodic cells, one gets firstly the bondonic energy per cell, under grand canonical conditions, as (Putz & Ori, 2012):

$$N^{-1}\left\langle E_{Bond}^N\right\rangle[kcal/mol] = -\frac{\partial}{\partial\beta}\ln\left\{\frac{z_B^N(\beta)_{\langle X_{bond}\rangle, \Xi, \Xi^{[2]}}}{N!}\right\} = \frac{3}{2\beta} + \Xi + \frac{1}{\beta + 2\beta^2\Xi}$$

(1.84)

having the same thermodynamic limits as in Eqs. (1.77) yet with 3/2 correction on the kinetic contribution as well as with more complexity on the topological dependency for finite temperature regime.

In the same grand canonical framework, the bondonic caloric capacity writes as (Putz & Ori, 2012):

$$C_B(N,\beta) = k_B\beta^2 \frac{\partial^2}{\partial \beta^2} \ln\left\{ \frac{z_B^N(\beta)_{\langle X_{bond}\rangle,\Xi,\Xi^{[2]}}}{N!} \right\} = \frac{5 + 4\beta\Xi(5 + 3\beta\Xi)}{2(1 + 2\beta\Xi)^2} Nk_B$$

(1.85)

Remarkably, the expression (1.85) recovers the Debye energy limit $5Nk_B/2$ for conducting electrons in solids either for high temperature limit and or for the no-topological structure ($\beta\vee\Xi\to0$), while having minimal limit as $3Nk_B/2$ energy for the bondons in most stable state or with maximum topological potential ($\beta\vee\Xi\to\infty$) corresponding with their freely spanning over the net considerate as a quantum box with infinite potential.

Employing the form (1.85) to phase transition modeled by Eq. (1.79) the absolute finite critical temperature is predicted with the form (Putz & Ori, 2012)

$$|\beta_{CRITIC}| = \frac{\Xi_0 + \Xi_D}{4\Xi_0\Xi_D} = \begin{cases} \dfrac{1}{2\Xi_0} & ...\Xi_0 = \Xi_D \\ \infty & ...\Xi_0 \to 0 \vee \Xi_D \to 0 \end{cases}$$

(1.86)

From the limits of Eq. (1.86) one observes the major role the topological index plays in predicting the characteristic temperature $T = 2\Xi/k_B$ at which a given nanosystems (topology) is stabilized (either as ideal or with defects) by the bondonic motion through it; otherwise, for an absent topology the bondonic motion is purely entropic at $T = 0K$, without observable character, according with Eq. (1.86) for the indefinite limit $\beta\Xi \to \infty \times 0$.

Worth noting that, through introducing the relative temperature and topological parameters respecting the critical and the defect ones,

$$\delta = \beta/\beta_{CRITIC}$$

(1.87)

$$\sigma = \Xi_0/\Xi_D$$

(1.88)

respectively, in Eq. (1.86) and then in Eq. (1.85) one obtains the predicted normalized bondonic caloric capacity per periodic cell (Putz & Ori, 2012)

$$\frac{C_B(\sigma,\delta)}{Nk_B} = \frac{3}{2} + \frac{1 + \delta\sigma(1 + \sigma)}{(1 + 0.5\delta\sigma(1 + \sigma))^2} \xrightarrow{\delta\&\sigma\to1} \frac{43}{18} \cong 2.38889$$

(1.89)

representing a sort of universal value for two fold critical capacity of nanosystems, i.e., at critical temperature and coexistence of ideal-with-defect structures. However, in the light of above discussion, the closer value of Eq. (1.89) with Debye value of Eq. (1.85), $C_B(N, \beta \to 0)/Nk_B \to 2.5$, indicates that at critical regime ($\delta\&\sigma \to 1$) the bondons behave like a Debye conducting electronic gas in solids at high temperature.

1.4.2 FOURTH ORDER IMPLEMENTATION

Now, the facing with superior potential first, second, and fourth order derivatives, can be systematically treated through replacing them with associate topological invariants and higher orders over the concerned bonds, networks or lattices, i.e., (Putz & Ori, 2014)

$$V(\bar{x}) \to \Xi = \Xi^{[0]}$$
$$\nabla V(\bar{x}) \to \Xi^{[1]}$$
$$\nabla^2 V(\bar{x}) \to \Xi^{[2]}$$
$$\nabla^4 V(\bar{x}) \to \Xi^{[4]} \tag{1.90}$$

Nevertheless, attention should be paid at this passage from physical to topological quantities since it actually replaces *electronic interactions* with *topology-based interactions*, being therefore restricted to those topological invariants bearing the energetic meaning, as is the case with the Wiener index, for instance.

Returning to the full partition function now the bondonic periodicity information on length and energy action maybe included to rewrite Eq. (1.71) to the actual form (Putz & Ori, 2014)

$$z_B^{[IV]}\left(\beta, \langle X_{Bond}\rangle, E_{Bond}, \Xi^{[0]}, \Xi^{[1]}, \Xi^{[2]}, \Xi^{[4]}\right) = \frac{1}{2\langle X_{Bond}\rangle}\sqrt{\frac{1}{\pi E_{Bond}\beta}}\exp\left[-\beta\Xi^{[0]}\right]$$

$$\times \int_{-\infty}^{+\infty} \exp\left[-\beta^2 E_{Bond}\frac{\Xi^{[2]}}{12}\bar{x}^2 - \beta^3 E_{Bond}^2\left(\frac{3\Xi^{[4]}}{160} + \beta^3\frac{29\left[\Xi^{[1]}\right]^4}{8640}\right)\bar{x}^4\right]d\bar{x}$$

$$= \frac{3}{a\beta}\sqrt{\frac{5\Xi^{[2]}}{\pi\left[29\beta^3\left(\Xi^{[1]}\right)^4 + 162\Xi^{[4]}\right]}}BesselK\left[\frac{1}{4}, \frac{15\beta\left(\Xi^{[2]}\right)^2}{58\beta^3\left(\Xi^{[1]}\right)^4 + 324\Xi^{[4]}}\right]$$

$$\times \exp\left[-\beta\Xi^{[0]} + \frac{15\beta\left(\Xi^{[2]}\right)^2}{58\beta^3\left(\Xi^{[1]}\right)^4 + 324\Xi^{[4]}}\right] \tag{1.91}$$

However, for workable measures of macroscopic observables, one employs the partition function of Eq. (1.91) to compute the canonical associated partition function according with the custom statistical rule assuming the N-periodic cells in the network

$$Z_\beta^{[IV]}(N,\beta)_{\langle X_{bond}\rangle, E_{bond}, \Xi^{[0]}, \Xi^{[1]}, \Xi^{[2]}, \Xi^{[4]}} = \frac{\left\{z_\beta^{[IV]}\left(\beta, \langle X_{bond}\rangle, E_{bond}, \Xi^{[0]}, \Xi^{[1]}, \Xi^{[2]}, \Xi^{[4]}\right)\right\}^N}{N!}$$

(1.92)

with the help of Eq. (1.92) one is provided with the canonical (macroscopic) internal energy contributed by N-bondons from the N periodic cells, through considering further thermal derivation (Putz & Ori, 2014):

$$E_{Bond}^{N[IV]}(N,\beta) = -\frac{\partial \ln\left\{Z_\beta^{[IV]}(N,\beta)_{\langle X_{bond}\rangle, E_{bond}, \Xi, \Xi^{[1]}, \Xi^{[2]}, \Xi^{[4]}}\right\}}{\partial \beta}$$

$$= \frac{N}{2\left(29\beta^3\left(\Xi^{[1]}\right)^4 + 162\Xi^{[4]}\right)^2}$$

$$\times \left\{ \begin{array}{l} 81\left(29\beta^2\left(13 + 8\beta\Xi^{[0]}\right)\left(\Xi^{[1]}\right)^4 - 30\left(\Xi^{[2]}\right)^2\right)\Xi^{[4]} \\ + \dfrac{1682\beta^6\left(2 + \beta\Xi^{[0]}\right)\left(\Xi^{[1]}\right)^8 + 870\beta^4\left(\Xi^{[1]}\right)^4\left(\Xi^{[2]}\right)^2 + 13122\left(5 + 4\beta\Xi^{[0]}\right)\left(\Xi^{[4]}\right)^2}{\beta} \\ + 30\left(\Xi^{[2]}\right)^2\left(81\Xi^{[4]} - 29\beta^3\left(\Xi^{[1]}\right)^4\right) \\ \times BesselK\left[\dfrac{3}{4}, \dfrac{15\beta\left(\Xi^{[2]}\right)^2}{58\beta^3\left(\Xi^{[1]}\right)^4 + 324\Xi^{[4]}}\right]\left(BesselK\left[\dfrac{1}{4}, \dfrac{15\beta\left(\Xi^{[2]}\right)^2}{58\beta^3\left(\Xi^{[1]}\right)^4 + 324\Xi^{[4]}}\right]\right)^{-1} \end{array} \right\}$$

(1.93)

Finally, by continuing with the inverse thermal energy derivatives, the internal energy of bonding of Eq. (1.93) may be employed also for estimating the allied caloric capacity (Putz & Ori, 2014)

$$C_\beta^{[IV]}(N,\beta) = -k_B\beta^2 \frac{\partial E_\beta^{N[IV]}(N,\beta)}{\partial \beta}$$

$$= \frac{Nk_B}{\left(29\beta^3\left(\Xi^{[1]}\right)^4 + 162\Xi^{[4]}\right)^4}$$

$$\begin{aligned}
&\left[225\left(\Xi^{[2]}\right)^4\left(29\beta^4\left(\Xi^{[1]}\right)^4 - 81\beta\Xi^{[4]}\right) \right. \\
&+ \left(29\beta^3\left(\Xi^{[1]}\right)^4 + 162\Xi^{[4]}\right) \begin{pmatrix} 841\beta^7\left(\Xi^{[1]}\right)^8\left(58\beta^2\left(\Xi^{[1]}\right)^4 + 45\left(\Xi^{[2]}\right)^2\right) \\ + 14094\beta^4\left(\Xi^{[1]}\right)^4\left(29\beta^2\left(\Xi^{[1]}\right)^4 - 30\left(\Xi^{[2]}\right)^2\right)\Xi^{[4]} \\ + 1712421\beta^3\left(\Xi^{[1]}\right)^4\left(\Xi^{[4]}\right)^2 + 5314410\left(\Xi^{[4]}\right)^3 \end{pmatrix} \\
&\times \left\{ -15\beta\left(\Xi^{[2]}\right)^2 \begin{array}{l} \left(29\beta^3\left(\Xi^{[1]}\right)^4 + 162\Xi^{[4]}\right)\left(1682\beta^6\left(\Xi^{[1]}\right)^8 - 23490\beta^3\left(\Xi^{[1]}\right)^4\Xi^{[4]} - 6561\left(\Xi^{[4]}\right)^2\right) \\ \times BesselK\left[\dfrac{1}{4}, \dfrac{15\beta\left(\Xi^{[2]}\right)^2}{58\beta^3\left(\Xi^{[1]}\right)^4 + 324\Xi^{[4]}}\right] \\ + 15\beta\left(\Xi^{[2]}\right)^2\left(29\beta^3\left(\Xi^{[1]}\right)^4 - 81\Xi^{[4]}\right)^2 \\ \times BesselK\left[\dfrac{3}{4}, \dfrac{15\beta\left(\Xi^{[2]}\right)^2}{58\beta^3\left(\Xi^{[1]}\right)^4 + 324\Xi^{[4]}}\right] \end{array} \right\} \\
&\left. \times BesselK\left[\dfrac{3}{4}, \dfrac{15\beta\left(\Xi^{[2]}\right)^2}{58\beta^3\left(\Xi^{[1]}\right)^4 + 324\Xi^{[4]}}\right]\left(BesselK\left[\dfrac{1}{4}, \dfrac{15\beta\left(\Xi^{[2]}\right)^2}{58\beta^3\left(\Xi^{[1]}\right)^4 + 324\Xi^{[4]}}\right]\right)^{-1} \right]
\end{aligned}$$

$$(1.94)$$

The treatment of pristine ("0")-to-defect ("D") networks goes now by equating the respective formed caloric capacities from Eq. (1.94) towards searching for the β-critic through the phase-transition equation

$$C_B^{[IV]}\left(\beta_{CRITIC}, \Xi_0^{[0]}, \Xi_0^{[1]}, \Xi_0^{[2]}, \Xi_0^{[4]}\right)_{IDEAL} = C_B^{[IV]}\left(\beta_{CRITIC}, \Xi_D^{[0]}, \Xi_D^{[1]}, \Xi_D^{[2]}, \Xi_D^{[4]}\right)_{DEFECTS}$$

$$(1.95)$$

Now one may use the above mentioned high temperature regime, $(\beta \to 0)$, see Eqs. (1.85) and (1.86), in accordance with the present semiclassical approach, to find the critical phase-transition "temperature" to be

$$\beta_{CRITIC} = \frac{216}{5}\frac{\Xi_0^{[4]}\Xi_D^{[4]}}{\left(\Xi_D^{[2]}\right)^2\Xi_0^{[4]} + \left(\Xi_0^{[2]}\right)^2\Xi_D^{[4]} + \Xi_0^{[2]}\Xi_D^{[2]}\sqrt{\Xi_0^{[4]}\Xi_D^{[4]}}}\left(\frac{Gamma\left[\dfrac{5}{4}\right]}{Gamma\left[\dfrac{3}{4}\right]}\right)^2$$

$$(1.96)$$

This algorithm will be next unfolded for the present honey0comb systems referenced in the graphene nanoribbons with Stone-Wales defects.

1.5   BONDONS ON NANO-RIBBONS WITH STONE-WALES DEFECTS

1.5.1   SECOND ORDER EFFECTS OF BONDONS ON GRAPHENIC FRAGMENTS

In this section topological modeling techniques are applied to describe the bondons role during the formation of Stone-Wales defects (SW defects) in graphene fragments, including boundary conditions and defect-defect interactions, without making assumptions on the electronic distribution density in the system, all specific physical information being conveyed by connectivity topological data; the present discussion follows (Putz & Ori, 2012).

Topological potential will Ξ operate over the graphene nanoribbons GNR^{z4} with fixed height ($z=4$) and variable lengths L whose super-cell is shown in Figure 1.7 for the $L=6$ case. First-principles calculations (Dutta & Manna, 2009) confirm defective hybrid nanoribbons systems having

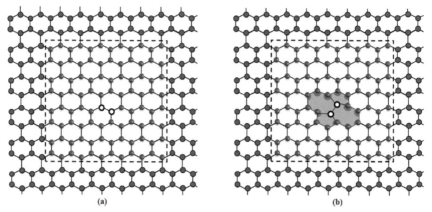

FIGURE 1.7   Framed super-cell of graphenic nanoribbon GNR^{z4} for the pristine condition (a) and after the creation of a single SW defect 5/7/7/5 by rotating the diagonal bond between the two circled carbon atoms (b). Horizontal size $L=6$ has been chosen. Bonds crossing the dashed line periodically close the structure in both directions (Putz & Ori, 2012).

limited dependence from ribbon width z in respect to their electronic properties, which will be characterized by z-dependent specific topological potential Ξ in ongoing studies on bondon behavior in 2D graphenic fragments. On the set of expanding GNR^{z4} structures, with periodic boundary conditions imposed in both directions, each super-cell presents a single central SW defect, obtained by rotation of one *diagonal* bonds of the net.

There is known that reversible SW rotations applied to *vertical* bonds generate, over finite graphene fragments, different defective configurations characterized by higher formation energies (Samsonidze et al., 2002, Ertekin & Chrzan, 2009). This anisotropy represents a peculiar topological feature of SW rotation on graphenic lattices, deeply encoded in their connectivity properties as reported in extensive investigations (Chen et al., 2010, Ori-Cataldo-Putz, 2011). Similarly, the multiplicity of ways SW defects have to combine and diffuse in sp^2 carbon structures arises from combining both types of bond rotations, each step involving four atoms. A very interesting topological effect is realized when a 5|7 pair (that may also come from an original SW double-pair), under the effects of iterated SW rotations, drifts in the hexagonal network and generates the so-called *SW wave*, a characteristic linear rearrangement of the hexagonal mesh (Ori-Cataldo-Putz, 2011).

As in previous topological modeling studies that have been recently devoted to graphenic layers with nanocones (Cataldo et al., 2010), C_{66} fullerene stability (Vukicevic et al., 2011) or schwarzitic nanoribbons conformations (De Corato et al., 2012), the topological potential derives from the Wiener index W of the chemical structure (Estrada & Hatano, 2010).

Topological invariants moreover offer an intriguing explicit dependence from system dimensionality D that fixes the leading exponent of the polynomial forms that express the indices as a function of the lattice size; in case of $D=1$ structures like GNR^{z4}, the following general laws hold: $W(L) \approx L^s$ and $w_1(L) \approx L^{s-1}$, being $s=2D+1$ (Ori-Cataldo-Putz, 2011; Ori et al., 2010).

Regardless to computational simplicity, topological potentials $\Xi^w = w$ expresses the necessary influence of L on the 5/7/7/5 defects when they interact with the surrounding hexagonal network and similar defects located in a different super-cell along the ribbon. The creation of SW defects in

GNR^{z4}, when $L = 4, 6, 8,...$ drives the lattice expansion, correctly reproduce ab-initio formation energy values E_{SW} extracted from Ertekin and Chrzan (2009), evidencing the role of long-ranged effects in generating topological defects in graphenic systems. Table 1.1 lists the variations of the topological descriptor W and the trends of the formation energies E_{SW} when single SW pentagon–heptagon double pairs are generated in the GNR^{z4} super-cell of size L when periodical stack is imposed in both directions (see, Figure 1.7).

Table 1.1 data allow the interpolation of the previously introduced parameters affecting the nature of the bondonic particles of these GNR^{z4} systems, that after calibration data one gets the 83% correlated Wiener topological index with the working pristine and SW defect graphene's potentials (Putz & Ori, 2012)

$$W \xrightarrow{\quad 83\% \quad} \Xi^W : \begin{cases} \Xi_0^W = 105.6658 + 0.0025 \times \left(320L + 256L^2 + 64L^3\right)[kcal/mol] \\ \Xi_D^W = 105.6658 + 0.0025 \times \left(69 + 333L + 249L^2 + 64L^3\right)[kcal/mol] \end{cases}$$

$$(1.97)$$

With the L-dependent polynomials of Eq. (1.73) the above critical temperature of pristine-defect coexistence in graphene as given by Eq. (1.86), along side the related bondonic action distance (1.76), bondonic N-normalized grand canonical internal energy (1.84) and the associate absolute caloric capacity (1.85) are, respectively, computed and

TABLE 1.1 A Single SW Defect Placed in the Mid of the Super-Cell of a GNR^{z4} Lattice with $N=16L$ Atoms Induces Polynomial Variations of the Topological Invariant W (*).

Lattice	L	N	W	W_0	R	B	E_{SW}
(4 × 4)	4	64	9472	9472	8	96	98.735
(6 × 4)	6	96	24855	24960	10	144	105.425
(8 × 4)	8	128	51437	51712	12	192	106.809
(10 × 4)	10	160	92299	92800	14	240	107.271
(12 × 4)	12	192	150513	151296	16	288	107.271

* E_{SW} represents the formation energies (kcal/mol) of the defect derived from recent literature (Ertekin & Chrzan, 2009). Parameters R and B give the lattice topological extension ($R=L+4$) and the number of bonds, respectively. Ideal graphenic closed structures have lower Wiener index W_0, sharing with GNR^{z4} the same R and B values. $L=6$ ideal and defective structures are represented in Figure 1.7 (Putz & Ori, 2012).

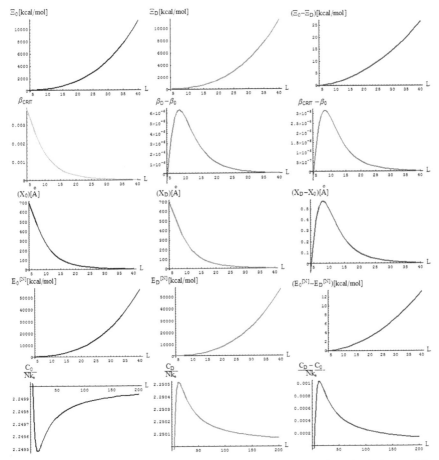

FIGURE 1.8   From top to bottom and left to right: the Wiener based topological potentials of Eq. (1.97) and their pristine (0) to SW defect (D) difference, critical temperature (1.86) along its pristine-defect induced differences, the distance of bondonic actions upon Eq. (1.76) and their D-to-0 difference, the bondonic N-normalized grand canonical internal energies of Eq. (1.84) and the 0-to-D difference, as well as the absolute caloric capacity from Eq. (1.85) and the D-to-0 difference (Putz & Ori, 2012).

represented in Figure 1.8. The results evidence the gaps that, at critical regime, generally affect all the physical quantities described by the bondonic model. This fact mainly originates from the different behaviors of the topological potential between pristine and defective lattices. Moreover, the curves suggest that (Putz & Ori, 2012):

i) the bondonic lengths in both lattices parallels the critical behavior of temperature, while the bondonic energies in grand canonical net mirrors the topological potential profiles, this being consistent with the bondonic energy-length relationship (1.2) and critical equation (1.86);

ii) the caloric capacity always displays a certain super-cell dimension L^* (in the range between 5 and 10Å) associated with its minimum or maximum values in case of ideal or defective structure, respectively. Same condition is emphasized by bondonic length difference at critical regime. That regime occurs when both caloric capacities equalize according with Eq. (1.79) valid in the grand canonical system. However, Eq. (1.86) rules the overall effects of the topological potentials on β_{CRITIC} that in turn, once applied in the Eq. (1.76), produces the peculiar peak in the X_D–X_0 gap, evidencing the same characteristic size threshold L^* for the super-cell edge at which the gain in energy caused by the formation of SW defects reaches, for the family of understudy nanoribbons GNR^{z4}, the maximum value of 107.217 kcal/mol (see Table 1.1); at this point bondons corresponding to the pristine lattice (with wave-length X_0) coexists with similar particles of the defective lattice having longer X_D.

1.5.2 FOURTH ORDER EFFECTS OF BONDONS ON HONEYCOMB FRAGMENTS

Next we will progress on the investigations of SW defects in graphene and related layers, as silicene germanene, and stannene, by analyzing the *propagation in the hexagonal nanoribbons* of the 5|7 pairs according to the wave-like topological mechanism originally introduced in Ori-Cataldo-Putz (2011) and baptized as Stone-Wales waves (SWw) along the effect such a drifting effect may have on the long-range electronic properties of such monolayers with the aid of bondonic path integral formalism, just exposed in its formal way in Section 1.3; the present discussion follows (Putz & Ori, 2014).

The topological skeletons of the systems considered in the present article are basically represented by a mesh of fused hexagons entirely paving the nanoribbons (Figure 1.9); closed boundary periodic conditions are

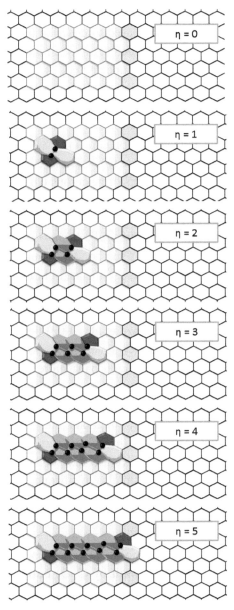

FIGURE 1.9 Propagation of the Stone-Wales wave-like defect along the zig-zag direction caused by the insertion of pairs of hexagons at $\eta=1$, corresponding to the SW defect generations step; the size of this dislocation dipole ranges from $\eta=0$ (pristine lattice) to $\eta=5$; pristine (rearranged) hexagons are in blue (and orange); pentagons and heptagons are in red and green, respectively (Putz & Ori, 2014).

imposed to form *nanotori* of carbon or silicon. Present section introduces to the graph-theoretical methods used to describe the *generation* and the *propagation* of the Stone-Wales defects in such a kind of *cubic* lattices, i.e., planar structures made of 3-connected nodes.

From the topological perspective, the key concepts applied in comparing graphene with silicone and related honey-comb networks' properties are basically two.

First, the two 5|7 pentagon-heptagon units SW constituting the SW defect, also called 5|7|7|5 dipole, are *considered free to migrate* in the hexagonal lattice by inserting η-1 pairs of hexagons 6|6.

Such a structural modification reflects *an universal topological property of the hexagonal meshes*. In this way, extended linear defects are created, keeping the modified structures *fully isomeric* to the initial one (i.e., conserving number of atoms, rings and bonds). Figure 1.9 shows, from the graphical point of view, the iterative sequence of bond rotations producing the *SW dislocation dipole* $5|7\{6|6\}_\eta 7|5$ with size η also called *SW wave* (SWw) (Ori-Cataldo-Putz, 2011).

DFT computations (Samsonidze et al., 2002) demonstrate that strain-induced local forces energetically favor the topological swap of two hexagons with the pentagon-heptagon pair; more details on topological dislocations are provided in the original paper on SWw (Ori-Cataldo-Putz, 2011). Although SWw defective configurations are not yet investigated in SiNR systems by mean of ab-initio methods, the introduction-mentioned $E_b \approx 2.8$ eV low values featured by the energy-barrier for normal SW rotations in Si hexagonal layers (see, Kara et al., 2009) encourage the search for that defect diffusion mechanism also in silicene.

Topological invariants Ξ are computed for the defective isomeric configurations illustrated in Figure 1.9. The nanoribbon building unit is made of $n_0 = 84$ atoms constituting the colored rings. In order to avoid long-range self-interactions, topological potential are computed in a *periodically closed* supercell E built by (3×3) building units. Supercell E has therefore a grand-total of $n = 756$ nodes and $B = 1134$ chemical bonds (or graph edges), the $B=3n/2$ relation being valid for other cubic graphs like the fullerene ones. At the center of that supercell, the $n_0 = 84$ array will hosts the generation and the propagation of the η-sized Stone-Wales wave for $\eta = 0, 1, 2, 3, 4, 5$ the $\eta = 1$ step corresponding to the generations of the standard SW defect 5|7|7|5. In Figure 1.9 the black-circled atoms mark the bonds rotated during the expansion of the SWw dislocation dipole. For the nanoribon fragments of Figure 1.9, through employing the pristine-to-defective steps $\eta=0\div5$, the topological potentials need in Eqs. (1.90) are generated by the associate polynomials, respectively (Putz & Ori, 2014):

$$\begin{cases} W^{[0]}(\eta) = W_0^{[0]} + \dfrac{3433}{60}\eta^5 - 732\eta^4 + \dfrac{37151}{12}\eta^3 - 3776\eta^2 - \dfrac{92777}{15}\eta \\ W_0^{[0]} = W^{[0]}(\eta = 0) = 4467960 \end{cases} \quad (1.98)$$

$$\begin{cases} W^{[1]}(\eta) = W_0^{[1]} + \dfrac{148111}{12}\eta^5 - \dfrac{31583}{2}\eta^4 + \dfrac{1605473}{24}\eta^3 - \dfrac{165319}{2}\eta^2 - \dfrac{3862409}{30}\eta \\ W_0^{[1]} = W^{[1]}(\eta = 0) = 40453182 \end{cases}$$

$$(1.99)$$

$$\begin{cases} W^{[2]}(\eta) = W_0^{[2]} + \dfrac{823187}{60}\eta^5 - \dfrac{2103211}{12}\eta^4 + \dfrac{8887421}{12}\eta^3 - \dfrac{10931693}{12}\eta^2 - \dfrac{7189376}{5}\eta \\ W_0^{[2]} = W^{[2]}(\eta = 0) = 267185898 \end{cases}$$

$$(1.100)$$

$$\begin{cases} W^{[4]}(\eta) = W_0^{[4]} + \dfrac{2083933}{20}\eta^5 - 1327800\eta^4 + \dfrac{66997285}{12}\eta^3 - 6750571\eta^2 - \dfrac{168681041}{15}\eta \\ W_0^{[4]} = W^{[4]}(\eta = 0) = 1410134950 \end{cases}$$

$$(1.101)$$

with the specialization for each instant nanoribon isomeric defective instants depicted in the Figure 1.9 and reported in Table 1.2.

The supercell in Figure 1.9 shows *two distinct topological regimes* according to the selected topological potential. Considering $\varXi^W = \alpha W$ as the potential energy of the system, see Eq. (1.73) and Table 1.3, the generation and the propagation of the SWw dipole results in a topologically favored condition.

The system evolves in such a way the Wiener index (1.72) decreases with $\eta = 1$ by reducing the chemical distances in the graph in the 7-rings region. Only the $W^{[1]}$ presents an anomaly to this behavior as illustrated in Table 1.2, starting for the steps $\eta = 4$ and 5; this justifies the present fourth

TABLE 1.2   Numerical Values Abstracted From Topological Potentials of Eqs. (1.98) – (1.101) Then Used to Generate the Interpolations Polynomials of Eqs. (1.104) – (1.111) As a Function of the η-step of the Forming (η=0, 0.2, 0.4, 0.6, 0.8, 1) and Propagation (η = 0, 1, 2, 3, 4, 5) of the SWw Dipole in the Periodic Nanoribon Supercell E of the of Figure 1.9, Respectively, see, Putz & Ori (2014)

η	$W^{[0]}$	$W^{[1]}$	$W^{[2]}$	$W^{[4]}$
0	4,467,960	40,453,200	267,186,000	1,410,130,000
0.2	4,466,600	40,424,600	266,868,000	1,407,660,000
0.4	4,465,060	40,392,500	266,508,000	1,404,880,000
0.6	4,463,470	40,359,500	266,134,000	1,402,000,000
0.8	4,461,900	40,329,100	265,764,000	1,399,170,000
1	4,460,420	40,305,200	265,416,000	1,396,500,000
2	4,455,370	40,542,500	264,226,000	1,387,400,000
3	4,453,620	42,849,300	263,807,000	1,384,160,000
4	4,452,140	51,493,100	263,439,000	1,381,240,000
5	4,450,930	74,805,700	263,132,000	1,378,770,000

TABLE 1.3   Synopsis of the Topo-Reactive Parameters for the Defects Instances (Starting From Pristine Net at the Step η = 0) From the SW Propagations in Graphene Sheet*

Defect Step	Instant Structure	Electronic Energy (eV)	Total Energy (eV)	Binding Energy (eV)	Parabolic Energy (eV)
$\eta = 0$		2858.69979	2595.306	7308.17	13063.1207
$\eta = 1$		2425.90314	2409.47	7494.0069	102344.109
$\eta = 2$		2641.35644	2410.023	7493.4534	100091.3384
$\eta = 3$		90063.404	10331.43	-427.9522	6770.427006

TABLE 1.3   Continued

Defect Step	Instant Structure	Electronic Energy (eV)	Total Energy (eV)	Binding Energy (eV)	Parabolic Energy (eV)
$\eta = 4$		2428.23769	2408.129	7495.3472	102353.338
$\eta = 5$		2484.84517	2468.133	7676.8912	107394.1399
Correlation Slope, α in Eq. (1.73)	$W^{[0]}(\eta)$	0.00384593	0.000846	*0.0013853*	0.01614977
	$W^{[1]}(\eta)$	0.00029961	7.01×10^{-5}	*0.000123*	0.001488221
	$W^{[2]}(\eta)$	6.4681×10^{-5}	1.42×10^{-5}	*2.335×10^{-5}*	0.000271766
	$W^{[4]}(\eta)$	1.2299×10^{-5}	2.71×10^{-6}	*4.445×10^{-6}*	5.16926×10^{-5}
Correlation Factor R^2	$W^{[0]}(\eta)$	0.21643315	0.622413	**0.813889**	0.727612957
	$W^{[1]}(\eta)$	0.16519269	0.538169	**0.8067669**	0.777077187
	$W^{[2]}(\eta)$	0.21568411	0.621567	**0.8144879**	0.725935217
	$W^{[4]}(\eta)$	0.21522938	0.621048	**0.8148337**	0.724864925

*As described in Figure 1.9, namely: electronic, total, binding and parabolic energy – the last one computed upon Eqs. (1.102) and/or (1.103) with the total number of pi-electrons $N_\pi = 82$ for the steps $\eta = 0$–4 and $N_\pi = 84$ for the last instant case $\eta = 5$, within the semi-empirical AM1 framework, respectively; the bottom of the table reports the free intercept correlation slopes and the associate correlation factors for each set of structural energies respecting the topological defective Wiener potential values of Table 1.2, providing the actual hierarchy (bolded values) and the calibration recipe (bolded italic) then used to generate the working potential polynomials of Eqs. (1.104)–(1.111) (Putz & Ori, 2014).

high order approach in order to properly describe complex topological electronic bonding features as well, within the frame of the bondonic formalism.

Nevertheless, other terms in the topological potential, coming from $W^{[2\&4]}$ descriptors, follow the $W^{[0]}$ behavior and they do not alter therefore this compactness-driven propagation effect along the zig-zag edge of the nanoribbons; numerically, topological distances span the $d_{ij} = 1, 2, \ldots, 29,$ 30 range in all the lattice configurations with $n = 756$ nodes whose central defective regions (having $n_0 = 84$ atoms) are step-by-step reproduced in Figure 1.9. The appropriate α values in Eq. (1.73) are determined by a specific interpolation process that is described in the following.

To model chemical reactivity, one considers various energetic quantities (such as the total energy, electronic energy or binding energy) alongside the celebrated parabolic form of the pi-energy (Putz 2011c, 2012a-b) computed by mean of a polynomial combination of *electronegativity* and *chemical hardness* for frontier orbitals (such as HOMO-highest occupied molecular orbital and LUMO-lowest unoccupied molecular orbital) respecting the number of pi-electrons engaged in the molecular reactivity; as such it runs upon the Mulliken-type formula (Putz 2012a-b)

$$E_\pi = \underbrace{\left(\frac{\varepsilon_{LUMO} + \varepsilon_{HOMO}}{2} \right) N_\pi}_{-Electronegativity} + \frac{1}{2} \underbrace{\left(\frac{\varepsilon_{LUMO} - \varepsilon_{HOMO}}{2} \right) N_\pi^2}_{Chemical\ Hardness} \qquad (1.102)$$

Equivalently, within the frozen core approximation or by Koopmans' theorem (1934), see also Volume I of the present five-volume set (Putz, 2016b), it is rewritable in terms of the ionization potential (IP) and electronic affinity (EA)

$$E_\pi = -\frac{IP + EA}{2} N_\pi + \frac{IP - EA}{4} N_\pi^2 \qquad (1.103)$$

Accordingly, Table 1.3 displays, respecting the defect-step evolution, the numerical values of these energies during the propagation of SWw defects in graphenic nanoribbons. The "best" free intercept correlation, as in Eq. (1.73), with the corresponding series of Wiener topological indices of Table 1.2 is then computed for each energetic frameworks considered in Table 1.3, deriving the related correlation factor hierarchy. One notes that, in line with above observations, only the correlation output in the step η=1 is systematically spurious in respect to the remaining correlation factors, most probably due to the dispersive effect present in the first order derivatives of the topological potential, the same dispersive effect being also present in the physical picture of dissipation phenomena, see also the Markovian and localization/delocalization effects described in the Chapter 5/Volume II (Putz, 2016) of the present five-volume set (Putz, 2016c). Also interesting, the parabolic based chemical reactivity analysis furnishes the second best results after the pure binding energy correlations; this behavior justifies both the pro and contra regarding its use in modern chemical reactivity theory, namely (Putz & Ori, 2014):

- the pro-argument, largely advocated by Parr works in last decades of conceptual chemistry research with application in inorganic and organic reactivity alike (Parr & Yang, 1989, Parr, 1983, Parr & Pearson, 1983, Chattaraj & Parr, 1993), while being recently employed by present authors in "coloring" chemical topology with chemical reactivity electronic frontier information of atoms in molecules

(Putz et al., 2013);

- the contra-argument, defended by late (Szentpaly, 2000) works on various inorganic systems, according which the parabolic description is slightly non-realistic neither for ground nor fro valence state of atoms and molecules since actually not having the minimum of the parabola on the right realm of exchanged electrons in bonding or in ionization-affinity processed; this limitation was also conceptually discussed by one of the present authors in a recent paper advancing the cubic form of chemical reactivity as a better framework for conceptual treatment for electronic exchange as driven by electronegativity and chemical hardness, with an universal (and also Bohmian) value, see (Putz, 2012e) and the Volume II of the present five-volume work (Putz, 2016c).

Therefore, although valuable, the parabolic reactivity calibration is also by this approach taken over by the cute binding energy for the correlation coefficients with topological potentials in Table 1.3, even at semi-empirical level – nevertheless in the line with the present semiclassical methodology. Once the correlation framework was established for according the topological with energetically passage of Eqs. (1.73) at its turn completing the recipe of Eqs. (1.90), one may further interpolate the creation and propagation of the SWw in the honey-comb nanoribbons of Figure 1.9 by employing the data of Table 1.2 and then appropriately calibrating the fifth order polynomials for the two cases, respectively (Putz & Ori, 2014):

- the energetically calibrated topological potentials for the forming SW defect instance (still corresponding to the "0" structure) within $[0,1]$ range of the η "steps":

$$\Xi_0^{[0]}(kcal/mol) = 142819 - 27.3514\eta - 7.43179\eta^2 + 0.947067\eta^3$$
$$- 0.0403542\eta^4 + 0.00058512\eta^5 \qquad (1.104)$$

$$\Xi_0^{[1]}(kcal/mol) = 114847 - 49.4016\eta - 14.4795\eta^2 + 1.91735\eta^3$$
$$- 0.127728\eta^4 + 0.011207\eta^5 \qquad (1.105)$$

$$\Xi_0^{[2]}(kcal/mol) = 144055 - 105.458\eta - 30.1329\eta^2 + 3.81941\eta^3$$
$$- 0.16288\eta^4 + 0.0023649\eta^5 \qquad (1.106)$$

$$\Xi_0^{[4]}(kcal/mol) = 144796 - 160.619\eta - 42.7699\eta^2 + 5.48559\eta^3$$
$$- 0.234943\eta^4 + 0.00341904\eta^5 \qquad (1.107)$$

• The polynomials for the topological potentials describing the SW waves still corresponding to the defective "D" structures) within [0,5] range of the η steps:

$$\Xi_D^{[0]}(kcal/mol) = 142738 + 443.209\eta - 576.127\eta^2 + 210.794\eta^3 \\ - 32.5354\eta^4 + 1.8285\eta^5 \tag{1.108}$$

$$\Xi_D^{[1]}(kcal/mol) = 114647 + 1027.55\eta - 1423.05\eta^2 + 719.265\eta^3 \\ - 219.918\eta^4 + 35.0219\eta^5 \tag{1.109}$$

$$\Xi_D^{[2]}(kcal/mol) = 143706 + 1818.3\eta - 2327.9\eta^2 + 850.485\eta^3 \\ - 131.361\eta^4 + 7.39031\eta^5 \tag{1.110}$$

$$\Xi_D^{[4]}(kcal/mol) = 144339 + 2546.85\eta - 3333.49\eta^2 + 1223.97\eta^3 \\ - 189.577\eta^4 + 10.6845\eta^5 \tag{1.111}$$

It is worth evidencing the advantage of this procedure which effectively allows an easy energetic calibration and a separate description of the 0-froming and D-propagating steps of the SWw defect by providing the associate polynomials that are (i) energetically realistic and (ii) with "equal importance" despite the different information contained: see for instance the numeric form of the fourth order topological potential Eq. (1.101) respecting those provided by Eqs. (1.107) and (1.111). This computationally-convenient method assures that higher order topological potentials will contribute in providing the bondonic related quantities of Eqs. (1.93), (1.94) and (1.96).

Nevertheless, all the present computational algorithm was implemented for graphenic structures, having the carbon atom as the basic motive; however they can be for further used in predicting similar properties also for similar atomic group like Si, Ge, Sn, through appropriate topological potential factorization depending on the displayed reactivity differences; since such differences are usually reflected in gap band or bonding distance differences, one may recall again the electronegativity as the atomic measure marking the passage from an atomic motive to another keeping the honeycomb structure. The influence of the lattice will be implemented by considering the (electronegativity dependency) function of the fermionic statistical type with 2-degeneracy of states spread over the graphenic type lattice – taken as a reference. Such a function accounts for the electronic pairing in chemical bonding is analytically taken as (Putz & Ori, 2014):

$$f_{X-Y} = \frac{2}{1+\exp\left(\dfrac{\sqrt{\chi_X \chi_Y} - \chi_C}{\chi_G}\right)} = \frac{2}{1+\exp\left(\dfrac{\sqrt{\chi_X \chi_Y} - 6.24}{5.1}\right)} \xrightarrow{X=Y=C} 1 \quad (1.112)$$

Numerically, Eq. (1.112) features the factorization with unity for C-C bonding, while departing to fractions from it when the Group A-IV of elements are considered as motives for honey-comb lattice with graphenic reference: Si-Si honey-comb bonding will carry the statistically Si atomic electronegativity $\chi(Si)=4.68$ [eV] in Eqs. (1.112) with X=Y=Si, and successively for Ge-Ge with $\chi(Ge)=4.59$ [eV], and Sn-Sn with $\chi(Sn)=4.26$ [eV] for the corresponding silicone as well as for similarly designed germanene and stannene nanoribbons' structures. Note that atomic electronegativity were considered within the Mulliken type formulation of IP and EA as in the first term of Eq. (1.103); furthermore, their geometric mean was "measured" against the referential graphenic C-C chemical bonding, while their difference was normalized under exponential of Eq. (1.112) to the so-called "universal" geometrical averaged form of Parr and Bartolotti, $\chi_G = 5.1[eV]$, at its turn obtained within the electronegativity geometric equalization framework (Parr & Bartolotti, 1982). Note that the present approach may allow for further extension towards XY hetero-bondings arranged in honey-comb lattice in which cases the mixed combinations C-Si, C-Ge, C-Sn, Si-Ge, Si-Sn, and Ge-Sn are implemented following the same formalism. Yet, here we will be restricted to homo-bondings in nanoribbons only due to their specific van-der-Waals interaction.

Going to have the final and most important part of discussion of the obtained bondonic observable properties through the present fourth order topological-potential formalism, they will be displayed through jointly implementing the short and medium ($\eta=0$–7) to long ($\eta=0$–10…50) range effect of the present interpolation-calibrated potentials (1.104)–(1.111); in other words, although having obtained the working topological potentials over a finite "movie" of forming and propagating SW defects in Figure 1.9, by letting "free" the step argument η in the actual evaluated quantities of Eqs. (1.93), (1.94) and (1.96) one actually will explore how much the short range behavior will echo into the long range as well, or whether this echo will feature some distortions, peaks or valleys, equivalently with a predicted signal to be recorded on extended nanosystems of graphenic type.

One starts with the topo-energetic Wiener potentials of Eqs. (1.104) – (1.111) with representations in Figure 1.10 noting that (Putz & Ori, 2014):

- The topological potentials modeling the forming ("0-to-1") step of SWw are all monotonically descending, meaning their eventual release into defective structures;
- The topological defective potentials have quite constant behavior over the entire computational η=0–5 plateau, while recording quasi-critical rise for the potential "echo" spanning the long range behavior, meaning the rising of the defective potential barrier in fact (with more emphasize for the first order potential, as expected from previous discussion, see Table 1.2, for instance); this strongly suggest the real finite range for the defective SWw, i.e., the annihilation long-range stage that came after their creation on the short-range realm;
- The D-to-0 difference shows some short range fluctuations for all potentials unless the first order one, yet ending into the D-potential definite rising barrier on the long range behavior; the difference for

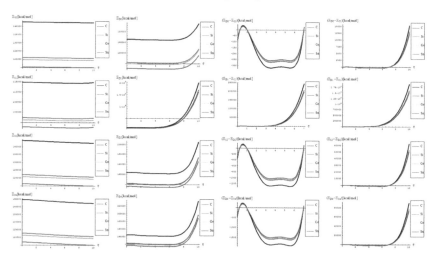

FIGURE 1.10 The Wiener based topological potentials of Eqs. (1.104)–(1.111): from top to bottom in successive orders and from left to right for the forming ("0") SW and for propagating ("D") of the SW defects on the medium (η=0–7)-to-long (η=0–10) range, respectively (Putz & Ori, 2014).

the first order potential displays such energetic barrier rise just from the short range, paralleling the defective "D" shape;

- Concerning the C>Si>Ge>Sn potential (paralleling the electronegativity) hierarchy, one sees that the C-to-Si large energetic gap for pristine "0" structure is considerably attenuated for the "D" propagation of the SWw defect, manifested especially for second and 4th order, while the Si-to-Ge energetic curves almost coincides for these orders; even more, all Si, Ge, and Sn shapes are practically united under first order defective potential "D1".

These topological potential features stay at the foreground for further undertaking of the remaining observable properties in a comparative analysis framework. As such, when analyzing the critical "temperature" through the inverse of the thermal energy of Eq. (1.96) one actually gets information on the phase transition *specific time* (via the celebrated statistical-to-quantum mechanics equivalence facilitated by the Wick rotation, $\hbar\beta \leftrightarrow \tau_{\beta-}$) for SW forming, propagating and disappearing through the topological isomeric forms of the considered structures, nanoribbons of Figure 1.9, with the dynamic representations of the Figure 1.11 (Putz & Ori, 2014):

- The general feature for the β signals is that it is constantly for the short range in phase transition (critical curve) between forming ("0-to-1") and transforming "D" of SW waves and along the predicted inverse C<Si<Ge<Sn signal (pulse) hierarchy;
- The situation changes on the long range "echo" when the critical signal may be recorded closer to the defective than to the pristine structures, when one should records also a shrink signal pulses gap between the C-to-Si-to-Ge-to-Sn;
- The differences between the critical-to-defective-to-pristine structures' pulses shapes follows on the short range the generally recorded topological potential difference on that range, while noticing definite cupolas for the long range behavior – especially on the critical regime, meaning that indeed the SWw echo is disappearing after about 50 isomeric topological transformation of the considered honey-comb nanosystems (see the right bottom line picture of Figure 1.11).

Worth noting that the β signals of Figure 1.11, when measured in seconds, through the [kcal/mol]-to-[Hz] transformations by Physical units (2013)

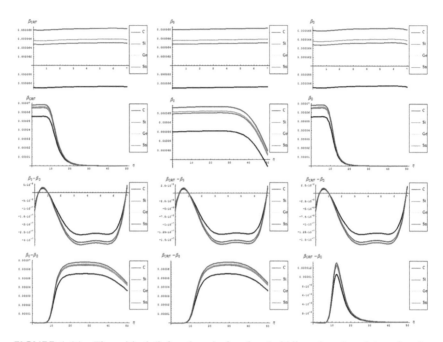

FIGURE 1.11 The critical (left column), forming (middle column) and transforming (right column) of SWw in Figure 1.9, as based on Eq. (1.96) on the short (first upper row) and long range (second upper row), and of their respective differences (Putz & Ori, 2014).

since the Eq. (1.96) relationship with topological potentials expressed in [kcal/mol] surpass the nowadays femtosecond limit (suitable only by synchrotron measurements); however, one can equally asses that these shorter times are specific to isomerization or topological rearrangements processes; they are however no more "theoretical" having by the present study an associate scale and algorithm of prediction. There is equally possible to imagine other nanosystems for which β having higher and therefore shorter times of detection for topological isomers.

Even more, these times are in fact the bondonic times for concerned lattices whose periodic radii of action is determined upon considering the β information into the specific Eqs. (1.96) featuring the Figure 1.12 representations and the following characteristics (Putz & Ori, 2014):

- The identical boning length for pristine and defective structures on the short range transformations (up to seven topological rearrangements paralleling the SW wave propagation into extended lattice); notably, the bonondic lengths are correctly shorter than the detected

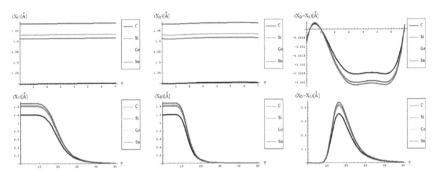

FIGURE 1.12   The bondonic "length" for SWw in forming (left column), defective propagation (middle column) along their differences (right column) behavior: on short (upper row) and long (lower row) ranges, upon considering critical information of Eqs. (1.96) into Eqs. (1.96) (Putz & Ori, 2014).

or previously estimated bonding length for C-C bond (in graphene) and elongated in Si-Si bond (in silicene), see the Section 1.2.1, since the bondonic agent nature, in assuring the bonding action for the basic atomic pairing in honey-comb nanoribbons.

- The decreasing of boning length and of the consequently action on the long range dynamics, paralleling the decreasing of the inter-elongation difference in bonding for C-to-Si-to-Ge-to-Sn;
- The prediction on the bondonic longest "echo" in an extended lattice, limited to 50 transformations, or dipole extensions steps continuing the Figure 1.9; this information has the practical consequence in predicting the longest nano-fragment still chemical stabilized by the bondonic "echo" by its long range action (sustained by its inner Bohmian nature, see Putz 2010a,b);
- The D-to-0 bondonic length differences parallels those recorded for critical-to-D one found for the β signal in Figure 1.11, this way confirming the finite (non-zero nor infinite) physical length over which the phase transition from pristine to Stone-Wales topological isomer is taking place; it also offers an spatial alternative to temporal (quasi-inaccessible) scale of measuring and detecting the bondonic effects on topological transformations and isomeric rearrangements.

Passing to the canonical measures one has in Figure 1.13 the representations of the sample total energy representations in "0" and "D" sates, and their behavioral difference, side by side for the actual fourth order algorithm

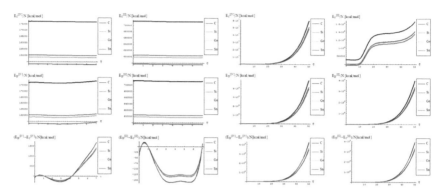

FIGURE 1.13　Side-by-side canonical internal energies of bondons in honey-comb supercells of Figure 1.9, as computed with the fourth order formulation of Eqs. (1.93) side by side with the former second order formulation of Putz & Ori (2012), for pristine "0" (upper row), defective "D" (middle row) and their differences (lower row), respectively (Putz & Ori, 2014).

with the previous second order restricted treatment (Putz & Ori, 2012). Accordingly, the specificities can be listed as following (Putz & Ori, 2014):

- No shape differences other than the overall scales along a quasi invariant gap energetics between C-to-Si-to-Ge-to-Sn related lattices are recorded between the [II] and [IV] order path-integral bondonic formalisms, with natural higher energetic records for the later approach since more interaction/interconnection effects included, for the internal energies of honey-comb lattice without and with topological defects as triggered on the short range as in Figure 1.9;
- The previous situation changes for the long range SW dipole transformations, noticing the same type of energetic rising as for the forth order topological potential/barrier in Figure 1.10, together with quasi-unifying the C with Si-to-Ge-to-Sn behaviors (being the last three atomic based lattices quite unified in total bondonic energetic shape); the peculiar behavior is noted just to [II] order treatment and in the long range pristine super cell self-arranging, when the energy rising displays two plateaus as well as still a C respecting Si-to-Ge-to-Sn energetic gaps for their honey-comb lattices; however, the energetic rising even for the so called pristine structures is in accordance with the bohemian quantum nature of the bondon which accounts for the self-arrangements of a quantum structure even when self-symmetric, being this in accordance with quantum

vacuum energy which is non-zero due to the energy required for internal self-symmetry eventually broken in the spring isomerization of space, here represented by the SW topological defects and of their (also finite) propagations.

- The D-to-0 differences are nevertheless replicating the defective behavior for both the [II] and [IV] order analysis on the long range, while showcasing some different types of fluctuations and inversions along the C-to-Si-to-Ge-to-Sn honeycomb nanosystems for the short-range of SW dipole evolution;

- with special reference to [IV] short range behavior one notes theta the pristine "0" state is still present as an "echo" over the defective "D" state, due to its energetic dominance, that nevertheless has the contribution in replicating "the learning" mechanism of generating SW defects in between each short range steps as was the case in between $\eta=0$ and $\eta=1$; remarkably, this may have future exciting consequences in better understanding the cellular morphogenesis by "replicating the learning" machinery of the Stone-Wales transformation, found to be present also at the cell-life-cycle phenomenology, see Guillot and Lecuit (2013).

Going to the last but the most "observable" quantity as it is the caloric capacity of Eqs. (1.94) within the present [IV] order path integral – bondonic approach, one has the results, and comparison with eth previous [II] order formalism of Putz and Ori (2012), exposed in Figure 1.14, with the notable characters (Putz & Ori, 2014):

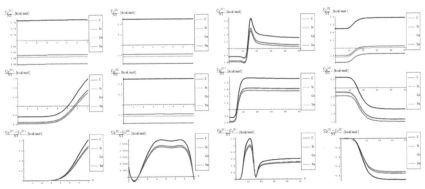

FIGURE 1.14 The same type of representations as in Figure 1.13, here for caloric capacity of Eq. (1.94) and of former formulation of Putz and Ori (2012), in the fourth and second order path integral of bondonic movement, respectively (Putz & Ori, 2014).

- form the scale values, one obtains actual quite impressing accordance with the previously calculated or predicted values, at the graphene and silicone networks: take for instance just the pristine "0" output, in [IV] other environment; for it one notes the constant results about $C_0^{[IV]}/(NT)[kcal/mol] \sim 0.77$ for graphene and $C_0^{[IV]}/(NT)[kcal/mol] \sim 0.65$ for silicene; when taking account of the units transformations by Physical units (2013), such as 1 [kcal/mol] = 503.228 [K], one arrives that, for instance, for room temperature of T~300K, and for short range transformation (say N=7 bondons involved, one created per each step of topological transformation) one gets $C_0^{[IV]}(C)[hartree] \sim 0.185$ and, respectively, $C_0^{[IV]}(Si)[hartree] \sim 0.156$, which in [eV] will, respectively, give about $C_0^{[IV]}(C)[eV] \sim 5.09$ and $C_0^{[IV]}(Si)[eV] \sim 4.24$ which nevertheless are quite close with previous estimations for SW rotation barriers as $E_b \approx 5$ eV for graphene and $E_b \approx 2.8$ eV for silicene (see Section 1.2.1); the discrepancy may be nevertheless avoided while considering the semiconductor properties of Si which requires more bondons being involved such that the SW rotational barrier to be passed and the defective dipole triggered; as such for N=10 created bondons for Si – SW super cell one refines the above result to $C_0^{[IV]}(Si, N = 10)[eV] \sim 2.96$ which fits quite well with literature results, see Sahin et al. (2013). On the other side, there is also apparent that for [II] order treatment the data of Figure 1.14 implies that more bondons are required to fit with the right observed or by other means estimated data, which leaves with the important conceptual lesson: more bondons – less connectivity relationship, very useful in addressing other fundamental chemical problems like crystal field theory and aromatic compounds, just to name a few.
- As previously noted the "D" effect is to shrink the energetic gap between C-to-Si-to-Ge-to-Sn lattice structural behavior, respecting the "0" pristine or defect forming transition state;
- The short range D-to-O differences closely follows the previously internal energy shapes of Figure 1.13, yet with less oscillations for the [IV] treatment, thus in accordance with more observable character of the caloric capacity;
- For the long range behavior, instead, what was previously parabolic increasing in internal energy acquires now a plateau behavior in [IV] and [II] order representations: "0", "D", and their "D-to-0" differences; nevertheless there seems that the "echo"/signal about η=10

is particularly strong in [IV] order modeling of D-to-0 differences in caloric capacities, while noticing for the "D" state the graphenic apex curvature about $\eta=7$ followed by that of silicone at $\eta=10$, in full consistency with above bondonic energetic analysis (N=7 for C-lattice, and N=10 for Si lattice), thus confirming it. Further signals are also visible, for accumulation of bondons (as the SW dipole evolve and extends over the nanostructure) at $\eta=15$ in pristine structure, as well as for the further plateaus within the [II] order analysis, again in accordance with the above discovered rule of more bondons required for acquiring the same effect with less connectivity (long-range-bonding neighboring) analysis.

The present results fully validate the bondonic analysis as a viable tool in producing reliable observable characters, while modeling and predicting the complex, and subtle, chemical phenomenology of bonding in isomers and topological transformations in the space of chemical resonances. Further works are therefore called in applying the present algorithm and bondonic treatment for other nanosystems as well as in deep treatment for the symmetry-breaking in chemical bonding formation of atoms-encountering in molecules and in large nanosystems.

1.6 CONCLUSION

Bondon's nature has roots in Bohmian theory on one hand and carries its bosonic character to model the inter-electronic repulsion in bonding on other hand, being therefore capable of exploiting the role of long range interactions in low dimensionality nanosystems. Present study advances a consistent picture to describe extended systems based on bondonic mass, length and energy ad especially at critical regime associated with caloric capacity of grand canonical samples associated to nanosystems, here applied at the graphene's nanoribbons. In particular, it was shown that in graphene nanoribbons the typical bondon energy E_{Bond} scales the delocalization X_{Bond} [Å] to a value L^* above the edge of lattice super-cell, providing in such a way the physical ground of an effective long-range mechanism capable to arrange order transitions in such a kind of 1D systems. Further studies enlarging the current approach may be done in many directions as such:

(i) considering excited states of bondonic mass, i.e., by further modes of Eq. (1.1) that may insight to the so called optical branches of extended systems, including the transition phase regime;

(ii) investigating the consequence of including the fourth-order effects of quantum propagator of Eq. (1.80) that may unveil more details on phase transitions features;

(iii) larger density of defects will also imply strongest signals about order-disorder transitions as a function of defects interactions; finally these models will also benefit from other topological long-range indices extending the simulations of bondon propagations on two-dimensional honeycomb lattices that are rich of critical effects produced by low-energy interacting electrons in a long-range Coulomb potential represented by a massless scalar gauge field (Herbut, 2006), a characteristic that fits in the bondons quasi-bosonic nature; equally, phase transitions earlier signaled by some non-analyticity of the free energy may also be explored.

Nevertheless, highly intriguing novel properties are theoretically derived here, namely: the topological potentials upto the fourth order, the so called beta-signal accounting for the time scale of bondonic pulses in a lattice supercell of graphenic type, the associate length of action, along the total internal energy of topological isomers and the remarkable behavior of the caloric capacity of the nanosystems. These quantities were evaluated and discussed for critical regime modeling the phase transition from a pristine to defective Stone-Wales topological transformations in the nanoribbon structures, either on short and long range of propagation, paralleling the creation of isomeric bondons, while identifying also the anti-bonding particle creation though following the dissipation echoes of the first over the lattice.

As an overall conclusion, the ability to indicate peculiar scale-threshold (like the η scale) for a given process in a given nanosystems, represents an important computational result, which provides an increasing importance to the topological simulation algorithms. Ab-initio models, hardly compete with topological modeling in selecting/proposing interesting configurations in extended nanosystems with hundreds or thousands atoms like the structure studied here; they have nevertheless the key role of refining the theoretical physical characterization of those proposed configurations, and asses their possibly physical-chemical relevance. The typical example is the

Stone-Wales wave isomeric mechanism that produces the η-extended dislocation dipole originally applied here to graphene-to-silicene nanoribbons; future computational studies, especially at DFT level will be necessary to describe and cross-check the actual bondonic findings regarding the energetic barriers and thermodynamic stability of presently considered SW topological defects in IVA elemental honeycomb and related hetero structures.

Moreover, the applications triggered by the present graphenic reference are envisaged as further projections towards the design of reactive supports for the pharmaceutical and cosmetic compounds, due to the conductor properties, magnetic and of flexibility and chemical saturation – unique for these new types of materials (as recognized also by the Nobel Prize in Physics in 2010), while the technological passage downward IVA elements as the basic constituents of nanosystems is expected to enrich the molectronic field with new semi-conductor and quantum electronic exotic properties.

Finally, the bondonic quantum condensate distribution picture for computing the energetic (observable) energies involved in isomeric series of nanostructures is formulated and applied at its general formalism with the exciting premises of simulating the creation and dissipation of SWw defective configurations in the graphenic-like regions which are preset on the surface of the *large* fullerenes, by modeling also the creation and the annihilation of bondons. Defective fullerenes of such a kind, modified by sequences of isomeric topological transformations are totally unexplored by scientific literature so far and their bondonic modeling appears as a forthcoming promising perspective and will be for sure investigated. The isomerization of honeycomb nanostructures by the application the generalized Stone-Wales transformations formalism is fully extendable to the case of non-spiral fullerenes as the demonstrated in the recent study (Berge, 1957; Ori, Putz et al., 2014).

KEYWORDS

- **conditional density**
- **extended nanostructures**
- **generation and propagation**
- **graphene**

- **honeycomb fragments**
- **partition function**
- **ribbons**
- **second and fourth orders of topo-defective graphenes**
- **Stone-Wales defects**
- **topo-bondonic algorithm**

REFERENCES

AUTHOR'S MAIN REFERENCES

Putz, M. V. (2016a). *Quantum Nanochemistry. A Fully Integrated Approach: Vol. III. Quantum Molecules and Reactivity.* Apple Academic Press & CRC Press, Toronto-New Jersey, Canada-USA.

Putz, M. V. (2016b). *Quantum Nanochemistry. A Fully Integrated Approach: Vol. I. Quantum Theory and Observability.* Apple Academic Press & CRC Press, Toronto-New Jersey, Canada-USA.

Putz, M. V. (2016c). *Quantum Nanochemistry. A Fully Integrated Approach: Vol. II. Quantum Atoms and Periodicity.* Apple Academic Press & CRC Press, Toronto-New Jersey, Canada-USA.

Putz, M. V. (2012a). Quantum Theory: Density, Condensation, and Bonding, Apple Academics, Toronto, Canada.

Putz, M. V. (2012b). Chemical Orthogonal Spaces, Mathematical Chemistry Monographs, Vol. 14, University of Kragujevac, Kragujevac.

Putz, M. V. (2012c). Nanoroots of Quantum Chemistry: Atomic Radii, Periodic Behavior, and Bondons. In: Nanoscience and Advancing Computational Methods in Chemistry: Research Progress, Castro, E. A., Haghi, A. K. (Ed.) IGI Global (formerly Idea Group Inc.), Hershey (PA) Chapter 4, pp. 103–143 (DOI: 10.4018/978–1-4666–1607–3.ch004).

Putz, M. V. (2012d). Density functional theory of Bose-Einstein condensation: road to chemical bonding quantum condensate. Structure and Bonding 149, 1–50 (DOI: 10.1007/978–3-642–32753–7_1).

Putz, M. V. (2012e). Valence atom with Bohmian quantum potential: the golden ratio approach. Chemistry Central Journal, 6, 135/16 pages (DOI: 10.1186/1752–153X-6–135).

Putz, M. V., Ori, O. (2014). Bondonic effects in group-iv honeycomb nanoribbons with Stone-Wales topological defects. Molecules 19(4), 4157–4188 (DOI: 10.3390/molecules19044157).

Putz, M. V., Ori, O., Cataldo, F., Putz, A. M. (2013). Parabolic reactivity "coloring" molecular topology: Application to carcinogenic PAHs. Curr. Org. Chem. 17(23), 2816–2830 (DOI: 10.2174/13852728113179990128).

Putz, M. V., Ori, O. (2012). Bondonic characterization of extended nanosystems: application to graphene's nanoribbons. Chem. Phys. Lett. 548, 95–100 (DOI: 10.1016/j.cplett.2012.08.019).

Putz, M. V. (2011a). Hidden side of chemical bond: the bosonic condensate. In Advances in Chemistry Research. Volume 10, Taylor, J. C. (Ed.), NOVA Science Publishers, Inc., New York, Chapter 8, pp. 261–298.

Putz, M. V. (2011b). Electronegativity and chemical hardness: different patterns in quantum chemistry. Current Physical Chemistry 1(2), 111–139 (DOI: 10.2174/1877946811101020111).

Putz, M. V. (2011c). Quantum parabolic effects of electronegativity and chemical hardness on carbon π-systems. In: Putz, M. V. (Ed.), Carbon Bonding and Structures: Advances in Physics and Chemistry, Carbon Materials: Chemistry and Physics series Vol. 5, Springer Verlag, London, Chapter 1, pp. 1–32.

Putz, M. V. (2010a). The bondons: the quantum particles of the chemical bond. Int. J. Mol. Sci. 11(11), 4227–4256.

Putz, M. V. (2010b). Beyond quantum nonlocality: chemical bonding field. Int. J. Environ. Sci. 1, 25–31.

Putz, M. V. (2009). Path integrals for electronic densities, reactivity indices, and localization functions in quantum systems. International Journal of Molecular Sciences 10(11), 4816–4940 (DOI: 10.3390/ijms10114816).

De Corato, M., M. Bernasconi, L. D'Alessio, O. Ori, M. V. Putz and, G. Benedek, (2013). Topological Versus Physical and Chemical Properties of Negatively Curved Carbon Surfaces. In: Ashrafi, A. R., Cataldo, F., Iranmanesh, A., Ori, O. (Eds.) Topological Modeling of Nanostructures and Extended Systems, Springer Verlag, Dordrecht, Chapter 4, pp. 105–136 (DOI: 10.1007/978–94–007–6413–2_4).

De Corato, M., Benedek, G., Ori, O., Putz, M. V. Topological Study of Schwarzitic Junctions in 1D Lattices, International Journal of Chemical Modeling 4(2/3) (2012). 105–113.

Ori, O., Putz, M. V., Gutman, I., Schwerdtfeger, P. (2014). Generalized Stone-Wales Transformations for Fullerene Graphs Derived from Berge's Switching Theorem. In: "Ante Graovac – Life And Works" Gutman, I., Pokrić, B., Vukičević, D. (Eds.), Mathematical Chemistry Monographs No. 16, University of Kragujevac, Chapter 5/Part C, pp. 259–272.

Ori, O., Cataldo, F., Putz, M. V. (2011). Topological anisotropy of Stone-Wales waves in graphenic fragments. Int. J. Mol. Sci. 12(11), 7934–7949 (DOI: 10.3390/ijms12117934).

SPECIFIC REFERENCES

Aufray, B., Kara, A., Vizzini, S., Oughaddou, H., Léandri, C., Ealet, B., Le Lay, G. (2010). Graphene-like silicon nanoribbons on Ag(110), A possible formation of silicene. Appl. Phys. Lett. 96, 183102–183103.

Babic, D., Bassoli, S., Casartelli, M., Cataldo, F., Graovac, A., Ori, O., York, B. (1995). Generalized Stone-Wales transformations. Mol. Simul. 14, 395–401.

Basheer, E. A., Parida, P., Pati, S. K. (2011). Electronic and magnetic properties of BNC nanoribbons: a detailed computational stud. New, J. Phys. 13, 053008(11pp).

Berge, C. (1957). Two theorems in graph theory. Proc. Nat. Acad. Sci. USA 43, 842–844.

Bhowmick, S., Waghmare, U. V. (2010). Anisotropy of the Stone-Wales defect and warping of graphene nano-ribbons: a first-principles analysis. Phys. Rev. B. 81, 155416, 1–155416, 7.

Britnell, L., Gorbachev, R. V., Jalil, R., Belle, B. D., Schedin, F., Mishchenko, A., Georgiou, T., Katsnelson, M. I., Eaves, L., Morozov, S. V., Peres, N. M. R., Leist, J., Geim, A. K., Novoselov, K. S., Ponomarenko, L. A. (2012). Field-effect tunneling transistor based on vertical graphene heterostructures Science 335, 947–950.

Butler, S. Z., Hollen, S. M., Cao, L., Cui, Y., Gupta, J. A., Gutiérrez, H. R., Heinz, T. F., Hong, S. S., Huang, J., Ismach, A. F.; Johnston-Halperin, E., Kuno, M., Plashnitsa, V. V., Robinson, R. D., Ruoff, R. S., Salahuddin, S., Shan, J., Shi, L., Spencer, M. G., Terrones, M.; Windl, W., Goldberger, J. E., (2013). Progress, challenges, and opportunities in two-dimensional materials beyond graphene. ACS Nano 7, 2898–2926.

Cahangirov, S., Topsakal, M., Aktürk, E., Sahin, H., Ciraci, S. (2009). Two- and one-dimensional honeycomb structures of silicon and germanium. Phys. Rev. Lett., vol. 102, 236804(4).

Carpio, A., Bonilla, L. L., de Juan, F., Vozmediano, M. A. H. (2008). Dislocations in graphene. New, J. Phys. 10, 053021, 1–053021, 13.

Cataldo, F., Ori, O., Iglesias-Groth, S. (2010). Topological lattice descriptors of graphene sheets with fullerene-like nanostructures. Mol. Simul. 36, 341–353.

Chattaraj, P. K., Parr, R. G. (1993). Density functional theory of chemical hardness. Struct Bond 80, 11–25.

Chen, S., Ertekin, E., Chrzan, D. C. (2010). Plasticity in carbon nanotubes: Cooperative conservative dislocation motion. Phys. Rev. B 81, 155417–8.

Chuvilin, A., Meyer, J. C., Algara-Siller, G., Kaiser, U. (2009). From graphene constrictions to single carbon chains. New, J. Phys. 11, 1–083019, 10.

Collins, P. G. (2011). Defects and Disorder in Carbon Nanotubes; In: Narlikar, A. V., Fu, Y. Y. (Eds.) Oxford Handbook of Nanoscience and Technology: Frontiers and Advances. Oxford University Press: Oxford.

Davydov, S. Y. (2010). On the elastic characteristics of graphene and silicene. Phys. Solid State. 52(1), 184–187.

De Padova, P., Quaresima, C., Ottaviani, C., Sheverdyaeva, P. M., Moras, P., Carbone, C., Topwal, D., Olivieri, B., Kara, A., Oughaddou, H., Aufray, B., Le Lay, G. (2010). Evidence of graphene-like electronic signature in silicene nanoribbons. Appl. Phys. Lett. vol. 96, 261905(3).

Dinadayalane, T. C., Leszczynski, J. (2007). Stone–Wales defects with two different orientations in (5, 5) single-walled carbon nanotubes: A theoretical study. Chem. Phys. Lett. 434, 86–91.

Dutta, S., Manna, A. K., Pati, S. K. (2009). Intrinsic Half-Metallicity in Modified Graphene Nanoribbons. Phys. Rev. Lett. 102, 096601(4).

Ertekin, E., Chrzan, D. C., Daw, M. S. (2009). Topological description of the Stone-Wales defect formation energy in carbon nanotubes and graphene. Phys. Rev. B 79, 155421–17.

Estrada, E., Hatano, N. (2010). Topological atomic displacements, Kirchhoff and Wiener indices of molecules. Chem. Phys. Lett. 486, 166–170.

Evans, M. R. (2000). Phase Transitions in One-Dimensional Nonequilibrium Systems. Brazil. J. Phys. 30, 42–57.

Ewels, C. P., Heggie, M. I., Briddon, P. R. (2002). Adatoms and nanoengineering of carbon. Chem. Phys. Lett. 351, 178–182.

Geim, A. K., Novoselov, K. S. (2007). The rise of graphene. Nature Materials 6, 183–191.

Guillot, C., Lecuit, T., (2013). Mechanics of epithelial tissue homeostasis and morphogenesis. Science 34, 1185–1189.

Guzman-Verri, G. G., Lew Yan Voon, L. C. (2007). Electronic structure of silicon-based nanostructures. Phys. Rev. B 76, 075131.

Hashimoto, A., Suenaga, K., Gloter, A., Urita, K., Iijima, S. (2004). Direct evidence for atomic defects in graphene layers. Nature 430, 870–873.

Herbut, I. (2006). Interactions and phase transitions on graphene's honeycomb Lattice. Phys. Rev. Lett. 97, 146401.

Huang, B., Liu, M., Su, N., Wu, J., Duan, W., Gu, B.-L., Liu, F. (2009). Quantum manifestations of graphene edge stress and edge instability: A first-principles study. Phys. Rev. Lett. 102, 166404, 1–166404, 4.

Im, J., Kim, Y., Lee, C.-K., Kim, M., Ihm, J., Choi, H. J. (2011). Nanometer-scale loop currents and induced magnetic dipoles in carbon nanotubes with defects. Nano Lett. 11, 1418–1422.

Iranmanesh, A., Ashrafi, A. R., Graovac, A., Cataldo, F., Ori, O. (2012). Wiener Index Role in Topological Modeling of Hexagonal Systems-From Fullerenes to Graphene; In: Gutman, I., Furtula, B. (Eds.) Distance in Molecular Graphs-Applications, Mathematical Chemistry Monographs Series, University of Kragujevac, Kragujevac, Volume 13, pp. 135–155.

Jeong, B. W., Ihm, J., Lee, G.-D. (2008). Stability of dislocation defect with two pentagon-heptagon pairs in graphene. Phys. Rev. B. 78, 1–165403, 5.

Jose, D., Datta, A. (2011). Structures and electronic properties of silicene clusters: a promising material for FET and hydrogen storage. Phys. Chem. Chem. Phys. 13, 7304–7311.

Kara, A., Léandri, C., Dávila, M. E., De Padova, P., Ealet, B., Oughaddou, H., Aufray, B., Le Lay, G. (2009). Physics of silicene stripes. J. Supercond. Nov. Magn. 22, 259–263.

Kleinert, H. (2004). Path Integrals in Quantum Mechanics, Statistics, Polymer Physics, and Financial Markets 3rd ed., World Scientific, Singapore.

Koopmans, T. (1934). Uber die zuordnung von wellenfunktionen und eigenwerten zu den einzelnen elektronen eines atoms. Physica1(1–6), 104–113.

Kotakoski, J., Krasheninnikov, A. V., Kaiser, U., Meyer, J. C. (2011b) From point defects in graphene to two-dimensional amorphous carbon. Phys. Rev. Lett. 106, 105505, 1–105505, 4.

Kotakoski, J., Meyer, J. C., Kurasch, S., Santos-Cottin, D., Kaiser, U., Krasheninnikov, A. V. (2011a) Stone-Wales-type transformations in carbon nanostructures driven by electron irradiation. Phys. Rev. B 83, 1–245420/6.

Krasheninnikov, A. V., Nordlund, K., Sirviö, M., Salonen, E., Keinonen, J. (2001). Formation of ion-irradiation-induced atomic-scale defects on walls of carbon nanotubes. Phys. Rev. B 63, 1–245405/6.

Kumeda, Y., Wales, D. J. (2003). Ab initio study of rearrangements between C60 fullerenes. Chem. Phys. Lett. 374, 125–131.

Lahiri, J., Lin, Y., Bozkurt, P., Oleynik, I. I., Batzill, M. (2010). An extended defect in graphene as a metallic wire. Nat. Nano. 5, 326–329.

Leandri, C., Le Lay, G., Aufray, B., Girardeaux, C., Avila, J., Dávila, M. E., Asensio, M. C., Ottaviani, C., Cricenti, A. (2005). Self-aligned silicon quantum wires on Ag(110). Surface Science 574(1), L9–L15.

Liu, Y., Yakobson, B. I. (2010). Cones, pringles, and grain boundary landscapes in graphene topology. Nano Lett. 10, 2178–2183.

Lusk, M. T., Carr, L. D. (2008). Nanoengineering defect structures on graphene. Phys. Rev. Lett. 100, 175503, 1–175503, 4.

Ma, J., Alfè, D., Michaelides, A., Wang, E. (2007). Stone-Wales defects in graphene and other planar sp2-bonded materials. Phys. Rev. B 80, 075131(4).

Manna, A. K., Pati, S. K. (2011). Tunable electronic and magnetic properties in BxNyCz nanohybrids: effect of domain segregation. J. Phys. Chem. C 115(21), 10842–10850.

Martin, J.-L., A. Migus, G. A. Mourou, A. H. Zewail (Eds.) Ultrafast Phenomena VIII, Springer-Verlag, Berlin-Heidelberg, 1993.

Meyer, J. C., Kisielowski, C., Erni, R., Rossell, M. D., Crommie, M. F., Zettl, A. (2008). Direct imaging of lattice atoms and topological defects in graphene membranes. Nano Lett. 8, 3582–3586.

Meyer, J. C., Kurasch, S., Park, H. J., Skakalova, V., Künzel, D., Groß, A., Chuvilin, A., Algara-Siller, G., Roth, S., Iwasaki, T., et al. (2011). Experimental analysis of charge redistribution due to chemical bonding by high-resolution transmission electron microscopy. Nat. Mater. 10, 209–215.

Nakano, H., Mitsuoka, T., Harada, M., Horibuchi, K., Nozaki, H., Takahashi, N., Nonaka, T., Seno, Y., Nakamura, H. (2006). Soft synthesis of single-crystal silicon monolayer sheets. Angew. Chem. Int. Ed. 45, 6303–6306.

Nordlund, K., Keinonen, J., Mattila, T. (1996). Formation of ion irradiation induced small-scale defects on graphite surfaces. Phys. Rev. Lett. 77, 699–702.

Novoselov, K. S., Falko, V. I., Colombo, L., Gellert, P. R., Schwab, M. G., Kim, K. (2012). A roadmap for graphene. Nature 490, 192–200.

Novoselov, K. S., Geim, A. K., V. Morozov, S., Jiang, D., Zhang, Y., Dubonos, S. V., Grigorieva, I. V., Firsov, A. A. (2004). electric field effect in atomically thin carbon films. Science 306, 666–669.

Ori, O., Cataldo, F., Vukicevic, D., Graovac, A. (2010). Wiener way to dimensionality. Iran. J. Math. Chem. 1, 5–15.

Parr, R. G. (1983). Density functional theory. Annu. Rev. Phys. Chem. 34, 631–656.

Parr, R. G., Bartolotti, L. J. (1982). On the geometric mean principle of electronegativity equalization. J. Am. Chem. Soc. 104, 3801–3803.

Parr, R. G., Pearson, R. G. (1983). Absolute hardness: companion parameter to absolute electronegativity. J. Am. Chem. Soc. 105, 7512–7516.

Parr, R. G., Yang, W. (1989). Density Functional Theory of Atoms and Molecules. Oxford University Press, New York.

Physical units (2013). http://users.mccammon.ucsd.edu/~blu/Research-Handbook/physical-constant.html (accessed August, 2013).

Pyshnov, M. B. (1980). Topological solution for cell proliferation in intestinal crypt. J. Theor. Biol. 87, 189–200.

Rutter, G. M., Crain, J. N., Guisinger, N. P., Li, T., First, P. N., Stroscio, J. A. (2007). Scattering and interference in epitaxial graphene. Science 317, 219–222.

Sahaf, H., Masson, L., Léandri, C., Aufray, B., Le Lay, G., Ronci, F. (2007). Formation of a one-dimensional grating at the molecular scale by self-assembly of straight silicon nanowires. Appl. Phys. Lett. 90, 263110(3).

Sahin, H., Sivek, J., Li, S., Partoens, B., Peeters, F. M. (2013). Stone-Wales defects in silicene: Formation, stability, and reactivity of defect sites. Phys. Rev. B 88, 045434(6).

Samsonidze, Ge. G., Samsonidze, G. G., Yakobson, B. I. (2002). Energetics of Stone–Wales defects in deformations of monoatomic hexagonal layers. Comput. Mater. Sci. 23, 62–72.

Stone, A. J., Wales, D. J. (1986). Theoretical studies of icosahedral C60 and some related species. Chem. Phys. Lett. 128, 501–503.

Takeda, K., Shiraishi, K. (1994). Theoretical possibility of stage corrugation in Si and Ge analogs of graphite. Phys. Rev. B 50, 14916–14922.

Terrones, M., Botello-Mendez, A. R., Campos-Delgado, J., Lopez-Urias, F., Vega-Cantu, Y. I., Rodriguez-Macias, F. J., Elias, A. L., Munoz-Sandoval, E., Cano-Marquez, A. G., Charlier, J. C. et al. (2010). Graphene and graphite nanoribbons: morphology, properties, synthesis, defects and applications. Nano Today 5, 351–372.

Todeschini, R., Consonni, V. (2000). Handbook of Molecular Descriptors, Wiley-VCH, Weinheim.

Vogt, P., De Padova, P., Quaresima, C., Avila, J., Frantzeskakis, E., Asensio, M. C., Resta, A., Ealet, B., Le Lay, G. (2012). Silicene: compelling experimental evidence for graphene-like two-dimensional silicon. Phys. Rev. Lett. 108(15), 155501(5).

von Szentpály, L. (2000). Modeling the charge dependence of total energy and its relevance to electrophilicity. Int. J. Quant. Chem. 76, 222–234.

Vukicevic, D., Cataldo, F., Ori, O., Graovac, A. (2011). Topological efficiency of C66 fullerene. Chem. Phys. Lett. 501, 442–445.

Zeng, H., Leburton, J. P., Xu, Y., Wei, J. (2011). Defect symmetry influence on electronic transport of zigzag nanoribbons. Nanoscale Res. Lett. 6, 254, 1–254/6.

Zhang, R. Q., Lee, S. T., Law, C.-K., Li, W.-K; Teo, B. K. (2002). Silicon nanotubes: Why not? Chem. Phys. Lett. 364(3–4), 251–258.

Zhoua, L. G., Shib, S. Q. (2003). Formation energy of Stone–Wales defects in carbon nanotubes. Appl. Phys. Lett. 83, 1222–1224.

CHAPTER 2

GEOMETRICAL CRYSTALLOGRAPHY

CONTENTS

ABSTRACT

The crystal (motif + lattice) system is a analyzed on its stable ground state by means of geometrical characterization of axes, inversion center and mirroring planes, as the main symmetrical elements driving the allowed operation in such ordered giant molecular structure: they where so provided the crystallographic systems (7), then extended to Bravais lattice (14), then enlarged to the point groups (32) when all elements of symmetry that intersects on a point are comprehensibly combined, and finally to the spatial groups (230) when the translation comes into the play too; these classification schemes correspond in fact with the most resumed way of chemical classification of substances, by their crystalline groups (points or spatial) and corresponds with the maps obtained by X-ray diffraction or by other optical action (optical activity by laser action or by magnetization) especially through the involved reciprocal space (with the allied theorems of faces represented by normal axes and axes represented by nodes in reciprocal space and of its projective maps) – there where the quantum behavior is recovered by means of wave vector, reciprocally related with the acting yet in resonance with electronic wave functions, specific to living electrons in crystals and ordered solids.

2.1   INTRODUCTION

The Mathematics is a formidable science and the mathematicians are very happy people! This is because, despite the fact that the mathematics proposes equations of whose solutions must be sought, there is a case even worst. Is that of the natural sciences, especially physicists, chemists, biologists, and geologists, who must find the solution of equations without not even knowing them.

In other words, the researchers in natural sciences, observing the objective reality, should propose equations by which linking the causes (and they must often be intuited) with the observed effects, then to solve these equations (and often this can not be in an exact way), then check the solutions obtained in concrete cases (and an equation with a viable solution have to cover a large number of phenomena observed), and finally to compare the numerical values obtained with those observed (in fact measured by experiment-and there appeared again a big problem, that of the measured phenomenon objectivity when it is "isolated" or "challenged in vitro" in the laboratory by the experimenters) – and in any case, the difference between these values must be find a reasonable margin of error (acceptable: less than 10%, optimal: less than 1%). Otherwise, everything should be reformulated or corrected, from the beginning (including the equations and/or the proposed base model).

Therefore, the natural science and their researchers have a truly difficult task, at the every step of their investigations, either experimental or theoretical.

However, paradigms, with general and even universal value, can be made in order to support the enunciation of the working models in the investigation of the Nature mysteries. Further will be explored the crystalline "organization" of the matter, based on which will be introduced its constituent schemes and in the following sub-chapters will be analyzed the consequences.

This is because all the chemical compounds can be brought in crystalline form under appropriate conditions of temperature and pressure. Further, such a possibility allows the investigation of the chemical compounds structure by the X-ray diffraction method. Therefore, the study of the crystalline form of matter (Crystallography) it becomes a necessary to elucidate the structures of the substances having as a result the prediction of the structure and the transformation of the chemical compounds.

At this point it is worth underline the fact that the names "chemical and physical crystallography" are arbitrary, rather serving to a methodological ordering, not fundamentally different.

In the favor of this vision, is sufficient to recall that in the middle of the XVIII-century Desaguliers suggested that the forces that keep stable the molecules are electrical, idea retaken by Berzelius a century later, long before the clarification of the electronic structure of the molecules, based on Bohr's findings and the X-ray diffraction

In the early twentieth century, the distinction between the physical forces and the chemical forces, at the molecular level, was superfluous under the conjugation of the atomic theories, coming from the quantum physics, with those chemical, of the valence theory. In the modern science, it was ultimately imposed the idea that the forces which hold the atoms in molecules or in crystals are physical, but expresses the reality of the chemical bondings.

2.2 A BRIEF HISTORY OF CRYSTALLOGRAPHY: THE INFLUENCE IN ART AND PHILOSOPHY

The solids and especially the crystals have always been fascinating; the present discussion follows (Chiriac-Putz-Chiriac, 2005).

Before proceeding to expose the crystalline paradigm of the matter, let's make a brief incursion in the flow of the events, which marked the history of science in terms of finding crystals and their legalities.

In Table 2.1 are shown the periods and the important events which marked the science and the solid state and of the crystals studies.

Moreover, even in the art, in the modern art, the study of the solid matter and of the geometrical structures associated had constituting a starting point, a prerequisite.

A famous example in the "crystallographic art" is M.C. Escher (1898–1972), Belgian artist and designer, who caused a perpetual fascination for the modern and postmodern generation, and whose work is recognized for the spatial illusions created, for impossible construction imagined, for the repetitive geometric subjects (the so-called tessellations), and no less for his incredible techniques in the images compilation and lithography, see World of Escher (2014).

Both artist and mathematician, M.C. Escher received a great appreciation from both communities, being considered in equal measure a humanist and a scientist, a mathematician and s crystallography, a visionary and a pragmatist.

Escher's work, representing periodic structures, complex mathematical structures, spatial perspectives requiring a "second look", leaves open the path between science and art in a spontaneous way, ingenious and challenging

Moreover, in the interrogations on the original principles, we reveal the deep connections between philosophy (seen as an attempt of explanation of the world based on the ideational and ideal forms) and the solid structures, as is mentioned in the ancient Greece in Plato's work. Based on philosophy, he introduced which has remained consecrated as the Platonic solids, see Table 2.3. The platonic solids (Wolfram Research, 2003), are regular solids under the form of the convex polyhedral with equivalent faces congruently located one to each other. There are exactly

TABLE 2.1 A Brief "Story of the Crystals", After Chiriac-Putz-Chiriac (2005)

Date	Event
ca. 6000 BC	Egyptian Turquoise mines are active.
Antiquity	The shiny stones (especially diamond, sapphire, emerald and the ruby) are very researched, making it magical and curative properties attributed!
ca. 350 IC	Theophrastus describes the regular form of the Garnet crystals.
ca. 30 IC	Strabo calls quartz (κρψστλλοζ = *crystallum* in Latin), which today we call "crystal".
1597	Alchemist Libavius noted that the geometric shape of the crystals is common to the sea salt.
Cen. XVII	Boyle, Leeuwenhoek, Kepler, Hooke and other make multiple observations about the crystals with the aid of the recently invented microscope.
1611	Kepler suggested that the hexagonal symmetry of the snowflakes is due to "the regular packing of constituent particles."
1665	Hooke advancing the hypothesis of the "spheroids" as components of the crystals.
1669	Steno observed as the crystals, whatever their origin, always keep some features related to the angles between the sides.
1780	Carangeot invents the goniometry with contact, thus managing the calculation of the angles between the sides.
1783	Bergman's studies on cleavage (split) in crystals suggest the hypothesis of the units rhombohedrally packed in crystals.
1783	De l'Isle makes the law of the "constant of the angles between faces."
1801	Haüy proposed the law of the "indices of the rationalized" in the crystal, thus formulating the fundamental law of crystal's morphology.
1808	Malus observed the polarization of the light produced by certain crystals.
1809	Wollaston invents the Goniometry with Reflection-helping to increase the accuracy of the data measured for the angles between the crystal faces.
1815	Biot discovered the rotator forms levo- and dextrose of the quartz.
1819–22	Mitscherlich discovered the isomorphism (the crystals with various composition but with the same form) and the polymorphism (crystals with different shapes but with the same chemical composition) – equivalent to the alotrophism of the chemical elements.

TABLE 2.1   Continued

Date	Event
1839	Miller uses the *Miller* indices to designates the crystal faces.
1848	Pasteur discovered the enantiomorphism crystals.
1880'-90'	Sohncke, Federov, Schönflies and Barlow develop theories about the internal symmetry of the crystals – but still with no experiments that confirm their theories.
1906–19	The Groth's work entitled "Chemische Krystallographie" (The Chemical Crystallography), based on the morphological properties, optical and other properties, puts in tables over 7000 crystalline substances, but still without specifying any information about the internal structure, in the absence of the experimental techniques!
1907	Barlow and Pope suggest the model through which the ionic crystals are composed by hard spheres in contact.
1912	Friedrich, Knipping and von Laue discover the X-ray diffraction.
1913	W.H. (father) and W.L. (son) Bragg uses the dependence of orientation of the X-ray diffraction on a monocrystal to solve (determine) the structure of the state NaCl (further also of the diamond, etc.).
1913	Ewald introduces the concept of reciprocal network.
1914	Debye's theory about the thermal movement of the atoms in solids predicts the appearance of the Debye-Waller's factor in the structures recorded by the methods of the X-ray diffraction.
1916	Debye and Scherer test the diffraction produced by the crystalline powders.
1924	Bernal et al. determine the structure of the graphite.
1926	The investigations of Frenkel reveals the existence of the punctual defects in the crystalline structures.
1926	Goldschmidt makes the hypothesis of the crystalline structures based on the spherical atoms.
1927	Pauling, starting from the ionic model of Goldschmidt, formulate the so-called rules of *Pauling* for the ions combinations in the crystals.
1929	It had been discovered how, rotating the anode of a X-ray generator – is obtained an increase of the intensities of the X-ray and a better vision of the diffraction figures on the crystals.
1934	Patterson introduces the *Patterson* function as a solution for determining the structure of the crystals based on the X-ray diffraction.

TABLE 2.1   Continued

Date	Event
1934	Ruska produces structure images of the crystals using the first electronic microscope.
1936	Halaban and Preiswerk – highlights the diffraction with neutrons on the crystals.
1941	Hughes uses the method of interpolation of the smallest (least) squares to obtain the most likely crystalline structure from a set of diffraction data.
1944	Buerger invents the precession camera.
1948	Exposed the direct methods for solving the crystalline structure, based on the X-ray diffraction.
1950s	Appear the automatically diffractometers, and the early computational age dramatically increase the ability to solve the crystalline structures.
1951	Bijvoet uses the anomalous scattering to determine the crystalline chirality (the absolute configuration).
middle of '50 years	The computers are used for the first time for complete solving of the crystalline structure of the x-ray diffraction data.
1955	The formulation of the Principles of Laves – explains the spatial occupation in crystalline structures.
1956	Menter creates the first image of crystalline network using the Transmission Electronic Microscopy (TEM)
1957	Müller – using the Field Ionic Microscopy: FIM notes the individual atoms in metals.
1970	Crewe, Wall and Langmore – develop the Darkfield Scanning Electron Microscopy, the first experimental method for the individual view of the heavy atoms in the crystalline structures.
1971	Formanek et al. – create the first detections of an individual atom using the High Resolution Electron Microscopy (HREM).
1974	Ijima – makes the first observation of the punctual defects using the electronic microscopy.
1980s	Becomes accessible the synchrotron radiation – which allows the massive increase of the x-ray intensity and of the quality of the recorded diffraction figures (e.g., the Laue diffraction figure in X-ray for the crystals had been obtained in the scale of milliseconds).
1982	The surface detectors are used to obtain the diffraction figures in x-ray, with the consequence of the considerable decrease of the time required for the reordering of the diffraction figures for the crystals.

TABLE 2.1 Continued

Date	Event
1982	Binnig and Rohrer – develop the Scanning Tunneling Microscopy (STM), which allows the detection even of the atoms of lighter surface.
1984	Schechtman et al. discover the *quasicrystals.*
1984	Binnig et al. – elaborate the Atomic Force Microscopy (AFM), which allows the detailed view of the surfaces, even more accurate than the one by STM.
1990'	Are recorded over 200.000 of crystalline structures (with total indexing for the specification of the coordinates of the internal atoms), stored in the databases.

5: the tetrahedron (with each three faces as adjacent triangles), the cube (with each three faces as adjacent square), the octahedral (with each four faces as adjacent triangles), the dodecahedron (with each three adjacent pentagonal faces) and the icosahedron (with each five triangles as adjacent faces), the first demonstration for their number and form being exposed by Euclid in his paper Elements.

Platon had been introduced this solids for the first time in *Timaios* (about 350 BC) as cosmic figures, trying to present an axiomatic cosmology.

Thus, the platonic universe is made by forms, particular objects, the absolute space, the Creator God and the brute matter.

As creator, God is unable to make a perfect world with an imperfect material. Thus, Platon believed that all the substances are composed by: air (octahedral units), earth (cubic units), fire (tetrahedral units) and water (icosahedron units).

The dodecahedron units were associated with the material composing the constellations and the paradise. In this universe, the Earth was the center, and the planets motions take place on crystalline spheres.

The human knowledge were "lost" at birth, but could be regained through a careful search and understanding of the "pure forms" (Republic (see, Platone 1996), Eikos Mythos – A true story).

One way, in this search, is the study of the bodies in terms of mathematics, based on the movements and regular forms. The truth must be found behind the appearances, or shapes, even if the absolute truth can not be completely elucidated!

Thus, had been announced the first major crystalline paradigm of the matter, by philosophical insight.

No less admirable, the atom solid paradigm, as the fundamental entity of the modern science on the Matter, will further reveal how rich conceptual is the putting of the programmatic knowledge of the nature structures on the geometric and symmetry bases.

2.3   THE SYMMETRY OF THE ISOLATED BODIES

The symmetry is a body property to reach an undifferentiated position respecting the initial one when it was "moved" in some way. To be emphasized the symmetry, the body movement must be made in relation to a geometric element that requires the type of the movement and is called the symmetry element. It is obvious that the symmetry elements of the isolated body can be; the present discussion follows (Chiriac-Putz-Chiriac, 2005):

- the *point*: the center of symmetry or of inversion;
- the *line*: the axis of symmetry or of rotation;
- the *plane*: planes of symmetry or of mirroring;

The movement that must be done by the body in relation with the element of symmetry is imposed by its nature and it is called symmetry operation. To illustrate the elements and the symmetry operations we will use the representation of some "candelabra" with numbered candles.

Thus towards the symmetry center must be accomplished the inversion ("the giving over head"). A body has his center of inversion if any point has an equivalent to the opposite side, on the same straight line passing through the center, Figure 2.1.

For this proof, the reticular string A_1, A_2, ..., A_n, with the parameter of the string equal with p will be considered in Figure 2.3.

If through the points of the string passes an axis of n order, then such an axis will pass through any of the other points.

The rotation around the axis that passes through A_2 with an angle $\alpha=2\pi/n$ will bring the point A_1 in the point B_1. The rotation around the axis, which passes through AI with the same angle α, must bring the point A_{i+1} in the point B_{i+1}. the points B_1, B_2, ..., B_n must compose a

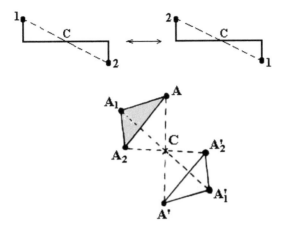

FIGURE 2.1 The illustration of the inversion operation, after Chiriac-Putz-Chiriac (2005).

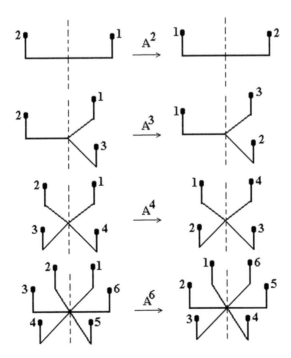

FIGURE 2.2 The representation of the operations of rotation allowed in a crystal, after Chiriac-Putz-Chiriac (2005).

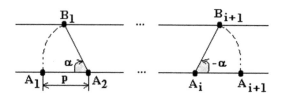

FIGURE 2.3   The construction helpful for the demonstration of the types of the possible axes of symmetry in a crystal, after Chiriac-Putz-Chiriac (2005).

string parallel with the string A_1, A_2, ..., A_n, having the same parameters of the string.

As long as the segment B_1B_{i+1} must be an integer multiple of the parameter of the network,

$$B_1B_{i+1}=N_1p \tag{2.1}$$

the same for the segment A_2A_i,

$$A_2A_i=N_2p \tag{2.2}$$

which, in terms of geometry, according to the Figure 2.3 means:

$$B_1B_{i+1}=A_2A_i + 2p\cos\alpha \tag{2.3}$$

wherefrom results:

$$\cos\alpha = \frac{N_1 - N_2}{2} = k\frac{1}{2} \tag{2.4}$$

Since N_1 and N_2 are integers also k must be integer. Thus, the values of cosine $\cos\alpha$ must be integer multiples of ½. Since the values of $\cos\alpha$ are expanded in the interval $[-1, +1]$ only five possible solutions of the Eq. (2.4) are obtained, such as:

$$
\begin{aligned}
\cos\alpha &= -1 &\Rightarrow& \quad 2\pi/n = \pi &\Rightarrow& \quad n = 2 \\
\cos\alpha &= -1/2 &\Rightarrow& \quad 2\pi/n = 2\pi/3 &\Rightarrow& \quad n = 3 \\
\cos\alpha &= 0 &\Rightarrow& \quad 2\pi/n = \pi/2 &\Rightarrow& \quad n = 4 \\
\cos\alpha &= \tfrac{1}{2} &\Rightarrow& \quad 2\pi/n = \pi/3 &\Rightarrow& \quad n = 6 \\
\cos\alpha &= 1 &\Rightarrow& \quad 2\pi/n = 2\pi &\Rightarrow& \quad n = 1
\end{aligned} \tag{2.5}
$$

If a body has axes of symmetry of various orders, the axis of the superior order is considered the main axis. The axes or rotation are also called "gyres" and are graphically symbolized such as: A^2: ◖, A^3: ▲, A^4: ■, and A^6: ◆, Figure 2.4.

Regarding the geometric demonstration above, is worth mentioned that it can be reopened to the phenomenological level. Thus, obviously results that the whole space (here planar, but the 3D case works similarly) can be completely covered (no gaps) by the successive rotation of the hexagons, of the equilateral triangular structures, and quadratic but not also the pentagonal one, Figure 2.5.

Toward a symmetry plan, the property is underlined by the operation of reflection. The point change its sign only for one coordinate. The notation of the plan of symmetry in the didactical system is P, if it contains an axis of symmetry or is perpendicular on an axis of symmetry of (at most) second order, or by Π if it is perpendicular on an axis of symmetry of an order superior to 2.

Any plane of symmetry is a plane of halving of the body, but not any plane of halving is a plane of symmetry. This fact is illustrated in the top side of the Figures 2.6a and 2.6b. In the representation Figure 12a the plane, which divides the points 1 and 3, is just a plane of halving; in the representation Figure 12b, it is also a reflection plane, of symmetry.

The inversion, the rotation and the reflection are simple symmetry operations; each of them is performed in relation with a single element of symmetry. A crystalline polyhedron can have more axes, more planes, but a single center of inversion.

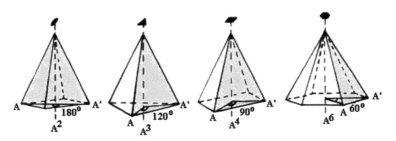

FIGURE 2.4   The presentation of the action of the gyres A^2, A^3, A^4, and A^6 toward the isolated bodies, after Chiriac-Putz-Chiriac (2005).

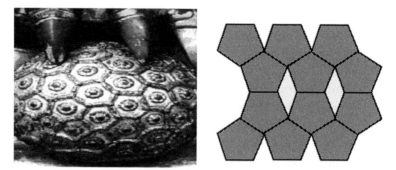

FIGURE 2.5 Left: lion claw holding a globe plated with gold the hexagons from the face "Paradise Purity Gate" work dated from the time of the Qing Dynasty (1736–1796), of the Forbidden City, Beijing; right: combined pentagons overlapping yet continued completely covering the planar surface; after Shubnikov and Koptsik (1977).

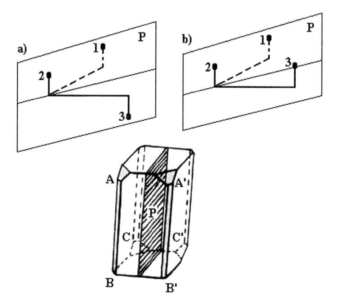

FIGURE 2.6 Up: the illustration of a plan P of non-symmetry (a) and of symmetry (b); down: the illustration of a plan of symmetry in a crystal, after Chiriac-Putz-Chiriac (2005).

The number and the type of the symmetry elements that character-ize a crystalline polyhedron are not random, as long as the association of the symmetry elements has a mandatory character. This will be further illustrated.

- If inside a polyhedron which has a gyre of even order (A^n, n=2k) is identified one of the elements of symmetry C (the inversion center) or (the plane of reflection $\perp A^n$) then it has all the three elements of symmetry, such as:

$$A^n + C \Rightarrow A^n C \Pi, \quad n=2k$$
$$A^n + \Pi \Rightarrow A^n \Pi C, \quad n=2k \qquad (2.6)$$

- If inside of a polyhedron there is a gyre of odd order (A^n, n=2k+1) is identified one of the elements of C (the center of symmetry) or Π (the plane of symmetry $\perp A^n$), then the non-identified is excluded from the system, i.e.:

$$A^n + C \Rightarrow A^n C, \quad n=2k+1$$
$$A^n + \Pi \Rightarrow A^n \Pi, \quad n=2k+1 \qquad (2.7)$$

If a gyre of n order is perpendicular on a gyre of second order ($A^n \perp A^2$) then inside the perpendicular plane will exist other (n–1) gyres of second order, i.e.:

$$A^n + A^2 \Rightarrow A^n n A^2 \qquad (2.8)$$

- If a gyre is parallel or contained in a reflection plane P, then the polyhedron has other (n–1) planes that contain the gyre crosses along its, i.e.:

$$A^n + P \Rightarrow A^n n P \qquad (2.9)$$

The combination of two operations of symmetry leads to the definition of complex elements of symmetry. For the isolated polyhedron, can be defined the following combinations:

- rotation + reflection;
- rotation + inversion;
- inversion + reflection.

A complex (composed) element of symmetry is self-consistent if it cannot be reduced to the succession of some simple operations of symmetry, with a real result (an undifferentiated position from the initial one).

Thus, the combination between the inversion and the reflection is equivalent with the existence of a gyre of second order and a reflection plane perpendicular on it. The combination does not lead to another element of symmetry.

By combining the rotation with the reflection are obtained the *axes of rotoreflexion* or the *gyroids*, and combining the rotation with the inversion are obtained the *axes of inversion or rotoinversion*. These are complex axes (elements) of symmetry. Are noted by the barred order of rotation ($\bar{n}$) or A_n for the gyroids, respectively with A_i^n for the inversion axes. It will be demonstrated that only one of those operations of symmetry is self-consistent, all others may be reduced to the association of simple operations; is about the tetra-gyroid A_4 or the fourth order axis of inversion A_i^4. Let's follow the effect of the rotoinversion axes.

- (A_i^1) For the rotation with 360° combined with the inversion, results the succession of the stages of Figure 2.7, wherefrom the identity is obtained:

$$A_i^1 = C \tag{2.10}$$

- (A_i^2) For the rotation with 360°/2 = 180° combined with the inversion, results the sequence of the hypostases of Figure 2.8, wherefrom is obtained the identity:

$$A_i^2 = \Pi(P) \tag{2.11}$$

- (A_i^3) For the rotation with 360°/3 = 120° combined with the inversion, results the sequence of the hypostases of Figure 2.9, wherefrom

FIGURE 2.7   The illustration of the operation of first order rotoinversion, after Chiriac-Putz-Chiriac (2005).

FIGURE 2.8   The illustration of the operation of second order rotoinversion, after
Chiriac-Putz-Chiriac (2005).

is obtained the identity:

$$A_i^3 = A^3 C \qquad\qquad (2.12)$$

The system has both A^3 and C.

- (A_i^4) For the rotation with $360°/4 = 90°$ combined with the inversion,
results the succession of the hypostases of Figure 2.10.

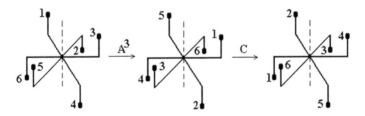

FIGURE 2.9   The illustration of the operation of the third order rotoinversion, after
Chiriac-Putz-Chiriac (2005).

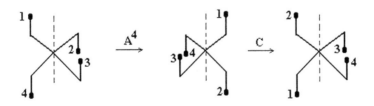

FIGURE 2.10   The illustration of the operation of the fourth order rotoinversion, after
Chiriac-Putz-Chiriac (2005).

The initial structure has not a fourth order axis of rotation or inversion center! Therefore, this compound symmetry operation is a self-consistent operation.

- (A_i^6) For the rotation of $360°/6 = 60°$ combined with the inversion, results the succession of the stages of Figure 2.11, wherefrom is obtained the identity:

$$A_i^6 = A^3 \Pi \qquad (2.13)$$

The structure has a gyre of the third order and the reflection plane perpendicular on it. Let's now follow the effect of the gyroids.

- (A_1) For the rotation with $360°$ combined with the reflection, results the succession of the hypostases of Figure 2.12, wherefrom results that the gyroid A_1 is equivalent with the reflection plane, i.e.:

$$A_1 = P(\Pi) \qquad (2.14)$$

- (A_2) For the rotation with $360°/2 = 180°$ combined with the reflection, results the succession of the hypostases of Figure 2.13, wherefrom results that the operation A_2 is reduced to the existence of the inversion center:

$$A_2 = C \qquad (2.15)$$

- (A_3) For the rotation with $360°/3 = 120°$ combined with the reflection, results the succession of the hypostases of Figure 2.14, obtaining the identity:

$$A_3 = A^3 \Pi \qquad (2.16)$$

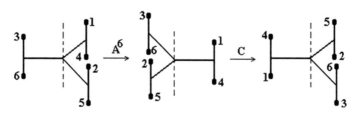

FIGURE 2.11   The illustration of the operation of the sixth order rotoinversion, after Chiriac-Putz-Chiriac (2005).

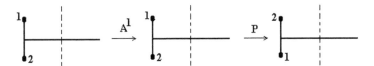

FIGURE 2.12 The illustration of the first order rotoreflection operation, after Chiriac-Putz-Chiriac (2005).

FIGURE 2.13 The illustration of the second order rotoreflection operation, after Chiriac-Putz-Chiriac (2005).

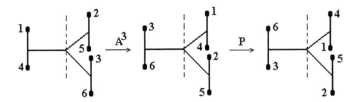

FIGURE 2.14 The illustration of the third order rotoreflection operation, after Chiriac-Putz-Chiriac (2005).

The gyroid A_3 is equivalent with the simple elements A^3 and Π.

- (A_4) For the rotation with $360°/4 = 90°$ combined with the reflection, results the hypostases of Figure 2.15.

The operation A_4 cannot be reduced to a succession of simple operations of symmetry because the system hasn't any gyre A^4 and neither a reflection plane on it.

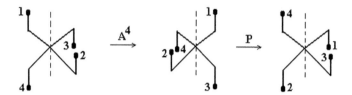

FIGURE 2.15   The illustration of the fourth order rotoreflection operation, after Chiriac-Putz-Chiriac (2005).

- (1.46) For the rotation with 360°/6 = 60° combined with the reflection, there results the succession of the stages of Figure 2.16, from where the identity is obtaining:

$$A_6 = A^3C \qquad (2.17)$$

In conclusion, from the analysis of the compound symmetry operations, Figure 2.17, there result the identities:

$$A_i^1 = A_2 = C$$
$$A_i^2 = A_1 = P$$
$$A_i^3 = A_6 = A_3C$$
$$A_i^6 = A_3 = A_3\Pi \qquad (2.18)$$

The fourth order symmetry axis (A_i^4) has the same effect with the tetra-gyroid (A_4). Both complex operations require the existence of a second order gyre. For this reason, the tetra-gyroid is noted by A_4^2.

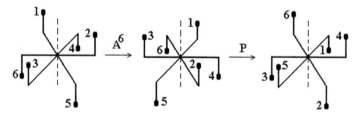

FIGURE 2.16   The illustration of the sixth order rotoreflection operation, after Chiriac-Putz-Chiriac (2005).

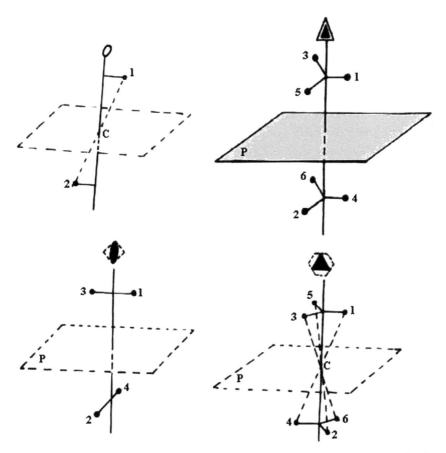

FIGURE 2.17 The illustration of the action of the compound axes of symmetry for the isolated bodies, after Chiriac-Putz-Chiriac (2005).

2.4 THE CRYSTALS' METRICS

2.4.1 MATHEMATICAL POINT NETWORKS: IDEAL CRYSTALS

First, one will proceed to the introduction of some specific definitions which will facilitate the explanation of the differences between the concepts below. Until now one has been generally spoken about symmetry. From now on it will be even more concretized this denomination by the one of *crystalline structure*, or simply, crystal. It must however be stated the difference between the crystal and the lattice (HyperPhysics, 2010).

A *network or lattice* is an infinite arrangement of mathematical points (infinitely small, unincorporated) into space, each being surrounded in the same way by the rest of them; the present discussion follows (Chiriac-Putz-Chiriac, 2005).

A *structure* (crystalline or solid, anyway regular) is formed by a lattice through the personalization of its points, that can be identified as atoms, molecules and assemblies of molecules, or other "species", all generically called motif or base, Figure 2.18.

In short: network (or lattice) + motif (or base) = crystalline structure.

As an observation, the lattice points should not necessarily correspond with the center of the base or motif. In addition, the lattice is the one who allows the identification of the unit cell (or elementary), which, by pure translations, and so by infinite periodic repetitions generating the whole crystal. In fact, in its ideal meaning the crystal is based on the infinite lattice. As illustrated in Figure 2.18 the unit cell can be chosen in several ways, identifying constructions with periodic symmetry from the lattice. Of these selections, the unit cell having a single point from the lattice will be called the primitive unit cell or simple primitive (P).

Therefore, to describe the crystal structure of the solid there is a model: the mathematical network of points, or the lattice.

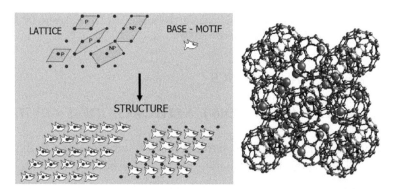

FIGURE 2.18   Left: the paradigm of generating a periodic structure by the combination between Base and Reason; is noted the possibility of identify inside the lattice of non-primitive unit cells (NP) and those primitive (P); right: a structural realization formed by the cubic lattice with the faces centered with the fullerene base (C60); after Heyes (1999).

A mathematical network of points is an ordered distribution of them, such that each point to enjoy the same neighborhood. The network of points can be: linear (one-dimensional), planar (two-dimensional) or spatial (three-dimensional). The real network of particles has the properties analogous to the mathematical network of points, except that point, being a particle, has a size.

The linear network/lattice of points can be completely defined by a single linear parameter called the reticular string parameter, equal with the distribution between two points (nodes) of the successive string, Figure 2.19.

In order to define the flat network of points – the reticular plane – are required three parameters: two linear parameters, the parameters of two reticular strings which intersect between them (a and b), and the angle between the directions of those reticular strings (γ), Figure 2.20. In a network of points we can identify more unparalleled reticular strings, Figure 2.21, so that in the network are formed flat closuring-cells of the network, of different forms and sizes. From these local rings, in order to define the network must be chosen the elementary cell. In the flat network, the elementary cell is the polygon with the minimal surface, minimum content of points and the maximum number of right angles.

For the examples of Figure 2.21, the III closure (the polygon $A_1A_2C_3C_2$) is excluded as elementary cell because of the number of the points that contains.

Inside the network, any corner point of an eye is common of four polygons, i.e., belongs in proportion of ¼ of the eye-closure. A point which stands on the side of the polygon, being in common for two polygons, belongs to the eye-closure in proportion of ½. Only an internal point fully belongs to the closure (in proportion of 1/100%).

Thus, to the III polygon belongs two points unlike the II closure (the polygon $A_1A_2B_3B_2$) and I (the polygon $A_1C_2D_2B_1$), either of them providing one single point.

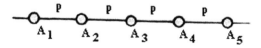

FIGURE 2.19   The linear network points, with p – the parameter of the reticular string, after Chiriac-Putz-Chiriac (2005).

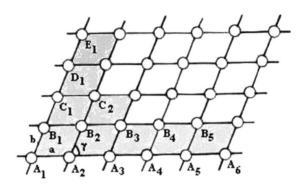

FIGURE 2.20 The flat network of points, with (a, b and γ) – the lattice parameters, after Chiriac-Putz-Chiriac (2005).

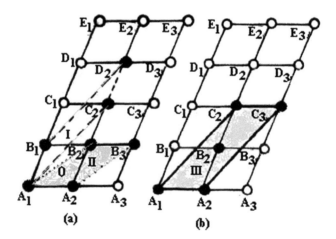

FIGURE 2.21 The planar network of points, with the reticular strings (A_1, A_2, A_3, ...), (B_1, B_2, B_3, ...), ...; in (a) and (b) are identified I, II, III as the closures of the network; the II closure is the elementary cell of the planar network, after Chiriac-Putz-Chiriac (2005).

Between the closures II and I there is no area difference, but the II closure is realized by the angles γ closer of the right one. Yet, the planar network proposed as example is open by the elementary cell 0 (the polygon $A_1A_2B_2B_1$) with the angle γ closest to 90^0.

To better understand the notion of primitive closure, the notion of effective point (or effective points) on the unit cell must be explained. To achieve that the Figure 2.22 is used. The effective point on the unit cell summarizes the total contribution to the cell of the whole points included in that cell and from all of its borders (faces, edges and tops/vertices).

Thus, in the 2D lattice of Figure 2.22, the superior unit cell actually has 4 (1/4)=1 effective point provided by the tops contributions, and one other effective point from the internal contribution, and with a total of two effective points on the cell it cannot be elementary or primitive. Instead, the inferior unit cell of Figure 2.22 is a primitive because the two atoms (physical) of the middle of the cell are assimilated with a motif (base of the lattice), i.e., generating 1 *inferior effective motif-point on the cell*, and there are not other. It's easy to imagine that the translating of this primitive cell in either direction along the parallels of its sides reproduces the planar structure of graphite.

Therefore, in judging the elementary or primitive cells one should take into account two related aspects: can not admit to the rotation axis of symmetry, other than those of the orders 1, 2, 3, 4 and 6, as previously shown, yet, at the same time, one must be able to generate the entire lattice (and hence also the structure) crystalline only through the successive (periodical) translations.

How does one to look to a unit cell "type" such that to reconcile the two geometric conditions: the rotations limited inside the cell, with

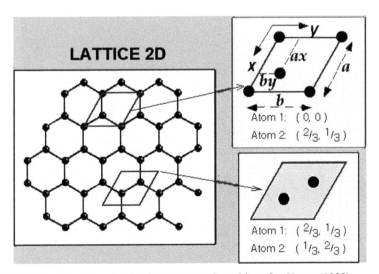

FIGURE 2.22   The lattice 2D for the structure of graphite; after Heyes (1999).

the pure translations with the periodical outside of it? One single spatial type can satisfy these two structural requirements, i.e. the parallelepiped unit cell, its general appearance being illustrated in Figure 2.23. Therefore, to characterize a spatial network points six parameters are required: three linear parameters linear, the parameters of the reticular strings (*a*, *b* and *c*) defining the so-called primitive vectors and which, by intersecting in the space, defining the characteristic polyhedron of the network; along with the angles that are formed between the three directions of the reticular string (*α, β, γ*). By convention, in crystallography, the linear parameters are denoted as following: *a* – on the direction *Ox*; *b* – on the direction *Oy* and *c* – on the direction *Oz*. For the angular parameters have been established: *γ* – the angle between the directions of the reticular strings characterized by the parameters *a* and *b*; *β* – the angle between the directions of the reticular strings characterized by the parameters *c* and *a*; *α* – the angle between the directions of the reticular strings characterized by the parameters *c* and *b*, Figure 2.22.

Accordingly, the elementary or primitive cell of a crystalline network must be: a convex polyhedron, with the faces parallel and equal two by two, having all corners filled with particles, the minimum volume and the maximum number of right angles, Figure 2.24.

The polyhedron should also satisfy the condition that the infinite network is generated only by translation operations on the three spatial directions.

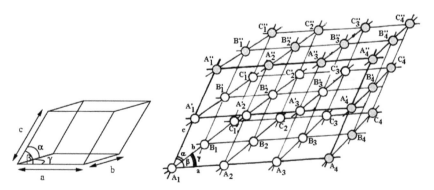

FIGURE 2.23 The parallelepiped unit cell (left) and the spatial network that it generates (right), after Chiriac-Putz-Chiriac (2005).

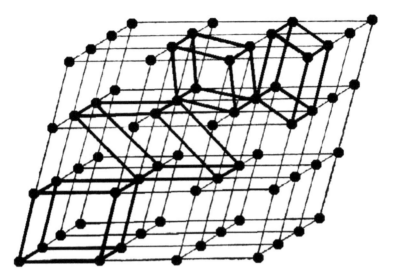

FIGURE 2.24 Various forms of some non-primitive tridimensional unit cell, after Chiriac-Putz-Chiriac (2005).

2.4.2 CRYSTALLOGRAPHIC SYSTEMS

Starting from the generic unit cell of Figure 2.23, the length (a, b, c) and the unit angles (α between b and c; β between a and c; γ between a and b), called *fundamental axes and angles*, can therefore be made in proportional relationships in a limited number of options, relative to the symmetry conditions at the internal rotations, the translation being already satisfied by the by choosing the rectangular coordinates (Verma & Srivastava, 1982).

Moreover, from these combinations only those that will fully satisfy the conditions of elementary and primitive polyhedron will be retained.

However, the crystallographic classification of the system of reticular points involves some definitions; the present discussion follows (Chiriac-Putz-Chiriac, 2005).

This will be understand by choosing the unique direction, a particular or, more precisely, a single direction. The singular direction can not be repeated by any of the symmetry element of a crystalline polyhedron will be retained.

By equivalent directions or directions of symmetry we will understand those directions which can be obtained one from one another, by a symmetry operation in relation with an element of symmetry of a polyhedron.

For example, the axes A^4 of the cube are transformed one into one another by the reflection respecting a diagonal plane.

The number of single directions and the axes of symmetry define the crystalline categories:

- The superior category has no singular directions and possesses more higher-order symmetry axis, where the cube belongs.
- The middle category is characterized by having one particular direction, an axis of symmetry of type A^3, A^4, or A^6, corresponding to the trigonal prism, tetragonal and respectively hexagonal symmetries.
- The inferior category contains several individual directions and axis of order 2, where rhombic prism belongs.

Based on the characteristic symmetry criterion and on the combination between the symmetry axes, the categories are divided into crystallographic systems as following.

The superior category has a single system – the cubic system.
The middle category contains:

- the rhombic or trigonal system: a single A^3 axis
- the tetragonal system: a single A^4 axis;
- the hexagonal system: a single A^6 axis.

The inferior category is divided in three systems:

- orthorhombic system: with more axes of 2nd order or more planes of reflection;
- the monoclinic system: one single A^2 axis, and one single plane of reflection or A^2P
- the triclinic system: an infinite cardinal of A^1 axes

The notion of syngony coincides with that of system for the cubic system, for the inferior category and for the quadratic (tetragonal) one. The concepts are not coincidental for the trigonal and hexagonal systems. The two systems are one single syngony as long as the crystals of these systems can be described in a single coordinate system, the hexagonal system. Therefore, the syngony is represented by the coordinate system that allows the description of the crystals with a particular symmetry. Thus, there are seven crystallographic systems and only 6 syngonies.

For each crystallographic system there can be identified the primitive elementary cell – the polyhedron, which has only the corners filled with particles, so that to the cell one single nodal particle belongs (each corner being common to 8 polyhedral neighbors in space, the ownership of sharing the particle to the cell is 1/8).

Given the possible relationship between the linear and angular parameters of an elementary cell, there can be defined the crystallographic systems as follows:

(1) $a=b=c$; $\alpha=\beta=\gamma=\pi/2$ CUBIC or TESSERAL

(2) $a=b\neq c$; $\alpha=\beta=\gamma=\pi/2$ QUADRATIC or TETRAGONAL

(3) $a\neq b\neq c$; $\alpha=\beta=\gamma=\pi/2$ ORTHORHOMBIC

(4) $a=b=c$; $\alpha=\beta=\gamma\neq\pi/2$ TRIGONAL or RHOMBOHEDRIC

(5) $a=b\neq c$; $\alpha=\beta=\pi/2$; $\gamma=2\pi/3$ HEXAGONAL

(6) $a\neq b\neq c$; $\alpha=\gamma=\pi/2$; $\beta\neq\pi/2\wedge2\pi/3$ MONOCLINIC

(7) $a\neq b\neq c$; $\alpha\neq\beta\neq\gamma\neq\pi/2$ TRICLINIC

$$(2.19)$$

The primitive cells of the seven crystallographic systems and their corresponding symmetry classes are presented in Table 2.2.

We can note that all the primitive polyhedrons come from the deformed cube – the polyhedron with maximum symmetry.

Thus, the stretching or the compressing of the cube following the direction 0z leads to the *quadratic* prism.

If the quadratic prism is elongated after one of the horizontal direction there is obtained the *orthorhombic* prism.

By "pushing" of two opposite edges, parallel to the 0z axis of the orthorhombic prism, we get the *hexagonal* primitive polyhedron when $\gamma=2\pi/3$. By "pushing" a face $z0y$ from the orthorhombic prism the *monoclinic* prism is reached. By "pressing" the edges of two diagonally opposite corners of the cube results the trigonal prism, and finally, by the total deformation of the cube there is reached the general case of *triclinic* prism.

However, relative to the hexagonal system, while considering the primitive cell as a rectangular form is sufficient to establish the crystallographic

TABLE 2.2 The Crystalline Systems and Their Description, After Chiriac-Putz-Chiriac (2005)

Crystallographic System	Geometry of Primitive Cell	Class of Symmetry Comments
Triclinic $a{\neq}b{\neq}c$, $\alpha{\neq}\beta{\neq}\gamma$	P	A^1 The polyhedron possesses an infinity of axes of first order (monogyre). Is the parallelepiped completely asymmetric. The only operation that allows is the identity.
Monoclinic $a{\neq}b{\neq}c$, $\alpha{=}\gamma{=}\pi/2$, $\beta{\neq}\pi/2;\ 2\pi/3$	P	A^2PC The digyre (A^2) is parallel with the axis of (b), and the reflection plane (P) is parallel with the axis 0z passing through the inversion center.
Orthorhombic $a{\neq}b{\neq}c$, $\alpha{=}\beta{=}\gamma{=}\pi/2$	P	$3A^23PC$ The three digyre pierce by two opposite faces of the parallelogram. Their intersection determines the center of inversion and the planes of reflection are, per head, perpendicular on a digyre.
Tetragonal $a{=}b{\neq}c$, $\alpha{=}\beta{=}\gamma{=}\pi/2$	P	$A^4A^24P\Pi C$ Tetragyre is perpendicular on the plane of bases, in their center ($\|0z$). The digyres out by the middles of the edges and lateral faces (placed in the plane x0y). The horizontal plane of reflection (Π) is perpendicular on the tetragyre A^4 and the vertical reflection planes contain, each of them, a gyre.

TABLE 2.2   Continued

Crystallographic System	Geometry of Primitive Cell	Class of Symmetry Comments
Cubic a=b=c, $\alpha=\beta=\gamma=\pi/2$	P	$3A^44A^36A^26P3\Pi C$ The tetragons pass through the centers of two opposite faces. The trigyres coincide with the large diagonal of the cube. The digyres pass by the centers of the opposite edges of the two parallel faces.
Trigonal a=b=c, $\alpha=\beta=\gamma\neq\pi/2$, $\alpha=\beta=\gamma<2\pi/3$	P	A^33A^23PC The trigyre is coincident with the large diagonal of the primitive parallelepiped.
Hexagonal a=b≠c, $\alpha=\beta=\pi/2$, $\gamma=2\pi/3$		$A^66A^26P\Pi C$ The conventional cell is the right prism with a rhomb base with the angle of, 120°, but with the symmetries at the level of the hexagonal prism: the sexagyre by the opposite central faces, and the 6 digyres passing through the lateral opposite faces and the edges.

system to which it belongs, i.e. just by fixing the relations between the fundamental parameters.

Thus, as shown in Table 2.2, to each crystallographic system is assigned a unique combination between axes and fundamental angles – that, once determined, involves the identification of the associated system type and the network symmetries.

A practical method for fixing the elementary cell type will be presented in the following.

2.4.3 THE REDUCING CELL METHOD

One raises the question: there is possible to find a lattice symmetry, and respectively, the system to which it belongs, from an arbitrary set of vectors of the unit cell? In other words, given a set of arbitrary fundamental parameters can these be reduced to the basic cellular level?

The adopted method, also called the reduction cell method introduced by Niggli, see Franzen (1994), is based on the reduction principle of the polyhedron crystalline bases built on the initial fundamental parameters, until finding the cell with the same area but with angles as close to 90°, as in Figure 2.25. As a working principle in plane, for example, once giving two fundamental vectors, a_0 and b_0, and the angle between them, all the combinations of scalar products, are calculated resulting in the quantities; the present discussion follows (Chiriac-Putz-Chiriac, 2005):

$$\begin{cases} r_{11} = \vec{a}_0 \cdot \vec{a}_0 = a_0^2 \\[2mm] r_{22} = \vec{b}_0 \cdot \vec{b}_0 = b_0^2 \\[2mm] r_{12} = \vec{a}_0 \cdot \vec{b}_0 = a_0 b_0 \cos(\hat{a_0, b_0}) \end{cases} \qquad (2.20)$$

Based on the values of Eq. (2.20), we will consider that b_0 is reducible to a_0 if the next inequalities occur:

$$\frac{\left| \vec{a}_0 \cdot \vec{b}_0 \right|}{a_0} \geq \frac{1}{2} a_0 \Leftrightarrow \left| \vec{a}_0 \cdot \vec{b}_0 \right| \geq \frac{1}{2} a_0^2 \Leftrightarrow \left| r_{12} \right| \geq \frac{1}{2} r_{11} \qquad (2.21)$$

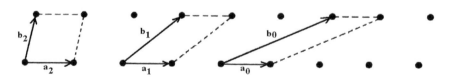

FIGURE 2.25   The illustration of the reduction of the unit cell until to the elementary cell, after Chiriac-Putz-Chiriac (2005).

i.e., the projection of b_0 on a_0 is higher than ½ of the length of a_0.

Thus, the length of b_0 can be modified as

$$\vec{b}_0 \to \vec{b}_1 = \vec{b}_0 - n\vec{a}_0, \quad n \in Z \tag{2.22}$$

such as, the angle toward a_0 will be implicitly modified by the appropriate finding of the parameter n of the "final point" inequality:

$$\left| \vec{a}_0 \cdot \vec{b}_1 \right| \leq \frac{1}{2} a_0^2 \tag{2.23}$$

In tridimensional case, one applies and repeats the above procedure for all the coordinates combinations, until the general conditions are satisfied:

$$\left| r_{ij} \right| \leq \frac{1}{2} r_{ii}, \quad \forall i, j \tag{2.24}$$

As an application, one requires to unfold the lattice symmetry from the fundamental (crystallographic) parameters relationships, for a network characterized by an arbitrary unit cell, with the set of fundamental parameters taking the values:

$$\begin{cases} a_0 = 5.000[\overset{0}{A}] \\ b_0 = 8.660[\overset{0}{A}], \\ c_0 = 6.403[\overset{0}{A}] \end{cases} \begin{cases} \alpha = 47.4^0 \\ \beta = 67.0^0 \\ \gamma = 30.0^0 \end{cases} \tag{2.25}$$

According to the above procedure, one firstly computes the quantities:

$$\begin{cases} r_{11} = \vec{a}_0 \cdot \vec{a}_0 = a_0^2 = 25.0[\overset{0}{A}]^2 \\ r_{22} = \vec{b}_0 \cdot \vec{b}_0 = b_0^2 = 75.0[\overset{0}{A}]^2, \\ r_{33} = \vec{c}_0 \cdot \vec{c}_0 = c_0^2 = 41.0[\overset{0}{A}]^2 \end{cases} \begin{cases} r_{12} = \vec{a}_0 \cdot \vec{b}_0 = a_0 b_0 \cos \gamma = 37.5[\overset{0}{A}]^2 \\ r_{13} = \vec{a}_0 \cdot \vec{c}_0 = a_0 c_0 \cos \beta = 12.5[\overset{0}{A}]^2 \\ r_{23} = \vec{b}_0 \cdot \vec{c}_0 = b_0 c_0 \cos \alpha = 37.5[\overset{0}{A}]^2 \end{cases} \tag{2.26}$$

Hence, there is the inequality:

$$|r_{12}| > \frac{1}{2}r_{11} \ (37.5 > 12.5) \tag{2.27}$$

which implies the reduction of the vector $\vec{b}_0$ with the vector $\vec{a}_0$ through the integer number n,

$$\vec{b}_1 = \vec{b}_0 - n\vec{a}_0, \quad n \in Z \tag{2.28}$$

searched so that it satisfies the inequality:

$$\left| \vec{a}_0 \cdot \vec{b}_1 \right| \leq \frac{1}{2}a_0^2 \tag{2.29}$$

Based on the transformation (2.28) the relation (2.29) can be rewritten as:

$$\left| \vec{a}_0 \cdot \left(\vec{b}_0 - n\vec{a}_0 \right) \right| \leq \frac{1}{2}a_0^2 \tag{2.30}$$

hence the values searched for the transformation parameter n turn out as

$$|r_{12} - nr_{11}| \leq \frac{1}{2}r_{11} \Leftrightarrow 25 \leq 25n \Rightarrow n = 1 \tag{2.31}$$

Under these conditions, the new fundamental vectors have the form:

$$\begin{cases} \vec{a}_1 = \vec{a}_0 \\ \vec{b}_1 = \vec{b}_0 - \vec{a}_0 \\ \vec{c}_1 = \vec{c}_0 \end{cases} \tag{2.32}$$

which implies the new scalar quantities and values:

$$\begin{cases} r'_{11} = \vec{a}_1 \cdot \vec{a}_1 = a_1^2 = r_{11} = 25.0[\overset{0}{A}]^2 \\ r'_{22} = \vec{b}_1 \cdot \vec{b}_1 = b_1^2 = b_0^2 - 2a_0b_0 \cos\gamma + a_0^2 = r_{22} - 2r_{12} + r_{11} = 25.0[\overset{0}{A}]^2 \\ r'_{33} = \vec{c}_1 \cdot \vec{c}_1 = c_1^2 = r_{33} = 41.0[\overset{0}{A}]^2 \end{cases}$$

$$\Leftrightarrow \begin{cases} r'_{12} = \vec{a_1}\cdot\vec{b_1} = \vec{a_0}\cdot\left(\vec{b_0}-\vec{a_0}\right) = r_{12}-r_{11} = 12.5[\overset{0}{A}]^2 \\ r'_{13} = \vec{a_1}\cdot\vec{c_1} = \vec{a_0}\cdot\vec{c_0} = r_{13} = 12.5[\overset{0}{A}]^2 \\ r'_{23} = \vec{b_1}\cdot\vec{c_1} = \left(\vec{b_0}-\vec{a_0}\right)\cdot\vec{c_0} = r_{23}-r_{13} = 25.0[\overset{0}{A}]^2 \end{cases} \tag{2.33}$$

where we notice that now occurs the inequality:

$$|r'_{23}| > \frac{1}{2}r'_{22} \; (25>12.5) \tag{2.34}$$

which recommends the reducing of the vector c_1 with b_1:

$$\vec{c_2} = \vec{c_1} - n'\vec{b_1}, \; n'\in Z \tag{2.35}$$

by finding the reduction parameter n', which implies:

$$\left|\vec{b_1}\cdot\vec{c_2}\right| \le \frac{1}{2}b_1^2 \tag{2.36}$$

With the transformation (2.35) the inequality (2.36) is firstly written as:

$$\left|\vec{b_1}\cdot\left(\vec{c_1}-n'\vec{b_1}\right)\right| \le \frac{r_{22}-2r_{12}+r_{11}}{2} \tag{2.37}$$

wherefrom the searched value of n' results:

$$|r'_{23}-n'r'_{22}| \le \frac{1}{2}r'_{22} \Leftrightarrow 12.5 \le 25n' \Rightarrow n'=1 \tag{2.38}$$

From now on, the new fundamental vectors will be written as:

$$\begin{cases} \vec{a_2} = \vec{a_0} \\ \vec{b_2} = \vec{b_1} = \vec{b_0}-\vec{a_0} \\ \vec{c_2} = \vec{c_1}-\vec{b_1} = \vec{c_0}-\vec{b_0}+\vec{a_0} \end{cases} \tag{2.39}$$

for which, by calculating all the possible scalar products,

$$
\left\{
\begin{array}{l}
r''_{11} = \vec{a}_2 \cdot \vec{a}_2 = a_2^2 = r_{11} = 25.0[\overset{0}{A}]^2 \\[2mm]
r''_{22} = \vec{b}_2 \cdot \vec{b}_2 = b_2^2 = b_1^2 = r'_{22} = 25.0[\overset{0}{A}]^2 \\[2mm]
r''_{33} = \vec{c}_2 \cdot \vec{c}_2 = c_2^2 = c_1^2 + b_1^2 - 2\vec{c}_1 \cdot \vec{b}_1 = r'_{33} + r'_{22} - 2r'_{23} = 16.0[\overset{0}{A}]^2
\end{array}
\right.
$$

$$
\left\{
\begin{array}{l}
r''_{12} = \vec{a}_2 \cdot \vec{b}_2 = r'_{12} = 12.5[\overset{0}{A}]^2 \\[2mm]
r''_{13} = \vec{a}_2 \cdot \vec{c}_2 = \vec{a}_1 \cdot \left(\vec{c}_1 - \vec{b}_1 \right) = r'_{13} - r'_{12} = 0[\overset{0}{A}]^2 \\[2mm]
r''_{23} = \vec{b}_2 \cdot \vec{c}_2 = \vec{b}_1 \cdot \left(\vec{c}_1 - \vec{b}_1 \right) = r'_{23} - r'_{22} = 0[\overset{0}{A}]^2
\end{array}
\right. \qquad (2.40)
$$

there is evident that the inequality type (2.24) is now satisfied for all the fundamental directions.

Therefore, the iterative process of reducing the cell stops here, while the geometric information of the resulting primitive cell is now extracted.

Accordingly, the values of the metric parameters of the elementary cell will be obtained by employing the values obtained in (2.40), firstly as:

$$
\begin{array}{l}
a_2 = \sqrt{r''_{11}} = 5[\overset{0}{A}] \\[2mm]
b_2 = \sqrt{r''_{22}} = 5[\overset{0}{A}] \\[2mm]
c_2 = \sqrt{r''_{33}} = 4[\overset{0}{A}]
\end{array} \qquad (2.41)
$$

and then, with the companion values of the angular parameters:

$$
r''_{12} = a_2 b_2 \cos \gamma_2 \Rightarrow \cos \gamma_2 = \frac{r''_{12}}{a_2 b_2} = \frac{12.5}{25} = \frac{1}{2} \Rightarrow \gamma_2 = 60^0
$$

$$
r''_{13} = a_2 c_2 \cos \beta_2 \Rightarrow \cos \beta_2 = \frac{r''_{13}}{a_2 c_2} = 0 \Rightarrow \beta_2 = 90^0
$$

$$
r''_{22} = b_2 c_2 \cos \alpha_2 \Rightarrow \cos \alpha_2 = \frac{r''_{23}}{b_2 c_2} = 0 \Rightarrow \alpha_2 = 90^0 \qquad (2.42)
$$

One is centralizing the relations:

$$a_2 = b_2 \neq c_2 \ \& \ \alpha_2 = \beta_2 = 90^0, \gamma_2 = 60^0 \qquad (2.43)$$

from where, as based on the classification of Table 2.2, the elementary cell in question is identified as belonging to the hexagonal system.

In the same way, one approaches any similar problem, until the corresponding reduction to one of the existing crystallographic system is achieved.

2.4.4   THE BRAVAIS LATTICE

Often, describing the crystal structure is easier if we rely on non-primitive elementary cell. To such a cell belong more particles whose position in the polyhedron is conditioned by the elements. For this reason, the non-primitive cells are not compatible with any crystallographic system (Goodhew, 2003).

A non-primitive cell can be *double primitive* (two particles belong to it). Depending on the position of the additional particles, the double primitive cells can be of two types; the present discussion follows (Chiriac-Putz-Chiriac, 2005):

- *Body (Internal) Centered Cells*, denoted by I, with one additional particle, in the center of symmetry of the primitive polyhedron. As an observation, every internal particle has a unitary participation to the polyhedron to which it belongs, Figure 2.26-left.
- *Double primitive cells* with two centered opposite faces. The specific notation of such a cell yet depends on the position of the centered faces, respectively on the direction where the particles are placed. Therefore, there are cells of type C (the particles are placed in the *c*-direction) Figure 2.26-right, of type A (the particles are placed on the *a*-direction) and of type B (the particles are placed on the *b*-direction).

The double primitive cells of type (I) are allowed by the crystallographic systems such as: cubic, tetragonal and orthorhombic, and those with two centered faces are compatible with the orthorhombic crystallographic systems (of type A) and monoclinic (of type C).

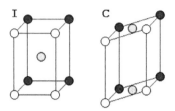

FIGURE 2.26   Left: double primitive cell of (I)-type; right: double monoclinic primitive cell of (C)-type, after Chiriac-Putz-Chiriac (2005).

Also, the quadruple primitive elementary cell (tetra primitive) can be identified and are denoted by (F) from *face*-centered cells: four particles/ motifs belong to them. Besides the eight particles from the corners of the polyhedron, other six particles are placed in the centers of all the faces, see Figure 2.27. As an observation, since the polyhedron face is common only for two neighbors of polyhedron, the participation quota for the "fill" of the polyhedron of a particle placed in a face is ½. The elementary cells of type F are allowed only by two crystallographic systems: cubic and orthorhombic.

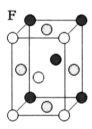

FIGURE 2.27   The quadruple primitive cubic cell, of F-type, after Chiriac-Putz-Chiriac (2005).

The primitive polyhedra define the maximum symmetry at rotation of the crystalline network. A crystalline system allows a non-primitive cell if three conditions are fulfilled:

(a) It does not cancel the symmetry at the specific rotation of the system;

(b) It can not be turned into another non-primitive cell, belonging to the same syngony, by changing the axes;

(c) It can turn into a primitive cell belonging to another syngony, of a lower class.

Let's illustrate the three conditions.

The cubic system does not allow a cell of C (A,B) type because the centering of only two opposite faces cancels the 4 axes of A^3 characteristic to the syngony, contrary to the a) condition above.

The tetragonal system does not allow a cell of type F because it turns into a cell of type I, of the same syngony, and the same for the cell of type C, thus contravening to the condition (c) above, see Figure 2.28. For the cubic system, the non-primitives of type A, B or C cancel the symmetries at the rotations around the trigyre through the opposite vertices (along the main diagonals). So, the non-primitive of type A, B or C can not be in the cubic system. Instead, the non-primitive I and those of F generate or are reduced to other primitive (i.e., tetragonal), different from the initial cubic, satisfying all the conditions a)-c) above, see Figure 2.29. Similar analysis at the unit cells level can be performed also for the other crystallographic

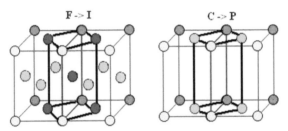

FIGURE 2.28   The F and C types of non-primitive tetragonal and their transformations, after Chiriac-Putz-Chiriac (2005).

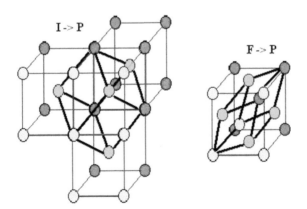

FIGURE 2.29   The I and F types of non-primitive cubic and their transformations, after Chiriac-Putz-Chiriac (2005).

systems. Accordingly, the combination between the non-primitive cells I, F and A, B, or C types with the crystalline systems of Table 2.2 generates in the total *14 lattices*, the so-called Bravais lattice of Figure 2.30.

2.5 POINT GROUPS

2.5.1 SYMMETRY FORMULAS: SYMMETRY CLASSES

All symmetry elements of a crystal polyhedron constitute its symmetry formula imposing the appurtenance to a symmetry class. Since in a crystal the rotation axis can only be by 1, 2, 3, 4 and 6 orders, and their association with other symmetry elements have strict rules (laws) is natural that the number of symmetry formula is perfectly defined, as will be listed below; the present discussion follows (Chiriac-Putz-Chiriac, 2005).

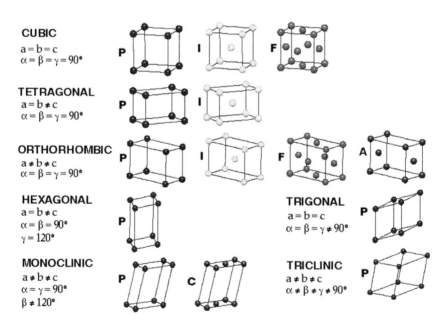

FIGURE 2.30   The type of cells of the Bravais lattices; after Heyes (1999).

The simplest symmetry formula meets the cyclic classes. In this case, a polyhedron has only a gyre of a particular order. Therefore, there are five cyclic classes:

$$(1) A^1; (2) A^2; (3) A^3; (4) A^4; (5) A^6 \qquad (2.44)$$

Given the rule (2.8) of the association of a gyre with the axis perpendicular to it we can deduce the axial classes' formulas. Since mono-gyre cannot multiply the axis A^2 we obtain only four formulas:

$$
\begin{array}{lllll}
(6) & A^2 & + & 2A^2 & = & 3A^2 \\
(7) & A^3 & + & 3A^2 & = & A^3\,3A^2 \\
(8) & A^4 & + & 4A^2 & = & A^4\,4A^2 \\
(9) & A^6 & + & 6A^2 & = & A^6\,6A^2
\end{array}
$$

In accordance with the general rule:

$$A^n + nA^2 = A^n nA^2 \qquad (2.45)$$

Because the tetra-gyre is the only self-consistent symmetry axis, the associated formula corresponds to the fifth axial symmetry classes:

$$(10)\ A_4^2 \qquad (2.46)$$

By combining the gyres with the inversion center, according to the rules (2.6) and (2.7), we obtain centered symmetry classes' formulas:

$$
\begin{array}{lllll}
(11) & A^1 & + & C & = & C \\
(12) & A^2 & + & C & = & A^2\Pi C \\
(13) & A^3 & + & C & = & A^3 C \\
(14) & A^4 & + & C & = & A^4\Pi C \\
(15) & A^6 & + & C & = & A^6\Pi C
\end{array}
$$

which can be resumed by the general formula:

$$A^n + C = A^n \Pi C \qquad (2.47)$$

The gyres can be associated also with the perpendicular or parallel reflection planes. Thus the formulas for planar symmetry classes are obtained.

The association axes with perpendicular plane of reflection leads to distinct formulas only for odd-order gyres as prescribed by Eq. (2.7). One yields:

$$
\begin{array}{llll}
(16) & A^1 + & \Pi = & P \\
(17) & A^3 + & \Pi = & A^3\Pi
\end{array}
\qquad (2.48)
$$

Any gyre (A^n) combined with a parallel reflecting plane multiplies it of n times, according to the rule (2.9). This way, one will obtain the other formulae of the planar classes:

$$
\begin{array}{llll}
(18) & A^2 + & P = & A^2 2P \\
(19) & A^3 + & P = & A^3 3P \\
(20) & A^4 + & P = & A^3 4P \\
(21) & A^6 + & P = & A^4 6P \\
(22) & A_4^2 + & P = & A_4^2 2P
\end{array}
\qquad (2.49)
$$

Associating the main symmetry axis with the binary symmetry axis, with the reflection planes and, when is possible, with the center of inversion, leads to symmetry classes of plane + axial symmetry whose formulae are:

$$
\begin{array}{lll}
(23) & A^2 + A^2 + P & = 3A^2 3PC \\
(24) & A^3 + A^2 + P + C & = A^3 3A^2 3PC \\
(25) & A^3 + A^2 + P + \Pi & = A^3 3A^2 3P\Pi \\
(26) & A^4 + A^2 + P + \Pi & = A^4 4A^2 4P\Pi C \\
(27) & A^6 + A^2 + P + \Pi & = A^6 6A^2 6P\Pi C
\end{array}
\qquad (2.50)
$$

Symmetry formulas derived so far contain a single main axis (A^n, $n \geq 3$). Group theory demonstrates that, in crystalline polyhedrons a unique association of a superior order $4A^3$ axis is possible. Such an association never appears alone and only with bi- gyre or tetra- gyre and leads to special classes of symmetry. The simplest formula corresponds to the tetrahedral group (T) with the formula:

$$
(28)\ 4A^3\ 3A_2
\qquad (2.51)
$$

Continuing the association of the tetra- gyre the octahedral group (O) is obtained:

$$(29)\ 3A^4\ 4A^3\ 6A_2 \qquad\qquad (2.52)$$

In the absence of the inversion center, along with the parallel reflection planes (P), group $4A^3$ implies tetra- gyres (A_4^2) and thus the obtained formula will be:

$$(30)\ 3A_4^2\ 4A^3\ 6P \qquad\qquad (2.53)$$

Involving the inversion center one is leaved with the last two possible crystallographic formulae:

$$(31)\qquad 4A^33A^23\Pi C$$
$$(32)\qquad 3A^44A^36A^26P3\Pi C \qquad\qquad (2.54)$$

There are therefore 32 point symmetry formulae, namely: 5 cyclic classes, 5 axial classes, 5 central classes, 7 plane classes, 5 plane-axial classes and 5 special classes.

2.5.2 SCHOENFLIES NOTATION

The Schoenflies system denotes with C_n the symmetry class with a single A^n axis. So, C_1 is the symbol for class with A^1, C_2 is for the class with A^2 as the only element of symmetry, and analogous C_3, C_4 and C_6; the present discussion follows (Chiriac-Putz-Chiriac, 2005)

With C_i one notes the class with the symmetry center as a single element, with C_s the P class, with C_{3i} the A_6^3 class, with S_4 the A_4^2 class (S is from "Spiegel" = mirror, in German language). With C_{mh} one notes the classes which results from combining the simple axes with a perpendicular plane (h – from "horizontal") on their direction, for example, $C_{2h} = A^2\pi C$, $C_{3h} = A^3\pi$, etc.

By combining the simple symmetry axes with (vertical) planes containing the axes the classes C_{nv}. So: $C_{nv} = A^22P$; $C_{3v} = A^23P$, etc. are resulting. The classes resulted by combining the simple axis of symmetry with the A^2 axes perpendicular on their direction are noted by Schoenflies with D_n, for example, $D_2 = 3A^2$; $D_3 = A^33A^2$; $D_4 = A^44A^2$.

If on these classes the P planes are added (as diagonal between secondary axis) there results the classes noted with $D_{nd} \cdot D_{2d} = A_4^22A^22P$; $D_{3d} = A_6^33A^23PC$.

Equally one has the classes D_{nh}:D_{2h} = $3A^2ePC$; D_{3h} = $A^33A^23P\pi$; D_{4h}= $A^44A^24P\pi C$; D_{6h}= $A^66A^26P\pi C$.

Cubic system uses the symbol T (from tetrahedron) for the $3A^24A^3$ class and O (from octahedron) for the $3A^44A^36A^2$class. For the rest of the classes: T_h will be the $3A^24A^43\pi C$ class, T_d the $3A_4^24A^36P$ class, and O_h the $3A^44A^36A^26P3\pi C$ class, Table 2.3.

At this point one will distinguish the symmetry operations associated to the symmetry elements (reflection, inversion, gyres/axes, gyroids, and identical operation), which in the notation Schoenflies are as shown in Table 2.4.

TABLE 2.3 The Platonic Solids From Top-to-Down: Tetrahedron, Cube, Octahedron, Dodecahedron and Icosahedron, With Specific Number of Faces (f), Vertices (v), and Edges (e) and the Associated Symmetry Operations*

Platonic Polyhedron	Number of			Symmetry operations
	f	v	e	
	4	4	6	i). Orthogonal bisecting axes of the edges: $6S_4$, $3C_2$;
				ii). The axes through vertices and the center of opposite faces: $8C_3$;
				iii). Diagonal planes to faces by edges: $6\sigma_d$.
				Total:
				$Td = \{T, 6S_4, 6\sigma_d\}$;
				Pure rotations:
				$T= \{E, 4C_3, 4C_3^2, 3C_2\}$.
	6	8	12	i). The axes through the center or the opposite faces (for
	8	6	12	cube)/ and vertices (for octahedron): $6S_4$, $3C_2$, $6C_4$;
				ii). The bisecting axes through the opposite edges: $6C_2$;
				iii). Axes through opposite vertices (for cube)/ and the center of the opposite faces (for octahedron): $8S_6$, $8C_3$, i;
				iv).Planes of 4 edges (cube)/vertices (octahedron): $3\sigma_h$;
				v). Planes of 4 opposite vertices (cube)/2 opposite vertices and 2 opposite edges (octahedron): $6\sigma_d$.
				Total:
				$Oh = \{O, i, 6S_4, 8S_6, 6\sigma_d, 3\sigma_h\}$;
				Pure rotations:
				$O= \{E, 8C_3, 6C_4, 6C_2, 3C_2\}$.

TABLE 2.3 Continued

Platonic Polyhedron	Number of			Symmetry operations
	f	v	e	
	12	20	30	i). Axes through the center of opposite faces (dodecahedron)/opposite vertices (icosahedron): $12C_5$, $12S_{10}$, $12C_5^2$, $12S_{10}^3$, i;
	20	12	30	ii). Bisecting axes through the opposite edges: $15C_2$;

iii). Axes through the opposite vertices (dodecahedron)/ center of the opposite faces (icosahedron): $20S_6$, $20C_3$;

iv). Planes containing 2 C_2 axes and 2 C_5 axes: $15\sigma_h$.

Total:

Ih={O, i, $12S_{10}$, $12S_{10}^3$, $_{20S6}$, $15\sigma_h$};

Pure rotations:

I= {E, $12C_5$, $12C_5^2$, $20C_3$, $15C_2$} .

*After Chiriac-Putz-Chiriac (2005).

TABLE 2.4 Types of Symmetry Elements and Associated Operations*

Symmetry elements		Associated Symmetry Operations
Inversion center (C)	i	Inversion transformation
Gyre (A^n)	C_n	Rotation (positive in clockwise direction) with the angle of $2\pi/n$ radians, with n natural number.
	C_n^p	Rotation (positive in clockwise direction) with angle of $2p\pi/n$ radians, with n natural number and p integer.
Gyroid (A$_n$)	S_n	Rotation with angle $2\pi/n$ radians followed by reflection in perpendicular plane on the rotation axis.
Mirroring plane (P,Π)	σ_h	The horizontal reflection – in the plane that passing by the origin and is perpendicular on the main axis of rotation.
	σ_v	Vertical reflection – in the plane that passes through the origin and contains the main axis of rotation.
	σ_d	Dihedral or diagonal reflection – a special case of the vertical reflection σ_v, but towards the plane that bisects in addition also the angle between the A^2 axes perpendicular on the main axis of rotation.
-	E	Identical transformation

*After Duch (2003) and Chiriac-Putz-Chiriac (2005).

More generally, since the need of systematizing the symmetry informa-tion from the group notion perspective worth noting the fact that all the symmetry operations introduced, exemplified and analyzed are executed in relation to the symmetry elements (center, axes, and planes) which are intersecting in a point.

Thus, the groups which summarize these symmetry operations will be named point symmetry groups. These point groups can be of three types: containing symmetry operations with respect to a single symmetry ele-ment, with respecting more than a symmetry element, and the so-called "Special" groups.

Listing and description of the groups contained of the each point groups is shown in Table 2.5 in Schoenflies notation. This notation has at least two advantages but also a disadvantage: the fact that entire group is noted with symbols that often coincide with one operation (object) from the group; yet this is also an advantage because, this way, it sym-bolizes the general features of the group and what symmetry elements are contained.

The second advantage of Schoenflies notation comes from the didacti-cal perspective, allowing the unitary (re-)introduction of crystallographic classes, see Table 2.6.

TABLE 2.5 Point Groups Characterization by Types and Schoenflies Notation*

Type	Notation	Feature
Groups only one element of symmetry	C_1	just E
	C_s	E & σ
	C_i	E & i
	C_n	E & C_n
	S_n	n = 2k, k natural
Groups with more than one element of symmetry	D_n	C_n & n $C_2 \perp C_n$
	C_{nh}	C_n & $\sigma \perp C_n$
	C_{nv}	C_n and two or more σ (Π) containing A^n (C_n).
	D_{nd}	C_n, $nC_2 \perp C_n$, n dihedral planes parallel with A^n (C_n) and which bisect the angles between those n axes $\perp A^n$ (C_n)
	D_{nh}	C_n, $nC_2 \perp C_n$, $\sigma \perp C_n$

TABLE 2.5 Continued

Type	Notation	Feature
Special groups	$C_{\infty v}$	Linear structures without i *(inversion)*
	$D_{\infty h}$	Linear structures with i
	T_d	Tetrahedral groups, including T_h and T
	O_h	Octahedral groups, including O
	I_h	Icosahedron groups, including I

*After Duch (2003) and Chiriac-Putz-Chiriac (2005).

TABLE 2.6 The 32 Crystallographic Classes/Groups in Schoenflies Notation (*Non-Axial Groups)**

Symmetry Groups		Rotation Order				
Combination of elements	*Description of the groups*	*1*	*2*	*3*	*4*	*6*
C_n	Cyclic	C_1^*	C_2	C_3	C_4	C_6
$\exists n\sigma_v \supset C_n$	Cyclic with vertical planes		C_{2v}	C_{3v}	C_{4v}	C_{6v}
$\sigma_h \perp C_n$	Cyclic with horizontal planes	C_s^*	C_{2h}	C_{3h}	C_{4h}	C_{6h}
i, S_{2n}	Improper rotation		C_i		S_4	S_6
$\exists nC_2$ axes $\perp C_n$	Dihedral		D_2	D_3	D_4	D_6
$\exists nC_2$ axes $\perp C_n$ and $n\sigma_d$	Dihedral with planes between axes		D_{2d}	D_{3d}		
$\exists (nC_2$ axes and $\sigma_h) \perp C_n$	Dihedral with horizontal planes		D_{2h}	D_{3h}	D_{4h}	D_{6h}
Special Groups	Cubic groups	T	T_h	O	T_d	O_h

**After Golden & Bernard (2002) and Chiriac-Putz-Chiriac (2005).

Thus, if the rotation, inversions and planes of reflections are combined, keeping as sub- indices the indicative of the rotation axes symmetry order in the group, there is naturally deduced the complete list of crystallographic class-groups relations in Schoenflies notation.

Note also that, although they are often named as "crystallographic classes" the combinations from Table 2.6 correspond in fact to group notion, since classes are only sub-units of the mathematical groups. Therefore, the name of "symmetry classes" doesn't refer to crystallography in strict mathematical meaning of the classes, but to the classifying of the symmetries considering the combinations of the symmetry operations.

2.5.3   THE INTERNATIONAL (HERMANN-MAUGUIN) NOTATION

Besides the Schoenflies notations, were imposed more compact notations (even in meaning of hermetic) for the elements and the symmetry operations (unique) trans-writing; the present discussion follows (Chiriac-Putz-Chiriac, 2005).

For example, symmetry compact international rotation (notation Hermann-Mauguin) uses the combinations between the symmetry axes (noted with X=1, 2, 3, 4, 6) and reflections planes (noted with "*m*", from the English word "mirror"). Writing this compact symbols follow some simple conventions:

(i) $\boxed{X/m}$ represents a mirroring plan perpendicular on the rotation axis of X order;

(ii) $\boxed{Xm}$ instead represents a mirroring plan containing the axis of X order;

(iii) $\boxed{X2}$ represents an axis of second order (a dyad) perpendicular to the axis of X order.

Regarding the five special groups, i.e., the cubic groups from Table 2.6, there is a separate convention:

(iv)   Occurrence of number $\boxed{3}$ refers to four trigyres equally inclined that any cubic system must have.

In addition, longer added the conventions:

(v) $\boxed{mm}$ represents two mutually perpendicular reflection planes;

(vi) $\boxed{mmm}$ represents three mutually perpendicular reflection planes;

(vii) $\boxed{X/mm}$ represents the axis of X order contained in one of the mirroring planes (orthogonal) with the axis' direction, in turn, perpendicular to the other plane;

(viii) $\boxed{4/mmm}$ and $\boxed{6/mmm}$ are highest symmetry groups, and are equivalent to groups 4/*mm* and 6/*mm*, respectively, the third symbol "*m*" referring to an extra- set of vertical mirroring planes.

In another way of speaking, there is worth specifying for each crystalline system, the crystallographic group's symbols' significations associated to 3D axis' system associated to unitary cell (Pettifor, 1995).

In the cubic system the first element of the group symbol indicates the (x, y, z) axes' symmetry, followed by the three gyres' symmetry and, finally, indicates the parallels' symmetry to the faces diagonals of the unitary cell.

In orthorhombic system, the elements of the group symbols give information about the symmetry elements associated to x, y and z axes, in this order.

In the tetragonal, trigonal and hexagonal systems, the group symbol starts signifying the symmetry of the unique axis 0z, following one or two components indicating the additional symmetry of the perpendicular 0x and 0y.

In the monoclinic system, the group symbol simply refers to the symmetry elements associated with unique axis 0y.

Further, some elements of crystalline structures will be presented and analyzed from *symmetry operations transcriptions* point of view in symmetry compact international (Hermann-Mauguin) notation. These compact symbols follow some simple writing conventions:

(i)   the *existing symmetry elements* is identified: especially rotations and reflections planes;

(ii) *the unique rotation axis* is identified, that which cannot be obtained from another symmetry elements;

(iii) the orders corresponding to the unique rotation axis are written in consecutive order, starting with the biggest one;

(iv) for each reflection unique plane (in previous meaning) is written one "*m*";

(v)   if one of the unique rotation axes is perpendicular to a reflection unique plane, it will be written near, but separated by the symbol "/".

This way, in Figure 2.31-left the rectangular block has three 2nd-order axes of rotation, and all unique since perpendicular on different faces, so it will be written as 222; then, there are three reflections planes, again, all unique because they "cut" different faces, and so will be written $2m2m2m$, and finally the unique rotation axes are (one by one) perpendicular to unique reflection planes, resulting the symbol $2/m2/m2/m$.

In the Figure 2.31-right there is one single axis of 2nd-order, and each of the two existing planes is unique at its turn, so associated with the symbol $2mm$, without registering perpendicular cases between these

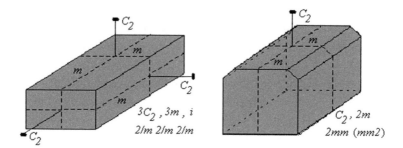

FIGURE 2.31 Symmetry operations and international (Hermann-Mauguin) notation determinations as for classes 2/m2/m2/m (on left) and 2mm (on right), after Chiriac-Putz-Chiriac (2005).

symmetry elements (often this symbol is written $mm2$, still without a precise motivation).

Instead, in Figure 2.32 more complex cases are analyzed. Crystalline structure from left presents an axis of the 4th order and is unique, four axes of 2nd order from which only two are unique, while other may be derived by the rotation around axis of order 4; so, till now one has the symbol 422; further, there are five reflection planes from which only 3 are unique, those perpendicular to unique rotation axis; so, one will write the symbol $4/m2/m2/m$.

On the right of Figure 2.32, however, the situation is more complex.

Of the three axis of 4th order, only one is unique (the other two may being respectively obtained by rotation around axes of 4th order), then from the six axis of 2nd-order again, only one is unique, because the rest are correlated with rotation of 4th order or/ and with the existing mirroring planes, and from the roto-inversion axis of 3rd order again only one is unique, the other being correlated by the corner of cube by the rotation of 4 order; so, until here it will be written the symbol 432.

Further on, from the nine mirroring planes three are perpendicular to axis of 4th-order (and if only one is unique results that there is associated an unique reflection plane) other six planes being perpendicular on axis of 2nd-order (and, of these only one axes in unique, the uniqueness is reflected on the correspondent perpendicular plane), and there is no reflection plane perpendicular on the roto-inversion axes of 3rd-order; finally the symbol will be $4/m\bar{3}2/m$.

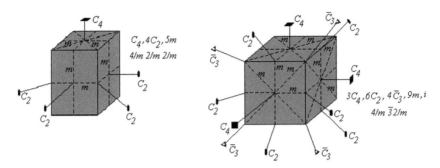

FIGURE 2.32 Symmetry operations and international (Hermann-Mauguin) notation determinations as for classes 4/m2/m2/m (on left) and 4/m$\bar{3}$2/m (on right), after Chiriac-Putz-Chiriac (2005).

One should note that, also very useful for condensing the elements and symmetry operations (unique) writing, international notation is not the most didactic in determination of all relation classes which can exist between the symmetry elements at unitary cells level of the solid bodies.

2.5.4 THE CRYSTALS MORPHOLOGY AND ITS STRUCTURAL LAWS

This section is dedicated to phenomenological correlation between the *internal structure* and the *external shape/form/morphology* of crystalline states; the present discussion follows (Chiriac-Putz-Chiriac, 2005).

2.5.4.1 The Crystalline Habitus

In the previous sections were analyzed and classified crystalline types based on their intrinsic unity: the unit cell, i.e., "to go" from the inside of crystal.

However, historically, the crystals were analyzed "seeing it" from the outside, i.e., "starting" from their faces.

Thus, over the time, the crystals' nature fascinated mainly due to the regular structures presented, with faces in geometric almost perfect shapes.

Not least intrigued is the fact that the assumed crystalline geometries were of the most diverse, the crystal being able to increase or split (cut) along its sides, however keeping its geometric regularity.

For example, in 1669, the Danish scientist Nicolaus Steno was the first who, studying the various forms of the quartz crystals by analyzing the sections of the crystals, had observed how, on matter what was the size and the shape of the crystal considered, the angles between the corresponding faces were kept constant. Then, in 1688, Guglielmini, while studying the calcite crystals, had observed how the cutting faces along certain planes is easier and, also, how the crystals thus obtained in turn keeps its faces plane, fine and with a regular geometry. In addition, this property shall be kept for any substance in crystalline state whatever would be its shape or size.

In 1784, Abbè Haüy, who could rightly be called the "father of crystallography", through studying the calcite crystals, by successive faces cuts, had tried to isolate the smallest structural block, concluding that the "rhomboid nucleus" is the smallest structural unit independent of the external form of the (calcite) crystals wherefrom it is obtained, see Figure 2.33. Then, Bravais was in turn who, in 1866, promoted the idea of reticular density, this way allowing the explanation of both the cleavage and

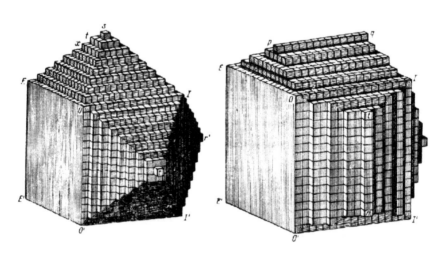

FIGURE 2.33 The relationship between the elementary building blocks form (identical on the left and on the right) and the external crystals form (differently developed on the left towards the right); after Haüy (1822).

the crystalline (faces) growth (the reverse phenomenon of the cleavage) as well.

Thus, there was historically concluded how the crystals grow in a perpendicular (normal) direction to its sides at a speed as much lower as the reticular density on that face is higher, naturally defining the crystal face as the reticular plane with maximum density (of its structural points, inhibiting the further advancement of the crystalline form in the normal direction at the plane-face in cause).

Conversely, the reticular planes with low density (compactization) either will allow an acute growth if they are at the surface of the crystal, or will allow the cleavage, and thus breaking the crystal along that plan, if they are inside the structure.

The relative developments (growths or cleavages) on the external aspect of a crystalline structure generate the so-called crystalline habitus (the set of shapes for a system), see Table 2.7. For example, the cube and the octahedron are two different forms of habitus, but belong to the same crystalline system, the cubic one, according with Figure 2.34.

The case of the NaCl crystals' growth is relevant in this regard. Usually, at the growth in aqueous solution, the crystalline salt cubical crystallized;

TABLE 2.7 Examples of Minerals, with Habitus Corresponding for the Crystalline Systems*

Examples of minerals	Syngony/Habitus
All the types of Garnet $A_3B_2(SiO_4)_3$, Diamond, fluorites, crystals of Gold, Sodalit $Na_4Al_3Si_3O_{12}Cl$, Lapis Lazuli (in Greek "lazakord"=paradise, due to the color of blue paradise) belonging to the class of solids with the formula $(Na, Ca)_8(Al,Si)_{12}O_{24}(S,SO)_4$, Pyrite, Silver crystals, Sphalerik, Spinel.	*Cubic* CUBIC PRISM OCTAHEDORN DODECAHEDRON
Apophylit $KCa_4Si_8O_2O(OH)_8H_2O$, Idokraz or Vezuvianit $Ca_{10}(Mg, Fe)_2Al_4(SiO_4)_5$ (SiO7)2(OH)4, Rutile, Scapolite $(Na,Ca,K)_4A_{13}(Al,Si)_3Si_6O_{24}$ (Cl,SO_4,CO_3), Wulfenite $PbMoO_4$, Zirconium $ZrSiO_4$.	*Tetragonal* TETRAGONAL PRISM BIPYRAMIDE PRISM WITH PYRAMIDE

TABLE 2.7   Continued

Examples of minerals	Syngony/Habitus
Andalusite Al_2SiO_5, Celestite $SrSO_4$, Chrysoberyl $BeAl_2O_4$, Cordierite $Mg_2Al_4Si_5O_{18}$, Danburite $CaB_2(Si_2O_4)_2$, Epidote $Ca_2(Al,Fe)_3Si_3O_{12}(OH)$, Enstatite $Mg_2Si_2O_6$, Hemimorphite $Zn_4Si_2O_7(OH)_2H_2O$, Fibrolite/Sillimanite Al_2SiO_5 (after the name of Benjamin Silliman, professor of Chemistry at Yale University in 1824), Olivine, sulfur crystals, Topaz $Al_2SiO_4(F,OH)_2$, Zoisite $Ca_2Al_3 (Si_3O_{12})(OH)$ (after the name of the Australian mineralogist Von Zois).	***Orthorhombic*** 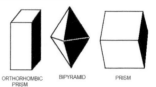
All variants of Quartz, Corundum Al_2O_3 (with the versions of Rubin and Sapphire), Hematite Fe_2O_3, Rhodochrosite CO_2Mn, Tourmaline (Na,Ca) $(Al,Fe,Li,Mg)_3A_{16}(BO3)_3 (Si_6O_{18})$.	***Trigonal (Rhombohedral)*** 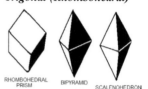
Apatite $Ca_5(F,Cl,OH)(PO_4)_3$, the crystals Beryl $Be_3Al_2Si_{16}O_{18}$ (with the versions cu variantele aquamarine, emerald), Zincite (Zn,Mn)O and others.	***Hexagonal*** 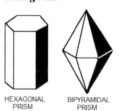
Azurite $Cu_3(CO_3)_2(OH)_2$, Crocoite $PbCrO_4$, Datolite $CaBSiO_4(OH)$, Diopside $CaMgSi_2O_6$, Jadeite $NaAl(SiO_3)_2$, Lazulite $MgAl_2(PO_4)_2(OH)_2$, Malachite $Cu_2(CO_3)(OH)_2$, Orthoclase Fel(d)spar (e.g., Albite $NaAlSi_3O_8$), Staurolite $(Fe,Mg,Zn)_2Al_9Si_4O_{23}(OH)$, Sphene $CaTiO(SiO_4)$, Spodumene $LiAlSi_2O_6$ and others.	***Monoclinic*** 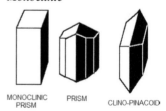
Amblygonite $(Li,Na)Al(PO_4)(F,OH)$, Axinite $Ca_2(Fe,Mn)Al_2(BO_3)(Si_4O_{12})(OH)$, Kyanite Al_2SiO_5, Labradorite $(Na,Ca)AlSi_3O_8$, Microline Fel(d) spar $KAlSi_3O_8$, Plagioclase Fel(d)spar (of type of Labradorite), Turquoise $(CuAl_6 [(OH)_3/PO_4]_{4.4}H_2O$ and others.	***Triclinic*** 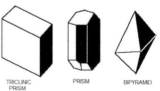

*After Chiriac-Putz-Chiriac (2005).

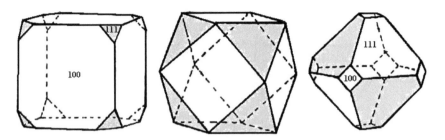

FIGURE 2.34 From left to right: cubic, cuboctahedron and octahedral habitus, after Chiriac-Putz-Chiriac (2005).

but if urea is added to this solution, the NaCl growth undergoes major changes in its external aspect, i.e., the cubical faces will change so that the cube's vertices will flatten more and more until the whole structure becomes octahedron.

2.5.4.2   The general Laws of the Crystalline Shape

The regular polyhedral shape of crystals is one of the most prominent characteristics which attracted the attention of the researchers since the ancient history of research.

The crystalline polyhedron is defined, as the geometric one, by its faces (f), edges (e) and vertices (v) between which number is *the Euler – Descartes relation*:

$$f + v = e + 2 \tag{2.55}$$

a relation easy to verify, for example, for the Platonic solids in Table 2.3. The identity between the geometric polyhedra and crystalline polyhedra is very rare, the crystalline polyhedra shape being influenced by the whole physical and chemical factors acting on the crystal growth. Thus, the crystals of the same mineral species may have faces of different shape and size and with different arrangement from one crystal to another, see Figure 2.35.

There is however a set of defining geometrical features, for a specific crystalline species, whose study is the subject of the geometrical crystallography.

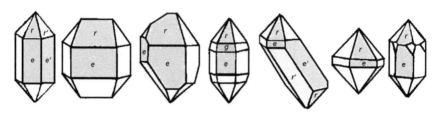

FIGURE 2.35   Various shapes of the quartz crystal, after Haüy (1822).

These features are included in the geometric crystallography's laws: the convexity law, the dihedral angles constancy law, the rationalizing law, the areas law and the symmetry principle. Explicitly, they are:

- The convexity law: a crystal is always a convex polyhedron, with outgoing angles between the adjacent faces. The appearance of an entering angle indicates an increase of two or more crystals, with the faces closing this angle belonging to different crystals.
- The dihedral angles constancy law: the crystalline shapes are characterized by the dihedral angles which the faces make between them. These angles are constant at constant temperature and constant pressure.
- The rationality law: all existing or possible faces of a crystal are found one towards each other in a geometrical dependence expressed by rational numbers (usually small integers).

In other words, the faces of a crystal are placed so that the crossing size for each of them on a reference is a commonly small integer multiple of a fundamental crystallographic parameter. The number with which are multiplied the fundamental parameter in order to obtain the parameter of a given face are called coefficients and their reciprocal (in inverse meaning) values are called indices of the face.

These laws are the result of a lengthy experimental observation on crystals. The first one is a definition of the crystal, and the other is a logical consequence of the reticular structure of the crystallized substance.

For instance, the experience knowledge said that the crystal faces correspond to some planes, usually of high reticular density, while the edges correspond to the dense reticular strings. As long as a certain plane or string has a fixed position in the lattice, the result is that also the angles between them are constant too.

Similarly, any reticular plane (so possible face of the crystal) encounters any direction in a node whose position towards the origin is expressed by an integer multiple of the parameter of the string parallel to the chosen direction. The relative position of two planes towards a direction will be explained by the relative position of two nodes, i.e., by the ratio of two rational numbers.

Using the law of rationality one can easily define the relative position of the various faces of a crystal. For this, one may choose a coordinate system of axes, defined by the directions of three non-parallel edges of the crystal, i.e., by the angles α, β and γ between them.

In relation to the chosen coordinate system, the position of a face is determined there is also known: the plane's equation which includes the face, the value and the sign of the parameters (i.e., indices) on these axes or, the position of the normal/versor as taken from the origin of the system of coordinates to the relative face.

The cosines of the angles between the normal/versor and the axis are called guiding/driving cosines of the face. For the current determinations the absolute values of the parameters are not relevant but their ratios.

To calculate the ratio between the parameters of a random face based on the guiding/driving cosines, a system of axes X, Y, Z is considered along an ABC face that cuts the three axes that at the distance OA = a, OB = b, OC = c (see Figure 2.36), which we will consider as parametric face.

Given OP the normal/versor at the ABC plane, it makes the guiding/driving angles with the reference system the angles POX, POY and POZ, respectively.

Then, from the Figure 2.36 there is directly written:

$$\frac{OP}{a} = \cos P\hat{O}X$$

$$\frac{OP}{b} = \cos P\hat{O}Y$$

$$\frac{OP}{c} = \cos P\hat{O}Z \qquad (2.56)$$

wherefrom the proportions results

$$a : b : c = \frac{1}{\cos P\hat{O}X} : \frac{1}{\cos P\hat{O}Y} : \frac{1}{\cos P\hat{O}Z} \qquad (2.57)$$

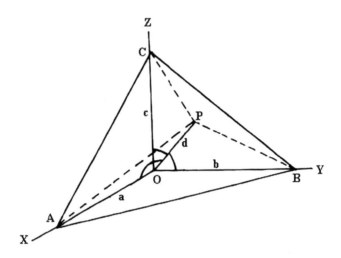

FIGURE 2.36   The definition of an ABC face by the guiding/driving angles, after Chiriac-Putz-Chiriac (2005).

These relationships between the parameters of the parametric face are representing the *axial relation* of the crystal.

On the other hand, from the constancy law there results that the guiding/driving angles are constant, so also the axial relation is constant. Consequently, the ratio a:b:c is a crystallographic constants for a certain crystalline substance.

If we consider a different face with the parameters (ma, nb, pc), similarly we have:

$$ma : nb : pc = \frac{a}{h} : \frac{b}{k} : \frac{c}{l} = \frac{1}{\cos P\hat{O}X} : \frac{1}{\cos P\hat{O}Y} : \frac{1}{\cos P\hat{O}Z} \quad (2.58)$$

In the relation (2.58) a constant part (*a:b:c*) for all the faces appears, which is usually irrational – since fixed by the crystal metrics– and a rational part (*h, k, l*) which determines the face position and may varying from one face to another.

However, in order to express the face position one can employ both numerical coefficients (*m, n, p*) of the parameters and the reciprocal indices (h, k, l): adopting the index notation, the symbol of a face (hence also of a reticular plane) will be written as (*hkl*).

Next, in determining the indices' values one will take into account that there are a total of 7 different ways to place a face with respect to the reference axes: when the face is parallel with two axes and cuts the third one (see Figure 2.37 top); when the face is parallel with an axis and cuts the other two (Figure 2.37 middle); when the face cuts all the three axes (Figure 2.37 below). The faces which are grouping in zones of faces are called *tauto-zonal, and have a common direction called zone axis. A face can belong of two or more zones* (Figure 2.38). A zone existence is determined by the intersection of at least two faces, the resulted edge from this intersection representing the zone axis.

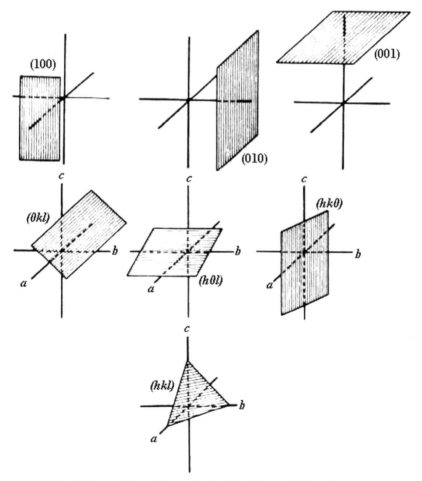

FIGURE 2.37   The possible positions of a face in relation with the reference axes.

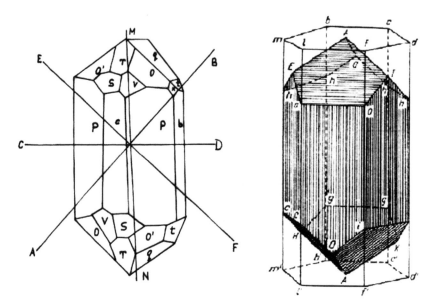

FIGURE 2.38   Left: AB, CD, MN and EF as zone's axes; right: zones of faces, after Haüy (1822).

Generally, any polyhedron can be obtained from a fundamental tetrahedron by successive truncating operations, i.e., by truncating the edges with the planes of their parallels, see Figure 2.39, this being a phenomenological form of the zones law, which will be discussed below in detail.

2.5.5 THE CRYSTALLOGRAPHIC INDEXING

Being exemplified and argued the independence of the unit cell in relation to the crystalline habitus, there remains just to correlate the unit cell with the faces planes under the normal physical conditions; the present discussion follows (Chiriac-Putz-Chiriac, 2005).

2.5.5.1 Miller Indices

Since the change in size and shape of a crystal does not alter the angles between the faces (experimental observation) and, moreover, as long as

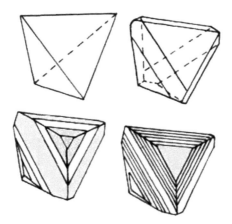

FIGURE 2.39   Determining the *tauto-zonal* faces from a fundamental tetrahedron.

the *crystal face* is a manifestation of the structural arrangement at the unit cell level from which it comes by growth, there is sufficient to specify the planes orientation of those faces, without localization their position in the space. Therefore, this orientation will be specified by specifying the *orientation of the parallel plane* to the unit cell level. Now, the unit cell is, in turn, characterized by the directions with fundamental lengths (a,b,c), thus being desirable the expression of the points and of the planes orientation faces or of any other direction in a crystal by the parameters of the unit cell. The unit cell, as a structural referential in a crystal was previously justified by its property of invariance at the modifications of the crystalline habitus. Thus, in order to determine the coordinates of a point in a network, one starts from a point of zero origin. This point is located in the left and posterior corner of the fundamental parallelepiped and is denoted by the symbol [[000]], with its higher symmetrical given by [[001]], and so on (see Figure 2.40).

Accordingly, the quantification of the orientation of the planes will also be related to the sizes of the fundamental lengths of the unit cell and of their fractions, Figure 2.41.

In this sense, the orientation of a crystallographic plane will be customized by a set of three numbers, called the *Miller indices*. They are successively written in the order corresponding to the fundamental axes, without

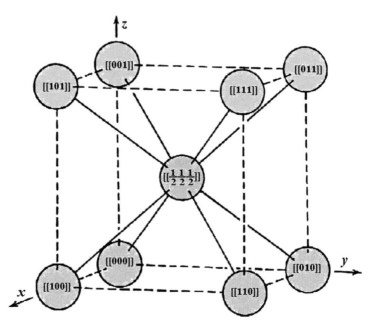

FIGURE 2.40   The coordination of the points in a lattice; the points which have a central or intermediate position are noted by the fractional indices, after Chiriac-Putz-Chiriac (2005).

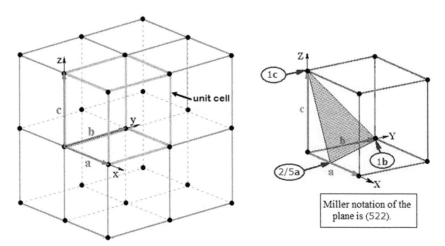

FIGURE 2.41   The unit cell (left) and the determination of the Miller indices associated to an arbitrary plane from the crystal (right) in relation with the parameters of the unit cell, after Chiriac-Putz-Chiriac (2005).

comma between them and inside the round brackets, (hkl), with h, k and l integer numbers.

Establishing the Miller indices assumes the following algorithm:

(i) consider the parallel plane with those under concern, at the level of the elementary cell;

(ii) note the fractions from the fundamental lengths of the unit cell by which the plane intersects the fundamental axes;

(iii) take the reciprocals of these fractions;

(iv) look for the lowest common multiple (l.c.m) of the denominators (this step is restricting the analysis to the unit cell);

(v) multiply the values of step (iii) and (iv) and getting the associated (hkl) string, see the example in Figure 2.41.

If the plane in focus is parallel with a fundamental axis, they will "intersect" at infinitum and the reciprocal number will be taken zero, so the steps (iv) and (v) not being affected by this convention.

The Miller method of definition of the crystallographic planes allows also the consideration of the intersections with the fundamental axes, in their negative part, generating negative integer numbers and the corresponding Miller indices re-noted, conventionally, with the positive values plus a top bar.

For example, the plane $(\bar{h}\bar{k}\bar{l})$ is obtained by the plane (hkl) by reversing (respecting the zero value) all the Miller indices, resulting that the two planes are equivalent, in the sense of the inversion symmetry.

A collection of equivalent planes, noted as $\{hkl\}$, includes the planes which can be obtained one from each other, by applying the symmetry operations allowed by the crystalline system or by the unit cell in focus.

Besides the planes, also the directions in a crystal are very important and should be rationalized to the fundamental referential system at the unit cell level.

In this case, any direction in the crystal can be specified by the associated direction vector, respecting to the fundamental axes and lengths, with the direction indices written along the axes, without comma, and between square brackets: $[uvw]$. The numbers (u, v and w) are integers and are calculated following the same procedure as for the indices Miller, see again the steps (i)–(v) above; also see the example of Figure 2.42.

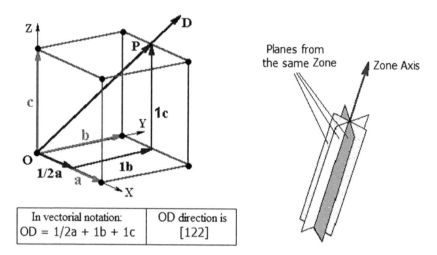

In vectorial notation:	OD direction is
OD = 1/2a + 1b + 1c	[122]

FIGURE 2.42 Left: the determination of the indices of an arbitrary direction of the crystal; right: the illustration of the direction of a zone axis for a family of equivalent planes, after Chiriac-Putz-Chiriac (2005).

Again, reversing the direction involves the inversion of the direction vector coefficients and thus also of the direction indices, so becoming $[\bar{u}\bar{v}\bar{w}]$, symbolically written – with the corresponding positive indices to which a top bar was added, as indicating the inversion.

Finally, a family of crystallographic (equivalent) directions, obtained each from other by symmetry operations allowed by the crystallographic system in cause is generally denoted by $\langle uvw \rangle$.

2.5.5.2   The Weiss Zone Law of Crystal Faces

The most crystals are characterized by the appearance of groups or sets of parallel edges, each of this group having its own crystallographic orientation. All faces of polyhedral shape that intersect after parallel edges and with a common direction form a *zone*.

The zone law summarizes, from the experimental findings, that all the possible faces at a crystallographic shape have between them *zonal bonds*.

The connection between planes and the directions in crystals consists in the fact that the intersection of two planes is made after a line, which, belonging to both planes, will be called the *zone axis*, i.e., of the zone where the planes in cause are intersecting (coexisting).

Thus, the zone planes will be those planes which either are all intersected upon the same zone axis, see Figure 2.42, or are all parallel with the zone axis in focus.

In terms of indices of planes and directions, the condition that the plane (hkl) is parallel with the (zone axis) direction [uvw] is analytically translated by the equation:

$$hu + kv + lw = 0 \qquad (2.59)$$

called the *Weiss* law of zones: it represents the fact that the zone axis and the normal/versor to the zone plane are perpendicular (with the null inner/dot/scalar product).

As long as only in the cubic system the normal direction to the plane (*hkl*) has the direction [*hkl*], with the direction vector $r_{hkl}=ha+kb+lc$, a general demonstration valid for the Weiss law for any crystallographic system can be given by speculating the condition of perpendicularity of two lines; alternatively it can be generated as based on the (arbitrary/general) construction of Figure 2.43.

In the Figure 2.43 the plane (*hkl*) which cut the fundamental axes at the coordinates (*a/h*, *b/k*, *c/l*) have been considered by omitting the common multiple, since the same for all the coordinates – and in any case also by the impossibility of its specification, in general, due to its variation from case to case.

The proof which follows uses only the basic vectorial notions that do not require a special introduction, however being necessary only a carefully analysis of the involved vectors' orientation.

Thus, for a vector to be in (or parallel to) the plane formed by the vectors (here) of basic directions AB (i.e., the vector from A to B) and AC (i.e., the vector from A to C) there is sufficient for the vector in question to be written as a linear combination of them such as:

$$r' = \lambda \, AB + \mu \, AC \qquad (2.60)$$

with λ and μ appropriately chosen.

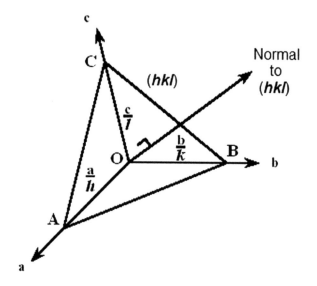

FIGURE 2.43   Construction for proofing the Weiss law of the zone faces, after Chiriac-Putz-Chiriac (2005).

Then, by re-expressing the vectors AB and AC in the unit cell system with the axes s of Figure 2.43, there is obtained:

$$
\begin{aligned}
AB &= b/k - a/h \\
AC &= c/l - a/h
\end{aligned} \tag{2.61}
$$

Note that while writing the relations (2.61) the directions of the vectors involved and the fact that the sum and the difference performed are expressed from the rule of vectorial composition in a parallelogram had been taken into account.

Under these conditions, with Eq. (2.61) in Eq. (2.60), the arbitrary vector r' is written at the level of the unit cell as follows:

$$
\begin{aligned}
r' &= \lambda(b/k - a/h) + \mu(c/l - a/h) \\
&= -(\lambda + \mu)a/h + \lambda\, b/k + \mu\, c/l
\end{aligned} \tag{2.62}
$$

If the vector r' of Eq. (2.62) is now identified with the direction vector $[uvw]$, $r_{uvw}=ua+vb+wc$, the next identities are obtained:

$$
\begin{aligned}
u &= -(\lambda+\mu)/h \\
v &= \lambda/k \\
w &= \mu/l
\end{aligned}
\tag{2.63}
$$

There is almost obvious that, as based on the relations (2.63), and by performing the sum of the products $uh + vk + wl$, the Weiss law is obtained under the relation (2.59).

The Weiss law (2.59) in fact expresses the relation between the indices of a face (hkl) and the zone axis with characterized by the $[uvw]$ indices.

Yet, some observations are still required: even if the $[uvw]$ symbol is a line, the u, v and w indices correspond to the $[[uvw]]$ node placed on a line that passes through the origin point $[[000]]$. Since also the (hkl) plane can be associated with the parallel plane which passes by the origin, there results that the $[uvw]$ zone plane and the zone axis to which it belongs have in common the nodal $[[uvw]]$ and $[[000]]$ points.

There is noteworthy how, in this proof, the individuation of any crystallographic system was not used in any step, since only the arguments of vectorial geometry were employed, thus validating the universality of the zones' law for any crystalline syngony (Nye, 1985).

Moreover, from the general zone equation some important equalities may result as follows.

(a) If two zone planes, for example $(h_1k_1l_1)$ and $(h_2k_2l_2)$ are known then, the direction of their zone axis $[uvw]$ can be determined using the formal vectorial product:

$$
(h_1k_1l_1)\times(h_2k_2l_2) =
\begin{vmatrix}
u & v & w \\
h_1 & k_1 & l_1 \\
h_2 & k_2 & l_2
\end{vmatrix}
\tag{2.64}
$$

wherefrom, each index of the zone axis with the vector $r_{uvw}=ua+vb+wc$ can be identified by barring the line and the column in cause and performing the remained determinant, with the results:

$$u = k_1 l_2 - k_2 l_1$$
$$v = l_1 h_2 - l_2 h_1$$
$$w = h_1 k_2 - h_2 k_1 \qquad (2.65)$$

(b) Reversely, one will tell that the plane (*hkl*) belongs to two zones, with the zone axes $[u_1 v_1 w_1]$ and $[u_2 v_2 w_2]$, if the relations analogous to those of Eq. (2.64) are satisfied:

$$[u_1 v_1 w_1] \times [u_2 v_2 w_2] = \begin{vmatrix} h & k & l \\ u_1 & v_1 & w_1 \\ u_2 & v_2 & w_2 \end{vmatrix} \qquad (2.66)$$

wherefrom the conditions satisfied by the Miller indices of the plane (*hkl*) immediately results:

$$h = v_1 w_2 - v_2 w_1$$
$$k = u_2 w_1 - u_1 w_2$$
$$l = u_1 v_2 - u_2 v_1 \qquad (2.67)$$

(c) The requirement that three faces with Miller indices $(h_1 k_1 l_1)$, $(h_2 k_2 l_2)$ and $(h_3 k_3 l_3)$, be part of the same area is that the indices' determinant to be null:

$$\begin{vmatrix} h_1 & k_1 & l_1 \\ h_2 & k_2 & l_3 \\ h_3 & k_3 & l_3 \end{vmatrix} = 0 \qquad (2.68)$$

The condition (2.68) is satisfied if a row or a column appears as the sum of the corresponding elements of other rows or columns, for example:

$$h_2 = h_1 + h_3,$$
$$k_2 = k_1 + k_3,$$
$$l_2 = l_1 + l_3 \qquad (2.69)$$

According to the equalities (2.69) the indices of a face contained between two other faces with known indices are obtained by summing indices of the last faces. Eventually, one will say that the plan $(h_2k_2l_2)$ belongs of the same zone as the planes $(h_1k_1l_1)$ and $(h_3k_3l_3)$ if the linear combinations are satisfied:

$$
\begin{aligned}
h_2 &= mh_1 \pm nh_3 \\
k_2 &= mk_1 \pm nk_3 \\
l_2 &= ml_1 \pm nl_3
\end{aligned}
\qquad (2.70)
$$

with m and n integer numbers.

2.5.5.3 Crystallographic Formulae

Another important property of the crystalline planes is that of the inter-planar distances d_{hkl}, defined by the perpendicular distance between the parallel and consecutive planes, starting with the closest ones to the unit cell origin.

If we consider the cuts of a (hkl) plan with the unit cell axes as having the coordinates $(a/h, b/k, c/l)$, then the $(nhnknl)$ plane will correspond to the coordinates by the cuts with the unit cell: $(a/(hn), b/(kn), c/(ln))$ which allows the writing of the relationship between these types of planes, in terms of the associated inter-planar distances:

$$
d_{nh\ nk\ nl} = \frac{d_{hkl}}{n}
\qquad (2.71)
$$

Of a particular practical importance is the calculation of the inter-planar distance d_{hkl} and the inter-planar angles $\cos\phi$, i.e., between two planes $(h_1k_1l_1)$ and $(h_2k_2l_2)$, of the $\cos\phi$ for the angles between two directions, say $[u_1v_1w_1]$ and $[u_2v_2w_2]$, as well as of the volume unit cell V, for the various crystalline systems.

Without going into the mathematical details, adjacent to the purpose of this chapter, the general formulae for these quantities will be exposed, with the working case of the (general) triclinic system, since the customizations for the rest of the crystallographic systems being performed in accordance with the basic parametric relationships between axes and angles, as shown in Table 2.2.

Thus, in the general case of the triclinic system, the quantities listed above are cast analytically as (Verma & Srivastava, 1982):

$$\frac{1}{d_{hkl}^2} = \frac{1}{V^2}\left(S_{11}h^2 + S_{22}k^2 + S_{33}l^2 + 2S_{12}hk + 2S_{23}kl + 2S_{31}hl\right)$$

(2.72)

$$\cos\theta = \frac{B}{C_{h_1 k_1 l_1} \cdot C_{h_2 k_2 l_2}}$$

(2.73)

$$\cos\phi = \frac{M}{I_{u_1 v_1 w_1} \cdot I_{u_2 v_2 w_2}}$$

(2.74)

$$V = abc\left(1 - \cos^2\alpha - \cos^2\beta - \cos^2\gamma + 2\cos\alpha\cos\beta\cos\gamma\right)^{1/2}$$

(2.75)

where:

$$S_{11} = b^2 c^2 \sin^2\alpha,$$

$$S_{22} = a^2 c^2 \sin^2\beta,$$

$$S_{33} = a^2 b^2 \sin^2\gamma,$$

$$S_{12} = abc^2\left(\cos\alpha\cos\beta - \cos\gamma\right),$$

$$S_{23} = bca^2\left(\cos\gamma\cos\beta - \cos\alpha\right),$$

$$S_{31} = acb^2\left(\cos\alpha\cos\gamma - \cos\beta\right),$$

$$B = \begin{bmatrix} h_1 h_2 b^2 c^2 \sin^2 \alpha + k_1 k_2 a^2 c^2 \sin^2 \beta + l_1 l_2 a^2 b^2 \sin^2 \gamma \\ + abc^2 (\cos \alpha \cos \beta - \cos \gamma)(k_1 h_2 + h_1 k_2) \\ + ab^2 c(\cos \gamma \cos \alpha - \cos \beta)(h_1 l_2 + l_1 h_2) \\ + a^2 bc(\cos \beta \cos \gamma - \cos \alpha)(k_1 l_2 + l_1 k_2) \end{bmatrix},$$

$$C_{hkl} = \begin{bmatrix} h^2 b^2 c^2 \sin^2 \alpha + k^2 a^2 c^2 \sin^2 \beta + l^2 a^2 b^2 \sin^2 \gamma \\ + 2hkabc^2 (\cos \alpha \cos \beta - \cos \gamma) + 2hlab^2 c(\cos \gamma \cos \alpha - \cos \beta) \\ + 2kla^2 bc(\cos \beta \cos \gamma - \cos \alpha) \end{bmatrix}^{1/2},$$

$$M = \begin{bmatrix} a^2 u_1 u_2 + b^2 v_1 v_2 + c^2 w_1 w_2 \\ + bc(v_1 w_2 + w_1 v_2)\cos \alpha \\ + ac(w_1 u_2 + u_1 w_2)\cos \beta \\ + ab(u_1 v_2 + v_1 u_2)\cos \gamma \end{bmatrix},$$

$$I_{uvw} = a^2 u^2 + b^2 v^2 + c^2 w^2 + 2bcvw\cos \alpha + 2acwu \cos \beta + 2abuv \cos \gamma$$

$$(2.76)$$

Thus, the Miller indices and the directions indices help in fixation-indexing of the orientation of the crystallographic planes and directions relative to the unit cell parameters.

Mostly, the Miller indices are integers and small numbers, rarely larger than 3 and even rarely greater than 5 (as based on empirical observations). These characteristics have substance in the Bravais observation that only the planes (reticular layers) with large reticular density will appear as natural faces, so these faces being simple planes in the unit cell with small Miller indices.

2.5.5.4 The Miller-Bravais Indices

The Miller indices, by their values indicate the planes and the equivalent directions, by combinations, permutations and inversion operations applied to the same numbers. However, such quality is missing in the *hexagonal system* because the model of the unit cell is a parallelepiped but the symmetry in the lattice is hexagonal (see Table 2.2). For example, in Figure 2.44, in terms of Miller indices, the faces parallel with 0z containing the segments AB, BC, CD, DE, EF and FA are indexed as planes (100), (010), $(\overline{1}10)$, $(\overline{1}00)$, $(0\overline{1}0)$ and respectively $(1\overline{1}0)$.

The problem which arises here is that, although those faces are equivalent – being in relationship of symmetry in the hexagonal symmetry, the Miller indexing produce *two* groups of faces: one with four faces indexed with a single index different from zero and another with two faces having two indices different from zero – those above emphasized.

Moreover, if the Miller indices are written as associated to the faces parallel with 0z containing the segments AC, BD, CE, DF, EA and FB the indexes: $(1\overline{1}0)$, $(\overline{1}20)$, $(\overline{2}10)$, $(\overline{1}10)$, $(1\overline{2}0)$ and $(2\overline{1}0)$ are obtained, respectively.

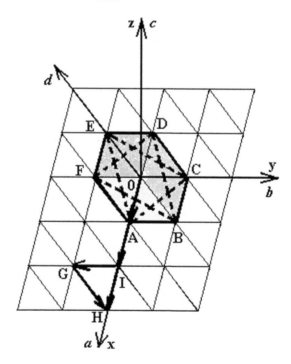

FIGURE 2.44 Introducing Miller-Bravais indices by special geometrical system reference, after Chiriac-Putz-Chiriac (2005).

This way, although again treated as equivalent planes, we note a different separation in groups of faces: those in which the index 2 appears and those in which such index is missing in the above emphasized ones.

Similar problems arise also in the directions indexing. As a result, Bravais proposed that, only for indexing, according to the identity principle of the indices type and of combinations for the equivalent planes and directions, to consider also the fourth direction as fundamental coordinated as denoted by d in Figure 2.44.

In these circumstances there is remarkable how, for example, following the Figure 2.44, we can write again in terms of vectors (with segment direction):

$$-0E = 0B = 0A + 0C, \tag{2.77}$$

which, in terms of Miller indices, produces the general relationship of bonding for all the planes, with the form

$$i = -(h+k) \tag{2.78}$$

being i the Miller index on direction d.

Thus, in the hexagonal system the indexing one will use 4, instead of 3, Miller indices, generating the so-called *Miller-Bravais indexing*: $\boxed{(hkil)}$, where, the forth index occupies the third position on the traditional Miller notation, as the minus sum of the first two.

In this new context, the Miller-Bravais indexing of the faces parallel to $0z$ containing the segments AB, BC, CD, DE, EF, and FA will produce the notation $(10\bar{1}0)$, $(0\bar{1}\bar{1}0)$, $(\bar{1}100)$, $(\bar{1}010)$, $(0\bar{1}10)$, and $(1\bar{1}00)$, and for the faces parallel to $0z$ containing segments AC, BD, CE, DF, EA, and FB will generate the new sets $(11\bar{2}0)$, $(\bar{1}2\bar{1}0)$, $(\bar{2}110)$, $(\bar{1}\bar{1}20)$, $(1\bar{2}10)$ and $(2\bar{1}\bar{1}0)$. This time, in all the cases, the notations are equivalent by symmetry operations. Also for the re-indexing of the directions of the hexagonal system one should follow the same procedure as for the planes case, so that the new Miller-Bravais index, from the third position, will be minus the sum of the first two Miller indices.

Again through calling the Figure 2.44, we notice how the segment 0A is directed as [100], but its bare translation on the same direction with a $1a$ unit generates the indexing [1000], where the third index does not appear as minus the sum of the first two; so, a proper "trail" to find the consistency Miller-Bravais construction should be built. Therefore, we will select the path:

$$0H=0I+IG+GH \tag{2.79}$$

equivalent to the translations

$$2a + \bar{1}b + \bar{1}d \tag{2.80}$$

which generates the indexing $[2\bar{1}\bar{1}0]$ on the 0A direction, observing how from now on the third index results from negative of the sum of the first two.

Generalizing the previous procedure, the transition from a Miller index of $[pqr]$ directions to associated $[uvtw]$ Miller-Bravais one it will be made by the indices' transformations:

$$
\begin{aligned}
u &= (2p\text{-}q)/3 \\
v &= (2q\text{-}p)/3 \\
t &= -(u+v), \quad w = r
\end{aligned}
\tag{2.81}
$$

However, there should be noted that, in terms of directions, many crystallographers prefer not to make the transformation within Miller-Bravais axes' system, while working with the Miller system for directions and with Miller-Bravais only for the planes indexing of the hexagonal system.

2.5.6 STEREOGRAPHIC PROJECTION. THE WULFF MAP

After indexing the crystal through the angles, faces and directions one should found the most appropriate representation for formulating their relationships. Firstly, one should choose a uniform representation for the planes and directions representation; a convenient choice is the normal lines to the planes as "representatives" of those planes. Then we chose a uniform representation system of various crystalline habitus, for any form and size. To this end, the ideal choice is the *spherical representation*: the crystal, with all the directions and the plans for analysis will be placed in the sphere center, with a unique radius for all the crystals from all the crystalline systems. This area will be called the *reference sphere* or *sphere projection*. In relation to this sphere, the crystalline planes and directions will be represented on the surface of the sphere; the present discussion follows (Chiriac-Putz-Chiriac, 2005).

Thus, the representation of the planes can be done in two ways:

(i) By considering the normal on the plan as crossing through the sphere and is extended until intersects the sphere in a point called pole;

(ii) By extended the plane (or the plane parallel to the one in cause) through the sphere center until intersects the sphere surface after the so called big circle for which the normal, passing through the center of the sphere, will find the pole plan, as in i).

Thus, until now, the planes have been reduced to poles (or points) on the sphere of projection, see Figure 2.45, while the relationships among angles of various crystallographic planes are represented by the poles' relationship on the projection's sphere.

For the case of crystalline directions, they are represented by drawing the associated line, eventually of its parallel that crosses the center of the sphere until will intersect the surface of the sphere in two points called *nodes*.

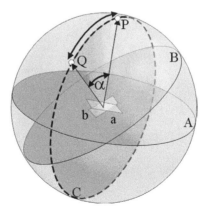

FIGURE 2.45 The illustration of the spherical projection; after Stereographic Projection (2003).

Applying this type of construction in the example of Figure 2.45, we note that the angle between the planes (a) and (b) is successively transferred to the level of the angle between the normal lines associated, and then to the poles P and Q on the sphere. Joining the poles P and Q on the sphere surface can be placed directly on a great circle of the sphere, and one can directly "read" the angle between them (the same as that of the associated planes) while the sphere surface was calibrated.

Further, as geographers transpose in plane the spherical coordinates, there is particularly useful to reduce the spherical projection to a two-dimensional one, or more simply, on a "paper surface." In this sense, very practical is the points projection on the equatorial plane of the sphere, called *stereographic projection*, this being achieved by joining each crystallographic pole of the sphere surface with the South pole (if the crystallographic poles are in the northern hemisphere) and, respectively, with the north pole (if the crystallographic poles are in the southern hemisphere), see Figure 2.46.

Remarkable how for the stereographic projection the angles between the crystallographic poles on the sphere are maintained, but the distance between them changes (this is more obvious since it can be imagined the projection plane perpendicularly sliding on the direction of the North-South axis).

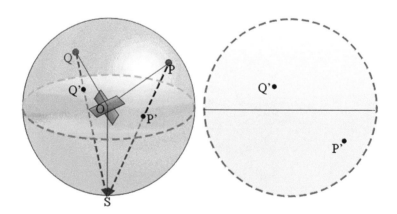

FIGURE 2.46   Illustration of the stereographic projection; after Stereographic Projection (2003).

The intersection of the equatorial plane (where is made the stereo-graphic projection) with the spherical surface generates the big circle of the sphere that is called *basic circle* or *primitive circle*. This name comes also from the fact that the stereographic projection of the points on the equator of the sphere results in themselves. In addition, the North Pole (for the stereographic projection of the northern hemisphere) has the stereo-graphically projection in the primitive circle.

At this point, one should consider the calibration of the sphere sur-face through an angular mapping (also called the "grid") of reference, so that, for example, considering the Figure 2.46, the angle between the poles P and Q on the sphere can be measured by measuring the angle between the projections P 'and Q' in the equatorial plane. Such a reference grid is obtained from the drawing in series of large circles (in a certain interval of each other) on the sphere so that to unite the North and South poles (forming the so-called *longitude lines*) and being then joined by a series of circles with smaller radii than the sphere one, called small circles (forming the so-called *latitude lines*).

Now, the angles measurement assume the poles alignment so that to be placed on the same big circle, operation performed by the rotation around an axis parallel to the equatorial plane of the stereographic projection, followed by the angular evaluation between the lines of latitude that cor-respond to them, see Figure 2.47.

Being fixed the spherical grid, further on, one should consider its flat projection to be used in conjunction with the stereographic pro-jection, through their poles. There are two types of flat projection of the spherical grid (always on the equatorial plane) as are shown in Figure 2.48.

(i) The first designs the latitude and longitude lines in a central plane, resulting concentric circles from the small circles, while having the large circles projected as diameter of the primitive circle, Figure 2.48-left. In any case, this projection is not practical because the North-South axis tilt is not take into account when the poles in question are aligned on the same large circle.

(ii) Conversely, if we take into account the poles aligning operation, the projection of the spherical grid can be considered *after* the

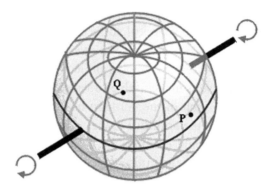

FIGURE 2.47   The spherical grid and the operation of poles' aligning; after Stereographic Projection (2003).

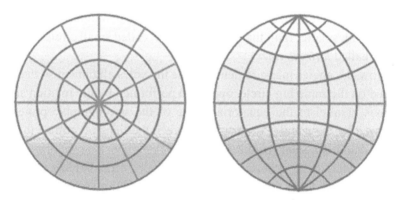

FIGURE 2.48   The concentric projection of the spherical grid (left) and on the Wulff map (right); after Stereographic Projection (2003).

projection sphere is rotated so that North-South axis to become horizontal (from vertical), resulting the so-called *Wulff map*, Figure 2.48-right.

In the Wulff map the latitude lines are horizontal arcs, while the longitudinal lines are vertical arcs connecting the North and South poles. The complete conjunction of the stereographic projection with the Wulff map is made considering the (Wulff) circle of the same radius with the projection sphere from the stereographic projection.

Usually the Wulff map is calibrated in angular intervals of 2° or 5°. Finally, by counting the lines of latitude between the poles aligned on the same longitudinal arc, the angle between them is measured.

In the Figure 2.49 the standard stereographic projection of the cubic crystal is considered oriented so that the direction [001] (thus along 0z) includes the North Pole of the sphere of projection. Only the information indicated in the northern hemisphere projection are required, those of the southern hemisphere being the duplicate with the opposite sign.

Worth remembering the fact that only for the cubic system the [hkl] directions are perpendicular to the (hkl) planes, so that the square brackets or round brackets can be omitted from the projections made. The standard projection indicates, in a clearly manner, all the important planes in the crystal and takes into account some rules, as follows.

(i)   the normal lines to the planes belonging to the same areas are coplanar and at right angles to the area axis.

(ii) all poles of the planes in the same area are on the arc (the projection or the trace) of the same large circle, whose pole is just the projection of the direction of the zone axis.

(iii) the planes' indices of a certain zone are correlated with the indices of the axis of the area in focus, by the Weiss zone law, see equation (2.59).

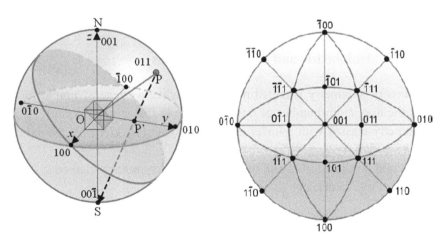

FIGURE 2.49 The standard projection of a cubic crystal: detail for the 001 plan projection (left) and of other planes and directions (right); after Stereographic Projection (2003).

(iv)   the indices of the planes belonging to the same area are related by linear relations, such as (2.70).

(v)   the directions indices are always presented in the lowest ratio $h{:}k{:}l$ of integers.

(vi)   the indices locations of stereographic projection must comply with the symmetry of perpendicular axes to the planes where were designed: see, for example, in Figure 2.49, how the symmetry of projection 001 is of 4th order, and how the projection 011 shows a symmetry of 2nd order in rotation in relation with the center of the Wulff map.

The last idea is crucial: the purpose of introducing the spherical and stereographic projection is to systematize in a planar representation the relations of tri-dimensional symmetry existing between the planes (associated to the faces) and between the important directions into a crystal.

2.5.7 THE CRYSTALLOGRAPHIC GROUPS

By comparison between various classes of a crystallographic system one notes that in each system there is a class which presents all the symmetry elements that can be associated according to the rules combination; the present discussion follows (Chiriac-Putz-Chiriac, 2005).

2.5.7.1 Holohedric and Merihedric Classes

A class of symmetry is called holohedric, when its general simple form has the maximum number of possible faces in this system in cause. The symmetry classes with a lower number of symmetry elements are called as *merihedric*.

Based on the number and types of deficient symmetry elements, three orders of merihedric are distinguished, as follows.

(i)   *hemihedric class* – with symmetry formula obtained by deleting one independent element of binary symmetry in the formula of holohedric class. The general simple form of the new class will be half of the number of faces of general simple shapes of holohedric class.

(ii) *tetartohedric class* – with formula of symmetry obtained by removing two independent elements of binary symmetry in the formula of holohedric class. The general simple form will have 1/4 of the number of the simple general faces of holohedric class.

(iii) *ogdohedric class* – with formula obtained by removing three independent elements of binary symmetry in the formula of holohedric class.

The general simple form of the new class will be 1/8 of the number of the simple general faces of holohedric class.

The three merihedric classes are exemplified in Figure 2.50. In fact, all the merihedric classes can be considered as hemihedric of a holohedric class, or of other hemihedric or even of a tetartohedric.

Based on the elements nature of binary symmetry kept constant in the formula of merihedric class, the following cases are distinguished:

(i) *holoaxial hemihedric* – when the classes in focus have only symmetry axes.

(ii) *anti-hemihedric* or *antimorphic hemihedric* – when the symmetry classes have only polar axes and symmetry planes parallel to these

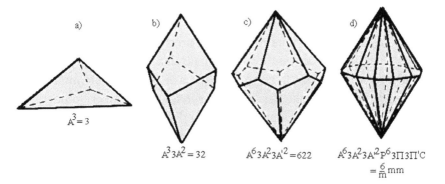

FIGURE 2.50 Examples of holohedric and merihedric with the three orders: (a) ogdohedric – trigonal pyramid; (b) tetartohedric – trigonal trapezohedron; (c) hemihedric – hexagonal trapezohedron; (d) holohedric – dihexagonal bipyramid, after Chiriac-Putz-Chiriac (2005).

axes (the polar axis means a symmetry axis connecting two non-equivalent geometry elements of a crystalline polyhedron). Most of anti-hemihedric forms are hemimorphic that means they present halves of holoedric forms.

(iii) *para-hemihedric* or *paramorphic hemihedric* – when the symmetry classes have a main plan of symmetry and center of symmetry.

Taking into account that a crystallographic face can have seven positions in relation to the crystallographic axes (see Figure 2.37) there results at most seven simple forms can be found for the systems with non-equivalent axes ($a{\neq}b{\neq}c$) on a given combination (thus, for a certain formula of symmetry).

For the medium and superior syngonies the identity of some axes ($a{=}b$ or $a{=}b{=}c$) makes that the ($h00$), ($k00$), and ($l00$) positions to become equivalent with ($h0l$), ($0kl$) and ($hk0$), respectively.

However, the number of the simple possible formulae for a combination still remains 7, as long as the equality of the axes distinguishes between the positions with equal indices and those of different indices ($hhl{\neq}hkl$).

2.5.7.2   The Crystallographic Forms

Crystallographic systems, symmetry classes and spatial groups are the general categories including and allowing the classification of the set of crystalline, natural and synthetic species. The same formula of symmetry may correspond to different polyhedra characterized by faces of different shapes, whose number and also the angle that closes them varies from polyhedron to another. All the possible faces of a crystal, determined by its symmetry formula, make the *crystallographic form*. The crystallographic forms can be simple or complex/composed. *The simple forms* consist of equivalent faces, i.e., all the forms can be obtained from one single face through multiplying it by the elements of symmetry. *Composed forms* (or combinations) consist of several simple crystallographic forms.

From the law of symmetry there results that a composed form is containing as many simple shapes as the types of faces are presented, i.e., the faces of a simple form on a combination are identical to each other. The simple forms may be open or closed, if they close or not a space (i.e., are capable or not of an independent geometric existence), respectively.

The open simple forms are classified as follows.

(a) Pedion – a crystal face that is not multiplied by any symmetry element;

(b) Pinacoid – 2 equivalent faces, symmetrical toward a center.

(c) Dome – 2 symmetrical faces toward a plane of symmetry and inclined to it.

(d) Sphenoid – 2 symmetrical sided towards a digyre and inclined to it.

(e) Pyramid – 3 or more equivalent faces equivalent symmetric toward a symmetry axis and inclined to it.

(f) Prism – 3 or more equivalent faces that overlapped by parallel edges.

The open simple forms are shown in Figure 2.51.

The simple forms differ from each other by the face position in relation to the symmetry's elements of the class to which they belong.

A crystal face can have an oblique position in relation to all the elements of symmetry. In this case, the faces' number of the corresponding simple shape will be maximum inside a given class, since one face will be multiplied by all elements of symmetry. This form it is called *general simple form* of the class, the name of the form also representing the name of the corresponding symmetry class. Unlike the general simple form with the indices (*hkl*) the other forms in a class occupy certain particular positions towards the symmetry axes (perpendicular, parallel or symmetric) being called as particular simple forms. The *closed simple form*

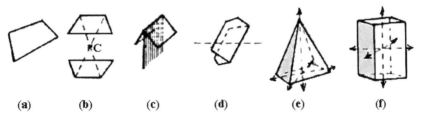

| (a) | (b) | (c) | (d) | (e) | (f) |

FIGURE 2.51 Open simple forms: (a) pedion; (b) pinacoid; (c) dome; (d) sphenoid; (e) pyramid; (f) prism.

completely circumvents the crystalline space with identical faces, forming a simple crystal, Figure 2.52.

In order to differentiate the simple forms among them one seems necessary that in addition of the symmetry formula and the number of equivalent faces the indices of the faces and the order of the form to be established, i.e., the position of the face in relation with the chosen system of coordinates. The system of coordinates used in geometric crystallography consists of three axes with axial and angular relationships as included in Table 2.2. Since the equilateral triangle and the regular hexagon shown in the plane of figure three equivalent directions, yet distanced at 120° for the trigonal and hexagonal systems, the coordinate system consists of four axes, where three are identical and situated in the same plane with the fourth (c axis) as perpendicular to the plane formed by the first 3 (Figure 2.44).

Coordinate axes 0x, 0y, 0z are oriented, by convention, as follows: 0x axis towards the observer and with positive way to it, 0y axis parallel to the observer and with positive way to the right and 0z axis in the vertical direction and with upward positive way. For the trigonal and hexagonal systems, the directions and the positive ways of the axes 0x, 0y, 0z are taken as before, while for the 0d axis (in Figure 2.44) the negative way is between the positive ways of the axes 0x and 0y. The assignment of

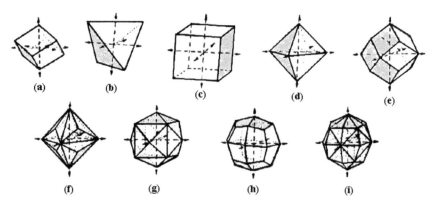

FIGURE 2.52 Simple closed forms: (a) rhombohedron; (b) tetrahedral; (c) cube; (d) octahedron; (e) dodecahedron; (f) triakis-octahedron; (g) cube pyramid; (h) trapezohedron; (i) hexakis-octahedron.

the indices and of the order for the simple forms presented on a crystal combination requires the orientation of the form in relation to crystallographic axes. For the simple forms orientation, the following rules are taking into account:

1. For the cubic system the crystallographic axes coincide with the 3-axes A^4, and in their absence with the three of those $4A^2$.
2. For the trigonal systems, tetragonal and hexagonal, the main symmetry axis is taken in coincidence with the crystallographic c axis, and the remaining axes (a = b) in a perpendicular plane – and coincide with the axes A^2 (if existing).
3. For the rhombic system, the crystallographic axes are taken into coincidence with the axes A^2. As an observation, for the four systems presented in the previous (2) and (3) step rules, for some hemimorphic forms, the axes choosing is performed as shown above, imagining that there is also the lower half.
4. For the monoclinic system, the crystallographic b axis is taken into coincidence with the axis A^2, and in its absence as perpendicular to the plane of symmetry. The other two axes are taken in the plane of symmetry (and in its absence in a plane perpendicular to the b axis) parallel with two more developed edges.
5. For the triclinic system, the crystallographic axes are chosen after three directions parallel to the more developed three edges.

For the oblique faces respecting to the crystallographic axes, an expeditious way for finding the indices is the visually assessment of the inclination of the face investigated towards the crystallographic axes. The visual assessment of the indices is based on the following rule: as much as bigger the inclination of the face (i.e., as much as the face intersects the axis considered to a lower distance), the higher the index will be. Basically, one proceeds as following: on the investigated surface a pencil is placed orienting the vertex respecting the considered axis; the size of the distance to which the imaginary extension of the pencil (i.e., extending face) intercepts the axis is estimated; the same operation is repeated for all the axes, resulting three segments where the longest corresponds to the lowest index. For the reference system with four axes, in Miller-Bravais indexing, there is not necessary to estimate the indices in relation to the three horizontal axes, because, taking into account the direction established by convention for d axis, the i-Miller index on this axis satisfies the relation (2.78):

$$h + i + k = 0 \qquad (2.82)$$

There is very important to establish correctly the indices in order to be identified the simple forms. For example, in the cubic system in each class we have at least two forms of the same number of faces which are equally inclined in relation with two axes and different towards the third one.

The differentiation is based on the assessment of the relative size of the third interception towards the two equal ones. If it is lower, then we have indices such as (112) and like (221) if it is higher.

Composed crystals often occur in nature, consisting of several simple crystallographic forms, in cases when constituent faces may have different developments, in terms of geometry and size. To treat the *combined forms*, one more developed face of a simple form is chosen as fundamental face, while the obtained segments are referred as above described at the interceptions of this face. To identify the simple forms, on a combination, there is not necessary to establish the indices of each face, but there is sufficient to consider the faces that appear in an octant. However, one always chooses the positive octant, and if in the same octant more equal faces appear, the closest face to the crystallographic a-axis is considered. The indices of this face are representative for the entire simple form as long as by a simple permutation and corresponding change of the indices' sign the indices of all the faces are obtained.

To indicate the representation of a simple form by indices, braces instead of parentheses are used for the notation of simple faces. For example {111} octahedron indicates the face (111) of octahedron of the positive octant, while {110} rhomboidal dodecahedron indicates the face (110) of rhomboidal dodecahedron of positive octant.

In the middle and lower syngonies, the different position of the representative face of a simple form (and hence of all other faces) in relation to the crystallographic axes are indicated by the order of the simple form in focus. Further on, the indices and the corresponding orders on crystallographic systems are presented. For the triclinic, monoclinic and rhombic systems, the pedions and pinacoids which are parallel with two axes are named as:

(100) – first pinacoid (pedion)
(010) – second pinacoid (pedion)
(001) – third pinacoid (pedion)

For the forms that are oblique towards the crystallographic axes the following denoting is in use:

($0kl$) first order (pinacoid, pedion, sphenoid, dome, prism)
($h0l$) second order (pinacoid, pedion, sphenoid, dome, prism)
($hk0$) third order (pinacoid, pedion, sphenoid, dome, prism)
(hkl) fourth order (pinacoid, pedion, sphenoid, dome, prism, except the rhombic system)

Regarding the trigonal and hexagonal systems, the order of simple forms (prisms, pyramids, bipyramids and rhombohedrics) is specify interested only towards the three equal axes of the horizontal plane, as:

($10\bar{1}0$) first order (trigonal or hexagonal prism)
($11\bar{2}0$) second order (trigonal or hexagonal prism)
($hk\bar{i}0$) third order (trigonal or hexagonal prism)
($h0\bar{h}l$) first order (pyramid, trigonal or hexagonal bipyramid, rhombohedron)
($h0\bar{2}l$) second order (pyramid, trigonal or hexagonal bipyramid, rhombohedron)
($hk\bar{i}l$) third order (pyramid, trigonal or hexagonal bipyramid, rhombohedron).

In the tetragonal system, the order is also specified only by the indices in relation to the two equal axes of the horizontal plane:

(110) prism of first order
(hhl) bisphenoids, tetragonal pyramids, enoids, tetragonal pyramids, tetragonal bipyramids
(100) prism of second order
($h0l$) bisphenoids, tetragonal pyramids, tetragonal bypiramids.

(*hk*0) prism of third order

(*hkl*) bisphenoids, tetragonal pyramids, tetragonal bypiramids.

All these informations will be centralized for the presentation of the correlated crystallographic classes and systems in the next section.

2.5.7.3   Correlated crystallographic systems and classes

Upon the classification in terms of primitives, which have been generated the seven crystalline systems, the enlargement of this perspective at the level of non-primitive, had generated the 14 Bravais networks, while the extension of the crystallographic perspective until the level of all the possible combinations between the symmetry's elements into a crystal, which, reflecting the set of possible relations of symmetry into the crystalline structures at the level of planes, directions and crystalline faces, generated the 32 crystallographic classes; there is now the time to put them in a correlated relation also with the geometry of the crystalline shapes. Thus, in Table 2.8 all these information for each of the 32 point groups are synthesized.

The classification of Table 2.8 by its double nature, for the tridimensional and bidimensional representations – by stereographic projection, wholly accomplishes the main purpose of this chapter: the unification of the crystalline description at the level of faces (from outside crystal) with those of the unit cell (from inside crystal), correlated as follows:

(i)   Each group is numbered and indexed as based on the international notation and using its symmetry formula.

(ii)   The class name is specified by rendering the model of the combined crystalline shape by detailed description of the simple forms of the faces that its possesses. The list of the forms presented in the class in focus is rendered in terms of Miller and Miller-Bravais indices, thus cumulating the types of allowed combinations. The

TABLE 2.8 Illustration of the Point Groups, Tridimensional and Stereographic Description for the Representative Faces-Planes (100), (010), (104), (024), and (214) for the Crystallographic Classes and Groups, with Examples*

Group	Class	Stereographic Projection	Crystals' Compounds
Triclinic System			
1: 1 A¹	Pedion (Monoedric)		Thiosulphate of Calcium $CaS_2O_3 \times 6H_2O$ Rubidium Ferrocyadine $Rb_4Fe(CN)_6 \cdot 2H_2O$

	(hkl)	Pedion 1
	(0kl)	Pedion 1
	(h0l)	Pedion 1
	(hk0)	Pedion 1
	(100)	Pedion 1
	(010)	Pedion 1
	(001)	Pedion 1

pedion

Group	Class	Stereographic Projection	Crystals' Compounds
2: ī C	Pinacoid		Copper Sulphate $CuSO_4 \cdot 5H_2O$, Nitrat de Bismut: Bismuth nitrate $Bi(NO_3)_3 \cdot 9H_2O$ Kianita: Kyanite Al_2SiO_5

	(hkl)	Pinacoid 2
	(0kl)	Pinacoid 2
	(h0l)	Pinacoid 2
	(hk0)	Pinacoid 2
	(100)	Pinacoid 2
	(010)	Pinacoid 2
	(001)	Pinacoid 2

pinacoid

Group	Class	Stereographic Projection	Crystals' Compounds
Monoclinic System			
3: 2 A²	Sphenoid (dihedric with axes)		Lithium Sulphate $Li_2SO_4 \cdot H_2O$ Halotrichite $FeAl_2(SO_4)_2 \cdot 22H_2O$ Sucrose $C_{12}H_{22}O_{11}$

	(hkl)	Sphenoid 2
	(0kl)	Sphenoid 2
	(h0l)	Pinacoid 2
	(hk0)	Sphenoid 2
	(100)	Pinacoid 2
	(010)	Pedion 1
	(001)	Pinacoid 2

dihedral (sphenoid)

TABLE 2.8 Continued

Group	Class	Stereographic Projection	Crystals' Compounds

4:

$m = \overline{2}$

P

Dome

(hkl)	Dome 2
(0kl)	Dome 2
(h0l)	Pedion 1
(hk0)	Dome 2
(100)	Pedion 1
(010)	Pinacoid 2
(001)	Pedion 1

dihedral (dome)

Sodium silicate Na_2SiO_3

Potassium nitrate KNO2

Hilgardite Ca8B18O33Cl4· 4H2O

5:

$\dfrac{2}{m}$

A^2PC

Prism

(hkl)	Prisms Ord. IV 4
(0kl)	Prisms Ord. I 4
(h0l)	Prisms Ord. II 4
(hk0)	Prisms Ord. III 4
(100)	Pinacoid 2
(010)	Pinacoid 2
(001)	Pinacoid 2

IV order prism

Potassium chlorate: $KClO_3$

Criolite: Na_3AlF_6

Anthracene: $C_{14}H_{10}$

Orthorhombic System

6:

222

$3A^2$

Rhombo-bisphenoid (Rombo-tetrahedric)

(hkl)	Bisphenoid 4
(0kl)	Prism Ord. I 4
(h0l)	Prism Ord. II 4
(hk0)	Prism Ord. III 4
(100)	Pinacoid 2
(010)	Pinacoid 2
(001)	Pinacoid 2

bisphenoid

Magnesium sulphate $MgSO_4 \cdot 7H_2O$

Calcomenite $CuSeO_3 \cdot 2H_2O$

Barium formate $Ba(HCOO)_2$

TABLE 2.8 Continued

Group	Class	Stereographic Projection	Crystals' Compounds

7:
mm2
$A^2 2P$

Rhombo-pyramid

(hkl)	Pyramid 4
(0kl)	Dome 2
(h0l)	Dome 2
(hk0)	Prism Ord. III 4
(100)	Pinacoid 2
(010)	Pinacoid 2
(001)	Pedion 1

pyramid

Bismuth Tycianate
$Bi(CNS)_3$
Sortit:
$Na_2Ca_2(Co_3)_3$
Resorcinol:
$C_6H_4(OH)_2$

8:
$\dfrac{2}{m} \dfrac{2}{m} \dfrac{2}{m}$
$3A^2 3PC$

Rhomobo-bipyramidal

(hkl)	Bipyramid 8
(0kl)	Prism 4
(h0l)	Prism 4
(hk0)	Prism 4
(100)	Pinacoid 2
(010)	Pinacoid 2
(001)	Pinacoid 2

bipyramid

Potassium Tycianate
KCNS
Stibnite Sb_2S_3
Oxalic acid
$(COOH)_2$

Rhombohedral-Trigonal System

9:
3
A^3

Trigonal-pyramid

(hkīl)	Trigonal pyramid 3
(h0h̄l)	Trigonal pyramid 3
(hh2h̄l)	Trigonal pyramid 3
(hkī0)	Trigonal pyramid 3
(101̄0)	Trigonal prism 3
(112̄0)	Trigonal prism 3
(0001)	Pedion 1

trigonal pyramid

Magnesium sulphite
$MgSO_3 \cdot 6H_2O$
Sodium Periodate
$NaIO_4 \cdot 3H_2O$

TABLE 2.8   Continued

Group	Class		Stereographic Projection	Crystals' Compounds
10: $\bar{3}$ A^3C	Rhombohedric			Ilmenite $FeTiO_3$ Dolomite $CaMg(CO_3)_2$
	$(hk\bar{i}\,l)$	Rhombohedron 6		
	$(h0\bar{h}\,l)$	Rhombohedron 6		
	$(hh\overline{2h}\,l)$	Rhombohedron 6		
	$(hk\bar{i}\,0)$	Hexagonal prism 6		
	$(10\bar{1}0)$	Hexagonal prism 6		
	$(11\bar{2}0)$	Hexagonal prism 6		
	(0001)	Pinacoid 2	rhombohedron	
11: 32 A^33A^2	Trigonal-trapezohedron			Potassium ditionate $K_2S_2O_6$ Alpha-quartz SiO_2 Cinnabar HgS
	$(hk\bar{i}\,l)$	Trigonal trapezohedron 6		
	$(h0\bar{h}\,l)$	Rhombohedron 6		
	$(hh\overline{2h}\,l)$	Trigonal bipyramid 6		
	$(hk\bar{i}\,0)$	Ditrigonal prism 6		
	$(10\bar{1}0)$	Hexagonal prism 6		
	$(11\bar{2}0)$	Trigonal prism 3		
	(0001)	Pinacoid 2	trigonal trapezohedron	
12: $3m$ A^33P	Ditrigonal-pyramid			Potassium bromate $KBrO_3$ Proustite Ag_3AsS_3
	$(hk\bar{i}\,l)$	Ditrigonal pyramid 6		
	$(h0\bar{h}\,l)$	Ditrigonal pyramid 3		
	$(hh\overline{2h}\,l)$	Hexagonal pyramid 6		
	$(hk\bar{i}\,0)$	Ditrigonal prism 6		
	$(10\bar{1}0)$	Trigonal prism 3		
	$(11\bar{2}0)$	Hexagonal prism 6		
	(0001)	Pedion 1	ditrigonal pyramid	
13: $\bar{3}\dfrac{2}{m}$ A^33A^23PC	Ditrigonal-scalenohedric			Corundum: Al_2O_3 Calcite: $CaCO_3$
	$(hk\bar{i}\,l)$	Ditrigonal scalenohedric 12		
	$(h0\bar{h}\,l)$	Rhombohedron 6		
	$(hh\overline{2h}\,l)$	Hexagonal bipyramid 12		
	$(hk\bar{i}\,0)$	Dihexagonal prism 12		
	$(10\bar{1}0)$	Hexagonal prism 6		
	$(11\bar{2}0)$	Hexagonal prism 6		
	(0001)	Pinacoid 2	ditrigonal scalenohedron	

TABLE 2.8 Continued

Group	Class	Stereographic Projection	Crystals' Compounds

Tatragonal System

14:
4
A^4

Tetragonal-pyramid

(hkl)	Quadratic pyramid 4
(0kl)	Quadratic pyramid 4
(h0l)	Quadratic pyramid 4
(hk0)	Quadratic prism 4
(100)	Quadratic prism 4
(010)	Quadratic prism 4
(001)	Pedion 1

tetragonal pyramid

Tartrate of Barium Antimony $Ba(SbO)_2$ $(C_4H_2O_6)_2 \cdot H_2O$

Metaldehide

$(CH_3CHO)_4$

15:
$\bar{4}$
A_4^2

Tetragonal-tetrahedric

(hkl)	Quadratic bisphenoid 4
(0kl)	Quadratic bisphenoid 4
(h0l)	Quadratic bisphenoid 4
(hk0)	Quadratic prism 4
(100)	Quadratic prism 4
(010)	Quadratic prism 4
(001)	Pinacoid 2

quadratic bisphenoid
(tetragonal thetrahedral)

Boron phosphate BPO_4

Canite:

$Ca_2B(AsO_4)(OH)_4$

16:
$\frac{4}{m}$
$A^4\Pi C$

Tetragonal-bipyramid

(hkl)	Quadratic bipyramid 8
(0kl)	Quadratic bipyramid 8
(h0l)	Quadratic bipyramid 8
(hk0)	Quadratic prism 4
(100)	Quadratic prism 4
(010)	Quadratic prism 4
(001)	Pinacoid 2

quadratic bipyramid

Schelite: $CaWO_4$

Lupinin nitrate $C_{10}H_{19}ON \cdot HNO_3$

17:
422
$A^4 4A^2$

Tetragonal-trapezohedron

(hkl)	Quadratic trapezohedron 8
(0kl)	Quadratic bipyramid 8
(h0l)	Quadratic bipyramid 8
(hk0)	Ditetragonal prism 8
(100)	Quadratic prism 4
(010)	Quadratic prism 4
(001)	Pinacoid 2

quadratic trapezohedron

Nickel sulphate $NiSo_4 \cdot 6H_2O$

Ammonia and methyl iodide $NH_3(CH_3)I$

TABLE 2.8 Continued

Group	Class	Stereographic Projection	Crystals' Compounds
18: *4mm* A⁴4P	Ditetragonal-pyramid 	 ditetragonal pyramid	Silver fluoride AgF·H₂O Diabolite Pb₂Cu(OH)₄Cl₂

For Group 18 inner table:

(*hkl*)	Ditetragonal pyramid 8
(0*kl*)	Quadratic pyramid 4
(*h0l*)	Quadratic pyramid 4
(*hk*0)	Ditetragonal prism 8
(100)	Quadratic prism 4
(010)	Quadratic prism 4
(001)	Pedion 1

For Group 19:

19: $\overline{4}2m$ $A_4^2 2A^2 2P$ — Class: Tetragonal-scalenohedron

(*hkl*)	Quadratic scalenohedron 8
(0*kl*)	Quadratic bisphenoid 4
(*h0l*)	Quadratic bipyramid 8
(*hk*0)	Ditetragonal prism 8
(100)	Quadratic prism 4
(010)	Quadratic prism 4
(001)	Pinacoid 2

quadratic scalenohedron

Crystals' Compounds: Mercuric cyanide Hg(CN)₂; Urea: CO(NH₂)₂

For Group 20:

20: $\frac{4}{m}\frac{2}{m}\frac{2}{m}$ $A^4 4A^2 4P\Pi C$ — Class: Ditetragonal-bipyramid

(*hkl*)	Ditetragonal bypiramid 16
(0*kl*)	Quadratic bipyramid 8
(*h0l*)	Quadratic bipyramid 8
(*hk*0)	Ditetragonal prism 8
(100)	Quadratic prism 4
(010)	Quadratic prism 4
(001)	Pinacoid 2

ditetragonal bipyramid

Crystals' Compounds: Rutile: TiO₂; Zircon: ZrSiO₄

Hexagonal System

For Group 21:

21: 6 A⁶ — Class: Hexagonal-pyramid

(*hki̅l*)	Hexagonal pyramid 6
(*h0h̅l*)	Hexagonal pyramid 6
(*hh2̅hl*)	Hexagonal pyramid 6
(*hki̅*0)	Hexagonal prism 6
(101̅0)	Hexagonal prism 6
(112̅0)	Hexagonal prism 6
(0001)	Pedion 1

hexagonal pyramid

Crystals' Compounds: Sulphate of Potassium Lithium LiKSO₄; Iodoform: CHI₃

TABLE 2.8 Continued

Group	Class	Stereographic Projection	Crystals' Compounds

22:

$\overline{6} = \dfrac{3}{m}$

A^3P

Trigonal-bipyramid

$(hk\bar{i}l)$	Trigonal bipyramid 6
$(h0\bar{h}l)$	Trigonal bipyramid 6
$(hh\overline{2h}l)$	Trigonal bipyramid 6
$(hk\bar{i}0)$	Ditrigonal prism 6
$(10\bar{1}0)$	Trigonal prism 3
$(11\bar{2}0)$	Trigonal prism 3
(0001)	Pinacoid 2

trigonal bipyramid

Unknown

23:

$\dfrac{6}{m}$

$A^6\Pi C$

Hexagonal-bipyramid

$(hk\bar{i}l)$	Hexagonal bipyramid 12
$(h0\bar{h}l)$	Hexagonal bipyramid 12
$(hh\overline{2h}l)$	Hexagonal bipyramid 12
$(hk\bar{i}0)$	Hexagonal prism 6
$(10\bar{1}0)$	Hexagonal prism 6
$(11\bar{2}0)$	Hexagonal prism 6
(0001)	Pinacoid 2

hexagonal bipyramid

Lanthanum Sulphate $La_2(SO_4)_3 \cdot 9H_2O$

Apatite: (CaF) $Ca_4(PO_4)_3$

24:

622

A^66A^2

Hexagonal-trapezohedron

$(hk\bar{i}l)$	Hexagonal trapezohedron 12
$(h0\bar{h}l)$	Hexagonal bipyramid 12
$(hh\overline{2h}l)$	Hexagonal bipyramid 12
$(hk\bar{i}0)$	Dihexagonal prism 12
$(10\bar{1}0)$	Hexagonal prism 6
$(11\bar{2}0)$	Hexagonal prism 6
(0001)	Pinacoid 2

hexagonal trapezohedron

Lithium Iodate $LiIO_3$

Quartz-beta SiO_2

Kalsilite $KAlSiO_4$

25:

$6mm$

A^66P

Dihexagonal-pyramid

$(hk\bar{i}l)$	Dihexagonal pyramid 12
$(h0\bar{h}l)$	Hexagonal pyramid 6
$(hh\overline{2h}l)$	Hexagonal pyramid 6
$(hk\bar{i}0)$	Dihexagonal prism 12
$(10\bar{1}0)$	Hexagonal prism 6
$(11\bar{2}0)$	Hexagonal prism 6
(0001)	Pedion 1

dihexagonal pyramid

Wurtzite: ZnS

Cloride of Triethyl ammonium $NH(C_2H_5)_3Cl$

Cadmium Selenide CdSe

Magnesium Telluride MgTe

TABLE 2.8 Continued

Group	Class	Stereographic Projection		Crystals' Compounds
26: $\bar{6}m2 = \dfrac{3}{m}m2$ $A^3 3A^2 3P\Pi$	Ditetragonal-bipyramid			Benitoite $BaTiSi_3O_9$
	(hkl)	Ditetragonal bypiramid 16		
	$(0kl)$	Quadratic bipyramid 8		
	$(h0l)$	Quadratic bipyramid 8		
	$(hk0)$	Ditetragonal prism 8		
	(100)	Quadratic prism 4		
	(010)	Quadratic prism 4		
	(001)	Pinacoid 2		

ditetragonal bipyramid

Group	Class	Stereographic Projection		Crystals' Compounds
27: $\dfrac{6}{m}\dfrac{2}{m}\dfrac{2}{m}$ $A^6 6A^2 6P\Pi C$	Dihexagonal-bipyramid			Beryl: $Be_3Al_2Si_6O_{18}$ Crystals of: Zn, Cd, Mg
	$(hki\,l)$	Dihexagonal bipyramid 24		
	$(h0\bar{h}l)$	Hexagonal bipyramid 12		
	$(hh\overline{2h}l)$	Hexagonal bipyramid 12		
	$(hk\bar{i}0)$	Dihexagonal prism 12		
	$(10\bar{1}0)$	Hexagonal prism 6		
	$(11\bar{2}0)$	Hexagonal prism 6		
	(0001)	Pinacoid 2		

dihexagonal bipyramid

Cubic System

Group	Class	Stereographic Projection		Crystals' Compounds
28: 23 $3A^2 4A^3$	Tri-tetrahedral			Sodium Chlorate $NaClO_3$ Sodium uranyl acetate: $Na(UO_2)(C_2H_3O_2)_3$
	(123)	Tetrahedral pentagonal dodecahedron 12		
	(221)	Dodecahedron deltoid 12		
	(211)	Triakis tetrahedron 12		
	(111)	Tetraedron 4		
	(120)	Pentagonal dodecahedron 12		
	(110)	Rhombic dodecahedron 12		
	(100)	Cube 6		

tetratoid
(pentagonal tetrahedral
dodecahedron)

Group	Class	Stereographic Projection		Crystals' Compounds
29: $\dfrac{2}{m}\bar{3}$ $3A^2 4A^3 3\Pi C$	Di-dodecahedron			Pyrite: FeS_2 Calcium Nitrate: $Ca(NO_3)_2$ Bixbite: $(Fe, Mn)_2O_3$
	(123)	Diakis dodecahedron 24		
	(221)	Triakis octahedron 24		
	(211)	Deltoidikosi tetrahedron 24		
	(111)	Octahedron 8		
	(120)	Pentagonal dodecahedron 12		
	(110)	Rhombic dodecahedron 12		
	(100)	Cube 6		

didodecahedron
(diakisdodecahedron)

TABLE 2.8 Continued

Group	Class		Stereographic Projection	Crystals' Compounds
30: 432 $3A^4A^3\,6A^2$	Tri-octhaedric			Unknown
	(123)	Icositetrahedron pentagonal (plagiedron) 24		
	(221)	Triakis octahedron 24		
	(211)	Deltoidikosi tetrahedron 24		
	(111)	Octahedron 8		
	(120)	Tetrakishexaedron 24		
	(110)	Dodecahedron rhomboidal 12		
	(100)	Cube 6	trioctahedral pentagon (pentagonal icositetrahedron) (plagiedron)	
31: $\bar{4}\,3m$ $3\,A_4^2\,4A^3 6P$	Hexa-tetrahedral			Sphalerite: ZnS Copper clorure: CuCl Tetrahedrite: Cu_3SbS_2 Cadmiu Telluride CdTe Selenide of Zinc: ZnSe
	(123)	Hexakistetrahedron 24		
	(221)	Dodecahedron deltoidal 12		
	(211)	Triakistetrahedron 12		
	(111)	Tetrahedron 4		
	(120)	Tetrakishexahedron 24		
	(110)	Dodecahedron rhomboidal 12		
	(100)	Cube 6	hexatetrahedron (hexakistetrahedron)	
32: $\dfrac{4}{m}\,\bar{3}\,\dfrac{2}{m}$ $3A^4 4A^3$ $6A^2 6P3\Pi C$	Hex-octahedral			Sodium chloride: NaCl Magnesia: MgO Diamond: C Crystals of: Ag, Au, Fe Hexamethylene-tetramine: $(CH_2)_6N_4$ Magnesium selenide: MgSe Lead Telluride: PbTE Garnetes: R3''R2''' $(SiO_4)_3$ Spinelli: R''R2'''O_4
	(123)	Hexakisoctahedron 48		
	(221)	Triakisoctahedron 24		
	(211)	Deltoidikositetrahedron (leucitohedron) 24		
	(111)	Octahedron 8		
	(120)	Tetrakishexahedron 24		
	(110)	Dodecahedron rhomboidal 12		
	(100)	Cube (hexahedron) 6	hexoctahedron (hexakisocthaedorn)	

*After (American Mineralogist Crystal Structure Database, 2002, Wolfram Research Crystalography, 2003, Bilbao Crystallographic Server, 2003).

number of faces of a certain type which the form under concern is also, in parenthesis, indicated.

(iii) For each group the stereographic projection specific for the general shapes (faces) is presented by the established symbols. Thus, for the reflection axes there is noted by ϕ for the axis of second order, $\blacktriangle$ for the axis of third order, $\blacksquare$ for the axis of fourth order and $\spadesuit$ for the axis of sixth order, while for the associated roto-inversions we will have $\triangle$ for the $\bar{3}$ axis, $\boxtimes$ for the $\bar{4}$ axis and $\leftmoon$ for the $\bar{6}$ axis, the case of $\bar{2} = m$ axis being associated with a reflection plane.

(iv) the examples of crystalline compounds characterized by the information of points i)-iii) are specified, when available.

2.5.8 GROUP SYMMETRY PERTURBATION

Beyond the crystallographic classification in terms of symmetry elements and operations, they have also a directly influence towards the structural properties to which they are associated; the present discussion follows (Chiriac-Putz-Chiriac, 2005).

2.5.8.1 The Crystallographic Anisotropy

The symmetry-properties correlation is summarized by the *Neumann principle*, according which "the symmetry elements associated to any physical property of a crystalline system must contain the symmetry elements of the point group to which the crystal in focus belongs."

Therefore, any physical property of any crystal may be grouped, classified and finally identified based on the elements (and the operations) of symmetry corresponding to the crystallographic group or class to which the crystal belongs.

Simply formulated, the crystallographic groups (classes) not only classify the crystal, but also fix (any) physical properties. Thus, the physical properties can be described by the elements of symmetry (and also by the point groups associated) – yet what is interesting here is the fact that through the elements of symmetry (especially by axes and planes) *the physical properties will depend on the symmetry directions.*

This kind of feature, i.e., the dependence on direction, will be generally called as *anisotropy*, while the isotropy corresponds to the particular case of uniformity, by the direction non-dependency. Therefore, based on the directionality of a physical system (crystalline in this case) and based on the type of the existing elements of symmetry one will determine the anisotropy degree, the homogeneity or the lack in homogeneity and the distribution of the physical properties of that system, Figure 2.53.

The anisotropic character is generally associated with the combination of the elements of symmetry (thorough the 32 point groups) and not (necessarily) only with a plan or a symmetry axis; therefore, the associated physical properties can be expressed throughout a more general way of measure than the dependence along one direction or belonging to a certain plan.

The scalar measures (as density, mass, volume, temperature) may represent the examples of the isotropic properties, which characterize a system globally, regardless the direction of analysis. On the other hand, the vector measures (as the forces, the electrical field, and the fluxes) fix the distribution's orientation of that physical property on the vector direction, since they are already representing the anisotropic materials.

Furthermore, the physical properties dependence according to the combination of axes (directions), plans and the inversion center can be expressed through more general measures than the vectors. These measures, called as *tensors*, are physical – mathematical objects which express

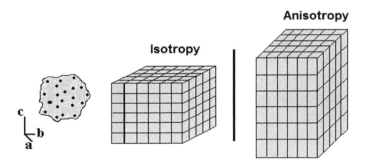

FIGURE 2.53   Isotropy vs. anisotropy; after Goodhew (2003).

in the highest degree the anisotropic properties of the (crystalline) complex ordered matter.

Therefore, one can say that the scalar quantities are zero degree tensors (and are written in a simple way, as symbols), while the vectors are first degree tensors (and are represented throughout some symbols that use a sub-index in order to specify their direction in space).

Nevertheless, the tensors may be, generally, of different kinds, for each kind being possible to add a sub-index that is able – regardless the others – to indicate any direction in space. When tensors of different kinds combine themselves, in order to "descent" the resulted tensor, one should proceed to the "so called" *contraction* (or *diagonalization*) of the indexes.

For example, if by applying a mechanical pressure (represented by a second kind tensor, marked with σ_{jk}) to a physical system's ending faces an electric charge of the system will appear (represented through the electric induction, a vector or a first degree tensor, marked with D_i); then, the measure which correlates the applied cause (the pressure) – to the effect (the electric charge moment) produced, must be a third degree tensor (with three sub-indexes, marked with d_{ijk}); yet, through the product that has been applied or contracted with the pressure tensor, the apparition of the electric charge vector is generated:

$$D_i = d_{ijk}\sigma_{jk} \qquad (2.83)$$

The phenomenon which is described by the third degree tensor d_{ijk} is called piezo-electric effect and it is manifested, for instance, by many quartz crystals, tourmaline with the following formula: (Na,Ca) $(Al,Fe,Li,Mg)_3Al_6(BO_3)_3(Si_6O_{18})$ and others.

Anyway, if a charging effect at the temperature variation (represented by scalar T) of the physical system appears at that moment, the cause (T) – effect (D_i) correlation will be made throughout a tensor, but of a first degree (marked with p_i), the described effect being called pyro-electricity:

$$D_i = p_i T \qquad (2.84)$$

Also in more general way, if we consider as causes the cumulated (infinitesimal) actions of a second degree scalar (e.g., the temperature T) with that one of a vector (e.g., the electric field symbolized with (E) with its components E_i and with the ones of a second degree tensor (e.g., the mechanical tension symbolized with [σ] with its' components σ_{ij}), then the registered (infinitesimal) effects will be able to jointly correlate with a scalar (as entropy, S), with a vector (as electric induction symbolized with (D) with its components D_i) and also with a second degree tensor (as the mechanical pressure symbolized with [ε] with its components ε_{ij}), through different combinations which involve other tensors having degrees varying from 0 to 4, each describing the corresponding effects, as represented in Table 2.9.

In Table 2.9 the tensorial definitions of the effects which connect the temperature variations, the electric field and the mechanical pressure have been introduced, around the equilibrium states of a certain physical system, with the thermodynamic (expanded) effect of the *state functions*

TABLE 2.9 Tensorial Variations of the Thermodynamic Equilibrium*

	T		*(E)*	*[σ]*
S	$dS = \left(\dfrac{\partial S}{\partial T}\right)_{[\sigma],(E)}$	$dT + \left(\dfrac{\partial S}{\partial E_i}\right)_{[\sigma],T}$	$dE_i + \left(\dfrac{\partial S}{\partial \sigma_{ij}}\right)_{(E),T}$	$d\sigma_{ij}$
	Heat capacity (C)	Electrocaloric effect (p_i)	Piezocaloric effect (α_{ij})	
(D)	$dD_i = \left(\dfrac{\partial D_i}{\partial T}\right)_{[\sigma],(E)}$	$dT + \left(\dfrac{\partial D_i}{\partial E_j}\right)_{[\sigma],T}$	$dE_j + \left(\dfrac{\partial D_i}{\partial \sigma_{jk}}\right)_{(E),T}$	$d\sigma_{jk}$
	Pyro-electricity (p_i)	Permittivity (κ_{ij})	Direct Piezoelectricity (d_{ijk})	
[ε]	$d\varepsilon_{ij} = \left(\dfrac{\partial \varepsilon_{ij}}{\partial T}\right)_{[\sigma],(E)}$	$dT + \left(\dfrac{\partial \varepsilon_{ij}}{\partial E_k}\right)_{[\sigma],T}$	$dE_k + \left(\dfrac{\partial \varepsilon_{ij}}{\partial \sigma_{kl}}\right)_{(E),T}$	$d\sigma_{kl}$
	Thermal expansion (α_{ij})	Reverse piezolectricity (d_{ijk})	Elasticity (s_{ijkl})	

*After Nye (1985).

(by quantities that depend only on the initial and final state of the evolving system) associated with entropy S, with the D_i -components of the electric induction vector (D), and with the ε_{ij} components of the mechanical tension tensor $[\varepsilon]$.

There can be observed the way the effects and the associated tensors are symmetrical towards the Table 2.9's main diagonal, thus identifying the actual tensors of the equivalent effects.

This fact can be easily proved as based on the state functions properties and also on the basic thermodynamic postulates.

One can start with the first law of the thermodynamics, which says that an internal energy variation of a system consists of the mechanical work variation (W) and of the heat variation (Q) which is realized/ received by the considered system

$$dU=dW+dQ \tag{2.85}$$

where, in this case:

$$dW=E_i\,dD_i + \sigma_{ij}\,d\varepsilon_{ij} \tag{2.86}$$

and:

$$dQ=TdS \tag{2.87}$$

as grounded on the second thermodynamic law.

If we continue to consider the state function $\Phi=\Phi(T,\,E_{ij},\,\sigma_{ij})$, this also admits the thermodynamically expansion (as an exact total differential) in the same dependence as those from Table 2.9:

$$d\Phi=\left(\frac{\partial\Phi}{\partial T}\right)_{[\sigma],(E)}dT+\left(\frac{\partial\Phi}{\partial E_i}\right)_{[\sigma],T}dE_i+\left(\frac{\partial\Phi}{\partial\sigma_{ij}}\right)_{(E),T}d\sigma_{ij} \tag{2.88}$$

In addition, if we consider the transformation:

$$\Phi=U-TS-E_jD_j-\sigma_{ij}\,\varepsilon_{ij} \tag{2.89}$$

then, through its total differentiation and by counting also the relations (2.85)–(2.87), we arrive at the following statement:

$$d\Phi = -SdT - D_i \, dE_i - \varepsilon_{ij} \, d\sigma_{ij} \tag{2.90}$$

from which, by identifying the corresponding terms with Eq. (2.88), the following statements also result:

$$-S = \left(\frac{\partial\Phi}{\partial T}\right)_{[\sigma],(E)} \, , \, -D_i = \left(\frac{\partial\Phi}{\partial E_i}\right)_{[\sigma],T} \, , \, -\varepsilon_{ij} = \left(\frac{\partial\Phi}{\partial\sigma_{ij}}\right)_{(E),T} \tag{2.91}$$

Finally, if we consider again the state function quality Φ, according to which the mixed derivatives of the second order are equal (the Cauchy condition of exact total difference),

$$\left(\frac{\partial^2\Phi}{\partial E_i \partial T}\right)_{[\sigma],(E)\backslash E_i} = \left(\frac{\partial^2\Phi}{\partial T\partial E_i}\right)_{[\sigma],(E)\backslash E_i} \tag{2.92}$$

the mixed identities can be formulated:

$$\left(\frac{\partial\varepsilon_{ij}}{\partial E_k}\right)_{[\sigma],T} = \left(\frac{\partial D_i}{\partial\sigma_{jk}}\right)_{(E),T}$$

$$\left(\frac{\partial\varepsilon_{ij}}{\partial T}\right)_{[\sigma],(E)} = \left(\frac{\partial S}{\partial\sigma_{ij}}\right)_{(E),T}$$

$$\left(\frac{\partial D_i}{\partial T}\right)_{[\sigma],(E)} = \left(\frac{\partial S}{\partial E_i}\right)_{[\sigma],T} \tag{2.93}$$

in agreement with the tensorial definitions equalization and of their corresponding effects in Table 2.9.

Yet, from a practical point of view, because of the fact that the quantitative values registered by the described effects by the tensors of the at least main order (the caloric capacity being excluded) are very low, as it is illustrated in Table 2.10, the existence or the appearance of such an effect can not be conclusive, as under no circumstances as an unique criterion to establish the existence of a certain type (class) of anisotropy should be considered.

Moreover, for the piezo- and pyro-electric effect, the physical nature (i.e., the appearance at the spatial ends of the system in focus, and so also for any crystal, of the opposite electric charges) is cancelled by any structure containing the inversion center, because its presence by the inversion of any pair of ends of the structure re-neutralizes it. Therefore, the appearance of such kinds of effects do not clearly indicate the crystallographic class (group), while existing 21 crystallographic groups without an inversion center, as shown in Table 2.11 (see also Table 2.8 for the presence of the symmetry elements and operations).

There is also a third type of action in Table 2.9, which generates the appearance of the electric induction (D) by the propagation of an electric field (E) through an anisotropic system (crystal),

$$D_i = \kappa_{ij} E_j \qquad (2.94)$$

this phenomenon being associated with the systems' permittivity, represented by the tensor of 2nd order, κ_{ij}.

The tensor of the permittivity anisotropic system (crystalline) can, in turn, be placed in relation with that of the vacuum (of zero order, represented by the scalar κ_0) by other tensor of second order, specific for the propagation environment, namely the refraction index n_{ij}, by the connection

$$\kappa_{ij} = \kappa_0 n_{ij} \qquad (2.95)$$

TABLE 2.10   The Magnitude of the Tensorial Effects at Thermodynamic Equilibrium*

(M.K.S.)	T	(E)	$[\sigma]$
S	$C/T \approx 10^4$	$p \approx 3 \times 10^{-6}$	$\alpha \approx 10^{-5}$
(D)	$p \approx 3 \times 10^{-6}$	$\kappa \approx 10^{-10}$	$d \approx 3 \times 10^{-12}$
$[\varepsilon]$	$\alpha \approx 10^{-5}$	$d \approx 3 \times 10^{-12}$	$s \approx 10^{-11}$

*After Chiriac-Putz-Chiriac (2005).

TABLE 2.11   Crystallographic Classed Without Inversion Center*

Syngony	Crystallographic Classes (groups)
Triclinic	1
Monoclinic	2, *m*
Orthorhombic	2*mm*, 222
Tetragonal	4, $\bar{4}$, 4*mm*, $\bar{4}$2*m*, 422
Trigonal	3, 3*m*, 32
Hexagonal	6, $\bar{6}$, 6*mm*, $\bar{6}$*m*2, 622
Cubic	23, –43*m*, 432

*After Chiriac-Putz-Chiriac (2005).

Due to the specific character of this refraction tensor, which correlates the propagation of the electromagnetic waves in vacuum – (*E*) with that of a crystalline environment – (*D*), the last statement can be written also for the propagation speeds in vacuum, represented by the scalar of the light speed *c*, respecting the velocity of the system in cause, represented by the vector (*v*). Thus, the tensor of the refraction index will not be of second order, but of the first order, to satisfy the contracting relation:

$$c = v_i \, n_i \qquad\qquad (2.96)$$

Thus, the propagation of the electromagnetic waves into a crystal will have three main directions of refraction or the *main optical axes* (for each of them being characteristic one refraction index as n_1, n_2, n_3), in a quadratic relation with the Cartesian coordinates associated to the crystalline system (*x*, *y*, *z*), through the form of a rotational ellipsoid which generates the so-called *surface of the refraction index* (indicatrix), with the equation:

$$x^2/n_1^{\,2} + y^2/n_2^{\,2} + z^2/n_3^{\,2} = 1 \qquad\qquad (2.97)$$

This way, for a direction or an optical (arbitrary) axis (O.A.) of incident radiation propagation in the considered anisotropic media, there can be always selected two refraction indices from the indicatrix, a phenomena called as *birefringence* – as shown in Figure 2.54, that, in turn, will impose two different types and two different directions of propagation: one *ordinary* and other *extraordinary*, see Figure 2.55.

However, since this surface of refraction indices applies to a crystalline system, it will be subjected to the same limitations of symmetry to which the crystallographic system belongs. If the orientation of the direction of propagation of the incident radiation (represented by the optical axis, O.A.) coincides with the main symmetry axis of the crystal, the birefringence effect is therefore canceled.

If from the main indices two are equal, i.e.,

$$n_1 = n_2 = w = n_{ordinary}, \; n_3 = e = n_{extra\text{-}ordinary} \tag{2.98}$$

the crystal in focus becomes *uniaxial* (with the supra- or under-unity ratio towards the unity, see upper of Figure 2.56), i.e., there is an only O.A. along which (aligned with the main axis of symmetry of the crystal) the

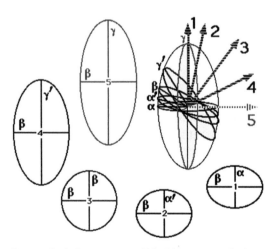

FIGURE 2.54   Geometrical phenomenon of birefringence; optical axes (O.A.) are in the direction of the 1–5 arrows; after (Optical Mineralogy, 2003, 2013).

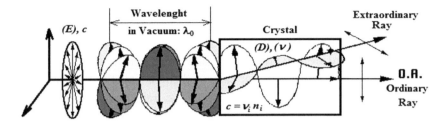

FIGURE 2.55 The birefringence phenomena (double refraction) into the crystal; O.A. corresponds to the optical axis, i.e., the propagation axis of the incident radiation (on the crystal); after (Optical Mineralogy, 2003, 2013).

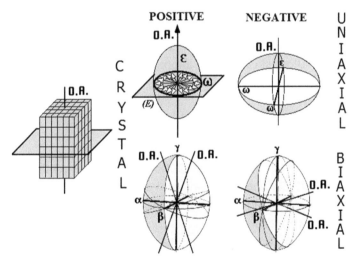

FIGURE 2.56 The alignment of the optical axes (O.A.) for the uniaxial (up) and biaxial (down) crystal, for the supra- and sub-unity ratios, respectively; after (Optical Mineralogy, 2003, 2013).

propagation and transmission are made in ordinary conditions, i.e., without recording two transmitted rays, the effect of birefringence being cancelled by the equalization of the refraction indices in the ellipsoid transversal section, i.e. perpendicular on O.A. This is the case of the crystals belonging to the trigonal, tetragonal and hexagonal systems, where each of them includes a single axis of third, fourth and sixth order, respectively.

For the crystals belonging to the orthorhombic systems, monoclinic and triclinic, there are two propagation directions (and alignment) of O.A. for which the crystals present, for each of them, a single refraction index, in the normal section plane of O.A. where the transmission waves are in an ordinary regime: this way the crystal is called *biaxial* (again with the supra- and under- ratio versions, according to the unity ratio of the refraction indices at the equatorial section of the indices surface as shown in bottom of Figure 2.56). The cubic system being isotropic it does not have the birefringence phenomenon; however the crystals belonging to this syngony would be oriented.

In conclusion, for the birefringence phenomena, its absence in any orientation or in one or two orientations of the O.A. can indicate the appurtenance of the crystal under concern to one of the set of the crystallographic systems that allow such a behavior; it is thus more an indicator of the system and not an absolute criterion of its selection.

2.5.8.2   The Curie Principle of Symmetry

From the analysis of the previous section we can also get the conclusion that by applying an operation (physical actions or perturbations) upon a group, there results the reduction of the symmetry operations towards the initial set (of symmetries).

Generalizing this observation on the principle grounds, there is worthy to note the idea through which under an external action upon a crystal, it will loses from its intrinsic symmetry operations that it had otherwise in the isolated state.

This way, there appears as particularly useful the power of prediction which, based on the nature of the external action, may anticipate the intrinsic symmetry operations of the crystal that will "survive" to that action and will be still active.

This kind of information exists, and, moreover, is included in a general principle at the groups' theory level, but which fully manifests its applicability in crystallographic analysis: it is called as (Pierre) Curie Principle

(1884). In phenomenology terms, the principle states that when different phenomena are combined, the resulted dynamical system will carry only symmetries common to the phenomena involved. At a crystalline level, this means that the state of a solid system depends both on the symmetry of the crystal free of any external influence and on the symmetry of the action externally applied and independent by the crystal presence.

In other words, a crystal found in an external field (or action) will manifest those symmetry elements and operations that are common both to the crystal without an external influence and to the external field in the absence of the crystal.

At a mathematical level of the groups, this principle is rewritten as follows: the symmetry group of a crystal under an external action $(\overline{K})$ is provided by the largest common symmetry sub-group of the crystal without external influence (K) and of the symmetry group of the external influence independently of the presence of the crystal (G):

$$\overline{K} = K \cap G \qquad\qquad (2.99)$$

A relevant example of this principle is given by Figure 2.57 for a crystal belonging to the $\overline{4}2m$ point group upon which an external field acts on the given direction, around which can be represented – in principle- an infinity of reflection planes.

This action may corresponds with the application of an electromagnetic field (or an electric plus a magnetic field) on a direction around which the intensities vectors of the electric fields (for example) oscillate as non-polarized, as shown in Figure 4.25.

These resulted phenomena specific to the system made by crystal + electromagnetic radiation, in terms of symmetry at the emerging polarization of radiation, have been already introduced as the birefringence or *double refraction* and will be more characterized in next.

2.5.8.3   Optical Activity by Crystals

The phenomenon of double refraction, firstly observed for the calcite crystals by Erasmus Bartholinus since 1669 is remarkable by the fact that

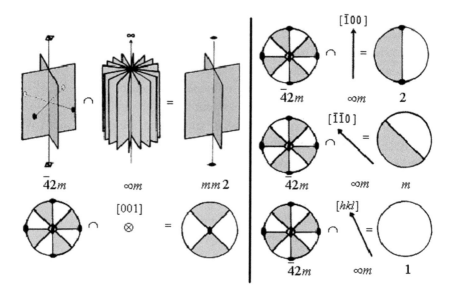

FIGURE 2.57 Application of Curie Principle for the point group $\bar{4}2m$ during the action of an external field, with the representative vector polarized in various directions; after (Hartmann, 2003, Commission on Crystallographic Teaching IUCR 2013).

from a non-polarized incident wave (with the electrical intensity vectors oscillating in all the directions, i.e., circularly respecting the propagation direction), after passing the crystalline environment, results into two flat polarized rays (with the intensity vectors oscillating only in a direction perpendicular to that of propagation). Which ones of the crystal symmetry elements will survive to this electromagnetic action?

Depending on the orientation of the electric field propagation direction, e.g. respecting the symmetry directions in the crystal belonging to the $\bar{4}2m$ point group, with the stereographic projection #19 of Table 2.8, – the crystal under concern, under electric field, will perceive one of the symmetries resulted in Figure 2.57, as based on the application of the relation (2.99).

Yet, the application of the Curie principle has another practical importance: due to the lowering of the crystal symmetry under the external influence, the anisotropy is practically increased, such that the initially uniaxial crystals becomes biaxial, thus increasing its *optical activity*.

This last optical effect, is also called as the *optical rotation power*, referring to the ability of (an)isotropic medium to rotate (to the left or to the right) the transmission polarization plane (of oscillation) of the electric intensity vector respecting the incident direction of polarization.

Yet, the optical activity must be carefully differentiated by the birefringence, but can accompany it as a "perturbation". Thus, the optical activity can be investigated based on the birefringence, when one of the monochromatic plane-polarized radiations continue to be used for a new incidence on a crystal, and recording, in transmission, the rotation of the polarization plane respecting the incident polarization direction.

Therefore, for all the crystals corresponding to the classes that do not contain symmetry operations of second order (mirroring planes, inversion center), as exposed in Table 2.12, and abstracted from Table 2.6, they can "exist" in *levo-* (L-) or *dextro-* (D-) *gyre* states, also called as *enantiomers' forms* (or *optical isomers*), they being able to rotate the incident monochromatic polarization plane to the left or to the right, respectively. Remarkably, the importance of the optical activity phenomenon and of the presence of the L- and D-states is extended also for the living organism, beyond the crystallographic classifications.

Thus, from the 20 amino acids, the 19 species can exist in both the L- and D-states, see the Alanine example of Figure 2.58; the exception

TABLE 2.12 Crystallographic Classes Without Inversion Center and Associated Reflection Planes*

Singony	Crystallographic Classes (Groups)
Triclinic	1
Monoclinc	2
Orthorhombic	222
Tetragonal	$4, \bar{4}, 422$
Trigonal	3, 32
Hexagonal	6, 622
Cubic	23, 432

*After Nye (1985).

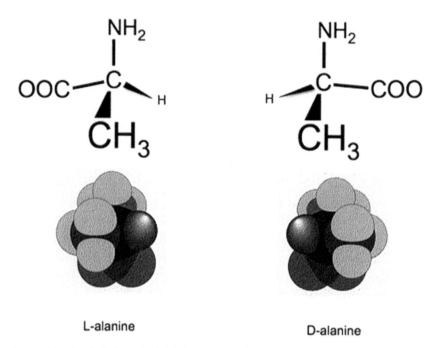

L-alanine

D-alanine

FIGURE 2.58   (left) L- and (right) D-enantiomeric forms of Alanine.

Glycine is a such because it has two indiscernible atoms of hydrogen attached to the alpha-carbon.

As an application, the amino acids in L-state are exclusively used for the protein synthesis, while determining the protein's function is essentially depending on its geometrical shape. Moreover, a protein with an D- instead of L-amino acid pushes the group R in a wrong direction.

Thus, the importance of the differentiation between the L- and D-forms is major for the living organism: if there is stated that one of the L- or D-forms plays a significant role in the biological structure (e.g., for the proteins with catalyzing role – as enzymes have, or in the receptors of the proteins surface), it is very likely that this will be not the case with the other form too.

In any case, in the laboratory chemical synthesis or in the pharmaceutical industry equal quantities of both the enantiomers, in the so-called racemic mixture are produced.

For example, the albuterol drug (Proventil®), which mimes the effect of adrenaline helping at the bronchi dilatation in an asthmatic act, contains an equal measure of enantiomers, of which only L- is active, the other one can causing unpleasant secondary effects; for this reason the levalbuterol (Xopenex®) has been subsequently synthesized, towards isolating the levo-effect.

The crystallographic analysis can be completed since based on the diffraction phenomenon with X-ray on crystals, due to the wavelength comparable with the dimensions of the inter-planar crystalline distances. Yet, even here a caution is required. If the incidence of the X-ray is not near the absorption edge in X-ray of the analyzed crystalline material, the X-ray diffraction automatically introduces a symmetry center (the so called *Friedel law*), thus not distinguishing between the planes (hkl) and ($\overline{hkl}$), i.e., not being able to distinguish between the crystals with or without inversion center. In such a case, from the 32 crystallographic groups, only 11 groups can be clearly individuated (the so-called *Laue groups*), Table 2.13.

Yet, even by appealing to other phenomena and interactions (e.g., the X-ray diffraction on crystals), the Curie Principle, by its various equivalent formulations, allows the individualization of the symmetry effect in Nature's phenomena and classification.

Finally, there is clear that the optical activity is able to distinguish between the various crystallographic classes, but it can not be used as the one and only criterion for classification, just because of its "perturbation"

TABLE 2.13   The Laue Groups*

Singony	Crystallographic Classes (Groups)
Triclinic	$\overline{1}$
Monoclinc	$2/m$
Orthorhombic	mmm
Tetragonal	$4/m, 4/mmm$
Trigonal	$\overline{3}, \overline{3}m$
Hexagonal	$6/m, 6/mmm$
Cubic	$m3, m3m$

*After Duch (2003).

feature, too weak to be decisive for a complete crystallographic analysis. For example, when the optical activity effects are very weak, it cannot provide any crystallographic indication.

Thus, beyond any crystallographic classification of symmetry, one can phenomenological and principally resumed that:

(i)   any symmetry-cause produces a symmetry-effect;
(ii) the cause of symmetries appears in effects;
(iii) the symmetry group of the cause is the sub-group of the symmetry group of the effect;
(iv) the effect is at least as symmetric as the cause;
(v)   if some effects exhibit an asymmetric presence, this asymmetry comes from causes.

2.6   SPACE GROUPS

2.6.1   INTRODUCTION TO THE SPACE GROUPS

In the previous sections an external perspective of the crystal (character-ized by its sides and morphology) with an internal one had been unified, as reduced to the unit cell (by the stereographic projection, for instance); the present discussion follows (Chiriac-Putz-Chiriac, 2005).

2.6.1.1   Helicals and the Glide Planes

However, the crystal, as an infinite network, can not be reduced to the unit cell. Moreover, the crystal symmetries analysis was previously considered only by the symmetry elements and operations associated to the reflection plan, to the rotation axis and to the inversion center.

On the other hand, translation, as an operation by which the unit cell expands to the infinite crystal (infinite lattice + its basis/motif), was not yet integrated into the possible combinations of symmetry. This inclusion, with the analysis of the geometric consequences for the crystalline charac-terization, is aimed for the present sections.

The combination of symmetry elements with the translations in the ele-mentary cell means the association of translation with rotation, reflection and inversion. The last combination, of translation with inversion center, can be embedded in turn in rotations, via rotoinversions.

But how does the combination of a symmetry element generally achieve the translation?

Firstly, there should be noted that the translations can not be considered as firstly applicable to a point (x, y, x), as long as the analysis of a crystal is initiated from the unit cell where, by priority, apply the restrictions to the specific rotational symmetry of the crystalline structure, to review the Table 2.6.

Thus, there appears that one cannot considered the translations "before" a mirroring (or reflection), since, in general, the translation can "hide" a reflection if this is made in the perpendicular direction on the considered mirroring plane and, moreover, also for the fact that the reflection can be associated also with a rotoinversion, for the second order axis $(\bar{2} = m)$.

Also for the inversion center the same reasoning applies: if a translation is effectuated along the directions that reverse the structural points, the existence of the inversion center "disappears". Therefore, the elements and the symmetry operations in the unit cell can not be reduced to translations. In other words, the translations must be considered *after* the application of all possible simple operations of symmetry.

This way, the first constraint in introducing the translations through the symmetry combination requires that the translation succeeds (and not precedes) the symmetry operations allowed in the unit cell and described by the point groups, see Table 2.6.

Further, we have to take into account that the constraints of the translational lattice, namely the translation T with n units along t-direction in a crystal should equal (or to express themselves through) m integer units of the vector t-size in the elementary cell:

$$nT = mt \qquad (2.100)$$

For example, if the fundamental axis $0x$ (with the fundamental vector of type a) is considered as the direction of translation, the previous lattice constraint recommends the translations:

$$T = (m/n)a \qquad (2.101)$$

This means that, for the case when $m=n$, the translation $T=a$, is recommended, i.e., through the translation the next structural point of the lattice on the direction $[a]$ is reached. In this case, the translation consideration did not bring any new element in the symmetry analysis of the unit cell, yet it has only translated this analysis.

But what happens if $m=n+1$? Then the translation becomes

$$T=(1+1/n)a=a+a/n \qquad (2.102)$$

which translates the analysis of a structural point of the unit cell outside the cell, in other words, in this case the translation does not correspond anymore to an internal element of symmetry of the elementary cell!

Through combination of the two constraints of the translation one concludes that the translations should be combined with the internal elements of symmetry of the elementary cell, by succeeding them.

For the axes of rotation, the combination with the translations will generate the so-called helical axes (or screw axes), through which, after a rotation of the axis order, along the axis, also a translation along the relative axis is effectuated, with the possible steps of translation, $T=(m/n)t$, so that $m<n$, and where t can take the values of the modules on the fundamental a, b or c directions. Table 2.14 details, for each of the symmetry axis from crystal, the possible screw axes, with the translations along the axis $0z$ (c), this axis being the only unique axis of rotation for many of the crystalline syngonies. In Figures 2.59–2.62 the helical axes of second order, of third order, of fourth order and of sixth order are exposed, respectively, with the considered "steps" of translation of Table 2.14, toward the action of the corresponding gyres (Chiriac-Putz-Chiriac, 2005).

Symbolically, the action (the axis and the direction) of the n-order helical with the step m ($<n$) along the direction q toward the point (x, y, q) will be indicated through the: $n_m[xyq]$.

Note how this notation combines the symbolism of the rotation axis (n) with the direction $[xyq]$ by the inferior m index.

In case the translations are combined with the reflection planes, the procedure presumes, firstly, the mirroring towards the reflection plane

TABLE 2.14 The Description of the Possible Helical Axes or the Screw Axes*

n-th order of rotation	The translation notations, types, and symbols				
2	2_1: a/2; b/2; c/2				
3	3_1: c/3	3_2: 2c/3			
4	4_1: c/4	4_2: 2c/4	4_3: 3c/4		
6	6_1: c/6	6_2: 2c/6	6_3: 3c/6	6_4: 4c/6	6_5: 5c/6

*After Chiriac-Putz-Chiriac (2005).

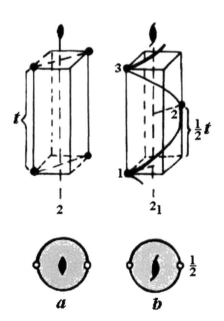

FIGURE 2.59 The action of the second order helical (b) towards the digyre (a).

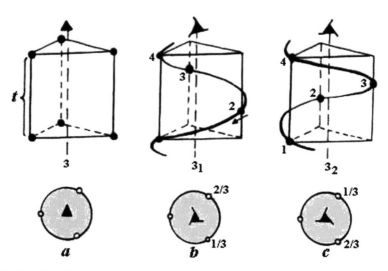

FIGURE 2.60   The action of the third order helical (b, c) towards the trigyre (a).

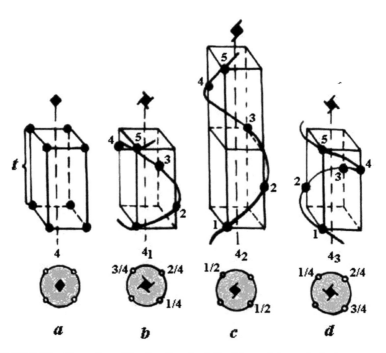

FIGURE 2.61   The action of the fourth order helical (b, c, d) towards the tetragyre (a).

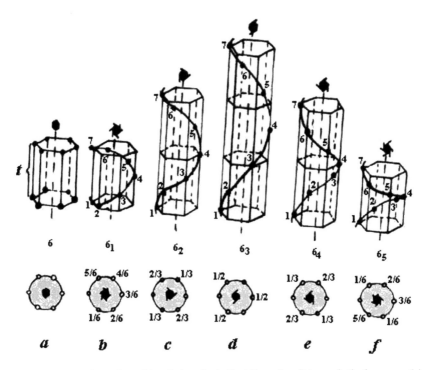

FIGURE 2.62   The action of the sixth order helical (b, c, d, e, f) towards the hexagyre (a).

execution, after which a translation inside the elementary cell in a plane or on a direction parallel with the reflection plane is considered, see Figure 2.63, thus defining the reflection plane as a glide plane.

Of course, also for the glide plane one should considered the translation effect of to be restricted to the unit cell. Accordingly, many combinations are possible, so generating more types of sliding glide planes, whose summary is included in Table 2.15.

In Table 2.15, the reflection planes, simple and combined with the translation are presented in the elementary cell.

In each symbolic diagram (of the third column) to the left side appears as the "R" the motif of the lattice, while the glide effect being given in a diagonal perspective in the right side of the column (unless for the planes of *m* and *c* types).

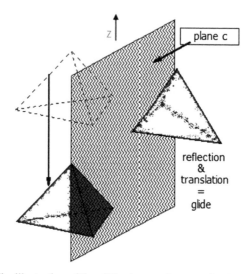

FIGURE 2.63 The illustration of the glide plane action; see also International Tables for Crystallography (2003).

TABLE 2.15 Types of Reflection and Glide Planes for the Unit Cell with R: the Motif of the Network/Lattice*

Symbol	Symmetry Action	Graphical Symbol of the Symmetry Action	The Nature of the Translation
m	Reflection plane	R / R̶ · R|	none
a, b	Reflection plane with axial glide	R / R̶ · R R̶ R|	$a/2$ along [100]; $b/2$ along [010] or <100>
c	Reflection plane with axial glide	R / R̶	$c/2$ along $0z$; $(a+b+c)/2$ along [111]
n	Reflection plane with diagonal glide	R R / R̶ · R R	$(a+b)/2$ or $(b+c)/2$ or $(c+a)/2$ or $(a+b+c)/2$ (tetragonal and cubic)
d	Reflection plane with diamond glide	R R R / R̶ R̶ · R R R	$(a\pm b)/4$ or $(b\pm c)/4$ or $(c\pm a)/4$ or $(a\pm b\pm c)/4$ (tetragonal and cubic)

*After International Tables for Crystallography (2003).

Moreover, in the right side of the diagrammatic symbols (unless for the c-plane) there appears the effect of the glide plane in the perpendicular view on the reflection plane. The arrows indicate the direction of translation.

As for the helical axis, we can symbolically write the symmetry action for the mirroring operations with a glide.

As an example, a reflection of a structural point with (x, y, z) coordinates towards a glide plane a (see the case of Table 2.15, of the third line) will be noted as $\boxed{a(x, y, z)}$ and will feature the point with $(1/2+x, -y, z)$ coordinates.

2.6.1.2   The Symbolistics for the Space Groups

Through combining:

- the seven crystalline syngonies (primitive cells of Table 2.2),
- with the 14 types of the Bravais lattice (the non-primitive cells of Figure 2.30),
- with the 32 point groups (representing the combinations of the symmetry elements of an elementary cell, review the Tables 2.6 and 2.8),
- and with the translations (the complex combinations of the symmetry elements inside the elementary cell, Tables 2.14 and 2.15),

there results the set of 230 crystallographic space groups (so called due to the intervention of the translation), as presented in Table 2.16, both in the international and Schoenflies notations.

Thus, the space groups describe the spatial distribution for the combinations of the symmetry elements of an elementary cell. Their importance is major, as long as they allow the identification of a crystalline structure of a chemical compound until the atomic level.

As a general guide in identifying the structure referred to in Table 2.16 symbols, one can synthesize some rules.

For their exposure, one should differentiate among the similar notation by using the notation of the planes and of the directions, in the same manner as a symbol, for example c, states as an indication for a direction [c], or for a plan, (ca), inside the symbols of the spatial groups.

TABLE 2.16   The Symbols and Classification of the Crystallographic Space Groups*

No.	International notation		Schoenflies notation	
	Point group	Spatial group	Group's symmetry formula	Spatial group
Triclinic System				
1	1	P1	A^1	C_1^1
2	$\bar{1}$	$P\bar{1}$	C	C_i^1
Monoclinic System				
3		P2		C_2^1
4	2	$P2_1$	A^2	C_2^2
5		C2		C_2^3
6		Pm		C_s^1
7	m	Pc	P	C_s^2
8		Cm		C_s^3
9		Cc		C_s^4
10		P2/m		C_{2h}^1
11		$P2_1/m$		C_{2h}^2
12	2/m	C2/m	$A^2\Pi C$	C_{2h}^3
13		P2/c		C_{2h}^4
14		$P2_1/c$		C_{2h}^5
15		C2/c		C_{2h}^6
Orthorhombic System				
16		P222		$D_2^1 = V^1$
17		$P222_1$		$D_2^2 = V^2$
18		$P2_12_12$		$D_2^3 = V^3$
19	222	$P2_12_12_1$	$3A^2$	$D_2^4 = V^4$
20		$C222_1$		$D_2^5 = V^5$

TABLE 2.16 Continued

No.	International notation		Schoenflies notation	
	Point group	**Spatial group**	**Group's symmetry formula**	**Spatial group**
21		C222		$D_2^6 = V^6$
22		F222		$D_2^7 = V^7$
23		I222		$D_2^8 = V^8$
24		$I2_12_12_1$		$D_2^9 = V^9$
25		Pmm2		C_{2v}^1
26		$Pmc2_1$		C_{2v}^2
27		Pcc2		C_{2v}^3
28		Pma2		C_{2v}^4
29		$Pca2_1$		C_{2v}^5
30		Pnc2		C_{2v}^6
31		$Pmn2_1$		C_{2v}^7
32		Pba2		C_{2v}^8
33	mm2	$Pna2_1$	A^22P	C_{2v}^9
34		Pnn2		C_{2v}^{10}
35		Cmm2		C_{2v}^{11}
36		$Cmc2_1$		C_{2v}^{12}
37		Ccc2		C_{2v}^{13}
38		Amm2		C_{2v}^{14}
39		Abm2		C_{2v}^{15}
40		Ama2		C_{2v}^{16}
41		Aba2		C_{2v}^{17}
42		Fmm2		C_{2v}^{18}

TABLE 2.16 Continued

No.	International notation		Schoenflies notation	
	Point group	Spatial group	Group's symmetry formula	Spatial group
43		Fdd2		C_{2v}^{19}
44		Imm2		C_{2v}^{20}
45		Iba2		C_{2v}^{21}
46		Ima2		C_{2v}^{22}
47		Pmmm		$D_{2h}^{1} = V_{h}^{1}$
48		Pnnn		$D_{2h}^{2} = V_{h}^{2}$
49		Pccm		$D_{2h}^{3} = V_{h}^{3}$
50		Pban		$D_{2h}^{4} = V_{h}^{4}$
51		Pmma		$D_{2h}^{5} = V_{h}^{5}$
52		Pnna		$D_{2h}^{6} = V_{h}^{6}$
53		Pmna		$D_{2h}^{7} = V_{h}^{7}$
54		Pcca		$D_{2h}^{8} = V_{h}^{8}$
55		Pbam		$D_{2h}^{9} = V_{h}^{9}$
56		Pccn		$D_{2h}^{10} = V_{h}^{10}$
57		Pbcm		$D_{2h}^{11} = V_{h}^{11}$
58		Pnnm		$D_{2h}^{12} = V_{h}^{12}$
59		Pmmn		$D_{2h}^{13} = V_{h}^{13}$
60	mmm	Pbcn	$3A^{2}3PC$	$D_{2h}^{14} = V_{h}^{14}$
61		Pbca		$D_{2h}^{15} = V_{h}^{15}$
62		Pnma		$D_{2h}^{16} = V_{h}^{16}$
63		Cmcm		$D_{2h}^{17} = V_{h}^{17}$

TABLE 2.16 Continued

No.	International notation		Schoenflies notation	
	Point group	Spatial group	Group's symmetry formula	Spatial group
64		Cmca		$D_{2h}^{18} = V_h^{18}$
65		Cmmm		$D_{2h}^{19} = V_h^{19}$
66		Cccm		$D_{2h}^{20} = V_h^{20}$
67		Cmma		$D_{2h}^{21} = V_h^{21}$
68		Ccca		$D_{2h}^{22} = V_h^{22}$
69		Fmmm		$D_{2h}^{23} = V_h^{23}$
70		Fddd		$D_{2h}^{24} = V_h^{24}$
71		Immm		$D_{2h}^{25} = V_h^{25}$
72		Ibam		$D_{2h}^{26} = V_h^{26}$
73		Ibca		$D_{2h}^{27} = V_h^{27}$
74		Imma		$D_{2h}^{28} = V_h^{28}$

Tetragonal System

75		P4		C_4^1
76		P4$_1$		C_4^2
77	4	P4$_2$	A^4	C_4^3
78		P4$_3$		C_4^4
79		I4		C_4^5
80		I4$_1$		C_4^6
81	$\bar{4}$	P$\bar{4}$	A_4^2	S_4^1
82		I$\bar{4}$		S_4^2
83		P4/m		C_{4h}^1

TABLE 2.16 Continued

No.	International notation		Schoenflies notation	
	Point group	Spatial group	Group's symmetry formula	Spatial group
84		$P4_2/m$		C_{4h}^2
85	4/m	$P4/n$	$A^4\Pi C$	C_{4h}^3
86		$P4_2/n$		C_{4h}^4
87		$I4/m$		C_{4h}^5
88		$I4_1/a$		C_{4h}^6
89		$P422$		D_4^1
90		$P42_12$		D_4^2
91		$P4_122$		D_4^3
92		$P4_12_12$		D_4^4
93	422	$P4_222$	A^44A^2	D_4^5
94		$P4_22_12$		D_4^6
95		$P4_322$		D_4^7
96		$P4_32_12$		D_4^8
97		$I422$		D_4^9
98		$I4_122$		D_4^{10}
99		$P4mm$		C_{4v}^1
100		$P4bm$		C_{4v}^2
101		$P4_2cm$		C_{4v}^3
102		$P4_2nm$		C_{4v}^4
103		$P4cc$		C_{4v}^5
104	4mm	$P4nc$	A^44P	C_{4v}^6

TABLE 2.16 Continued

No.	International notation		Schoenflies notation	
	Point group	Spatial group	Group's symmetry formula	Spatial group
105		$P4_2mc$		C_{4v}^7
106		$P4_2bc$		C_{4v}^8
107		I4mm		C_{4v}^9
108		I4cm		C_{4v}^{10}
109		$I4_1md$		C_{4v}^{11}
110		$I4_1cd$		C_{4v}^{12}
111		$P\overline{4}2m$		$D_{2d}^1 = V_d^1$
112		$P\overline{4}2c$		$D_{2d}^2 = V_d^2$
113		$P\overline{4}2_1m$		$D_{2d}^3 = V_d^3$
114		$P\overline{4}2_1c$		$D_{2d}^4 = V_d^4$
115		$P\overline{4}m2$		$D_{2d}^5 = V_d^5$
116	$\overline{4}2m$	$P\overline{4}c2$	$A_4^2 2A^2 2P$	$D_{2d}^6 = V_d^6$
117		$P\overline{4}b2$		$D_{2d}^7 = V_d^7$
118		$P\overline{4}n2$		$D_{2d}^8 = V_d^8$
119		$\overline{I4}m2$		$D_{2d}^9 = V_d^9$
120		$\overline{I4}c2$		$D_{2d}^{10} = V_d^{10}$
121		$\overline{I4}2m$		$D_{2d}^{11} = V_d^{11}$
122		$\overline{I4}2d$		$D_{2d}^{12} = V_d^{12}$
123		P4/mmm		D_{4h}^1
124		P4/mcc		D_{4h}^2
125		P4/nbm		D_{4h}^3

TABLE 2.16 Continued

No.	International notation		Schoenflies notation	
	Point group	Spatial group	Group's symmetry formula	Spatial group
126		P4/nnc		D_{4h}^4
127		P4/mbm		D_{4h}^5
128		P4/mnc		D_{4h}^6
129		P4/nmm		D_{4h}^7
130		P4/ncc		D_{4h}^8
131		P4$_2$/mmc		D_{4h}^9
132	4/mmm	P4$_2$/mcm	$A^44A^24P\Pi C$	D_{4h}^{10}
133		P4$_2$/nbc		D_{4h}^{11}
134		P4$_2$/nnm		D_{4h}^{12}
135		P4$_2$/mbc		D_{4h}^{13}
136		P4$_2$/mnm		D_{4h}^{14}
137		P4$_2$/nmc		D_{4h}^{15}
138		P4$_2$/ncm		D_{4h}^{16}
139		I4/mmm		D_{4h}^{17}
140		I4/mcm		D_{4h}^{18}
141		I4$_1$/amd		D_{4h}^{19}
142		I4$_1$/acd		D_{4h}^{20}

Trigonal System

No.	Point group	Spatial group	Group's symmetry formula	Spatial group
143		P3		C_3^1
144	3	P3$_1$	A^3	C_3^2

TABLE 2.16 Continued

No.	International notation		Schoenflies notation	
	Point group	Spatial group	Group's symmetry formula	Spatial group
145		$P3_2$		C_3^2
146		R3		C_3^4
147	$\bar{3}$	$P\bar{3}$	A^3C	C_{3i}^1
148		$R\bar{3}$		C_{3i}^2
149		P312		D_3^1
150		P321		D_3^2
151	32	$P3_112$	A^33A^2	D_3^3
152		$P3_121$		D_3^4
153		$P3_212$		D_3^5
154		$P3_221$		D_3^6
155		R32		D_3^7
156		P3m1		C_{3v}^1
157		P31m		C_{3v}^2
158		P3c1		C_{3v}^3
159	3m	P31c	A^33P	C_{3v}^4
160		R3m		C_{3v}^5
161		R3c		C_{3v}^6
162		$P\bar{3}1m$		D_{3d}^1
163		$P\bar{3}1c$		D_{3d}^2
164	$\bar{3}m$	$P\bar{3}m1$	$A_6^33A^23PC$	D_{3d}^3

TABLE 2.16   Continued

No.	International notation		Schoenflies notation	
	Point group	**Spatial group**	**Group's symmetry formula**	**Spatial group**
165		P$\bar{3}$c1		D_{3d}^4
166		R$\bar{3}$m		D_{3d}^5
167		R$\bar{3}$c		D_{3d}^6
Hexagonal System				
168		P6		C_6^1
169		P6$_1$		C_6^2
170	6	P6$_5$	A^6	C_6^3
171		P6$_2$		C_6^4
172		P6$_4$		C_6^5
173		P6$_3$		C_6^6
174	$\bar{6}$	P$\bar{6}$	A$^3\Pi$	C_{3h}^1
175		P6/m		C_{6h}^1
176	6/m	P6$_3$/m	A$^6\Pi$C	C_{6h}^2
177		P622		D_6^1
178		P6$_1$22		D_6^2
179	622	P6$_5$22	A^{6}6A^2	D_6^3
180		P6$_2$22		D_6^4
181		P6$_4$22		D_6^5
182		P6$_3$22		D_6^6
183		P6mm		C_{6v}^1
184	6mm	P6cc	A^{6}6P	C_{6v}^2

TABLE 2.16 Continued

No.	International notation		Schoenflies notation	
	Point group	Spatial group	Group's symmetry formula	Spatial group
185		P6$_3$cm		C_{6v}^3
186		P6$_3$mc		C_{6v}^4
187		$\overline{P}6m2$		D_{3h}^1
188	$\overline{6}m2$	$\overline{P}6c2$	$A^3\Pi3A^23P$	D_{3h}^2
189		$\overline{P6}2m$		D_{3h}^3
190		$\overline{P6}2c$		D_{3h}^4
191		P6/mmm		D_{6h}^1
192	6/mmm	P6$_3$/mcc	$A^66A^26P\Pi C$	D_{6h}^2
193		P6$_3$/mcm		D_{6h}^3
194		P6$_3$/mmc		D_{6h}^4
Cubic System				
195		P23		T^1
196		F23		T^2
197	23	I23	$3A^24A^3$	T^3
198		P2$_1$3		T^4
199		I2$_1$3		T^5
200		Pm3		T_h^1
201		Pn3		T_h^2
202		Fm3		T_h^2
203	m3	Fd3	$3A^24A^33\Pi C$	T_h^4
204		Im3		T_h^5
205		Pa3		T_h^6

TABLE 2.16   Continued

No.	International notation		Schoenflies notation	
	Point group	Spatial group	Group's symmetry formula	Spatial group
206		Ia3		T_h^7
207		P432		O^1
208		P4$_2$32		O^2
209		F432		O^3
210	432	F4$_1$32	$3A^44A^36A^2$	O^4
211		I432		O^5
212		P4$_3$32		O^6
213		P4$_1$32		O^7
214		I4$_1$32		O^8
215		P$\bar{4}$3m		T_d^1
216		F$\bar{4}$3m		T_d^2
217	$\bar{4}$3m	I$\bar{4}$3m	$3A_4^24A^36P$	T_d^3
218		P$\bar{4}$3n		T_d^4
219		F$\bar{4}$3c		T_d^5
220		I$\bar{4}$3d		T_d^6
221		Pm3m		O_h^1
222		Pn3n		O_h^2
223		Pm3n		O_h^3
224		Pn3m		O_h^4
225	m3m	Fm3m	$3A^44A^36A^2$	O_h^5
226		Fm3c	6P3ПC	O_h^6
227		Fd3m		O_h^7

TABLE 2.16   Continued

No.	International notation		Schoenflies notation	
	Point group	Spatial group	Group's symmetry formula	Spatial group
228		Fd3c		O_h^8
229		Im3m		O_h^9
230		Ia3d		O_h^{10}

*After (Verma & Srivastava, 1982).

Thus, the writing rules of symbols for the space groups claims the following conventions:

(i)   The first letter of the space symbols indicates the type of the lattice (P, C/A/B, I, F or R) and the suffixes (i.e., all that follow) indicate the symmetries.

(ii)   Because $\bar{2} = m$, and as long as the reflection planes are essential elements in the space groups, the symbol m will be always used instead the $\bar{2}$ axis in the suffixes of the symbols of the space groups.

(iii)   For the triclinic system, the most asymmetric one, there are only two space groups: P1 and P$\bar{1}$.

(iv)   For the monocyclic system, the unique axis of rotation in the system, 2 or $\bar{2}$, can be identified with the fundamental axis c (as the first choice), or with the fundamental axis b (as the second choice) at the level of the unit cell.

(v)   For the orthorhombic system there are three orthogonal axes and one element of 2 and $\bar{2} = m$ symmetry along each one; accordingly, the suffixes of the associated space groups refer to the symmetries along the axes $[a]$, $[b]$ and $[c]$, respectively. For example, Pcc2 means a primitive lattice (partitioned in primitive cells) with the glide planes as the (bc) and (ac) planes and with a helical axis of the second order along the direction $[c]$.

(vi)   In the tetragonal system, the $[a]$ and $[b]$ axes are equivalent, i.e., the symmetries in relation with them are also equivalent, and there is not required their separated specification in the symbol of group. Thus, the first symmetry in the symbol of the space group refers to the symmetry in relation to the axis [c], the next symmetry refers to the symmetry in relation with both the [a] and [b] axes, and the third symmetry refers to the symmetry in relation with the equivalent [110] and [1$\bar{1}$0] directions. All the space groups from this syngony contain the specification to the fourth order axis, of rotation or helical. For example, I4mm is a system with a tetragonal centered cell (due to the symbol I) having a rotational axis of fourth order in [c] direction, a normal reflection plane to the equivalent [100] and [010] directions, and another normal line to the equivalent [110] and [1$\bar{1}$0] directions.

(vii) For the hexagonal system, similar with the tetragonal one regarding the uniqueness of the axis along [c] direction, the first symbol of suffixes represents the symmetry along the [c] direction, the second and the third symmetry respectively refers to the one along and to the one on the normal line to the equivalent [a], [b] and [d] axes (see Figure 2.44 and the discussion of the Miller-Bravais indices); All the space groups of this system contain as a first symmetry a rotation or a helical axis of sixth order.

(viii) For the trigonal system the choice of the fundamental axes as in the hexagonal system is convenient, so that the space symbols have, also in this case, the same significance as those for the hexagonal system, having instead the third order axes of rotation and the helicals.

(ix) For the cubic system, the three fundamental crystallographic axes are equivalent. The first symmetry refers to the symmetry along the equivalent axes [a], [b] and [c], or of the set of equivalent <100> directions. The second symmetry refers to the one along the main diagonal, or to the set of equivalent <111> directions. The third symmetry refers to the symmetry of the <110> set of the diagonals of the faces. In this system two sets of planes are possible. The first set refers to those parallel with the faces of the unit cell and appear only for the m3 point group, with m before 3, therefore, also in the allied symbols of the group, the presence of a reflection (or glide) plane before 3 indicates how those planes are parallel with the faces of the unit cell. The second set of planes contains the planes which include the diagonal axes of third order and appears also for $43m$ and $m3m$ point groups, with m after 3; consequently, a sequence of reflection (or glide) planes relative to (before or after) the third order axis immediately indicates the location of the relative planes in the space group. Moreover, the presence of the 3 and $\overline{3}$ axes *on the secondary position of the symmetry* indicates the fact that the symbol belongs to the cubic syngony (unlike the trigonal-rhombohedral syngony where such axes appear only on the first position of symmetry!).

2.6.1.3   The Pearson Classification

At this point, there is worth, since having the geometric information characteristic to a crystalline solid structure, to give a uniform classification of the chemical compounds (inorganic, organic, and mineral) with respect to the space group where the crystalline solid of the compound can be represented.

This way, the set of 230 possible space groups corresponds to the broader minimal classification that can be given to the chemical compounds. It is minimal because having only 230 such possibilities is much less than the almost infinite set of chemical compounds (natural and synthetic). Then, it is broad because this classification is very rigorous for the space group perspective, while being also able to broaden the information to the physical, mathematical and quantum properties.

In other words, due to the aim to rigorously classify as conceptually-compact as it can be the chemical compounds, with the indexing given by spatial groups, one may introduce also an alternative classification which may be the terminus point of any crystallographic analysis.

This way, the set of the chemical compounds in the crystalline state can be described by indexing through a combination between the types of the primitive cell (Table 2.2) of the Bravais networks (Figure 2.30) and of the number (n) of atoms (all types), thus generating the specified so-called Pearson notations (Villars & Calvert, 1991), as listed in Table 2.17.

TABLE 2.17   The Categories of Pearson Classification With n-the Total Number of Atoms in the Unit Cell*

Nr.Crt	Pearson Symbols (Prototype/Syngony/Groups)	Unit cell's pattern
1.	Triclinic structures (#1-#2) aPn a: asymmetric	
2.	Monoclinic structures (#3-#15) mPn mCn m: monoclinic	
3.	Orthorhombic structures (#16-#74) oPn oFn oI n oCn o: orthorhombic	
4.	Tetragonal structures (#75-#142) tPn tIn t: tetragonal	
5.	Trigonal structures (#143-#167) hPn hRn h: Hexagonal/reference system	
6.	Hexagonale structures (#168-#194) hPn h: hexagonal /reference system	
7.	Cubic structures (#195-#230) cPn cFn cIn c: cubic	

*After (U.S. Naval Research Laboratory/Center for Computational Materials Science, 2003).

With the aid of the Pearson indexing the "inverse path" of the crystallographic classification is crossed, i.e., by reducing the classification of the chemical compounds to the set of 230 space groups to the number of 14 Bravais lattice, yet leaving open (like a variable) the total number of atoms present in the elementary cell, a number that varies from case to case.

Accordingly, the Pearson classification is merely reductive than complete, as long as no parameter is specified, for example, the number of atoms per unit cell, which however varies. Overall, the Pearson indexing serves for an immediate "view" of the type of structure, at the same time indicating stoichiometric information for the unit cell.

2.6.2 THE CRYSTALLOGRAPHIC DESCRIPTION OF THE SPATIAL GROUPS

In the previous section the 230 crystallographic space groups, with the symbols listed in Table 2.16, have been introduced. Yet, one should be aware that in the symbol of a space group are included the minimum symmetries necessary to deduce the other ones and does not reflect all the symmetries that a space group involves. Here we analytically unfold such reality; the present discussion follows (Chiriac-Putz-Chiriac, 2005).

2.6.2.1 The Symmetry Analysis of the Space Groups

To deduce all the symmetries of a space group actually it means to determine all the *equivalent points*, i.e., those points that can be mutually transformed one in other through the internal symmetry operations, the translations included.

This way, in order to deduce all the equivalent points in a space group a structural point of coordinates (x, y, z) is considered upon which the symmetry operation prescribed by the symbol of the spatial group are applied.

Then, all the transformations possible between the new points are applied until there are no longer obtained new structural points from the symmetry operations: as such the set of equivalent point had been determined. Further, the symmetry operations that transform the equivalent points into each other, previously exhausted, are examined toward possibly identifying new operations of symmetry, not specified in the symbol of group, thus determining the total set of symmetry operations allowed by that space group.

This is the general strategy in determining all symmetries of a given space group. Practically, however, a method by which to consider the transformations of symmetry of the symbol of group (then also the correlated ones) on the generic structural point (x, y, z) must be introduced. As long as the coordinates of this generic point are fixed in a crystallographic system associated to the concerned syngony, a transformation of symmetry applied to the point is equivalent to a symmetry transformation applied to the fundamental axes associated with $(a, b,$ and $c)$. Therefore, the new structural point obtained by the symmetry operation will have the coordinates (x', y', z'), which is equivalent to saying that, by the symmetry operation, the fundamental axes had been changed, so becoming (a', b', c').

As such, the problem of considering the symmetry operations applied to a structural point means considering the coordinate transformations, respectively the fundamental axes' changing, by the virtue of the considered symmetry: when passing from the system with the axes (a, b, c), through the symmetry operation, the axes (a', b', c') are generated to correlate with the initial ones through transformation matrix:

$$
\begin{array}{cccc}
\begin{array}{c} Symmetry \\ Operation \end{array} & a & b & c \\
a' & ? & ? & ? \\
b' & ? & ? & ? \\
c' & ? & ? & ?
\end{array}
\qquad (2.103)
$$

For example, for the identical and the inversion operations, the coordinates of the point (x, y, z) do not modify while all change the sign, respectively. This is equivalent with the fact that in the first case the fundamental axes do not changes at all, while in the second case all the axes inverse the sign. Thus, there result the matrix of transformation for the identical and the inversion operations, having a non-null completed matrix structure on the main diagonal, filed with the +1 and −1 components, respectively:

$$
\begin{array}{cccc}
E & a & b & c \\
a' & 1 & 0 & 0 \\
b' & 0 & 1 & 0 \\
c' & 0 & 0 & 1
\end{array}
\; ; \;
\begin{array}{cccc}
i & a & b & c \\
a' & -1 & 0 & 0 \\
b' & 0 & -1 & 0 \\
c' & 0 & 0 & -1
\end{array}
\tag{2.104}
$$

The immediately following case corresponds to the symmetry operations on reflection, through which, depending on the orientation of the mirroring plane respecting the fundamental axes, only the coordinate's sign is modified corresponding to the change of the direction of the perpendicular axis on the reflection plane. Depending on the coordinate that changes its sign and on the axis that changes its direction, the following variants for the transformation matrices at mirroring result:

$$
\begin{array}{cccc}
\sigma_{xy} & a & b & c \\
a' & 1 & 0 & 0 \\
b' & 0 & 1 & 0 \\
c' & 0 & 0 & -1
\end{array}
\; ; \;
\begin{array}{cccc}
\sigma_{xz} & a & b & c \\
a' & 1 & 0 & 0 \\
b' & 0 & -1 & 0 \\
c' & 0 & 0 & 1
\end{array}
\; ; \;
\begin{array}{cccc}
\sigma_{yz} & a & b & c \\
a' & -1 & 0 & 0 \\
b' & 0 & 1 & 0 \\
c' & 0 & 0 & 1
\end{array}
\tag{2.105}
$$

To evaluate the *rotation effect* on the coordinates and fundamental axes the construction of Figure 2.64 will be consider, depicting the equivalence between the proper rotation of a segment and its coordinates in plane $(x, y) \perp$ on the (main) axis of rotation $0z$.

The coordinates' transformation (and of the associated axes) follows form the summing of the projections' contributions on the orthogonal directions of Figure 2.64, analytically as:

$$
\begin{aligned}
x_2 &= x_1 \cos\phi + y_1 \sin\phi \\
y_2 &= -x_1 \sin\phi + y_1 \cos\phi
\end{aligned}
\tag{2.106}
$$

Further on, by taking into account the identity, respectively the inversion operations, along the rotation axis, also the transformation matrices of the improper rotations can be immediately written:

$$
\begin{array}{cccc}
C_n & a & b & c \\
a' & \cos\phi & \sin\phi & 0 \\
b' & -\sin\phi & \cos\phi & 0 \\
c' & 0 & 0 & 1
\end{array}
\; ; \;
\begin{array}{cccc}
S_n & a & b & c \\
a' & \cos\phi & \sin\phi & 0 \\
b' & -\sin\phi & \cos\phi & 0 \\
c' & 0 & 0 & -1
\end{array}
\tag{2.107}
$$

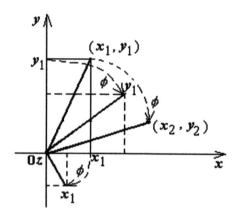

FIGURE 2.64 The analytical unfold of the (clockwise) rotation about the vertical axis (0z), after Chiriac-Putz-Chiriac (2005).

At this point one must notice that for the hexagonal and trigonal (rhombohedral) systems, there is preferred that the transformation of the axes at the rotational symmetry operations to not being performed according to the rules (2.107), due to the symmetries of the rotational axes of third and sixth order, but developing a construction (Figure 2.65) similar to that (Figure 2.44) used for the introduction of the indices Miller-Bravais, in previous Section 2.5. Actually, for a structural (x, y, z) point, which in the (a, b, c) system corresponds to the segment at the origin:

$$r = x\,a + y\,b + z\,c, \tag{2.108}$$

the rotational operation of third order will be described in a system generated by the rotation of the fundamental axes in the new set of (a', b', c') axes, correlated with the first set of axes by the relations abstracted from the vectorial relation of Figure 2.65, namely:

$$a'_{[3]} = b,\ b'_{[3]} = -(a+b),\ c'_{[3]} = c \tag{2.109}$$

For the system of axes (2.109) the relation (2.108) will be successively written:

$$r_{[3]} = x\,a'_{[3]} + y\,b'_{[3]} + z\,c'_{[3]}$$

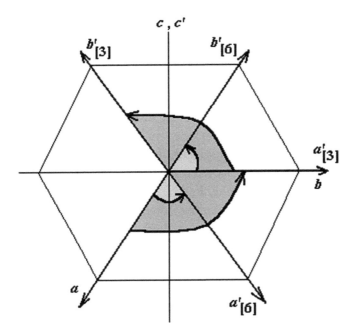

FIGURE 2.65 The rotations of sixth and third order in the hexagonal and trigonal syngonies, respectively, after Chiriac-Putz-Chiriac (2005).

$$= x\,b - y\,(a + b) + z\,c$$

$$= -y\,a + (x - y)b + z\,c \tag{2.110}$$

wherefrom, through the new coordinates' identification on the old system (a, b, c) axes, there results that the third rotation actually had transformed the (x, y, z) point in that of $(-y, x-y, z)$ coordinates, compare (2.110) with (2.108), in a system of coordinates trigonally ordered (such as in Figure 2.65).

The procedure can be repeated for the rotations of sixth order as applied to the a and b axes in Figure 2.65, with the successive results:

$$r_{[6]} = x\,a'_{[6]} + y\,b'_{[6]} + z\,c'_{[6]}$$

$$= x\,(a + b) - y\,a + z\,c$$

$$= (x - y) a + xb + z c \qquad (2.111)$$

Comparing the result (2.111) with (2.108) follows that, in the hexagonal system, respecting the fundamental axes in the hexagonal reference, the coordinates of the structural (x, y, z) point are changed to the rotation of the sixth order into the coordinates $(x - y, x, z)$.

The rotoinversion of the third and sixth orders can be obtained also by applying the inversion, after performing the required rotation, as applied on the sign of the involved coordinates.

Another essential fact should be stated and taken into account in the description of the space groups: as earlier noted in this Chapter the point groups are called "punctual" because all the included elements of symmetry intersect in a point.

Nevertheless, the spatial groups, although introduced as extensions of the point groups by considering the translations "in steps" inside the elementary cell, include elements of symmetry (helicals and glide planes) which, added to the symmetries of the basic point group, make that their intersection can not be reduced to a point any longer.

This fact has a major importance for the crystallographic description of the spatial groups, as long as the inversions can not be considered all performed respecting a single origin, i.e., the inversion center. Thus, also translations of the coordinates of a structural point respecting non-common origin points must be performed for the elements of symmetry in the group.

As example, the inversion of the (x, y, z) point towards the origin $(0, 0, 0)$ simply produces the coordinates $(-x, -y, -z)$. But what happens if the point about which the inversion is considered, is in turn, of the (X, Y, Z) coordinates?

In this case, one should firstly appeal to the translation of the coordinates (x, y, z) respecting the origin $(0, 0, 0)$ towards the inversion point (X, Y, Z) – the new origin, which produces the new coordinates $(x–X, y–Y, z–Z)$.

Then, the inversion towards this new point is considered, with the coordinates being reversed as usually and becoming $(X-x, Y-y, Z-z)$. The last operation will imply the translation of the coordinates in the inverted plane, back toward the origin (0, 0, 0), which produces the final coordinates $(2X-x, 2Y-y, 2Z-z)$.

The lesson is that, whenever in the space groups a coordinate to the origin (0, 0, 0) is reversed, the inversion towards the space point (X, Y, Z) will produce $+2X$ or/and $+2Y$ or/and $+2Z$ as a contribution to the reversed coordinate, the result being finally reported to the origin (0, 0, 0).

Altogether, in Table 2.18 the effects of the symmetry operations previously described are listed and exemplified, for the symmetry transformations with the inversion of the coordinates of a point (x, y, z) respecting the point (X, Y, Z), but finally reported to the reference origin (0, 0, 0).

2.6.2.1 Determination of a Space Group Symmetries

After the analytical and methodological clarifications from the previous section, one can now proceed to formulate the algorithm by which any space group can be completely described, starting from its symbol, as follows (International Tables for Crystallography, 2003).

(a) Given a symbol of a space group, the type of elementary cell (Bravais) and all the symmetries allowed are identified in the resumed suffixes of the symbol.

TABLE 2.18   The Coordinates Respecting the Origin (0, 0, 0) Upon the Symmetry Operations That Implies the Inversion Around the Points (0, 0, 0) and (X, Y, Z)*

Symmetry Element	(0, 0, 0)	(X,Y,Z)
i	$(-x, -y, -z)$	$(2X-x, 2Y-y, 2Z-z)$
$m\ (\sigma_{xy})$	$(x, y, -z)$	$(x, y, 2Z-z)$
$C_2(0z)$	$(-x, -y, z)$	$(2X-x, 2Y-y, z)$
$C_3(0z)$	$(-y, x-y, z)$	$(2Y-y, 2Y+x-y, z)$
$\bar{3}(0z)$	$(y, y-x, -z)$	$(y, 2X+y-x, 2Z-z)$
$C_4(0z)$	$(-y, x, z)$	$(2Y-y, x, z)$
$\bar{4}(0z)$	$(y, -x, z)$	$(y, 2X-x, z)$
$C_6(0z)$	$(x-y, x, z)$	$(2Y+x-y, x, z)$
$\bar{6}(0z)$	$(y-x, -x, -z)$	$(2X+y-x, 2X-x, 2Z-z)$

*After Chiriac-Putz-Chiriac (2005).

(b) The origin of coordinates on a symmetry element is chosen, preferably at the intersection of as many elements of symmetry.

(c) The identified symmetries are operated, respecting the chosen origin, for a generic (x, y, z) point and the equivalent points are obtained.

(d) The possible operations of symmetry are again applied to the new equivalent points, and one could thus generate the other new equivalent points.

(e) The previous steps are repeated until no new equivalent points are obtained.

(f) Additional elements of symmetry are explored from the presence in the set of equivalent points identified, for those not explicitly specified in the space group symbol.

(g) The diagram corresponding to the space group is represented.

We will further illustrate the case of the space group (#17)P222$_1$.

The symbol denotes a primitive lattice with a helical axis of second order along the direction [a], a second order helical axis along the [b] direction and a 2$_1$ helical in the [c] direction.

The origin (0, 0, 0) on the helical axis 2$_1$ is chosen.

With this choice one should fix the locations of the other helicals of second order: $2[xq_1q_2]$ along the [a] direction and respectively $2[q_3yq_4]$ along the [b] direction, with the q_1, q_2, q_3, q_4 coordinates to be next determined.

Through taking into account the prescriptions of Table 2.18, and the execution of the above operations of associated symmetry to a generic (x, y, z) point, the following transformations are obtained:

$$x,y,z \xrightarrow{2[xq_1q_2]} x,2q_1-y,2q_2-z \qquad (2.112)$$

$$x,y,z \xrightarrow{2[xq_3q_4]} 2q_3-x,y,2q_4-z \qquad (2.113)$$

$$x,y,z \xrightarrow{2_1[00z]} -x,-y,\frac{1}{2}+z \qquad (2.114)$$

The new coordinates of the right side of the transformations (2.112)-(2.114) are the new equivalent points.

Further, one should re-consider the existing symmetries between these equivalent points. There is noted how the points obtained in Eqs. (2.113) and (2.114) should be correlated by an axis as a helical of second order in the [c]-direction, due to the fact that in the coordinate 0z the same sign is recorded.

Thus, the points (2.112) and (2.113) are mutually transformed one into another by the effect of a second order helical in the [c]-direction, so identified with the helical $2_1[00z]$, while the next equivalences are valid:

$$\begin{cases} q_1 = q_3 = 0 \\ 2q_2 - 2q_4 = \dfrac{1}{2} \Rightarrow q_2 = q_4 + \dfrac{1}{4} \end{cases} \qquad (2.115)$$

Also, from the analysis of the Eqs. (2.112) and (2.114) transformations one notes how the obtained equivalent points should, at their turn, be correlated by a helical of second order, due to, in this case, the sign invariance on the coordinate 0y.

The helical of this case should be along the [b] direction, and moreover, to be of $2[0, y, q_2+1/4]$ type in order to be equivalent with the points of the right side of the transformations (2.112) and (2.114).

Yet, since on the [b]-direction already is a second order helical, namely $2[q_3yq_4]$, the corresponding identifications result as:

$$\begin{cases} q_3 = 0 \\ q_4 = q_2 + \dfrac{1}{4} \end{cases} \qquad (2.116)$$

Comparing the results (2.115) and (2.116), one yields that, excepting the unique solution $q_1 = q_3 = 0$, there are two sets of values for q_2 and q_4:

$$\begin{cases} q_2 = 0 \, , \ q_4 = \dfrac{1}{4} \\ q_2 = \dfrac{1}{4}, \ q_4 = 0 \end{cases} \qquad (2.117)$$

The solution (2.117) indicates the doubling for the possibilities of the equivalent points resulted from the transformations (2.112) and (2.113), respectively:

$$\begin{cases} q_1 = 0, q_2 = 0, q_3 = 0, q_4 = \dfrac{1}{4} \overset{(2.112),(2.113)}{\Rightarrow} Q(x,-y,-z); S\left(-x,y,\dfrac{1}{2}-z\right) \\ q_1 = 0, q_2 = \dfrac{1}{4}, q_3 = 0, q_4 = 0 \overset{(2.112),(2.113)}{\Rightarrow} T\left(x,-y,\dfrac{1}{2}-z\right); V(-x,y,-z) \end{cases}$$

$$(2.118)$$

The analysis of the new points (Q, S) & (T, V), by their coordinates, show that they are equivalent points by the already existed and identified symmetries.

Thus, applying a second order helical operation on the [c]-direction to the points (Q, S) the transformations of Q in V and of S in T are noted. Therefore, the minimal set of structural points contains four equivalent groups that exhausted the elements and the symmetry operations from the analyzed space group.

The last step is that of representation.

For this, a plane a perpendicular on the main axis (correlated with the most symmetry operations) is chosen, in this case the helical on the [c]-direction.

Furthermore, form the information that the considered space group belongs to the orthorhombic system, the projection on the (ab) plane of P222$_1$ primitive will be a rectangle, within which the origin (0, 0, 0) shall be considered and about which the equivalent structural points (x, y, z); (x, -y, -z); (-x, -y, 1/2+z); (-x, y, 1/2-z) can be represented along with the set of symmetry operations associated with the unit cell.

There results the diagram as that one represented in the left side of the Figure 2.66.

There is noted how, the four equivalent points are concentrated inside the area that configures the xy-projection of the unit cell (primitive), with the (0, 0, 0) center in the middle of the cell, on the [c]-axis.

The diagram shows at the same time that, for any other points than those provided above, the effect of applying the existing symmetry around each point produces results outside the unit cell. One finally notices how,

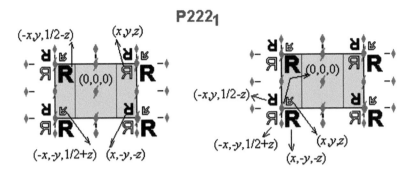

FIGURE 2.66 The diagrammatic representations of the crystallographic space group (#17) P222$_1$; after Verma and Srivastava (1982) and U.S. Naval Research Laboratory/ Center for Computational Materials Science (2003).

equivalently, the origin (0, 0, 0) can be reconsider in a corner of the primitive cell (e.g., on the bottom left of Figure 2.66-right).

According to this new choice of the reference point the generic structural point of coordinated (x, y, z) is re-defined and then, by applying the symmetry group, all other equivalent points are re-obtained.

Similarly, the diagram of the other space groups of Table 2.16 can be detailed and represented. Some results are shown in the Table 2.19. From there should be noted that, when appropriate, also the equivalent diagrams for the projections of the Bravais unit between the centered forms on the faces A, B, or C were shown in Table 2.19 too.

One such example is the case of space group #15 for which both the equivalent projections, C2/c and B2/b, were presented, depending on the considered various faces in the unit cell, as centered with the structural points.

As shown in the previous description of the space group #17 (P222$_1$), for the space group #15 the representation with the (0,0,0) in the down left corner of the diagram C2/c of Table 2.19 was directly taken, noting how the four equivalent points exhaust their space expanding it until the another point coordinate, here as (1/2,1/2,0), wherefrom the four equivalent points are again repeated, respecting this new origin, so that a structural (x, y, z) point becomes (1/2+x, 1/2+y, z), and so on. There will be said that the group #15 has, in this case, two coordinates origins, namely (0,0,0) and (1/2,1/2,0).

TABLE 2.19   The Diagrammatic/Projective Representation of the Space Groups and the Pearson Classification for Representative Examples of Crystals' Primitives*

No.	Crystallographic System/Syngony
Triclinic System	

1	
P1 $AsKSe_2$; aP16	

2	
P$\bar{1}$ Cf; aP4	

| *MONOCLINIC SYSTEM* | |

4	
P2$_1$ (>Pa)Te; mP4	

8	
Cm PZT: $O_3PbTi_{0.48}Zr_{0.52}$;mC10	

11	
P2$_1$/m $KClO_3$; mP10;	

12	
C2/m $AlCl_3$; mC16	

TABLE 2.19   Continued

No.	Crystallographic System/Syngony

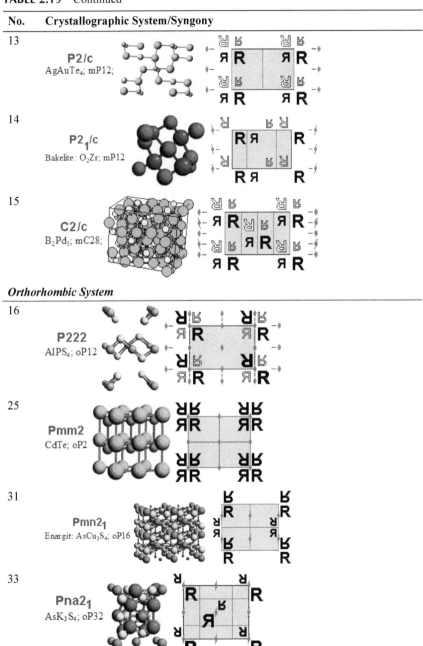

13   **P2/c**
AgAuTe$_4$; mP12;

14   **P2$_1$/c**
Bakelite: O$_2$Zr; mP12

15   **C2/c**
B$_2$Pd$_5$; mC28;

Orthorhombic System

16   **P222**
AlPS$_4$; oP12

25   **Pmm2**
CdTe; oP2

31   **Pmn2$_1$**
Enargit: AsCu$_3$S$_4$; oP16

33   **Pna2$_1$**
AsK$_3$S$_4$; oP32

TABLE 2.19   Continued

No.	Crystallographic System/Syngony

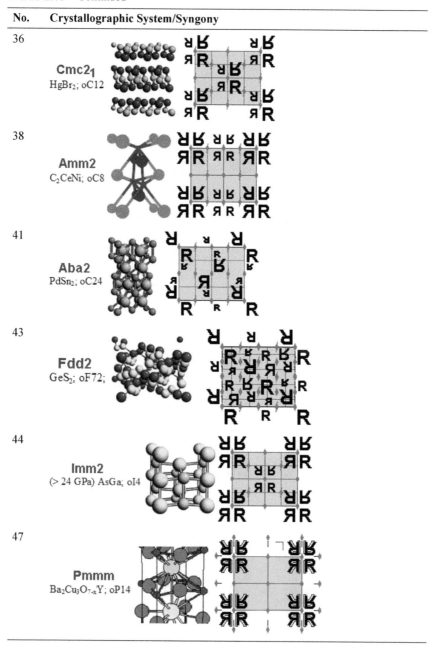

36	**Cmc2₁** HgBr₂; oC12
38	**Amm2** C₂CeNi; oC8
41	**Aba2** PdSn₂; oC24
43	**Fdd2** GeS₂; oF72;
44	**Imm2** (> 24 GPa) AsGa; oI4
47	**Pmmm** Ba₂Cu₃O₇₋ₓY; oP14

TABLE 2.19 Continued

No.	Crystallographic System/Syngony
51	**Pmma** AuCd; oP4
58	**Pnnm** CFe_2; oP6
59	**Pmmn** Cyanogen chloride CClN; oP6
61	**Pbca** CdSb; oP16
62	**Pnma** BFe; oP8
63	**Cmcm** BCr; oC8

TABLE 2.19   Continued

No.	Crystallographic System/Syngony
64	**Cmca** B_2C_2Mg; oC80
65	**Cmmm** Ga_3Pt_5; oC16
69	**Fmmm** FTl; oF8
70	**Fddd** $Pu(\gamma)$; oF8
71	**Immm** $MoPt_2$; oI6

TABLE 2.19   Continued

No.	Crystallographic System/Syngony

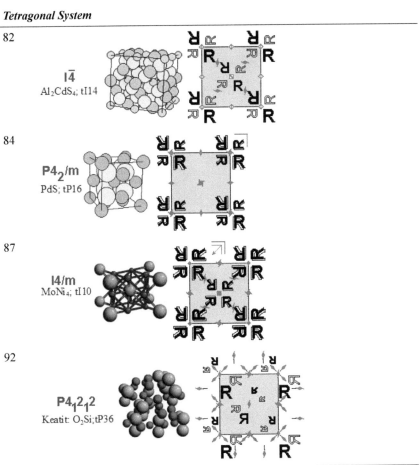

72   **Ibam**
     S_2Si; oI12

Tetragonal System

82   **$I\bar{4}$**
     Al_2CdS_4; tI14

84   **$P4_2/m$**
     PdS; tP16

87   **I4/m**
     $MoNi_4$; tI10

92   **$P4_12_12$**
     Keatit: O_2Si; tP36

TABLE 2.19   Continued

No.	Crystallographic System/Syngony

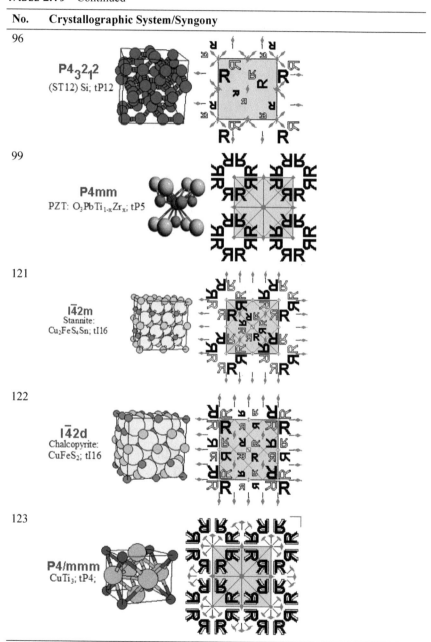

96

$P4_32_12$
(ST12) Si; tP12

99

P4mm
PZT: $O_3PbTi_{1-x}Zr_x$; tP5

121

$I\bar{4}2m$
Stannite:
Cu_2FeS_4Sn; tI16

122

$I\bar{4}2d$
Chalcopyrite:
$CuFeS_2$; tI16

123

P4/mmm
$CuTi_3$; tP4;

TABLE 2.19 Continued

No.	Crystallographic System/Syngony
127	**P4/mbm** Si_2U_3; tP10
129	**P4/nmm** Cu_2Sb; tP6
134	**P4$_2$/nnm** B; tP50
136	**P4$_2$/mnm** CrFe(σ); tP30

TABLE 2.19   Continued

No.	Crystallographic System/Syngony

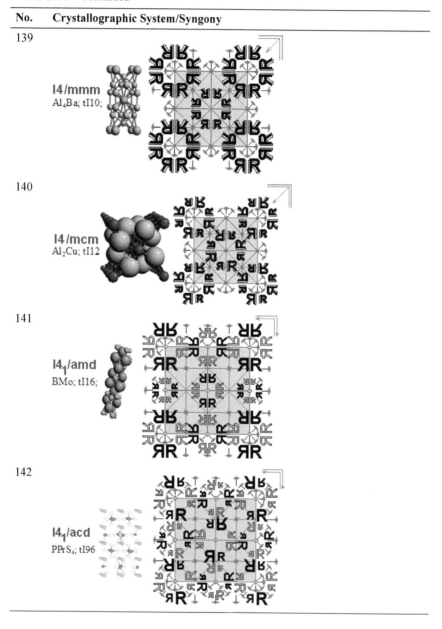

139   **I4/mmm**   Al₄Ba; tI10;

140   **I4/mcm**   Al₂Cu; tI12

141   **I4₁/amd**   BMo; tI16;

142   **I4₁/acd**   PPrS₄; tI96

TABLE 2.19   Continued

No.	Crystallographic System/Syngony

Trigonal System

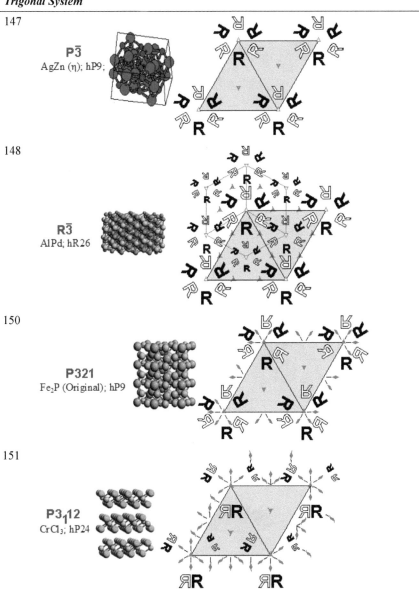

147   **P3̄**
AgZn (η); hP9;

148   **R3̄**
AlPd; hR26

150   **P321**
Fe₂P (Original); hP9

151   **P3₁12**
CrCl₃; hP24

TABLE 2.19   Continued

No.	Crystallographic System/Syngony

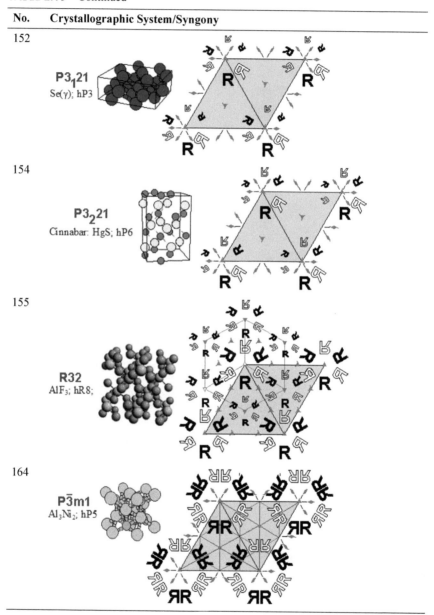

152   **P3$_1$21**   Se(γ); hP3

154   **P3$_2$21**   Cinnabar: HgS; hP6

155   **R32**   AlF$_3$; hR8;

164   **P$\bar{3}$m1**   Al$_3$Ni$_2$; hP5

TABLE 2.19   Continued

No.	Crystallographic System/Syngony
166	

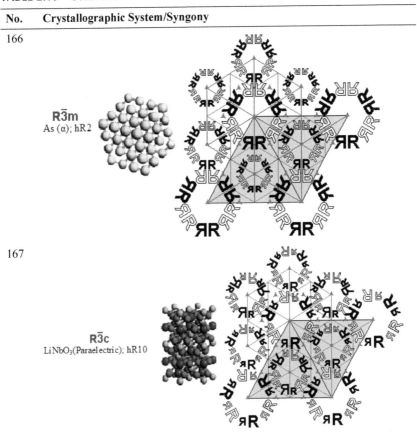

R3̄m
As (α); hR2

167

R3̄c
LiNbO₃(Paraelectric); hR10

Hexagonal System

180

P6₂22
CrSi₂; hP9

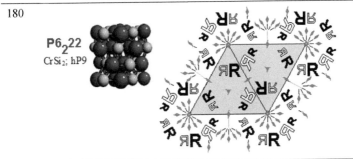

TABLE 2.19   Continued

No.	Crystallographic System/Syngony

182

P6₃22

Bainite: CFe₃; hP8

186

P6₃mc

Al₅C₃N; hP18

187

P6̄m2

BaPtSb; hP3

189

P6̄2m

Fe₂P (Modificat); Hp9

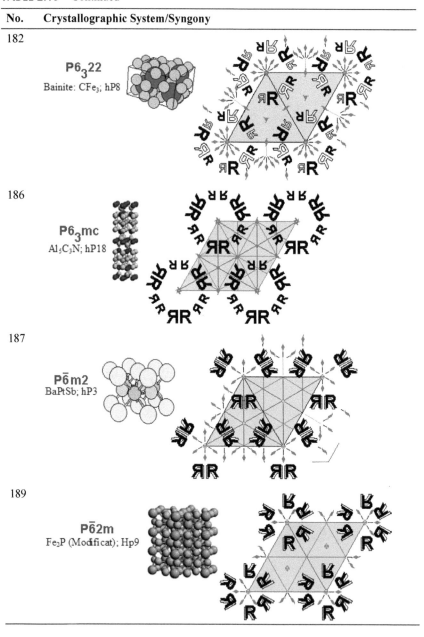

TABLE 2.19 Continued

No.	Crystallographic System/Syngony
191	

P6/mmm

AlB$_4$Mg; hP6;

| 194 | |

P6/mmc

AlCCr$_2$; hP8;

Cubic System

| 197 | |

I23

Ga$_4$Ni; cI40

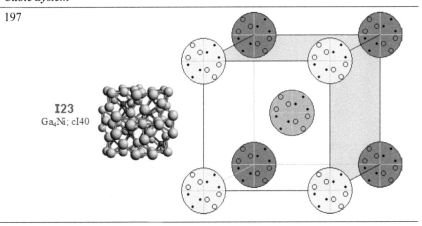

TABLE 2.19   Continued

No.	Crystallographic System/Syngony
198	
199	
204	

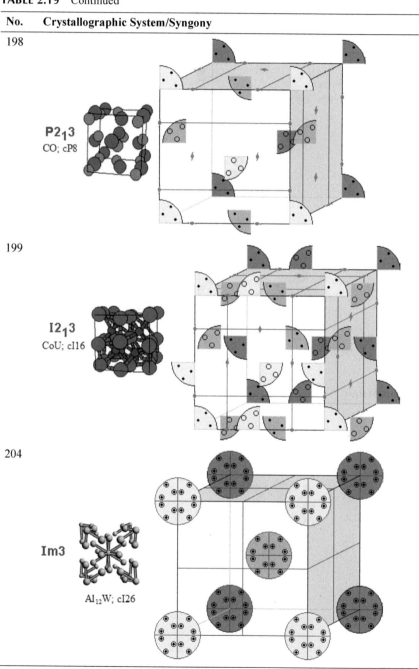

$P2_13$

CO; cP8

$I2_13$

CoU; cI16

Im3

$Al_{12}W$; cI26

TABLE 2.19   Continued

No.	Crystallographic System/Syngony
205	
206	

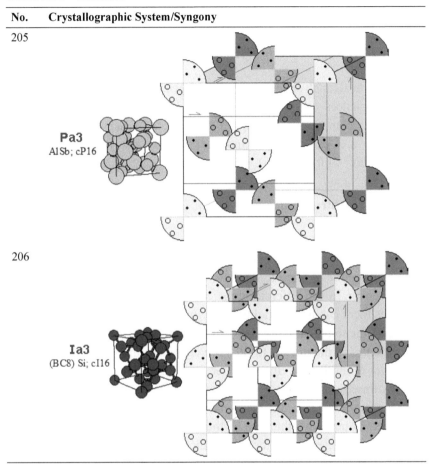

Pa3
AlSb; cP16

Ia3
(BC8) Si; cI16

*After (Goodhew, 2003; Commission on Crystallographic Teaching – IUCR 2003; International Tables for Crystallography, 2003; U.S. Naval Research Laboratory/Center for Computational Materials Science, 2003; Optical Mineralogy, 2003).

The complete list of the diagrams with the equivalent points associated is found in the *International Tables of Crystallography* (2003 and its further editions), with the interpretation as previous explained. However, in Table 2.19 not all the space groups are exposed from two certified reasons: the first refers to the fact that not all the space groups are actually identifiable, i.e., about 11 groups are enantiomers, i.e., they are equivalent by mirroring symmetry to the other space groups; the second argument consists in the fact that not all the space groups correspond to real crystals, i.e., there are space groups where no chemical compound is crystallized.

2.6.3 EXTENSIONS OF THE SPACE GROUPS

The statistical analysis (Zorky, 1996) revealed the fact that the inorganic substances crystallize in approximately 209 space groups, while the organic compounds overlap only 185 of the space groups. The consequences are particularly important; the present discussion follows (Chiriac-Putz-Chiriac, 2005).

2.6.3.1 The Structural Class

Firstly, there is noted how the set of 230 space groups largely overlap with almost all the geometric possibilities of crystallization of a chemical compound, inorganic or organic.

Moreover, by the different statistical overlay of the crystallization possibilities there can be highlighted also from the space groups perspective, the phenomenological unity between the inorganic and organic classes of the structure of the matter. In fact, the crystallography offers a geometric paradigm of unification of the matter, by the geometric view of the structural symmetries.

Thus, until this step, the classification of the chemical compounds was advanced in terms of: the type of the unit cell-crystalline system, of the Bravais lattice-elementary cells to which a crystalline state belongs, of the point group associated to the morphology and crystalline habitus and, finally, of the spatial group where the structural points associated with the motif/base of the associated lattice that can be noted for the concerned compound.

However, beyond the ability to "classify" the chemical compounds in terms of the space group that belongs to the associated crystalline state, a special case is represented by the molecular crystals of the organic compounds – also for the fact that they introduce new elements of symmetry, thus generalizing the consecrated crystallographic ones.

For a statistical overall view, there is instructive to note the organic crystals distribution in the crystalline systems, namely (Belsky & Zorky,

1977; Belsky et al., 1995): 14.0% in the triclinic system, 55.0% for the monoclinic, 27.7% for the orthorhombic, 1.0% in the trigonal, 0.3% in the hexagonal and cubic at only 0.2%.

The generalization of the space group concept, can be performed in a topological manner, for example by neglecting the coordinates of the atoms from the equivalent points, as well for the parameters of the lattice of the periodical objects. Which is left refers to:

(i)   the symmetry group (one is in the Table 2.8 presented), and
(ii) the specification of the equivalent structural points (in relation with the symmetry operations of the symmetry group of the previous point), thus defining the concept of *structural class* (SC).

This view can be applied to any arrangement of atoms in: molecules, reticular strings, and finally, in crystals. For a molecular crystal, its structural class – SC will be indicated by:

(i)   a symbol that includes the symbol of the associated space group,
(ii) to which is added the number of molecules (Z) on the unit cell (one notes a similarity with the previous Pearson notation, in this case at the molecular level),
(iii) and followed by a bracket where the point group to which the molecule (the motif in the crystal) belongs (as it would be independent).

As illustrative examples take the structural classes $P2_1/c$, $Z = 4(1)$ (this structure is also called "elephant"), being the most common SC of the homo-molecular crystals, or $P2_1/c$, $Z = 2(1)$, rendered in the projection on a base of the unit cell, as in Figure 2.67.

If the motif contains two or more molecules, the crystal will be called *polysystemic*, a case in which two or more symbols of punctual groups are indicated in the bracket of the symbol of the structural class associated; for example $P2_1/c$, $Z = 4(1, 1)$ is abbreviated as: $P2_1/c$, $Z = 4(1^2)$; or as $P2_1/c$, $Z = 6(1,1)$; $P2_1/c$, $Z = 8(1.86)$; $P2_1/c$, $Z = 12(1.89)$ and so on.

The classification of the crystals by the extended notion of SC allows the description of some molecular crystals which, otherwise, would not fit in any other spatial group.

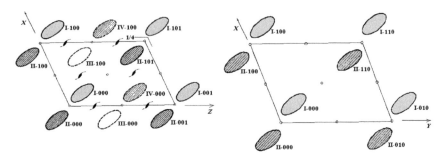

FIGURE 2.67   The representations of the unit cells for the structural classes P2$_1$/c, Z = 4(1) (left) and P2$_1$/c, Z = 2(1$\bar{1}$) (right); after Zorky (1996).

An already classical case (Belsky & Zorky, 1977; Belsky et al., 1995; Zorky, 1996) is that of the tolan molecule (C$_6$H$_5$–C≡C–C$_6$H$_5$) that belongs to the structural class P2$_1$/c, Z = 4(1^2) (see Figure 2.68). From the Figure 2.68, the independent molecules of type A and B are noted, i.e., those non-equivalent molecules/structural points in the associated space group, are yet transformed one in another, by a rotation-displacement operation of 2$_q$ type, with the axis of the 2nd order passing through the M (1/4,0,0) point to a ω = 6.3° angle towards the *(X0Z)* plane and parallel with the *(Y0Z)* plane.

This operation presumes, for example, a rotation of the A-type molecule with 180° around the direction of the axis, followed by a displacement (deliberately it is not called translation!) around the axis with the value δ = (a/2) cosβ cosω, so obtaining the "placement" of the molecules of B-type in the molecular crystal.

There is obvious that this kind of symmetry, i.e., the presence of the axis of 2$_q$ type, can not be found in none of the symmetries of the space groups, being this the reason for which it is introduced as a *hypersymmetry*, but still being a symmetry.

Moreover, the presence of the of 2$_q$ type axis is not accidental, since not being unique. Other similar can be identified as: the one which passes, in Figure 2.68; and both through the points M and N (1/4, 1/4, 1/4), in this case being the A and B' molecules the hyper-equivalent through the 2$_q$ axis.

Similar axes are found also in other molecular crystals from which, the most representative, are those of the structural P2$_1$/c, Z = 4(1^2) class, this differing only by the specific value of the angle ω, see Table 2.20.

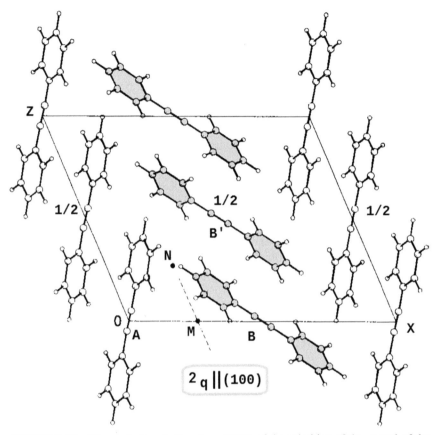

FIGURE 2.68   The projection in the plane (X0Z) of the primitive of the crystal of the Tolan molecule, after Zorky (1996).

TABLE 2.20   Examples of Organic Molecules, the Working Crystalline Motifs for the Molecular Hyper-Symmetric Crystals*

Organic Molecules	ω [^{0}]
Stilben/Stilbene	6.6
3,3'-Bi-isoxazol	40.7
Tetrabromo-ethylene	25.9
4,8-Dihidroxil-1,5-	26.0
Naphthaquinones	7.3
1,5-Difluoroantraquinona	−5.9
Isodibenzanthrone	77.0

*After Belsky & Zorky (1977); Belsky et al. (1995); Zorky (1996).

There can be noted how, with the axis of 2_q hyper-symmetry, the hyper-symmetric m_p (reflection + displacement) plane also coexists in the center-symmetric crystals, a plane that can be identified also for the tolan case of Figure 2.68 as being the normal plane on the 2_q directions.

At this step the difference between the axes of 2_q type that define the hyper-symmetry and the axes of the 2_1 type can be made, implying an operation that acts on an entire reticular (molecular) string.

Moreover, an intermediate situation between the hyper-symmetry of the independent molecules and the symmetry of the complete equivalent axes in a crystal generates a case that will be called as *pseudo-symmetry*.

Illustrative examples are presented in Table 2.21.

Finally, there is worthy to point out the fact that inside the molecular organic crystals the hyper-symmetry prevails, due to the frequency with which the independent molecules have similar conformations, as revealed from the analysis of the statistics of the Cambridge structural database (Sona & Gautham, 1992).

2.6.3.2   The Magnetic/Colored Groups

Another extension of the space groups can be performed when, besides the consecrated symmetries, the *anti-symmetry* is also considered, i.e., the possibility that the structural points to be endowed also with the *spin*, with the, "up" and "down" traditional possibilities.

Therefore, new positions in the space are generated, for which, the equivalences must be re-considered by applying of a new operation to re-symmetric the produced anti-symmetry.

The idea equivalents with a "coloring" of the equivalent points or of the crystalline region (the crystallography faces or planes, for instance) in complementary colors (in this case, by white-and-black) so that only certain symmetries from those initial will be further viable, while other will appear.

Then, the anti-symmetry operation will mean the mutually change of the colors. This "enrichment" with spin or color of a crystalline structure is not accidental; it resides in the magnetic quality (that depends on the spin orientations of the components of a crystal) that certain chemical compounds recorded.

TABLE 2.21 Organic Molecules' Example of and Their Pseudo-Symmetric Crystals*

Motif/Pseudosymmetry	The Projection of the Pseudosymmetric Unit Cell
2-deoxy-12-oxolemnacarnol $C_{15}H_{22}O_3$ molecular reticular string formed by H-bonds of the I-000, I'-000, I-010, and I'-010 molecules has the pseudo-symmetry $Pc(Y)2_1$	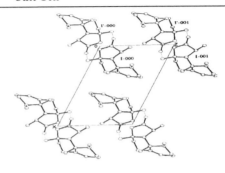
cis-4a,5,8,8a-tetrahydro-1,4-naphtoquinone $C_{10}H_{10}O_2$ molecular strings are formed by CH···O bonds and are correlated in pairs by the $P_{c(X)}2_1 11$ pseudo-symmetry	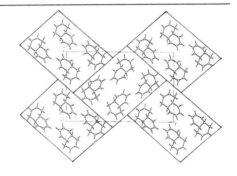
2-(2-hidroxi-3-fenil-2-propen-1-iliden)-3,3,5,5-tetramethyl-cyclopentanone $C_{18}H_{22}O_2$ CS: P(-1), Z = 6(1.89), with molecular strings with the pseudo-symmetry $P_{c(X)}3_1$, obviously non-included in the associated spatial group	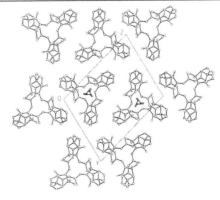

*After Belsky & Zorky (1977); Belsky et al. (1995); Zorky (1996).

Thus, the extension of the crystallographic space groups to those containing the anti-symmetry operation will be called *magnetic or colored groups* (Verma & Srivastava, 1982, Sona & Gautham, 1992).

As in the space groups introductory case, the magnetic groups characterization of the should relay on the level of the punctual crystallographic groups.

Accordingly, this will be exposed for the 4*mm* (C_{4v}) group of whose symmetry operations include (revise Tables 2.6 and 2.8): the inversion center, the rotations of 2nd and 4th (± clockwise or counterclockwise) orders, as also the reflections toward the m_1, m_2, m_1' and m_2' planes, see Figure 2.69 for the so called (*0*)-symmetric operations.

The representation of this punctual group can be affected by the magnetic influence through its "coloring" (in white-and-black) in two possible (*I, II*) variants, both based on the (0)-representation of Figure 2.69.

For (*I*) - case of Figure 2.69, the set of symmetry operations allowed by the (0) state are still available: the rotations of 1st order (the inversion center), the 2nd order rotation, and the mirroring on m_1, m_2 planes. The rest of the symmetries can be recovered only by the additional application of an anti-symmetry operation (*θ*), with the effect of the appearance of new anti-symmetry operations, namely: $θ4^{±}$, $θm_1'$ and $θm_2'$: the symmetries and the anti-symmetries together generate the new magnetic 4'*mm*' punctual group.

Analogously, for (*II*) - case of Figure 2.69, the symmetry operations that are "survived of the coloring" by anti-symmetry are: the rotations

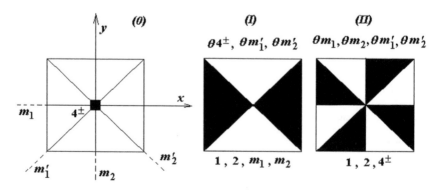

FIGURE 2.69 The simple (0) and magnetic (I, II) symmetry and anti-symmetry operations of the 4mm point group, respectively; after Verma and Srivastava (1982).

TABLE 2.22   The Point- and Generalized-Space Groups*

Shubnikov (Magnetic) Group	Point Group	Spatial Group
Type I	32	230
Type II	32	230
Type III	58	1191
Total	122	1651

*According to Shubnikov and Belov (1964).

of orders 1, 2 and $4^{\pm}$, to which those lost by considering anti-symmetry operation through the "inversion of the color" are added through the newly generated $\theta\, m_1$, $\theta\, m_2$, $\theta\, m_1$' and $\theta\, m_2$' operations. The two classes of symmetry operations compose the new $4m'm'$ point magnetic group.

The "survived" and resulted sub-groups of the symmetry operations from the consideration of the anti-symmetry, over the initial group, are called as *reduced groups* of the original point groups. Their identification, out of all the 32-point crystallographic possible groups, increases the number of the point groups, also by eliminating the identical cases, to 58-point magnetic groups, and which, through further combination with the translation operations further increases the number of the space group to 1191 (!) *magnetic space groups*.

These are the so-called *Shubnikov groups* (after the name of the initiator of the study of those types of groups) of Type III. The Shubnikov groups of Type I are those ordinary (32 punctual, with the 230 spatial), i.e., the "uni-colored" ones, while those of Type II being obtained form those of the Type I by applying (all along) the composition with the θ anti-symmetry operation, which results, again, in the 32-point groups and of the 230 spatial, "uni-colored" spatial groups, yet in the complementary color to those of Type I. The summary of these generalized groups is in Table 2.22 presented.

2.7   CONCLUSION

The main lessons to be kept for the further theoretical and practical investigations of the crystallography by geometrical principles that are approached in the present chapter pertain to the following:

- Identifying the main symmetries allowed in a crystal structure as an ordered chemical structure-a giant molecule with periodical structure;
- Employing the main symmetry elements and operations, especially the rotations, to characterize the possible mutual arrangements of structural particles/the motifs in the crystal lattice, including the restraining of the rotational axes of rotations to 1, 2, 3, 4, and 6, so excluding pentagons from geometries able to cover the extended spaces by themselves;
- Writing the geometrical length and angles relationships so driving the crystals classifications in syngonies and crystallographic systems;
- Dealing with specific rules of transformations and reducing between the unit cells of crystallographic systems towards expanding their number to those known as Bravais lattice so allowing also the inner or face-shared structural particles for single or neighboring unit cells of lattices;
- Characterizing the crystals by their point groups through systematically combining all rotational axes with mirroring planes and inversion centers towards the consecrated 32 classes;
- Understanding the symmetries features of crystals by their resumed notations (Schoenflies and international/Hermann-Mauguin) as well as by their specific habitus influencing their morphology (external shape) as paralleling the inner planes and symmetries at the unit cell level;
- Describing the nodes, axes and planes of crystals, at unit cell level, i.e., related with the fundamental referential (the crystallographic lengths and angles) system of specific crystal system through Miler indexing and by face and Weiss law of zones, so pre-introducing the reciprocal space framework in which the electronic behavior is modeled by allied quantum nature – the inverse of the wave function aka wave-vectors at their turn de Broglie quantified;
- Learning to recognize specific crystalline system by their projection in general and by stereographic and Wulff projections/maps in special, this way preparing the required tools in dealing with X-rays diffraction diagrams for crystal structure investigation, naturally provided in the reciprocal space – the quantum space of electronic behavior in solid states and systems;
- Treating the crystalline states by their polyhedral modeling, i.e., correlating the point group with the specific morphology created by the crystals' habitus;

- Solving the perturbation effects on crystals by systematic counting of mechanical, electrical, and optical effects;
- Formulating the requirements for anisotropy properties of a crystal structure, and illustrating it by the optical activity in crystal leaving with birefringence effects, and restraining groups of symmetry (Laue groups) not distinguishing upon inversion center of symmetry;
- Interpreting the Friedel effect/law of inducing of an inversion center in a crystal (unit cell) structure by optical perturbation, in special due to the X-ray near-edge of absorption of characteristic spectra;
- Connecting the physical perturbation of crystal symmetry with Curie principle that provides a practical receipt characterizing the system + perturbation common symmetry by considering their commonalities;
- Developing the space groups and the extended symmetries including the translation along axes (helicals) and planes (glides), including the Pearson classification and the in detail-characterization/identification of a given space group unit cell symmetries and geometrical ordering as starting from the symmetry space group formula;
- Finding applications for the space groups by further extending them by understanding the pseudo-symmetry specific to macromolecular chemistry as well as by coloring or (spin) magnetization specific to quantum action (by lasers, for instance) on concerned chemical solid state systems.

KEYWORDS

- **art and crystals**
- **Bravais lattice**
- **crystal's metrics**
- **Curie principle of symmetry**
- **Hermann-Maugin notation**
- **history of crystallography**
- **magnetic/colored groups**
- **Miller indexing**

- **Miller-Bravais indices**
- **Pearson classification**
- **point groups**
- **reducing cell method**
- **Schoenflies notation**
- **space groups**
- **stereographic projection**
- **symmetry perturbations**
- **Weiss zone law**
- **Wulff map**

REFERENCES

AUTHOR'S MAIN REFERENCE

Chiriac, V., Putz, M. V., Chiriac, A. (2005). *Crystalography* (in Romanian), West University of Timişoara Publishing House, Timişoara.

SPECIFIC REFERENCES

American Mineralogist Crystal Structure Database (2002, 2013) (http://www.minsocam.org/)

Belsky, V. K., Zorkaya, O. N., Zorky, P. M. (1995). Structural classes and space groups of organic homomolecular crystals: new statistical data. *Acta Crystallogr. A* 51:473–481.

Belsky, V. K., Zorkii, P. M. (1977). Distribution of organic homomolecular crystals by chiral types and structural classes. *Acta Crystallogr. A* 33:1004.

Bilbao Crystallographic Server (2003). http://www.cryst.ehu.es

Commission on Crystallographic Teaching – IUCR (2003, 2013). http://www.iucr.org/education/pamphlets

Duch, S. (2003). *Lecture Notes for Natural and Applied Sciences*, University of Wisconsin-Green Bay.

Franzen, H. F. (1994). *Physical Chemistry of Solids. Basic Principles of Symmetry and Stability of Crystaline Solids*, World Scientific, Singapore.

Golden, M., Bernard, D. (2002). *Lectures Notes on Condensed Matter Science*, Master program at University of Amsterdam and Free University of Amsterdam.

Goodhew, P. (2003). *Introduction in Crystallography* (MATS 110), Course Lectures, University of Liverpool.

Hartmann, E. (2003), *An Introduction to Crystal Physics* (Description of the Physical Properties of Crystals), Teaching Commission – IUCR Pamphlets.

Haüy, R. J. (A.) (1822). *Traité de cristallographie.* Bachelier & Huzard, Paris.

Heyes, S. J. (1999). *Four Lectures in the 1st Year Inorganic Chemistry Course*, Oxford University.

HyperPhysics (2010). http://hyperphysics.phy-astr.gsu.edu/Hbase/hframe.html

International Tables for Crystallography (2003, 2013). http://it.iucr.org/

Nye, J. F. (1985). *Physical Properties of Crystals*, Clarendon Press, Oxford.

Optical Mineralogy (2003, 2013). http://edafologia.ugr.es/optmine/indexw.htm

Pettifor, D. (1995). *Bonding and Structure of Molecules and Solids*, Oxford Science Publications, Clarendon Press, Oxford.

Platone (1996). *La Republica*, Arnaldo Mondatori Editore, Milano.

Shubnikov, A. V., Belov, N. V. (1964). *Colored Symmetry*, Oxford, Pergamon Press.

Shubnikov, A. V., Koptsik, V. A. (1977). *Symmetry in Science and Art*, Plenum Press, New York and London.

Sona, V., Gautham, N. (1992). Conformational similarities between crystallographically independent molecules in organic crystals *Acta Crystallogr. B* 48(1), 111–113.

Stereographic Projection (2002). Notes on Mineralogy, University of Würzburg.

U. S. Naval Research Laboratory/Center for Computational Materials Science (2003, 2013). http://cst-www.nrl.navy.mil/.

Verma, A. R., Srivastava, O. N. (1982). *Crystallography for Solid State Physics*, Wiley Eastern Limited, New Delhi.

Villars, P., Calvert, L. D. (1991). *Pearson's Handbook of Crystallographic Data for Intermetallic Phases*, 2nd Edition, ASM International, Materials Park, Ohio.

Wolfram Research 2003: http://mathworld.wolfram.com

Wolfram Research Crystalography (2003, 2013). http://mathworld.wolfram.com/CrystallographicPointGroups.html

World of Escher (2014). http:www.worldofescher.com

Zorky, P. M. (1996). Symmetry, pseudosymmetry and hypersymmetry of organic crystals. *J. Mol. Struct.* 374, 9–28.

CHAPTER 3

QUANTUM ROOTS OF CRYSTALS AND SOLIDS

CONTENTS

ABSTRACT

The quantum mechanics postulates are shortly revived towards of their application in providing the basic crystal Bloch theorem and of further specialization to the quantum modeling of crystals in reciprocal space, on various levels of electronic behavior from free to quasi-free, to quasi-binding, to tight-binding models, as well as for modeling solids by quantum statistical means, especially grounded on Fermi statistics driving the valence-to-conducting levels' displacements so providing the quantum framework for semiconducting electrons and to chemical related moletronics.

3.1   INTRODUCTION

In 1783, Immanuel Kant said that "human reason many times afforded to construct the science as a tower and then to completely demolish it to analyze its foundation; never it's too late for becoming reasonable and wise; but the more knowledge comes latter, the more difficult is to promote a reform!"

In a surprisingly way and unjustly passed in the science history shadow, there appears the concept of solid "state" of the atom as proposed by Lewis (1916) in his seminal article for the modern chemistry, in which the rule of the doublet and the octet based on the concept of the cubic atom are established; the present discussion follows (Chiriac-Putz-Chiriac, 2005).

In 1916, Lewis was expressing his criticism for the recent atomic model proposed by Niels Bohr, based on the Max Planck's quantum theory, which assumed the atomic system as composed of fixed spherical orbits, so quantified – because, Lewis argued that, the spherical atomic system "contains electrons whose motion is insensitive to the action of the external charges".

In other words, Lewis admitted that the electrons from an atomic layer, eventually the valence one, are equivalent (interchangeable), but felt the need to customize this symmetry, or intra-atomic equivalency, on the inter-atomic level.

The following geometric element with a high degree of symmetry after the Sphere is the Cube. Therefore, as a counterweight to the spherical *Bohrian atom*, Lewis proposed the cubic atom model. Moreover, and here lies the genius of Lewis, the cube, keeping its internal symmetry, allows its customization from the outside, when is differently occupied by electrons, to its corners.

The period between 1900–1916 was full of experimental evidences, based on the new methods of the X-ray structural investigations, see the Chapter 5 of the present volume (Putz & Lacrămă, 2005), and it was already known that, along a chemical period of elements, the atoms present the external electrons, as distinguished from the internal one, with those external being more free, less attracted by the atomic nucleus, and so the first likely to participate in chemical bonds.

One had concluded that the external electrons are indexed from 1 to 8, in a period. This experimental evidence combined with the symmetry idea, specialized for each atomic structure, finds its full expression in the Lewis theory of the cubic atom.

In this context, an atomic system is composed of a core (nucleus + internal electrons) found at the (virtual) cubic center and a layer of valence (the external atom) with the electrons symmetrically disposed toward the core in the cube's corners, so that a maximum symmetry to the cube center to be covered, see Figure 3.1.

From here, the next step was immediate: *the octet rule*.

Each atomic system is "oriented" to enter into combinations with other atomic systems, after which each of the participant cubic atoms will at

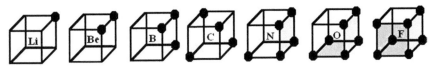

FIGURE 3.1   The cubic atom: the atomic structure is seen as a center-core equidistantly surrounded by the cubic corners by the valence electrons, so disposed to achieve a maxim symmetry in the cube; after Lewis (1916).

maximum symmetrize its valence in the cube, by giving/accepting electrons from/to the cube corners to achieve the "magical" cardinals of 0 and 8 electrons, with a maximum symmetry towards the cube center.

This way, one concluded that the full occupancy with 8 electrons in the cube corners corresponds to the noble gas configuration. But, as long as also He was considered as being part of the same class, despite the fact that the X-ray experiments have proved the total existence of only two electrons in the structure of He, Lewis felt the need to complete the octet rule with the one of the doublet, in order to explain the stable structure of He.

Remarkably, Lewis anticipated, by the atom cubic paradigm, even the fundaments of the quantum mechanics, much later fully accomplished.

For example, the brilliant observation that "the electron should be seen not just as a particle charged with electric charge, but also as a carrier of an own "magnetic moment", coupled with the Lewis interrogation, i.e., if the octet rule and of the doublet can be somehow reduced to a "basic rule", i.e., by forming pairs of electrons – anticipated the fundamental concept of electronic spin and the Exclusion Principle (Pauli principle).

Moreover, with an extraordinary power of prediction Lewis said how the "Coulomb's law of the attraction proportional to the inverse of the distance's square is no longer valid at small distances". And even here, Lewis was not wrong since he only omitted a single attribute: instead of "small distances" was proved that his intuition is applied at "*very* small distances".

Thus, the Coulomb forces take place on the atomic distances, when, instead, at nuclear distances they are renormalizing into the *weak forces*, one of the four fundamental forces existing in nature (the gravitational forces, the electromagnetic-Coulomb, the weak-nuclear and the strong-subnuclear forces) (Putz, 2010).

However, based on cubic atom paradigm, Lewis had formulated, in an original manner, a set of postulates of the chemical behavior of matter composed of atoms, that can be constituted, in fact, as the first postulates of the modern chemistry, as follows (Lewis, 1916).

1. Every atom is essentially composed of an *atomic core*, which remains unaltered in all the ordinary chemical processes, with a positive charge corresponding, as number, to the number of the periodic group to which the associated element belongs.

2. Any atom, besides the atomic core, has an *external shell/layer* or *shell/layer of valence*, which contains a number of electrons equal with the positive charge of the atomic core, and can vary between 0 and 8.
3. Any atom tends to fill the valence layer to an even number of electrons, and in particular to 8 electrons, arranged in a symmetrical manner in the 8 corners of a cube, consecrating *the octet rule*.
4. Two atomic layers of valence are mutually *interpenetrable*.
5. The electrons can freely move from one position to another in the valence layer, making the so-called *intra-atomic isomers*, which favor the creation of molecular compounds.
6. The electric forces between the particles close to each other do not satisfy the simple law of the inverse square of the distances between them, which, instead, are manifested at large distances. Even this postulate can be valid if it is interpreted as an "equivalent reformulation" of the Coulomb's law in an exponential form, current called as *the pseudopotential*, and which is an approximate analytical substitute of the Pauli principle.

However, based on the cubic atom paradigm, and hence with the octet rule, one can describe the atomic structures and the chemical bonds in a consistent and unified manner, with a high degree of causality and interpretation.

Although the Bohr's quantum paradigm have been imposed at the end and nowadays, worth mentioning that the chemical bond and its transformations have their origin, as a phenomenological projection, in the cubic paradigm of the atom, the Lewisian cubic atom.

In the other side, in modern age of quantum mechanics, Wigner and Seitz stated in 1955 in front of the quantum challenges of solid state of the matter and of its combinations and transformations: "if someone would have a huge computer, probable was able to solve the Schrödinger problem for each (crystal/solid/) metal and thus obtained interesting physical sizes... which most likely will be in concordance with the experimental measures, but nothing vast and new could not be obtained in such a procedure. Instead, there is preferable a lively picture of the behavior of wave functions, as basis for the description of the essence of the factors... and for understanding the origin in the variation of properties from a (crystal/ solid/) metal to another".

Without claiming to be closer, even at the begin of XXI, to the complete quantum solution of matter and of its properties, we continue to employ and explore the vivacity of the wave functions modeling a plethora of new scientific insights and technical opportunities! These present sections will guide us in this way. However, to comply also with Kant's requirement for a viable theory of matter, yet without "demolishing" the observational edifice of natural sciences (with Physics and Chemistry carrying the preeminent role of knowledge advancement), one can reanalyzing the characterization of crystalline structure from the quantum roots, directly related with the solid-state and crystalline specificity: the periodicity of crystallographic planes and the reciprocal lattice. This way, the "inductive" process of Chapter 2 will be here completed and confirmed by the current "deductive" one.

3.2   OVERVIEW ON THE QUANTUM POSTULATES OF MATTER

According with the visionary Gilbert Newton Lewis (1933): "There can be no doubt that the quantum mechanics gates the full solution of the chemistry problems!" Therefore, the structure of matter, including the solid state, along its transformations as well, found the logical-mathematical and physical explanation in the quantum mechanics formalism.

One starts from the representation of a physical system (atom, molecule, substance, or macroscopic body) in (temporal) evolution, as a dynamic system; the present discussion follows (Putz, 2006).

Thus, the problem of measuring the properties of this system shows up.

The classical mechanics prescribes a full knowledge of the system properties only if they can be measured in all instants or at any point of its evolution.

The quantum mechanics is a more general theory which does not require that at the repetition of an experiment for the measuring of a physical property (as length, speed, energy, etc.) this property to be reproduced *exactly* with the same value at any measurement.

The quantum mechanics is a statistical theory par' excellence. The famous experiments in the electron diffraction, as those of Davidson and Germer, Thomson, and Rupp, as well as those in the molecular beams of

Rabi had fully confirmed the validity of the quantum (statistical) description (upon measurement) of matter (Davisson & Germer, 1927; Thompson, 1928; Rupp, 1928; Rabi, 1936).

In the classical mechanics, the statistical approach is seen as a convenient tool for characterizing a system evolution, by means of probability, density, etc. Instead, in the quantum mechanics, the physical senses have only the averaged reality, to be confirmed by measured. Under these conditions, which are the dynamic laws of the quantum theory?

These will be synthesized in the so-called Postulates of the quantum mechanics. Here we just overview them, without excessively detailing the mathematical device and physical consequences, for which the Volume I of the present five-volume work is fully dedicated (Putz, 2016a).

3.2.1 POSTULATE OF OPERATORS AND STATE FUNCTIONS [P1]

The fundamental connection between the classical and quantum mechanics is achieved through the operators concept. They symbolize the processes, the operations, through which a (measurable) quantity/property or a function turns in another structural property or in another function. For example, the square root $\sqrt{}$ when "operates" to the function f turns it in a different function: $\sqrt{f}$. There is clear therefore how the quantum formalism is from the beginning based on the elements capable to represent a system's evolution, changelings and dynamic, wherefrom the statistical approach appears as natural, i.e., as a result of these operations.

The first postulate of the quantum mechanics refers thus to the "correspondence principle" by which the functions and the classical quantities (as coordinate, momentum, energy, or orbital momentum) become operators, having as the connection element (of correspondence) the Planck constant h or its reduced form $\hbar = h/(2\pi)$. The list of these correspondences is shown in Table 3.1.

In Table 3.1, the symbol • means the "object under focus upon which the quantum operation/operator is applied.

One also notes how only the space dependency function retains the function quality and not becomes a quantum operator (i.e., multiplies with the identity operator which leaves invariant any other function upon which is applied).

TABLE 3.1   The Classical (left) to Quantum (right) Correspondences for the Main Structural Properties of Matter (Putz, 2006)

Clasical Property	Significance	Associated Quantum Operator
$f(x)$	Any function of position, of coordinate, as the potential $V(x)$	$f(x)$
p_x	The component on (the axis) x of the impulse (likewise for the axis y and z)	$\dfrac{\hbar}{i}\dfrac{\partial}{\partial x}\bullet$
E	The Hamiltonian (representative of the total energy of the system) in the variant independent of time	$\dfrac{p_{operator}^2}{2m}+V(x)=$ $-\dfrac{\hbar^2}{2m}\dfrac{\partial^2}{\partial x^2}\bullet +V(x)$
E	The Hamiltonian in the variant independent of time.	$i\hbar\dfrac{\partial}{\partial t}\bullet$
L_z	The component of the axis z of the angular kinetic moment (orbital or of azimuth ϕ)	$-i\hbar\dfrac{\partial}{\partial \phi}\bullet$

These operators, and any operator in general $\hat{O}$, have the remarkable property to select from all the state functions in which a dynamic system can be found (i.e., the functions which describe the quantum states) those which exactly ensure the same values measured at an experiment repetition, despite the general (quantum) statistics of a measurement process.

These special state functions are called eigen-functions (in German: eigen = proper) of the system, arbitrarily noted with f, while their corresponding measured values are called eigen-values, arbitrarily noted by α. Thus, an operator $\hat{O}$ associates to its eigen-function their eigen-value α through the so-called eigen-value problems:

$$\hat{O}\bullet f = \alpha f \tag{3.1}$$

and which in the operators' language reads as follows: "at the action of the operator $\hat{O}$ toward the eigen-state f the associated eigen-value α is measured".

3.2.2 POSTULATE OF THE AVERAGED MEASURED VALUES [P2]

In the context in which the state function $f(x, y, z, t)$, in general a complex function, does not correspond to an eigen-function, the application of the operator $\hat{O} \bullet f$ does not produce the same measured value at reloading of the same experiment: what is actually measured is not an eigen-value of the system, but an averaged measured value $\langle \alpha \rangle_{\hat{O}}$.

This averaged value, being also a number, it can be obtained from Eq. (3.1), since proceeding to the replacement $\alpha \rightarrow \langle \alpha \rangle$, multiplying to the left the new equation with the complex conjugate function f^* and performing the average operation under the integral over the whole established space-time interval, so resulting the value (Putz, 2006):

$$\langle \alpha \rangle = \frac{\int f^* (\hat{O} \bullet f) d\tau}{\int f^* f d\tau} \tag{3.2}$$

Note how the space-time volume element $d\tau$ is defined as the product between the elementary temporal variation dt and the spatial volume element $dv=dxdydz$, which for a system composed of N particles become $3N$ dimensional, so called *the configuration space*.

Next, by comparing the Eqs. (3.1) and (3.2), one observes how the averaged measured value $\langle \alpha \rangle$ is identified with the eigen-value α if the condition according which the denominator from Eq. (3.2) is identical with the unity is satisfied, thus defining the *normalization condition* (Putz, 2006):

$$\int f^* f d\tau = 1 \tag{3.3}$$

In other words, the eigen-functions f which generate the eigen-values α to any measurement (through applying of the operator $\hat{O}$, for instance) have to be normalizable. Moreover, this condition implies the fact that the eigen-functions tends to zero when the space extends to the infinity, for example on the direction x: $f(x \rightarrow \infty) \rightarrow 0$, which further implies the "confining" of the system in a finite physical space through the eigen-functions f.

This idea has a meta-physical relevance in the relation with the fact that the eigen-function f has the role of associated-wave (even corpuscular) for the system it represents.

Furthermore, by interpreting the equation (3.3) in terms of the probabilities there is clear how the measure $f^*(x, y, z, t) f(x, y, z, t)$ can be interpreted as being the relative probability with which the system is found with the coordinated (x, y, z) at the moment t.

Alternatively, but with a highly phenomenological importance, the connection between the eigen-values and the averaged measured values of Eqs. (3.1) and (3.2), respectively, the variational theorem of the quantum mechanics can be established: the state functions f for which the variation of the average measured value is zero, $\delta\langle a \rangle = 0$, are the eigen-functions satisfying the Eq. (3.1). The reciprocal assertion is also valid.

3.2.3   POSTULATE OF WAVE FUNCTION BASIC SETS AND OPERATORS' COMMUTATIVITY [P3]

Because all the measured values are real, the operators allowed by the general equation (3.2) must satisfy the hermiticity relationship (Putz, 2006):

$$\int f * (\hat{O} \bullet f) d\tau = \int f(\hat{O} \bullet f) * d\tau \qquad (3.4)$$

these operators being called as *observables*.

An important theorem of the observables says that the spatial $f_1, f_2, \dots$ eigen-functions (meaning stationary, i.e., which do not depend by time) for any stationary hermitian operator (for which the stationary Hamiltonian operator in Table 3.1 is a typical case) form a complete orthogonal set of functions, also called *the basic functions set*.

This means that any other function of spatial state g can be written as a linear combination (by expansion) of the basic functions:

$$g = \sum_i a_i f_i \qquad (3.5)$$

with a_i called as the expansion Fourier coefficients.

Another important operators' property describes how two operators, for example $\hat{O}_1$ and $\hat{O}_2$, can have in common all the eigen-functions (i.e., can reciprocally "borrowing" them) if satisfying the condition of commutativity, i.e., (Putz, 2006):

$$\left(\hat{O}_1 \hat{O}_2 - \hat{O}_2 \hat{O}_1 \right) \bullet f = 0 \qquad (3.6)$$

for any state function f.

This theorem is particularly useful when dealing with operators who switch with a system's Hamiltonian, and for the operators which can assume all the stationary states of the system as eigen-functions, and so as the basic functions.

3.2.4 POSTULATE OF THE SCHRLIDINGER EQUATIONS [P4]

An elegant method to introduce the quantum particles motion's equation, called – at non-relativistic level – the *Schrödinger equation*, consists in employing the classical-quantum correspondences in Table 3.1. Thus, for the conservation of the total energy of a particle under the action of an external potential $V(x)$ there is equivalently obtained that the classical form of energy conservation (Putz, 2006)

$$-\frac{p^2}{2m} + E - V(x) = 0 \qquad (3.7)$$

is turned into the associated quantum form

$$-\left(\frac{\hbar}{i} \frac{\partial}{\partial x} \right)^2 \frac{1}{2m} + \left[E - V(x) \right] = 0 \qquad (3.8)$$

$$\Leftrightarrow \frac{\hbar^2}{2m} \frac{d^2}{dx^2} + \left[E - V(x) \right] = 0 \qquad (3.9)$$

If each term of the Eq. (3.9) is applied to a wave function independent of time, $\psi(x)$, the so-called stationary form of the Schrödinger equation is obtained (Putz, 2006):

$$\frac{\hbar^2}{2m} \frac{d^2 \psi}{dx^2} + \left[E - V(x) \right] \psi = 0 \qquad (3.10)$$

The form (3.10) allows introducing the total energy (Hamiltonian) operator of the system

$$\hat{H} = -\frac{\hbar^2}{2m}\nabla^2 + V \tag{3.11}$$

with which the Schrödinger equation (3.10) can be compactly rewritten as a problem of eigen-values and functions:

$$\hat{H}\psi = E\psi \tag{3.12}$$

with ψ as eigen-functions and E as eigen-values (or spectrum), the actual solutions of the quantum motion equation, respectively.

If the Hamiltonian a system is the considered observable with the operator general time-dependency $\hat{H}(x, y, z, p_x, p_y, p_z, t)$, one can further apply the correspondence principle (Table 3.1) resulting in the fundamental dynamical equation of the quantum mechanics for determining the evolutionary state functions (Putz, 2006):

$$\hat{H}\left(x, y, z, \frac{\hbar}{i}\frac{\partial}{\partial x}, \frac{\hbar}{i}\frac{\partial}{\partial y}, \frac{\hbar}{i}\frac{\partial}{\partial z}, t\right) \bullet f = -\frac{\hbar}{i}\frac{\partial f}{\partial t} \tag{3.13}$$

equation referred to as *the first Schrödinger equation*.

If, moreover, the state functions are also eigen-functions, i.e., they satisfy the eigen-values equation (3.1), this time with the total energy as a measured eigen-value (Putz, 2006):

$$\hat{H}\bullet f = -\frac{\hbar}{i}\frac{\partial f}{\partial t} = Ef \tag{3.14}$$

the factorized form of the eigen-function can be considered:

$$f = \psi(x, y, z)\exp\left(-i\frac{E}{\hbar}t\right) \tag{3.15}$$

with the spatial component of the eigen-function as the stationary solution of the eigen-problem (3.12); the Eq. (3.14) is known as *the second Schrödinger equation*.

However, the form (3.12) of the eigen-functions for the Hamiltonian of a system allows the normalization conditions (3.3) to be rewritten as the average measured values (3.2) as well by the employment of the variational theorem exclusively in terms of stationary functions, i.e., the solutions of the Eq. (3.10). Yet, it happens that just the finding of a general expression for the stationary solution of the Eq. (3.10) to generate the "drama" of the twentieth century quantum chemistry: so close to the stationary solution of the matter (as Lewis says) and yet inaccessible in a complete analytical version. From this on, the science intuition and creativity is called to give rise the phenomenological and analytical tools and models of the matter, structures of quantum states and dynamics, of atoms and molecules, see the Volumes I–III of the present five-volume work (Putz, 2016a-c) while the crystal and solid state quantum paradigm is to be in next unfolded in the present chapter.

3.3 BLOCH' THEOREM AND THE CRYSTAL ORBITALS

The quantum molecular model under Hückel approximation is quite general as to allow the enlargement of the discourse from molecules to the crystalline orbitals; the present discussion follows (Putz, 2006)

Firstly, on can formulate the situation of the linear polyenes of type C_nH_{n+2}, for which, the string of identical and periodical atoms forms a periodical fiend of nuclei, as prototype for a reticular arrangement (on one line only) of the atoms, as structural points, in a crystal.

Accordingly, the Hückel model applied for the general case of the (quasi) linear molecules will produce the orbital eigen-energies with the dependence (Putz, 2006):

$$E_j = \alpha + 2\beta \cos\left(\frac{j\pi}{n+1}\right), j = 1,2,3,\cdots,n \qquad (3.16)$$

easy to be verified (for instance) for the molecule of allyl with $n=3$, which for $j=1, 2, 3$; see the E_1, E_3 and E_2 levels' solutions presented in Volume III of the present five-volume work (Putz, 2016c). However, the spectral placement of the molecular orbitals for the molecules with identical atoms, linear and consequently bound, will have the general form presented in Figure 3.2 (Putz, 2006).

The Eq. (3.16) allows further generalization for molecules with an infinite number of atoms linearly arranged as in a reticular string, or a crystallographic direction in a solid body.

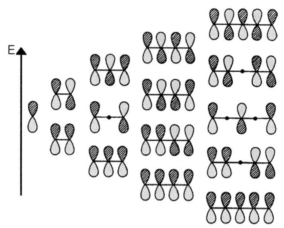

FIGURE 3.2 The spectral arrangement of the orbitals for the linear polyene molecules; see (Further Readings on Quantum Crystal 1940-1978).

So what is the limit of the expression (3.16) when n→∞?
Firstly, for *n* tending to the infinity an infinity of intermediate spectral levels also occur, which is the equivalent to an effective union of them in spectral bands.

How many are they?
Basically, there are three such bands, corresponding to the molecular types of bonds, schematically represented in Figure 3.3 (Putz, 2006):

- the *bonding band* (band of the lowest energy) corresponding to the so-called *band of valence* (abbreviated as BV);
- the *anti-bonding band* (the highest energy band) corresponding to the so-called *band of conduction* (abbreviated as BC);
- and the non-bonding band (intermediate band of energy) corresponding to a band of transition (abbreviated BT; originally in English is

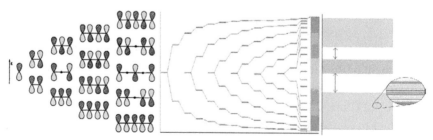

FIGURE 3.3 The generalization of the molecular orbitals to the electronic bands in a solid; see (Further Readings on Quantum Crystal 1940-1978).

"Band Gap" = word-for-word: the band of jump, but understood as a passing; sometimes it is also called *prohibited band*, but in a quantum meaning this can cause confusion, as long as there are accessible and allowed states even inside of it).

Secondly, which is the analytical generalization of n tending to the infinite of the expression (3.16)? The simplest case consist in considering the level with j-maximum, i.e., $j=n$, upon which performing limit $n\to\infty$ (Putz, 2006),

$$\lim_{n\to\infty} E_n = \alpha + 2\beta \cos\left(\pi \lim_{n\to\infty}\left(\frac{n}{n+1} \right) \right) = \alpha + 2\beta \cos(\pi) \qquad (3.17)$$

after which the re-consideration of the possibility of varying the expression (3.17) also for the other spectral levels will me made by introducing the additional k-(Brillouin) parameter,

$$k = \frac{\pi}{a} \qquad (3.18)$$

with which the general expression for the energy of the orbital levels in an infinite reticular string of bounded atoms writes as:

$$E(k) = \alpha + 2\beta \cos(ka) \qquad (3.19)$$

with the continuous representation of Figure 3.4 on a relevant interval of variation of the parameter k, $0 \le k \le \pi/a$, while taking into account for the parity property of the cosine function: $E(k)=E(-k)$ (Putz, 2006).

Who is "a" in the expressions (3.18) and (3.19)?
Actually, its physical nature does not matter as fart as it is absorbed by the k-parameter insertion of Eq. (3.18) in the expression (3.19).

This way, there is the freedom of choosing the suited meaning with the quantum physicochemical model approached. For the picture of the atoms placed in an infinite number of periodical places on a reticular string, the only unspecified information until now is the periodical distance between them.

In addition, if we choose the parameter a as a measure for the distance between two neighboring atoms – also the k parameter from Eq. (3.18) would gain the physical significance that the electromagnetic wave vectors have. This identification further allows the indication of the type of the sensitivity the electromagnetic waves manifest in scattering (interaction)

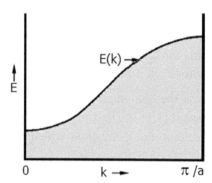

FIGURE 3.4   The variation of the spectral energy of an infinite reticular string of identical atoms periodically placed at the distance "a"; see (Further Readings on Quantum Crystal 1940-1978).

with the atoms of the reticular crystalline string considered: for the dimensions of the crystalline networks a is of the order Å ($10^{-10}m$), which in the electromagnetic spectrum corresponds to the X-radiation.

The last observation is essential to justify and motivate the X-ray diffraction on the crystals, as an efficient experimental tool for investigating the periodic lattice arrangement (associated solid state) of a substance or chemical compound.

Accordingly, for a reticular string composed of atoms periodically placed at the a distance one form another, the eigen-energy associated to this multi-atomic bounded system corresponds to the expression (3.19).

But what about he associated eigen-functions?
In order to answer at this question, one should firstly analyze the periodic potential produced by such atomic arrangement:

$$V(x) = V(x+a) \tag{3.20}$$

As represented in Figure 3.5 with continuous lines overlapped to the dotted lines of the Coulomb-potentials associated to the nuclei of the atoms in the string (Putz, 2006).

From the analytically point of view, the potential periodicity of the Eq. (3.20) allows the effective writing of the potential form in a general manner, known as Fourier expansion:

$$V(x) = \sum_n v_k \exp(ik_n x) \ , \ k_n = n\frac{2\pi}{a} \tag{3.21}$$

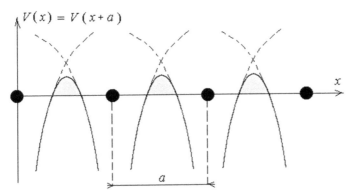

FIGURE 3.5 Periodical potential for a reticular atomic crystalline string; see (Further Readings on Quantum Crystal 1940-1978).

where v_k are the so-called Fourier coefficients, and k_n is in agreement with the parameter (3.18) and corresponds to the so-called reciprocal vector of the considered lattice (reticular string) for which a is defining the fundamental crystallographic distance on $0x$

With the Euler identity:

$$\exp(im) = \cos(m) + i\sin(m) \tag{3.22}$$

one can immediately verify how the solution (3.21) satisfies the Eq. (3.20). The solution meaning of Eq. (3.21) is that the periodical functions are transformed as driven by the terms of exponential $\exp(ik_n x)$ type.

On the eigen-function level one may have the question: for the reticular type as in this case or for the crystalline type in general, does the periodicity of the crystalline lattice imposing a transformation of the eigen-function such as is the potential of Eq. (3.20) or it is modified by other periodical terms?

The answer can be given through considering the periodical translation in the crystal as a specific symmetry operation, as are the symmetry operations in a molecule, yet here for modeling the crystalline eigen-function in general.

However, which is the translation operator $\hat{T}$ relationship with the Hamiltonian of the periodic system?

The crystalline Hamiltonian is periodic because the potential (3.20) upon the electronic evolution is periodic. On the other hand, the translation operator applied on the crystalline eigen-functions, by the very definition of the periodic translation, produces the necessary result (Putz, 2006):

$$\hat{T}_a \bullet \Psi(x) = \Psi(x+a) \tag{3.23}$$

While combining the relation (3.23) with the periodic Hamiltonian of the crystalline system further result the identities:

$$\hat{T}_a \left(\hat{H} \bullet \Psi(x) \right) = \hat{H}(x+a) \bullet \Psi(x+a) = \hat{H}(x) \bullet \Psi(x+a) = \hat{H}\hat{T}_a \bullet \Psi(x)$$

$$\tag{3.24}$$

which generate the commutativity relation between the translation operator and the Hamiltonian of the system (Putz, 2006):

$$\hat{T}_a \hat{H} = \hat{H}\hat{T}_a \tag{3.25}$$

Based on this commutativity, according to the postulate [P3] of the quantum mechanics, further follows that the Hamiltonian of the system and the translation operator have the same set of eigen-function Ψ, which implies, for the translation operator, the existence of the eigen-values equation (according to the quantum mechanics postulate [P1]):

$$\hat{T}_a \bullet \Psi(x) = t(a)\Psi(x) \tag{3.26}$$

with $t(a)$ being the eigen-values of the translation operator.

Now, by identifying the Eqs. (3.23) and (3.26) there result that the crystalline eigen-function must be transformed into the periodic lattice upon the expression (Putz, 2006):

$$\Psi(x+a) = t(a)\Psi(x) \tag{3.27}$$

So that, the periodicity is analytically extended at the level of the crystalline eigen-functions yet under different form from the periodic potential behavior, see the Eq. (3.20), but being correlated with the eigen-values of the translation operator.

Further on, the determination of these eigen-values will be done by appealing to the commutativity property and through composing of the successive translations, according to the opeatorial formula:

$$\hat{T}_a \hat{T}_{a'} = \hat{T}_{a'} \hat{T}_a = \hat{T}_{a+a'} \tag{3.28}$$

which produces the alternatively eigen-equations:

$$\hat{T}_{a'} \hat{T}_a \bullet \Psi(x) = t(a) \hat{T}_{a'} \bullet \Psi(x) = t(a)t(a')\Psi(x)$$
$$\hat{T}_{a'} \hat{T}_a \bullet \Psi(x) = \hat{T}_{a+a'} \bullet \Psi(x) = t(a+a')\Psi(x) \tag{3.29}$$

which results into the eigen-values equation

$$t(a+a') = t(a)t(a') \tag{3.30}$$

and whose solution can only be as of an exponential form. Now, in relation to the Fourier expansion prescription of the periodic potential (3.21) and of its interpretation, the natural choice of the exponential form for the eigen-functions of the translation operators will be:

$$t(a) = \exp(ik_a a) \tag{3.31}$$

with k_a the so-called reciprocal vector in the a-direction, in this case. For the three-dimensional analysis, the generalization immediately occurs, if we consider the lattice vectors written in the base of the crystallographic directions (a, b, c):

$$\vec{R} = m\vec{a} + n\vec{b} + l\vec{c} \tag{3.32}$$

for which the eigen-values at translation yield:

$$t\left(\vec{R}\right) = t\left(\vec{a}\right)^m t\left(\vec{b}\right)^n t\left(\vec{c}\right)^l = \exp\left(i\vec{k}\vec{R}\right) \tag{3.33}$$

with $\vec{k}$ the (integer) vector or the reciprocal lattice. Under these conditions, the crystalline eigen-function satisfies, based on the lattice periodicity, the general form (Putz, 2006):

$$\Psi\left(\vec{r} + \vec{R}\right) = \exp\left(i\vec{k}\vec{R}\right)\Psi\left(\vec{r}\right) \tag{3.34}$$

expression known as the *Bloch's theorem* for the crystalline eigen-functions, see also (Further Readings on Quantum Crystal 1940–1978), this way being deductively demonstrated here.

Returning to the reticular atomic string, the next question refers on how its eigen-function is written by the virtue of the already exposed Bloch's form of crystalline eigen-functions?

The answer can be eventually formulated by calling the basic functions postulate [P3], so admitting the atomic orbitals as a basis set for the crystalline eigen-functions, equivalently with the LCAO expansion (quantum superposition) for the molecular eigen-functions, yet considering also the Bloch's prescription (3.34), i.e., through additionally including the translation eigen-values (3.31) and (3.33), respecting the (isolated/in gas phase) molecular case.

The so-called crystalline orbitals are obtained:

$$\Psi_k = \sum_n \exp\left(inka\right)\phi_n \tag{3.35}$$

The connection between the orbital crystalline form (3.35) and the eigen-energies associated with Eq. (3.19) consists in customizations of the k values into the so-called k-points, which specifies the various types of crystalline orbitals. For example, one can immediately evaluate the crystalline orbitals corresponding to the extreme k-points for the zone of continuous variation in Figure 3.4, namely $k=0$ and $k=\pi/a$, which generates the crystalline orbital for the zone center (noted by Γ) with the superposition of atomic orbitals, all in phase (Putz, 2006):

$$\Psi_\Gamma = \sum_n \phi_n = \phi_0 + \phi_1 + \phi_2 + \phi_3 + ... \tag{3.36}$$

as well as the and the crystalline orbital on the zone frontier (noted by X) with the superposition of the atomic orbitals, all in opposition of phases:

$$\Psi_X = \sum_n (-1)^n \phi_n = \phi_0 - \phi_1 + \phi_2 - \phi_3 + ... \tag{3.37}$$

The visualization of these types of crystalline orbitals from the linear series of atomic orbitals in Eqs. (3.36) and (3.37), for various types of atomic orbitals, is shown in Table 3.2 (Putz, 2006).

From the analysis of Table 3.2 there is noted how, for example, the atomic orbitals such as s, $p_{x,y}$, d_{z^2}, d_{xy} and d_{x2-y2}, produce, by their arrangement as prescribed by Eq. (3.36), the bonding crystal orbitals in the center of k-zone (Γ), and respectively, by their arrangement in the anti-phase as

TABLE 3.2 Types of Crystalline Orbitals (CO) in the Center (X) and at the Frontier (Γ) of the k-Zone of the Crystalline Eigen-Function for the Various Types of Basic Atomic Orbitals (AO); after (Further Readings on Quantum Crystal 1940-1978)

$\Gamma\ (k=0)$		OA	$X\ (k=\pi/a)$	
CO	AOs in CO		AOs in CO	OC
σ		s		σ*
π		p_x, p_y		π*
σ*		p_z		σ
σ		d_{z^2}		σ*
π*		d_{xz}, d_{yz}		π
δ		$d_{xy}, d_{x^2-y^2}$		δ*

prescribed by the (3.37) expansion, the anti-bonding orbitals at the frontier of the k-zone (X). The situation is reversed for the crystal orbitals constructed from the arrangements in the phase/anti-phase of the atomic orbitals such as p_z, d_{xz}, and d_{yz}.

Figure 3.6 illustrates the layout of these crystalline orbitals of bonding/anti-bonding in relation with the associated eigen-energies of Figure 3.4 (Putz, 2006).

In this way, the crystal orbitals, as was the case with the molecular ones, are derived from the basic set of the atomic orbitals, i.e., of the constitutive atoms, within the quantum phenomenology.

We therefore employed the quantum hierarchy: from the atom to the molecule, to the solid, generalizing and preserving from each level of organization of the matter the information and the required features, along the common features that contribute in describing a more complex structural organization: this is the case of the expansion and specialization of the LCAO superimposing quantum picture of basic sets orbitals and of the Hückel approximation, from the molecular orbitals description to the eigen-functions and the crystal eigen-energies, taking into account the specificity of a lattice periodicity, hence the Bloch "correction" over the molecular case.

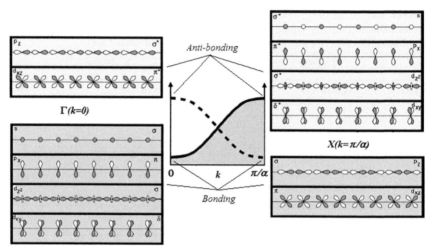

FIGURE 3.6   The crystalline orbitals in the center and at the frontier of the k-zone of the associated eigen-energies; see (Further Readings on Quantum Crystal 1940-1978).

However, the quantum description of the solid and crystal state can be also directly made, by considering the second Schrödinger equation (3.10) for the periodic crystal potential (3.20), when admitting approximations relatively to the form of this potential, within a period.

This kind of approach will be practiced also in the following sections of the present chapter, in order to show and argue, in more analytical details, the image of k-zones and of energetic bands here phenomenological deduced, as specific for the quantum description of the crystal and solid state.

3.4   QUANTUM MODELS FOR CRYSTALS

3.4.1   RECIPROCAL LATTICE

The main characteristic of a crystalline arrangement consists in the existence of crystallographic planes, which, implicitly "collect" de periodicity of engagement of structural dots (atoms and assemblies of atoms) in all the directions in which the crystal expands; the present discussion follows (Putz, 2006).

Miller notations gave the indices of equivalent crystallographic planes (*hkl*), belonging to the same family of planes, that have in common the same interplanar distance d_{hkl} measured on the normal direction between any two consecutive planes of the same family.

Then, the characterization of crystallographic planes and of the families of such planes by the normal directions to them generated the possibility of representation of all the planes families by the projections of the normal by stereographic projection method.

Yet, stereographic projection, also efficient for establish the orientations of families of planes, can not give indications towards the interplanar specific distance to that family.

This way, a new type of crystallographic projection should be introduced in relation to crystallographic planes, families of planes, normal directions but also to the interplanar distance.

A such projection had been introduced for the first time by P. P. Ewald in 1921 and then detailed by the applications of J. D. Bernal in 1927 and called as *reciprocal lattice* (abbreviated as RL), because it is formed by dots associated to the inverse $1/d_{hkl}$ of interplanar distances in the direct lattice (abbreviated DL). Thus, both crystallographic characteristics are kept (Putz, 2006):

1. the crystallographic families of planes characterization by normal directions on their planes (i.e., an entire series of parallel planes is represented in a one single normal direction) and
2. the characteristics of interplanar distance is included in the relation of inverse proportionality, though which, on the normal direction of family planes the characteristic dot is fixed (i.e., the characterization of family planes, from the direction of their common normal had been restricted even more to the dot associated to the inverse of interplanar distance specific to the family).

For a better fixation of the conceptual advantage introduced by the lattice reciprocal-projection of the direct lattice, in Figure 3.7 a triclinic system of crystallographic axes and the family of crystallographic axes (100) is considered.

The family of planes (100) is characterized by the common normal 0P. Moreover, the interplanar distance is d_{100}. At this distance the reverse distance $R_{100}=M/d_{100}$, will correspond (by projection) with M a factor of proportionality of projection (also called the factor of enlargement).

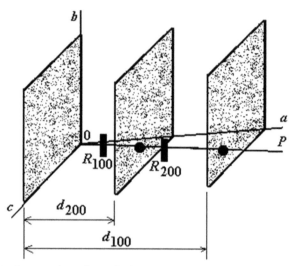

FIGURE 3.7 Construction of a reticular string in reciprocal lattice; after (Verma & Srivastava, 1982; Putz, 2006).

Further, from the definition of by Miller indexing, there is clearly how the family of planes (200) "cuts" the axis $0a$ at distance $\frac{1}{2}a$ and therefore will be characterized, on the same normal as that of the family of (100) planes, by the interplanar distance $d_{200}=d_{100}/2$. But this distance will be projected on the inverse distance $R_{200}=M/d_{200}=2M/d_{100}=2R_{100}$.

Thus, what appeared in Miller indexing a family of planes places in reverse order respecting the increasing order of the Miller indices, in reciprocal projection naturally, appears, as a string of consecutive dots.

Simply, one can say that the reciprocal projection reverse, once again (or gives replica as the reciprocally inversing) the Miller indexing of crystallographic planes.

Which have been accomplished by Figure 3.7, in fact projected, is just the crystallographic a-direction, along which crystallographic planes with inter-planar distances d_{h00} are recorded through reciprocally projected on $0P=a^*$ direction thus called as reciprocal crystallographic direction along which reciprocal distances R_{h00} are projected, with modulus of reciprocal vectors $d^*_{h00}=ha^*$ ($h=1,2,3,...$).

Thus there is noted how a^* direction (the $0P$ direction) is from construction (i.e., from the definition of reciprocal definition) perpendicular on the planes of family of planes ($h00$), and therefore it is perpendicular also on the crystallographic b and c directions of the direct space (that forms the plane parallel with the d_{h00} planes) (Putz, 2006).

Similar constructions and projections can be considered also for the planes of $(0k0)$ and $(00l)$ type with interplanar d_{0k0} and d_{00l} distances from which results, respectively, the reciprocal crystallographic $b*$ and $c*$ directions along the reciprocal vectors $d^*_{0k0}=kb*$ $(k=1,2,3,...)$ and $d^*_{00l}=lc*$ $(l=1,2,3,...)$.

The basic vectors of reciprocal lattice this way formed are orthogonal on the basic vector of direct lattice:

$$a* \cdot b = a* \cdot c = b* \cdot a = b* \cdot c = c* \cdot a = c* \cdot b = 0 \qquad (3.38)$$

forming the so-called *unit cell in reciprocal lattice*, see Figure 3.8 (Putz, 2006).

But which are vectorial expressions and the modules of crystallographic vectors of reciprocal lattice?

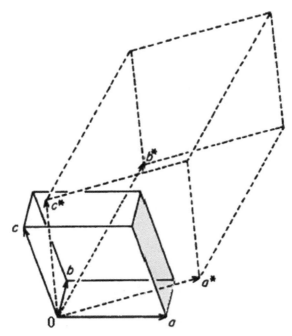

FIGURE 3.8 Construction of unit cell in reciprocal space (dotted line); after (Verma & Srivastava 1982).

For this the vectorial expression of volume of unit cell is called. In the direct space, the fundamental vectors (a, b, c) generate the volume of unit cell with the expression

$$V = a \cdot (b \times c) \tag{3.39}$$

with $(b \times c)$ corresponding to the vector n normal to area A_{bc} of the face of unit cell fixed by the vectors b and c.

In the same time, the same volume can be expressed considering the interplanar distance d_{100} on the direction of the normal n:

$$V = d_{100} A_{bc} = d_{100} n \cdot (b \times c) \tag{3.40}$$

The equality of the two equivalent expressions of the volume of the unit cell is firstly written such as:

$$n / [(n \cdot n) d_{100}] = (b \times c)/[a \cdot (b \times c)] \tag{3.41}$$

wherefrom, by taking into account the quality of the normal for A_{bc},

$$n \cdot n = 1 \tag{3.42}$$

and considering also the construction of Figure 3.7

$$n / d_{100} = n R_{100}/M = a* \tag{3.43}$$

the expression of fundamental vector $a*$ in reciprocal lattice there immediately results, and analogously for the other $b*$ and $c*$ directions, so obtaining the set of statements:

$$a* = b \times c/[a \cdot (b \times c)]$$
$$b* = c \times a/[a \cdot (b \times c)]$$
$$c* = a \times b/[a \cdot (b \times c)] \tag{3.44}$$

The statements (3.44) allow the scalar products calculation between the fundamental vectors of reciprocal lattice and those correspondents in direct lattice, with the results:

$$a* \cdot a = b* \cdot b = c* \cdot c = 1 \tag{3.45}$$

based on the cyclicity properties of the scalar product towards the vectorial one,

$$a \cdot (b \times c) = b \cdot (c \times a) = c \cdot (a \times b) \tag{3.46}$$

The results (3.45) are in accordance and in addition justifies the title of *reciprocal lattice* for the constructed lattice on the reciprocal a^*, b^*, *and* c^* vectors abstracted from the fundamental ones of the direct lattice according to the relations (3.44).

Moreover, with the aid of the statements (3.44) also the driving cosines, i.e., the cosines of the angles α^*, β^*, γ^* respectively between the (b^*, c^*), (a^*, c^*), (a^*, b^*) vectors of reciprocal lattice can be calculated as depending on the driving cosines of angles α, β, γ between the vectors of the direct lattice (b, c), (a, c), (a, b), respectively.

These general results correspond to the case of unit cell of triclinic syngony. The complete list of the results for the seven crystalline systems of Table 2.2 is presented in Table 3.3 (Putz, 2006).

Therefore, any d^*_{hkl} vector in reciprocal lattice can be written on the basic vectors of reciprocal lattice, generally as:

$$d^*_{hkl} = h\,a^* + k\,b^* + l\,c^* \tag{3.47}$$

Yet, which is the relation between the d^*_{hkl} vector of reciprocal lattice and the d_{hkl} vector of direct lattice?

TABLE 3.3 Relations Between the Modules and Driving Cosines in Reciprocal Lattices Towards the Direct Ones for the Seven Crystalline Systems; after (Verma & Srivastava, 1982).

	a^*	b^*	c^*	$\cos\alpha^*$ or α^*	$\cos\beta^*$ or β^*	$\cos\gamma^*$ or γ^*
Triclinic	$bc\sin\alpha/V$	$ca\sin\beta/V$	$ab\sin\gamma/V$	$(\cos\beta\cos\gamma-\cos\alpha)/(\sin\beta\sin\gamma)$	$(\cos\gamma\cos\alpha-\cos\beta)/(\sin\gamma\sin\alpha)$	$(\cos\alpha\cos\beta-\cos\gamma)/(\sin\alpha\sin\beta)$
Mono-clinic	$1/(a\sin\beta)$	$1/b$	$1/(c\sin\beta)$	$\alpha^*=\alpha=90°$	$\beta^*=180°-\beta$	$\gamma^*=\gamma=90°$
Ortho-rhombic	$1/a$	$1/b$	$1/c$		$\alpha^*=\beta^*=\gamma^*=90°$	
Tetragonal	$1/a$	$1/a$	$1/c$		$\alpha^*=\beta^*=\gamma^*=90°$	
Rhombohedral	$a^* = b^* = c^* = 1/[a\sin\alpha\sin\alpha^*]$				$\alpha^*=\beta^*=\gamma^*$, $\cos(\alpha^*/2)=1/[2\cos(\alpha/2)]$	
Hexagonal	$a^* = b^* = 2/[a\sqrt3]$		$1/c$		$\alpha^*=\beta^*=90°$	$\gamma^*=60°$
Cubic	$a^* = b^* = c^* = 1/a$				$\alpha^*=\beta^*=\gamma^*=90°$	

The question has two stages: the vectorial level and the level of modules. For the formulation of the answers one starts by considering the crystallographic (*hkl*) plane in direct lattice that cuts the fundamental crystallographic *a*, *b*, *c* directions at *a/h*, *b/k*, and respectively *c/l*, distances as in Figure 3.9 depicted.

For the answer at vectorial level, the vector *PQ* in construction of Figure 3.9, for example, is written in relation to the vectors of direct lattice:

$$PQ = b/k - a/h \tag{3.48}$$

Further on, the scalar product between this vector and that one of Eq. (3.47) that design the plane (*hkl*) in reciprocal lattice is performed, and while taking into account the relations (3.38) and (3.45) as well as the distribution of scalar product, the identity is obtained (Putz, 2006):

$$d^*_{hkl} \times PQ = (h\, a^* + k\, b^* + l\, c^*) \times (b/k - a/h) = 0 \tag{3.49}$$

which indicates the perpendicularity of vector d^*_{hkl} on *PQ* vector.

Similarly, there is shown how the d^*_{hkl} vector is perpendicular on the *QR* vector, i.e., finally d^*_{hkl} vector is perpendicular to the plane *PQR*, i.e., is oriented in the direction of the normal *n* at crystallographic plane (*hkl*) in Figure 3.9.

As such an answer for vectorial orientation of d^*_{hkl} was given.

For the issue of scalar measure, as a module of associated vector, the vector d^*_{hkl} will firstly be written as based on its module d^*_{hkl} and the normal along which is oriented that gives:

$$d^*_{hkl} = n\, d^*_{hkl} \tag{3.50}$$

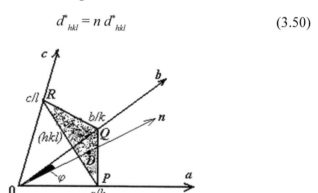

FIGURE 3.9 Crystallographic plane (hkl) and its cuts on the axes of reciprocal space; after Putz (2006).

The last statement combined with the general expression (3.47) result into an analytical expression for vector n:

$$n = (ha^*+kb^*+lc^*)/d^*_{hkl} \tag{3.51}$$

The expression (3.51) can be further used for rewriting the interplanar distance d_{hkl}, which in the direct space is measured on the normal direction to the crystallographic (hkl) plane, for example, through the identities based on Figure 3.9 (Putz, 2006):

$$\begin{aligned}
d_{hkl} &= 0D \\
&= 0Q \cos(\varphi) \\
&= (b/k) \cdot n \\
&= (b/k) \cdot (ha^*+kb^*+c^*)/d^*_{hkl} \\
&= 1/d^*_{hkl} \tag{3.52}
\end{aligned}$$

where again the properties (3.38) and (3.45) were used.

Thus, the vector d^*_{hkl} of reciprocal lattice results as perpendicular on crystallographic (hkl) plane and has the $1/d_{hkl}$ magnitude properly to the inverse of interplanar distance for family of (hkl) planes.

Further on, since based on the previous answers, there is the question: how are projected inside the reciprocal lattice the direct lattices and the Bravais unit cells of the seven crystalline syngonies and at which type of unit cell such projection corresponds?

To this aim, the Figure 3.10 explains how the unit cell of cubic F-type is transformed in unit cell of I-type in reciprocal lattice, by the transformation of the minimum interplanar distances that correspond to the (010) and (111) planes of F-cell in direct lattice, and to dots of I-cell in reciprocal lattice, with dots placed at equal distances with the inverse of interplanar distances of direct lattice (Putz, 2006).

In a similar manner, when one starts from a cubic cell of I-type in direct lattice a cubic cell of F-type will be obtained in reciprocal lattice.

The complete list of correspondences of Bravais unit cells of direct lattice to the reciprocal ones, is in Table 3.4 exposed (Putz, 2006).

Still, beside the Bravais lattices presented in the Chapter 2 of the presented volume, the crystalline space can be homogeneously partitioned

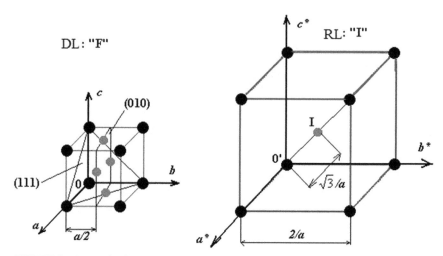

FIGURE 3.10   Projection of F-unit cell of direct lattice (DL) in I-unit cell of reciprocal lattice (RL); after (Kittel, 1996).

TABLE 3.4   Correspondences of Bravais Unit Cells Between the Direct Lattice (DL) and the Reciprocal Lattice (RL) Types; after (Verma & Srivastava, 1982; Putz, 2006).

Syngony	Type of Cell in DL	Type of Cell in RL
Triclinic	P(primitive)	P
Monoclinic	P	P
	C	C
Orthorhombic	P	P
	I	F
	F	I
	C	C
Tetragonal	P	P
	I	I
Rhombohedral	R	R
Hexagonal	P	P
Cubic	P	P
	I	F
	F	I

(and respectively re-composed) also by a special *Wigner-Seitz unit cells* construction (abbreviated as WSC).

This type of cell is not different by those of Bravais type, as long as it can be composed by any of them. The construction is based on the closed space of medians union (lines or planes) that cuts the middle of the short-est distances around a structural dot, so that the unit cell resulted, WSC, behave like a primitive.

Similar construction of reciprocal space is called *Brillouin zone* and can be either obtained by WSC projection in reciprocal space or by unit cell projection that is at the base of WSC from the direct space to the reciprocal one, following that in this reciprocal space the union of the medians between the closest dots around a reciprocal dot of interest is to be performed.

This way, using this method a direct primitive cell from the unit cell can be always constructed from any Bravais cell, without calling the neighboring cell for the "reduction" of the unit cell in focus at the aimed primitive cell. *Brillouin zones* are, therefore, primitive cells in reciprocal space, since directly obtained by projection of the Bravais cell.

In order to keep the close with the example of Figure 3.10, in Figure 3.11 the primitive cells constructed in direct lattice of I-cubic cell with recipro-cal F-projection are exposed (Putz, 2006).

There is noted how, indeed, the Brillouin Zone is the primitive cell in reciprocal space, being constructed around the central dot of I-cell (as reciprocal F-projection from direct lattice) however not including any other reciprocal dot in the closed space formed from the union of median planes at distance towards the central dot to the closest neighbor dots (from the corners of the cube).

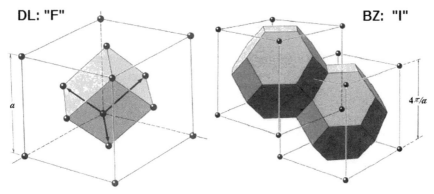

FIGURE 3.11   Cubic primitive cell of F cubic cell (left) in the direct lattice (DL) and the correspondent primitive cell (truncated octahedron) as Brillouin Zone (BZ) of I-cell type, appeared as the reciprocal primitive cell in reciprocal lattice (right); after (Kittel, 1996).

There is also noted how in Figure 3.11, unlike the case of Figure 3.10, the disproportion of the dimensions of cubic cells between direct and reciprocal lattices was abolished, since the reciprocal lattice being multiplied from now on with M=2π factor, an operation allowed by increase of M factor, so introduced from the basic reciprocal lattice construction (see Figure 3.7).

Beside the advantage of treatment of similar proportions, the introduction of M=2π factor for the projection of the distance $a/2$ on $2M/a$ (Figure 3.10) allows the identification of the measure $M/a=2\pi/a$ specifically to reciprocal lattice with a precise physical meaning: the wave vector $k=2\pi/\lambda$ modulus is indexing the number of nodes of a wavelength λ.

Making a step even further, with the aim to quantum characterizing a crystal, one notes that k can be correlated with the nodes of "wave" states for the associated Schrödinger-Bloch eigen-function and, as well, it can be correlated with the impulse of the de Broglie associated "corpuscular" motion in the crystal:

$$p=k\hbar \qquad (3.53)$$

The introduction of reciprocal space allows for both the characterization by unitary projection (planes are represented by dots) and completes projection (both information of orientation and interplanar distance are present) for the crystallographic planes of direct space as well as the possibility of their quantum representation by the characterization of Brillouin Zones of wave vectors associated to the crystal's eigen-states (Pettifor, 1995).

3.4.2   QUANTUM MODELS OF CRYSTALS

Typical models for illustrating the crystalline quantum states' characterization will be exposed, in terms of reciprocal lattice, as following; the present discussion follows (Putz, 2006).

3.4.2.1   Quantum Model of Free Electrons in Crystal

A crystal is composed from atoms or assemblies of atoms spatially ordered. The atoms are occupied by electrons in quantum shells, according to the rules of minimum energy occupancy and the maximum orbital penetration which confer them a classification after the complete occupied and the free shells in Periodical Table of associated elements.

Thus, quantum atom in any Ψ state (fundamental or excited) can by simplified treated as being composed by a core Ψ^m state that includes the occupied shells (of minimum energy and those orbital close to the nuclei) and a valence Ψ^v state with the farthest (complete or incomplete) shell from the nuclei influence. In quantum terms such a states separation requires the associated eigen-functions to be orthogonal

$$\int \Psi^m(r)\Psi^v(r)dv = 0 \tag{3.54}$$

in order to constitute the basis for any function of atomic state (postulate [P3] of quantum mechanics) (Slater, 1939).

Therefore, since the crystal is composed by bonded periodically atoms it can be, in turn, characterized by two basic orthogonal states, namely that of Ψ_k^m core and that Ψ_k^v of valence, being also characterized by the wave vector associated to the wave packets of those states, according to the relation (3.53), and intimate bonds (described in previous section) with crystallographic planes by reciprocal lattice and Brillouin zone(s).

The procedure of orthogonal construction of eigen-function of valence of k-modulation (of frequency or nodes) of core eigen-function is in Figure 3.12 illustrated (Putz, 2006). There results the idea that in order to satisfy the orthogonality, the eigen-function of valence should be modulated by large frequencies and many k-nodes, which is equivalent with the existence of an extra-energy of valence: high k means high p, according to Eq. (3.53), which means major energy $E=p^2/2m$. Yet, the extra-energy of valence has the role of shielding and cancellation of quasi-total of Coulombian potential of nuclei of atoms periodically disposed in crystalline lattice, Figure 3.13-left, resulting a net potential that in turn can be modeled by a gap potential type, Figure 3.13-right, for the entire crystalline potential (Putz, 2006).

This kind of treatment is based by the so-called pseudo potential theory, wherefrom, for the present case, the free electrons picture inside the crystal which are moving in a potential hole (of length L, crystal length) accordingly results.

The Schrödinger equation (3.10) for modeling free electrons in crystal (Putz, 2006):

$$-\frac{\hbar^2}{2m_e}\nabla^2 \bullet \Psi = E\Psi \tag{3.55}$$

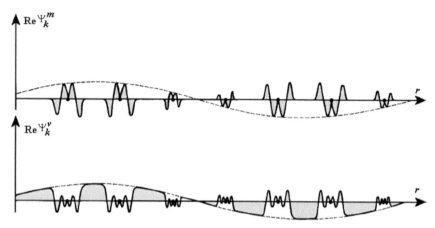

FIGURE 3.12 Construction of valence state (down) from the orthogonality condition towards the crystalline state of core (up), in relation to the modulations of wave vector associated to those states; see (Further Readings on Quantum Crystal 1940-1978).

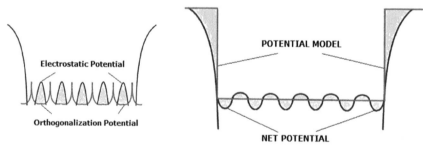

FIGURE 3.13 Construction of orthogonalization potential for the Coulombian periodical one (left), with the effect in the appearance of net potential and of the model of potential of free electrons in a crystal (right); see (Further Readings on Quantum Crystal 1940-1978).

should not be seen as a simple particularization of general Schrödinger equation for the external potential absence applied to the electrons in crystal, $V(r)=0$, but rather (and more coherent) as a re- energetic scaling:

$$E=E_{Coulombic(Core)+Orthogonal(Valence)}-V_{Coulombic} \tag{3.56}$$

The Eq. (3.55) can be rewritten, when reduced at crystallographic direction 0x, as a differential equation of 2nd order:

$$\frac{d^2}{dx^2}\Psi(x)+k\Psi(x)=0 \tag{3.57}$$

where the wave vector has been identified:

$$k = \sqrt{\frac{2m_e E}{\hbar^2}} \Leftrightarrow E = \frac{\hbar^2}{2m_e} k^2 \tag{3.58}$$

from combining the Eq. (3.53) with the first expression of the total energy of Table 3.1.

Expression (3.58) shows how the total electronic energy varies parabolically with the variation of wave vector associated in reciprocal space, Figure 3.14 (Putz, 2006). Thus, the total energy will be quantified if the wave vectors are quantified. Let's see next.

The eigen-solutions of Eq. (3.57) are such as:

$$\Psi_k^{1,2}(x) = c_{1,2} \exp(\pm ikx) \tag{3.59}$$

easily to verify by direct substitution of Eq. (3.59) in Eq. (3.55). Coefficients $c_{1,2}$ are determined from the condition of normalization of eigen-function, see the [P2] quantum mechanical postulate, here adapted to the conditions of present crystalline quantum model:

$$\int_0^L \left[\Psi_k^{1,2}(x) \right]^* \Psi_k^{1,2}(x) dx = 1 \Rightarrow c_{1,2} = \frac{1}{\sqrt{L}} \tag{3.60}$$

With the aid of coefficients (3.60), the eigen-functions (3.59) are rewritten:

$$\Psi_k^{1,2}(x) = \frac{1}{\sqrt{L}} \exp(\pm ikx) \tag{3.61}$$

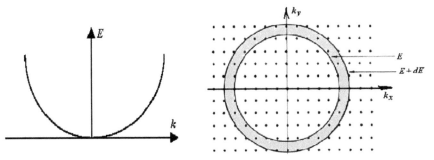

FIGURE 3.14   Parabolic E-k dependence (left) and its projection in reciprocal k-space (right); see (Further Readings on Quantum Crystal 1940-1978).

If also the condition of the crystal periodicity is employed then also the eigen-functions should be periodic at the ends of crystal, which generates the k-quantification in the crystal:

$$\Psi_k^{1,2}(x=0) = \Psi_k^{1,2}(x=L) \Leftrightarrow 1 = \exp(ikL) \Rightarrow k = n\frac{2\pi}{L} \qquad (3.62)$$

which means that the electronic states allowed in crystal are uniformly disposed in reciprocal k-space at $2\pi/L$ distance one from other. This result allows the introduction of the electronic states density quantity, as the inverse of their separation:

$$g(k) = \frac{L}{2\pi} \qquad (3.63)$$

so that the number of electronic states allowed in interval $(k, k+dk)$ will be given by $g(k)dk$.

If N cubical unit cells (of side a) are considered in crystal, then

$$L = Na \qquad (3.64)$$

and the total number of states allowed in first *Brillouin zone* $(-\pi/a \le k \le +\pi/a)$ will be:

$$\int_{-\pi/a}^{+\pi/a} g(k)dk = \frac{L}{2\pi}\frac{2\pi}{a} = \frac{L}{a} = N \qquad (3.65)$$

i.e., equal to the number of unit cells (not atoms!) from the system. This important result thus allows the quantum characterization of the entire crystal by reducing the analysis to the reciprocal space at the level of first Brillouin zone.

Moreover, for the 3D case for an isotropic crystal, there can be successively written:

$$g(k)dk = \frac{L_x L_y L_z}{(2\pi)^3} dk_x dk_y dk_z = \frac{v}{8\pi^3} 4\pi k^2 dk = \frac{v}{2\pi^2} k^2 dk \qquad (3.66)$$

If from the relations (3.58) the first derivative is considered for the dependence $E = E(k)$:

$$\frac{dE}{dk} = \frac{\hbar^2}{m_e}k = \frac{\hbar^2}{m_e}\sqrt{\frac{2m_e E}{\hbar^2}} = \hbar\sqrt{\frac{2E}{m_e}} \qquad (3.67)$$

and it is further combined with the expression (3.66), the allowed density of states as function of total energy can be rewritten, by taking into account also the factor 2 of multiplicity for the two states of spin-electrons (up and down)

$$g(E) = 2g(k)\frac{dk}{dE} = \frac{vm_e}{\pi^2 \hbar^3}\sqrt{2m_e E} \tag{3.68}$$

Eq. (3.68) says that the number of allowed electronic states with energy E on unit volume, $g(E)/v$, does not depend on the dimension of material, being an intrinsic property, specific to the electronic states in a solid.

The result is very important for the description of electric properties of solids and will be also in next used.

Moreover, the density of state $g(E)$ allows in evaluation for number of free electrons in crystal by the integration after the entire energetic spectrum.

Here another fundamental notion is introduced: the maximum energetic level that can be populated by electrons is a finite measure and is called *Fermi level* (energy) and is noted by E_F.

This level can not be infinite, otherwise it would create a picture of auto-ionized electrons from the crystal. Thus, the number of free electrons inside an isotropic crystal will be given as (Putz, 2006):

$$N_e = \int_0^{E_F} g(E)dE = \frac{v\sqrt{2m_e^3}}{\pi^2 \hbar^3}\int_0^{E_F}\sqrt{E}dE = \frac{v\sqrt{2m_e^3}}{\pi^2 \hbar^3}\frac{2}{3}E_F^{3/2} \tag{3.69}$$

a relation which, firstly, allows its reverse use, i.e., by expressing the Fermi energy based on the number of free electrons in crystal:

$$E_F = \frac{\hbar^2}{2m_e}\left(\frac{3\pi^2 N_e}{v}\right)^{2/3} \tag{3.70}$$

and then, the introduction of the wave vector associated to this energy based on the relation of type (3.58)

$$E_F = \frac{\hbar^2 k_F^2}{2m_e} \Rightarrow k_F = \left(\frac{3\pi^2 N_e}{v}\right)^{1/3} \tag{3.71}$$

which allows performing the ratio of Fermi energy to reciprocal k-space of dots.

In terms of Brillouin zones, the projection of Figure 3.15 (Putz, 2006), will be progressively completed until k_F fixes E_F when the occupancy with electrons of levels allowed in crystal is complete, so generating the so-called *Fermi surfaces* (in this case spheres).

In relation with the Fermi energy and wave vector one can introduces also other measures of interest, as Fermi velocity v_F, calling the relation between velocity v and impulse p through the mass m, $v = mp$, so having

$$v_F = \frac{p_F}{m_e} = \frac{\hbar k_F}{m_e} \qquad (3.72)$$

which results in practice in significant numerical values, but not relativistic.

For example, for a typical solid cube with $a=2.5\times10^{-10}$ [m], in the hypothesis that exist at lest an atom with one valence electron on unit cell, so with an electronic density N_e/v of about 6×10^{28} [m^{-3}] results with the aid of which the Femi energy $E_F \approx 9\times10^{-19}J=6$ [eV] is evaluated from Eq. (3.70), for the Fermi wave vector $k_F \approx 1.2\times10^{10}$ [m^{-1}] from Eq. (3.71) and consequently also Fermi velocity $v_F \approx 1.4\times10^6$ [m s^{-1}] from Eq. (3.72)!

Another important measure that can be calculated based on Fermi energy is just the total energy of electrons in a crystal: it is calculated

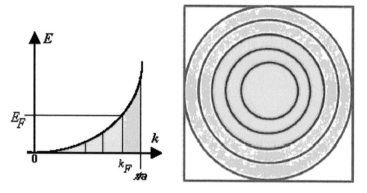

FIGURE 3.15 The occupancy of energetic levels allowed until Fermi level (E_F) by the free electrons of a crystal (left) and the Fermi surface-sphere with circle projection of k_F radius (full occupancy) in the first Brillouin zone (right); after (Further Readings on Quantum Crystal 1940-1978; Putz, 2006).

through the evaluated total number of electrons (3.69) by using the density of electronic states, but in regime of averaged measured value for the crystal state by the virtue of [P2] postulate of quantum mechanics (Putz, 2006)

$$\langle E \rangle = \int_0^{E_F} E g(E) dE = \frac{v\sqrt{2m_e^3}}{\pi^2 \hbar^3} \int_0^{E_F} E^{3/2} dE$$

$$= \frac{v\sqrt{2m_e^3}}{\pi^2 \hbar^3} \frac{2}{5} E_F^{5/2} = \frac{3}{5} N_e E_F = \left[\frac{\hbar^2}{2m_e} \left(\frac{3\pi^2}{v} \right)^{2/3} \right] \frac{3}{5} N_e^{5/3} \qquad (3.73)$$

and which, among others, indicated an average energy on electron to be with the $(3/5)E_F$ measure. Yet, beside the measures with a strong physical meaning, calculated in the framework of the model of free electrons, there still remains a fundament problem: the conciliation between continuous E-k dependence and the discreet k-quantification, Figure 3.14, for electronic states in crystal.

However, the impossibility of direct quantum prescription of the energy of Schrödinger equations (3.55) and (3.57), but only indirectly by the notation (3.58) combined with k-quantification of the crystalline eigenfunctions (3.62), indicates the fact that for obtaining a better picture of quantum treatment of the crystal a re-consideration of Schrödinger equation itself is required.

This issue and the generated quantum model will be referred in next section.

3.4.2.2   Quantum Model of Quasi-Free Electrons in Crystals

A more realistic model of the quantum crystal is here derived from the reconsideration of Schrödinger equations, (3.55) and (3.57), in the presence of a periodic potential with the periodicity of the lattice, say the uni-directional lattice with a-period, which in reciprocal lattice of wave vectors means $2\pi/a$ periodicity.

A periodical potential symmetrically respecting $x=0$ that acts towards the electrons that evolve in both the directions in any periodical interval of reciprocal lattice, see (Further Readings on Quantum Solid 1936–1967), assumes the form:

$$V(x) = \sum_n \left[V_n \exp(i2\pi nx/a) + V_{-n} \exp(-i2\pi nx/a) \right] = 2\sum_n V_n \cos(2\pi nx/a)$$

(3.74)

with V_n being the Fourier coefficients for expanding of $V(x)$ potential, also indicating the degree of perturbation of the evolution of electrons in crystal.

The model of free movement (in the absence of nuclei potential or in effectively zero nuclei potential) is thus improved, but here arise another problem: does the potential (3.74) introducing also the re-consideration of eigen-functions (3.61) for the free movement? Therefore, should again the Schrödinger equations (3.55) and (3.57) be solved in the absence of potential (3.74)? Not necessarily!

The eigen-functions (3.61) can be re-used in the conditions of potential (3.74) reduced at the first term, i.e., in conditions of a small perturbation of free movement of electrons.

Thus, the model of quasi-free electrons uses the eigen-functions of free electrons model, so without the constraint to solve the Schrödinger equation of the first order. The identification of the ingredients of this picture reduces the potential (3.74) at the first order as (Putz, 2006):

$$\hat{H}^{(1)} = 2V_1 \cos(2\pi x/a)$$

(3.75)

and the eigen-functions (3.61) of free electrons model at considered at the frontier of the first *Brillouin zone*, with $k=\pi/a$, in the version of stationary waves by normalized re-combinations:

$$\Psi_e^{(0)}(x) = \frac{1}{\sqrt{2}}\left[\Psi_{k=\pi/a}^1(x) + \Psi_{k=\pi/a}^2(x) \right] = \sqrt{\frac{2}{L}}\cos\left(\frac{\pi x}{a}\right)$$

$$\Psi_0^{(0)}(x) = \frac{1}{\sqrt{2}}\left[\Psi_{k=\pi/a}^1(x) - \Psi_{k=\pi/a}^2(x) \right] = \sqrt{\frac{2}{L}}\sin\left(\frac{\pi x}{a}\right)$$

(3.76)

Under these conditions, the eigen-functions, not being modified by the existence of the potential perturbation of first order, there remains the energetic correction (7.8) to be evaluate for the energy of free electrons (3.58) at the frontier of first *Brillouin zone*, i.e., in the dot of reciprocal space corresponding to entire family of crystallographic planes of direct space, with interplanar *a*-distance, thus evaluating the entire periodical potentials influence of type (3.75) upon the quasi-free electrons in crystal.

For each eigen-function (3.76) the correction (7.8) induced by potential (3.75) in dot $k=\pi/a$ can be calculated, with the focus on the difference between these corrections, equivalent to the energy that remove the degeneration (multiplication of eigen-functions that correspond to the same energy), here by de-doubling of the two eigen-functions (Putz, 2006):

$$E_0^{(1)} - E_e^{(1)} = \int_0^L 2V_1 \cos(2\pi x/a)\frac{2}{L}\left[\sin^2(\pi x/a) - \cos^2(\pi x/a)\right]dx$$

$$= -\frac{4}{L}V_1\int_0^L \cos(2\pi x/a)\cos(2\pi x/a)dx = -\frac{4}{L}V_1\left[\frac{L}{2} + \frac{a}{8\pi}\sin(4L\pi/a)\right]_{L=Na}$$

$$= -2V_1$$

$$(3.77)$$

The result (3.77), by the energetic (3.58) separation induced in $k=\pi/a$ dot of reciprocal space, at the frontier of Brillouin zone, actually equalizes with the separation of energetic bands, with a discretization, by the quantification of the energy of the quasi-free electrons in the crystal.

As an observation, the title of quasi-free is not equivalent with that of quasi-bonding, because the state that is perturbed is that of free electrons, not that of the bonding electrons, thus becoming quasi-free and not quasi-bonding picture!

Thus, the quantification in crystals is transposed in energetic bands, separated by an energetic gap, as indicated in Figure 3.16, based on stationary eigen-functions (3.76) and energetic gap (3.77) (Putz, 2006).

An important consequence of quasi-free electrons treatment in solid consists in the possibility of classification of solids as metals, insulating, and semiconductors based on the width of the separation of energetic bands in crystal.

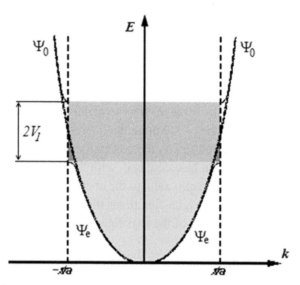

FIGURE 3.16   Energetic discretization at the frontier of the first Brillouin zone in the quantum model of quasi-free electrons in crystal; after (Further Readings on Quantum Solid 1936–1967; Putz, 2006).

Firstly, as based on the result (3.65) the energetic analysis can be reduced at the behavior of E-k dependence at the level of the first Brillouin zone, see Figure 3.17 (Putz, 2006).

From Figure 3.17, there is noted how, in projection, the E-k contours record a deformation of the frontier of the first Brillouin zone.

On the other side, because the energies of the first Brillouin zone before the reduction process, i.e., those corresponding to the scheme of extreme left of Figure 3.17, are the lowest ones, these will be the first occupied,

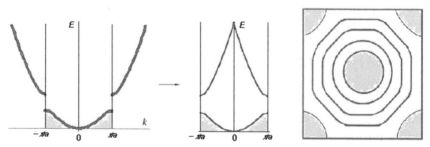

FIGURE 3.17   Reduction of E-k dependence at first Brillouin zone (left) and its projection in contours (right); after (Further Readings on Quantum Solid 1936–1967; Putz, 2006).

so that Fermi surface (indexing all the occupied electronic states) will be expanded towards the frontier of the Brillouin zone, as the allowed states of inferior energetic band of Figures 3.16 and 3.17-left are occupied.

For a large separation between the allowed energetic bands, the occupancy is continuously made only in the inferior band, of valence, which will also record the exactly Fermi level at the frontier of the first Brillouin zone, and correspond to the *E-k* representations of Figure 3.18.

The situation is changed if the quasi-free electrons have enough energy (and respectively high values of the wave vector) to exceed the potential gap of Figure 3.16 and to access to allowed states of second energetic band, actually to enter with the associated k-wave vector in second Brillouin zone, in E-*k* non-reduced version of extreme left of Figure 3.17.

How the *E-k* occupancy of electrons of energetic bands and respectively of successive Brillouin zones is represented in projection contours is exposed in Figure 3.19-left (Putz, 2006):

- it starts from the first Brillouin zone in non-distorted version of spherical model of free electrons (Figure 3.15-right);
- it is continued with the drawing of projection circles also outside the first Brillouin zone;
- the effect of distortion of projection circles of the first Brillouin zone is applied (as in Figure 3.17-right) towards the zone border;

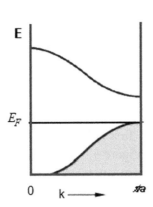
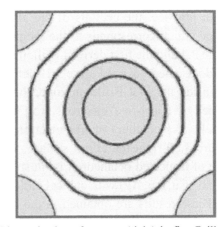

FIGURE 3.18   *E-k* diagram (left) and its projection of contour (right) in first Brillouin zone, with the indication of electronic occupancy until Fermi level, for an insulating crystal; after (Further Readings on Quantum Solid 1936–1967; Putz, 2006).

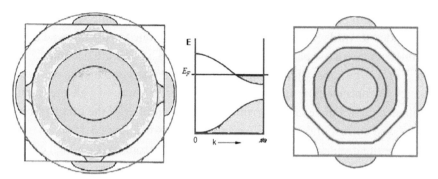

FIGURE 3.19   Construction of representation of electronic occupancy for energetic bands and respectively of Brillouin zones (left). *E-k* diagram (middle) and its contour projection (right) for the first Brillouin zone, with the indication of the electronic occupancies until Fermi level, for an insulating crystal; after (Further Readings on Quantum Solid 1936–1967; Putz, 2006).

- the deformation from the first zone inside the second one is continued also through circles;
- the second zone corresponding to the electronic states energetically occupied is completed by electrons and with relative wave vectors;
- the representations of the middle and the right side of Figure 3.19 are obtained.

This situation corresponds to the metals.

In practice, the crystals of the alkaline metals have cubic lattice of I-type which corresponds to the reciprocal lattice of F-type according to Table 3.4 and the first derived Brillouin zone is of type of a regular rhombic dodecahedron, see Figure 3.20-left (Putz, 2006).

Inside the first Brillouin zone, the Fermi surface corresponds to the model of quasi-free electrons, for which slight deformations without circular *E-k* contours of free electrons are recorded, Figure 3.20-left. In Figure 3.20-right the Fermi surface, found in first Brillouin zone of the type of truncated octahedron is illustrated in Figure 3.11-right, for noble metals (Cu, Au, Ag) with cubical lattice (in direct lattice) of F-type.

These results are obtained in the conditions of assuming a perturbed potential of first order upon the free motion of electrons in a crystal.

Does these resulted change (and if so, how) for an exactly treatment, the non-perturbed occupancy (and energetically by E) and the movement

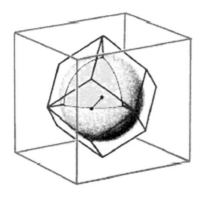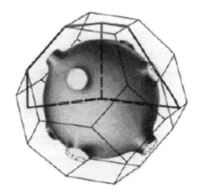

FIGURE 3.20   First Brillouin zone and associated Fermi sphere to the cubical crystals of I-type (in direct lattice) of alkaline metals (to the left) and of these of F type (in direct lattice) of noble metals (to the right); after (Kittel, 1996).

(associated with k-wave vectors) for the electrons, as long as they are bond by crystal's periodic potential? The next section is dedicated to the answer to this question.

3.4.2.3   Quantum Model of Bonding Electrons in Crystal

The model of a periodic crystalline potential is considered as a generalization of the effectively null potential model of Figure 3.13, by the alternation of the wells with potential barriers of a finite altitude V_0, as illustrated in Figure 3.21 (*Kronig-Penney model*), see (Further Readings on Quantum Solid, 1936–1967).

Thus, the crystalline potential satisfies the periodicity relation:

$$V(x+a) = V(a) \tag{3.78}$$

However, what can there be said about the eigen-function associated by general Schrödinger equation (Putz, 2006)

$$-\frac{h^2}{2m_e}\frac{d^2}{dx^2}\Psi(x) + V(x)\Psi(x) = E\Psi(x) \tag{3.79}$$

regarding the periodicity (3.78)?

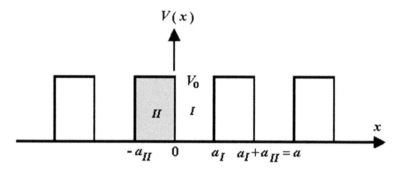

FIGURE 3.21   Model or periodical well in a Kronig-Penney crystal potential model; after Putz (2006).

Directly, without solving the Eq. (3.79), there can be said only that the associated density of probability according to the quantum mechanical postulate [P2], since associated with localization electrons in crystalline field, it should keep the same periodicity, so generating the equation:

$$\left|\Psi(x+a)\right|^2 = \left|\Psi(x)\right|^2 \tag{3.80}$$

From Eq. (3.80) there results that eigen-function $\Psi(x)$ varies with a basic factor from a unit cell to another, written in the direct space as:

$$\Psi(x+a) = e^{i\phi}\Psi(x) \tag{3.81}$$

which, by induction, traversing all the N-unit cells on $0x$ direction, the generated eigen-function based on the rule (3.81), namely

$$\Psi(x+Na) = e^{iN\phi}\Psi(x) \tag{3.82}$$

it should regain the initial eigen-function:

$$\Psi(x+Na) = \Psi(x) \tag{3.83}$$

being the last condition known as the Born-von Karman *boundary* condition. By identifying the terms of Eqs. (3.82) and (3.83) the equation and respectively the quantization of the phase ϕ is obtained:

$$\exp(iN\phi) = 1 \Rightarrow \phi = \frac{2n\pi}{N}, \, n{=}1, 2, ..., N \qquad (3.84)$$

The connection with the reciprocal lattice may be obtained by noting also that the phase induced by traversing of a unit cell by the eigen-function associated to the electrons in crystal can be written as:

$$\phi = ka \qquad (3.85)$$

wherefrom, by the equalization to the statement of Eq. (3.84) also the quantification of the wave vectors allowed in a crystalline periodical system with the length Na results as

$$k = \frac{2\pi}{a}\frac{n}{N}, \, n{=}1, 2, ..., N \qquad (3.86)$$

Moreover, with Eq. (3.85), the relation (3.81) is rewritten as:

$$\Psi(x+a) = e^{ika}\Psi(x) \qquad (3.87)$$

which allows the general writing of the crystalline eigen-function, such as (Putz, 2006):

$$\Psi(x) = U(x)\exp(ikx) \qquad (3.88)$$

as long as the function $U(x)$ satisfies the periodical relation:

$$U(x+a) = U(x) \qquad (3.89)$$

i.e., a relation exactly of type of that one assumed by the crystalline potential in Eq. (3.78).

Thus, the periodicity of the density of probabilities (3.80) was "lowered" at the level of eigen-function such as Eq. (3.88), regaining the celebrated *Bloch* theorem of Eq. (3.34), here in a generalized form: "the eigen-function of an electron in a periodic potential can be written as a product of a function carrying the potential periodicity and a basic exponential factor exp(ikx)".

The 3D extension is immediate.

With potential (3.78) as in representation from Figure 3.21 and with the form of eigen-functions of type (3.88) here separated on I and II regions, $\Psi_{I,II}(x) = U_{I,II}(x)\exp(ikx)$, the effective solving of Schrödinger equation (3.79) separated in the region I and II, can be settle by assuming the eigen-energies $E < V_0$ (Putz, 2006):

$$\begin{cases} \dfrac{d^2\Psi_I}{dx^2} + \dfrac{2m_e}{\hbar^2}E\Psi_I = 0 \\[3mm] \dfrac{d^2\Psi_{II}}{dx^2} - \dfrac{2m_e}{\hbar^2}(V_0 - E)\Psi_{II} = 0 \end{cases}$$

$$\Leftrightarrow \begin{cases} \dfrac{d^2U_I}{dx^2} + 2ik\dfrac{dU_I}{dx} - (k^2 - \delta^2)U_I = 0 \\[3mm] \dfrac{d^2U_{II}}{dx^2} + 2ik\dfrac{dU_{II}}{dx} - (k^2 + \gamma^2)U_{II} = 0 \end{cases} \qquad (3.90)$$

with notations:

$$\delta^2 = 2m_e E/\hbar^2$$
$$\gamma^2 = 2m_e(V_0 - E)/\hbar^2 \qquad (3.91)$$

The solutions of the system (3.90) have the general form:

$$\begin{cases} U_I(x) = A\exp[i(\delta - k)x] + B\exp[-i(\delta + k)x] \\[2mm] U_{II}(x) = C\exp[(\gamma - ik)x] + D\exp[-(\gamma + ik)x] \end{cases} \qquad (3.92)$$

of those coefficients A, B, C and D are found by the continuity and periodicity conditions of the $U_{I,II}(x)$ functions and of their derivatives:

$$\begin{cases} U_I(0) = U_{II}(0) \\[2mm] \dfrac{dU_I}{dx}\Big|_{x=0} = \dfrac{dU_{II}}{dx}\Big|_{x=0} \end{cases} \& \begin{cases} U_I(a_I) = U_{II}(-a_{II}) \\[2mm] \dfrac{dU_I}{dx}\Big|_{x=a_I} = \dfrac{dU_{II}}{dx}\Big|_{x=-a_{II}} \end{cases} \qquad (3.93)$$

With Eq. (3.92) in Eq. (3.93), the next system is formed:

$$\begin{cases}
A + B = C + D \\
iA(\delta - k) - iB(\delta + k) = (\gamma - ik)C - (\gamma + ik)D \\
A\exp[i(\delta - k)a_I] + B\exp[-i(\delta + k)a_I] = C\exp[-a_{II}(\gamma - ik)] \\
\qquad\qquad\qquad\qquad\qquad\qquad\qquad + D\exp[a_{II}(\gamma + ik)] \\
iA(\delta - k)\exp[i(\delta - k)a_I] - iB(\delta + k)\exp[-i(\delta + k)a_I] \\
\qquad = C(\gamma - ik)\exp[-a_{II}(\gamma - ik)] - D(\gamma + ik)\exp[a_{II}(\gamma + ik)]
\end{cases}$$

$$(3.94)$$

which allows the non-trivial solutions if and only if the associated determinant is vanishing, which, after an elementary (although lengthily) algebra, it reduced to the working equation:

$$\frac{\gamma^2 - \delta^2}{2\delta\gamma}\sinh(\gamma a_{II})\sin(\delta a_I) + \cosh(\gamma a_{II})\cos(\delta a_I) = \cos[k(a_I + a_{II})]$$

$$(3.95)$$

At this point, there is convenient, for avoiding the problems when $E > V_0$, to consider the case with $V_0 \to \infty$ (Figure 3.22) & $a_{II} \to 0$ so that the product $V_0 a_{II}$ tending to a constant. This way, the approximation:

$$\sinh(\gamma a_{II}) \cong \gamma a_{II} \text{ and } \cosh(\gamma a_{II}) \cong 1 \qquad (3.96)$$

is also generated. Moreover, in this case one gets the lattice constant limiting case:

$$a_I \to a \qquad\qquad (3.97)$$

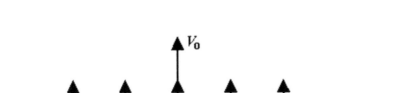

FIGURE 3.22 Potential dirac delta ("comb") function.

If one consider the notation

$$P = m_e a_I V_0 a_{II} / h^2 \qquad (3.98)$$

the Eq. (3.95) becomes (Putz, 2006):

$$f(\delta a) \equiv \frac{P}{\delta a} \sin(\delta a) + \cos(\delta a) = \cos(ka) \qquad (3.99)$$

of those solutions can be graphically evaluated, through identified the δa dots in the graphic of the function $f(\delta a)$ of left-hand side of Eq. (3.99), for which it equalized with the maximum limits of the cosine function associated, +1 and -1, of the right-hand side of the Eq. (3.99).

Such a graphical evaluation of the solutions of Eq. (3.99) for $P=3\pi/2$, which corresponds to a high limit of the barrier V_0 is in Figure 3.23 exposed.

Accordingly, from the analysis of Figure 3.23 there results how the dots (δa) can take values only on certain intervals, thus being discredited, so quantified. Next, by reminding the notations of Eq. (3.90), one can write for the eigen-energy of the electrons:

$$E = \frac{h^2}{2m_e} \delta^2 = \frac{h^2}{2m_e a^2} (\delta a)^2 \qquad (3.100)$$

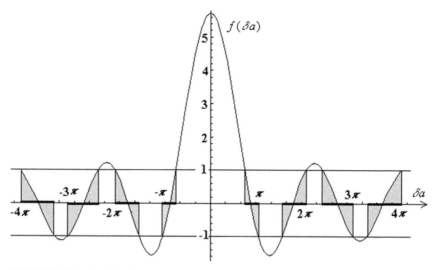

FIGURE 3.23   Graphical solution of Eq. (3.99) for $P=3\pi/2$, after Putz (2006).

Eq. (3.100) means the quantification also in the electronic eigen-energies of the crystal.

Regrouping the intervals allowed as solutions of Eq. (3.99) from the Figure 3.23 in terms of energy quantification (3.100), the representation of Figure 3.24 results (Putz, 2006); in its extended version (from the left side) it presents the discontinuities at the wave vectors values $k \equiv \delta = \pm n\pi / a$ and respectively the energetic bands in the reduced version at the first Brillouin zone (from the right side).

Therefore, the basic result of this section consists in the fact that the application of a model of non-perturbed potential to the quantum description of bonding electrons in a crystal do not change the picture of bands and energetic gaps already introduced by the model of quasi-free electrons, of previous section (see Figure 3.18), but just generalized it at even many energetic bands.

The next case, for which the thickness of the region II of Figure 3.21 is not zero, implies the case of *tight-binding electrons* between the potential fields of the structural consecutive positions (nuclei) in crystalline structure, and will be further analyzed.

3.4.3.4 Quantum Model of Tight-Binding Electrons in Crystal

Once again: "a crystal is a compound of atoms or assemblies of atoms ordered in space"; this because the atoms and the atomic orbital are the quantum fundament of the compounds of the super-atomic (yet nano-) matter.

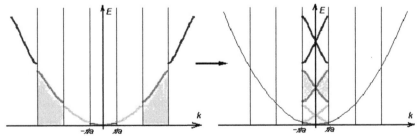

FIGURE 3.24 Extended quantification (left) of the eigen-energies of the bonding electrons of a crystal and their reduction to the first Brillouin zone (right); after (Further Readings on Quantum Solid 1936–1967; Putz, 2006).

Thus, the atomic orbital must be considered as a quantum basis also for the description of the electrons in crystals.

Postulate [P3] of quantum mechanical allows this picture, employed at the level of molecular orbital through the LCAO construction.

This way, the crystalline $\Psi_k(r)$ orbital should be constructed from the atomic ones, $\phi(r–R_{mnp})$, located at the R_{mnp} distance respecting the origin of the reference system of (direct) lattice, so that also the Bloch theorems (3.88) and (3.89) to be respected for including the periodicity of the lattice in crystalline eigen-function.

Under these conditions, the crystalline orbital can be written as a complete LCAO expansion of basic Bloch factors, see (Further Readings on Quantum Solid, 1936–1967)

$$\Psi_k(r) = \frac{1}{\sqrt{N}} \sum_{mnp} \exp\left(ikR_{mnp}\right)\phi\left(r - R_{mnp}\right) \qquad (3.101)$$

when the statements (3.61) and (3.64) were considered and the constant of the lattice a, was excluded from the normalization factor.

As such there can be immediately verified how the eigen-functions of type (3.101) satisfy the Bloch theorem under the general form:

$$\Psi_k(r+t) = \frac{1}{\sqrt{N}} \sum_{mnp} \exp\left(ikR_{mnp}\right)\phi\left(r + t - R_{mnp}\right)$$

$$= \frac{1}{\sqrt{N}} \exp\left(ikt\right)\sum_{mnp} \exp\left(ik(R_{mnp} - t)\right)\phi\left(r - (R_{mnp} - t)\right)$$

$$= \exp\left(ikt\right)\Psi_k(r)$$

$$(3.102)$$

based on the fact that the periodic t-vector produces the translation of the lattice R_{mnp} vector position in the $t + R_{mnp}$ position vector like another vector of the lattice.

Thus, the functions (3.101) are indeed eigen-functions of the crystalline periodical system, thus satisfying the Schrödinger equation associated (according with the [P1] postulate of quantum mechanics) under the form of unidirectional case on the direction $0x$:

$$\hat{H} \bullet \sum_{m} \exp\left(ikR_m\right)\phi\left(x - R_m\right) = E_k \sum_{m} \exp\left(ikR_m\right)\phi\left(x - R_m\right) \quad (3.103)$$

By multiplying at the left the Eq. (3.103) with conjugate atomic orbital $\phi^*(x-R_{m'})$ and by further integrating the result, the linear set of equations is obtained (Putz, 2006):

$$\sum_{m}\left(H_{m'm} - E_k S_{m'm}\right)\exp(ikR_m) = 0 \qquad (3.104)$$

where the consecrated notations have been considered:

$$H_{m'm} = \int \phi^*(x - R_{m'})\hat{H}\bullet\phi(x - R_m)dx \qquad (3.105)$$

$$S_{m'm} = \int \phi^*(x - R_{m'})\phi(x - R_m)dx \qquad (3.106)$$

In order to solve the Eq. (3.104) for the eigen-energies E_k (note that the corresponding eigen-functions are those of Eq. (3.101)!) the approximate forms for the interaction (3.105) and overlapping (3.106) terms of atomic orbital in crystalline structure should be specified (since the general analytical ones being inaccessible).

This aim can be fulfilled if there is reminded the construction of the bonding (symmetrical, low) and those from the anti-bonding (anti-symmetrical, the highest) states, see Figure 3.25-left, form the combination of the atomic orbital for two neighboring atoms and the variation of the relative energies based on the distance between atomic nuclei.

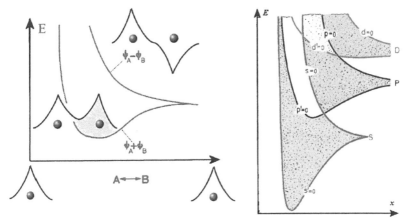

FIGURE 3.25 The Bi-atomic energetic levels (left) and their crystal bands counterpart (right); after (Further Readings on Quantum Solid 1936–1967; Putz, 2006).

The generalization of this picture to an entire string of atoms periodi-cally disposed under the influence of the neighboring nuclei emphasizes the formation of energetic bands, with very close energetic levels, from the removing of the degeneration of all the states located between bonding and those of anti-bonding states, by the combination of any type of atomic basic orbitals $(s, p, d,...)$ appearing to the corresponding terms of bands (S, P, D), but without energetically overlapping (atomic orbital s with those p, etc. ...) from their orthogonal quality, see Figure 3.25-right (Putz, 2006).

This way, the model of energetic bands (since one is looking for a right approximation for the eigen-energies of Eq. (3.104)), results according to the similar predictions of bands in the model of the quasi-free electrons and bonded-electrons from previous sections, only based on the interaction and combination of atomic orbitals of each pair of the neighbor atoms in crystal.

Therefore, energetically, the crystal can be quantum mechanically char-acterized within the crystalline electrons approximation (and respectively of the orbital) through the model of the first order interaction of the atomic orbitals (i.e., the most neighboring being the neighbors of first order); this picture being called the *tight-binding model of electrons* in crystal.

For the interactions (3.105) and of overlapping (3.106) terms this model prescribes the values (written symbolical as) (Putz, 2006):

$$H_{m'm} = \begin{cases} \alpha \ , \ m' = m \\ \beta \ , \ m' = m \pm 1 \\ 0 \ , \ in \ rest \end{cases}$$

$$S_{m'm} = \begin{cases} 1 \ , \ m' = m \\ 0 \ , \ in \ rest \end{cases} \tag{3.107}$$

under approximation which is recognized to be of Hückel type.

There is not less true that by applying the approximations (3.107) to the Eq. (3.104) there is firstly obtained:

$$\beta \exp(ikR_{m-1}) + (\alpha - E_k) \exp(ikR_m) + \beta \exp(ikR_{m+1}) = 0 \tag{3.108}$$

and with specific realizations

$$R_{m-1} = (m-1)a \ \& \ R_m = ma \ \& \ R_{m+1} = (m+1)a \tag{3.109}$$

Along the simplification of the common exp($ikma$) factor and through the application of the Euler decomposition (3.22) for the rest of terms, the last two equations provide the eigen-energies of the electronic states in the crystal, consistently as:

$$E_k = \alpha + 2\beta \cos(ka) \qquad (3.109)$$

with a form identically to that of Eq. (3.17), as abstracted from the generalization of the π-bonds of polyenes at the infinite reticular string in a crystal.

Still, in this case the relation (3.109) was directly obtained, without calling the intermediary molecular case, and even performed through a similar approximation to that one of molecular Hückel; the result (3.109) is not restricted by a certain type of atomic orbital (as was the case of the π- conjugated bond as the ground of generalization: polyenes chain→crystalline string). Besides, for the present approach, there appears as intrinsic feature the connection between the Bloch theorem, the form of crystalline eigen-functions (3.101) and the eigen-energies (3.108), a non-obvious connection by the inductive approach of previous sections.

This way, the dependence (3.109) can be rewritten for the 3D case (Putz, 2006):

$$E_k = \alpha + \beta \sum_{mnp} \exp(ikR_{mnp}) \qquad (3.110)$$

and which in 2D projection gives information about the E-k dependence at the level of first Brillouin zone.

In order to investigate the form and the eventual differences towards the treatment of quasi-free and free electrons in crystal, a cubical lattice will be considered, which contains, in projection the lattice vectors:

$$R_{mm} = \{(a,0); (-a,0); (0,a); (0,-a)\} \qquad (3.111)$$

and for which the eigen-energies (3.110) will be in relation with the (k_x, k_y) wave vectors associated by the relationship:

$$E_k = \alpha + 2\beta \left[\cos(k_x a) + \cos(k_y a) \right] \qquad (3.112)$$

One is interested in the behavior of the *E-k* dependence from the center (Γ: *k*=0) towards the border of the first Brillouin zone (X: *k*=π/a).

Around the center of the zone the cosine expansion $\cos\theta \approx 1 - \theta^2/2$ is used in (3.112) which results in the dependence:

$$E_\Gamma = \alpha + 4\beta - \beta\left[(k_x^2 + k_y^2)a^2\right] \tag{3.113}$$

i.e., a parabolic dependence (circular in projection) of quasi-free electrons type, Figure 3.19 (left or right in the middle of first Brillouin zone).

For the *E-k* behavior around the border of the first Brillouin zone, the respective wave vectors will be written with the truncated expanded forms:

$$k_x = \frac{\pi}{a} - \delta_x$$

$$k_y = \frac{\pi}{a} - \delta_y \tag{3.114}$$

which, along with the trigonometric identity

$$\cos(a - b) = \cos(a)\cos(b) + \sin(a)\sin(b) \tag{3.115}$$

generate from Eq. (3.112) the dependence:

$$E_X = \alpha - 4\beta + \beta\left[(\delta_x^2 + \delta_y^2)a^2\right] \tag{3.116}$$

i.e., again a parabolic dependence (circular in projection).

In the middle of the bands, the condition $E_k = \alpha$ in Eq. (3.112) results in the equation and the attached trigonometric solution:

$$\cos(k_x a) + \cos(k_y a) = 0 \Rightarrow k_x a = \pi - k_y a \tag{3.117}$$

corresponding to the right lines.

Cumulating the results (3.113), (3.116) and (3.117) in a single representation, the Figure 3.26 at the level of first Brillouin zone is generated (Putz, 2006).

Nevertheless, what is relevant from the previous analysis, although not demonstrated here for the most general crystalline lattice – the triclinic

FIGURE 3.26 Construction and energetic occupancy of the first Brillouin zone for the tight-binding electrons crystalline model; after (Further Readings on Quantum Solid 1936–1967; Putz, 2006).

one, refers on how the eigen-energies of the electrons in crystal have a parabolic dependence on the wave vectors associated, both the center of Brillouin zone (at the last level of band of valence, BV) and at the border/frontier of the Brillouin zone (at first level of band of conduction, BC), as based on the results of various levels of approximation considered, according to the statements (3.113) and (3.116), respectively.

The model of tight-binding electrons, therefore predicts the qualitative aspects similar with the model of the quasi-free electrons, but through an analysis including the orbital interactions of the neighboring atoms in the crystal.

Kant was wrong in this case.

The present chapter founds all the basic results for the crystalline quantum state characterized from previous section, but in a more consistent and unitary comprehension, yet without "demolishing" the molecular state theory, but only short-cutting it.

The results reconfirmation from the molecular extension by the present approach does not reformulate the molecular modeling, but even more confirms its validity, while also emphasizing the differences between molecular orbitals and those crystalline.

Reconciliation between the two "worlds", crystalline and molecular, from an even surprising point of view, will be treated in the following section, at its ending. The impact of this fusion, for life and future, is – we believe – most valued for the future nanoscale systems and functionalities.

3.5   MODELING QUANTUM SOLIDS

3.5.1   FERMI DISTRIBUTION

The previous section demonstrated *in extenso* how the quantum description of electrons in crystals requires a discretization of the eigen-energies of crystalline orbitals in energetic bands; the present discussion follows (Putz, 2006).

Therefore, the electrons showcase discontinuous energetic spectrum of bands in crystals; but how are they disposed in between these bands, if any?

An idea was already previously introduced, regarding the necessity of the existence of a maximum level of occupancy, i.e., the Fermi level.

The position of this level in the entire energetic spectrum of bands in crystals is essential, as long as it establishes the maximum limit of the possibilities of electronic occupancy.

But how are disposed the electrons and what is the probability with which the successive energetic bands are occupied until the maximum limit of Fermi level?

And even more: what is happening beyond the Fermi level?

The analysis will start with another interrogation: there is any connection between the electron's distribution in energetic spectrum in a solid body and Fermi level?

The answer is affirmative. And this is the so-called Fermi distribution $f(E)$!

However, as long as it is a distribution, we have to start from statistics, see the Volume I of the present five-volume set and (Putz, 2010): it prescribes that a total number of electrons N_e can occupy a total number of Z_E allowed states (inside a spectrum) with a probability:

$$\Omega = \frac{Z_E!}{N_e!(Z_E - N_e)!} \qquad (3.118)$$

Physical statistics further prescribes the occupancy with probability (3.118) to produce an entropy variation ΔS proportional with $\ln\Omega$ through the constant k_B of Boltzmann, such as:

$$\Delta S = k_B \ln \Omega = k_B \ln\left(\frac{Z_E!}{N_e!(Z_E - N_e)!}\right) \qquad (3.119)$$

The entropy variation (3.119) will subsequently combine with the variation of eigen-energies ΔE for the N_e electrons producing the variation of free energy ΔF necessary for the realization of such occupancy at thermodynamic T-temperature:

$$\Delta F = N_e\Delta E - T\Delta S = N_e\Delta E - k_BT \ln\left(\frac{Z_E!}{N_e!(Z_E - N_e)!}\right)$$

$$\cong N_e\Delta E - k_BT\left[Z_E \ln Z_E - N_e \ln N_e - (Z_E - N_e)\ln(Z_E - N_e)\right]$$
$$(3.120)$$

which will involve a variation of chemical potential $\Delta\mu$ of the system electrons-occupying-available energetic states:

$$\Delta\mu = \frac{\partial\Delta F}{\partial N_e} = \Delta E + k_BT \ln\left(\frac{N_e}{Z_E - N_e}\right) \qquad (3.121)$$

The dependency (3.121) allows the writing of the N_e/Z_E distribution of electrons/states available occupancy as:

$$\frac{N_e}{Z_E} = \frac{1}{1+\exp\left[(\Delta E - \Delta\mu)/(k_B T)\right]} \xrightarrow[\Delta\mu \equiv E_F]{\Delta E \equiv E} f(E)$$

$$= \frac{1}{1+\exp\left[(E - E_F)/(k_B T)\right]} \tag{3.122}$$

which generated Fermi distribution when the energetic variations with the energetic levels themselves have been identified, with E the eigen-level currently occupied and respectively E_F the maximum available eigen-level.

There is immediately noted from Eq. (3.122) how for $E=E_F$ the distribution becomes $f(E)=1$, i.e., properly corresponding to the maximum occupancy when all the Z_E available states are filled with the N_e electrons ($N_e = Z_E$).

Then, how is manifesting the temperature influence in Eq. (3.122) around the Fermi level?

For a typical cubic crystal there results the Fermi energy $E_F \approx 9 \times 10^{-19} J = 6$ [eV], which in terms of the temperature, $E_F = k_B T_F$, means a Fermi temperature T_F of about 70,000 [K] (i.e., orders of 10^5 [K])!

This means that at common temperatures a slight variation in distribution (3.122) around Fermi energy is barely recorded, as well as its transformation from a (discontinuous) step function into a continuous function having values between 0 and 1, see Figure 3.27 (Putz, 2006).

Yet, continuous form of Fermi distribution (3.122) allows the formulation of some important applications, just based on the model of free electrons in a solid body.

For example, for an extraction work (ϕ_0, also called extraction energy) of electrons from a small enough material, the electrons in a solid can be ionized by a heating of the solid so that the Fermi distribution to become continuous (as in the middle and the right side of Figure 3.27) allowing, by the Fermi "tail" distribution, the existence of the energetic levels even over the limit over the extraction work (ϕ_0) so obtaining the *thermo-emission* phenomenon, Figure 3.28-left.

Also, by applying intense fields (electric or magnetic) upon a solid the difference of solid-environment potential is changed so that tunneling

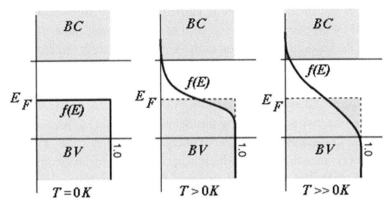

FIGURE 3.27 Illustration of Fermi function and Fermi level, (E)and E_F, respectively, for energetic bands (BV: band of valence and BC: band of conduction) for electronic occupancy in a solid body; after (Further Readings on Quantum Solid 1936–1967; Putz, 2006).

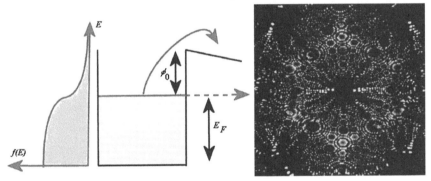

FIGURE 3.28 Illustration of thermo-ionic-emission (left) and the field-emission (left and right); after (Further Readings on Quantum Solid 1936–1967; Putz, 2006).

of electrons even at Fermi level are caused, see dotted representation of Figure 3.28-left, thus obtaining the so-called field emission.

For a polished metal the application of such fields produces a picture of localization of atoms in metal, due to fields' variation between atoms, Figure 3.28-right, the process representing the base of the methods of *atomic scanning microscopy*.

Regarding the electro-magnetic action upon solids, a photon with energy higher than the extraction work produces the electron propulsion from the occupied levels receiving this energy, so obtaining the *photo-emission*, Figure 3.29-left.

Yet, if the incident photon is carried an energy high enough to be absorbed by en electron from the core of the electronic states (the underneath of the valence states!) then such electron will be excited on a superior level, while other electron "falling" in its place from the valence band, and emitting photons corresponding to X-radiation, see Figure 3.29-right, being phenomenon underlying the *X-ray emission* and the *Auger spectroscopy*.

Moreover, two solids having the electronic contact (so forming a *junction*) tend to equalize their levels of energetic occupancy, i.e., to equal the Fermi levels in a single one, which is accompanied by an electronic flow, migration from the region with superior Fermi level (becoming the positively "ionized" region) to the region with inferior Fermi level (becoming the negatively "ionized" region), this way obtaining the so-called *contact potential*, see Figure 3.30.

The electrons migration between two solids in contact raises the question: do the electrons of crystalline orbital's have a motion inside the solid body that they belong, in the absence of any contact potential?

The answer is affirmative and generates an analysis both interesting and essential for the completion of the quantum characterization of electrons in crystals.

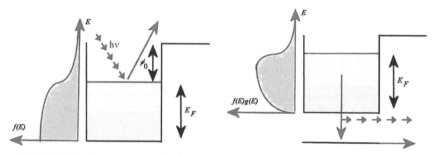

FIGURE 3.29 Representation of photoemission (left) and of the X-ray emission (right) phenomena; after (Further Readings on Quantum Solid 1936–1967; Putz, 2006).

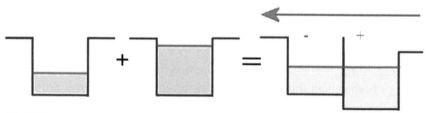

FIGURE 3.30 Illustration of the contact potential between two solids; after (Further Readings on Quantum Solid 1936–1967; Putz, 2006).

If the electrons would have their own motion on the valence band (VB), they should be characterized also by a velocity. However, since the electrons have a quantum nature depending on their occupancy on Bloch-Schrödinger crystalline orbitals (3.101) they should be represented by the associated wave package, further characterized by the group velocity correlated with the quantum energy as such (Putz, 2006):

$$\vee_k := \frac{d\omega_k}{dk} = \frac{1}{\hbar}\frac{dE_k}{dk} \tag{3.123}$$

If an electron from VB has the velocity (3.123) it also has an acceleration and thus a force (F^*) and a corresponding mass m^*. One may considering that the force, in turn, can be expressed as "momentum in time" and that the quantum momentum is associated with the wave vector of the electron according to the relation (3.53), so follows the equalities:

$$\frac{\left(F^*\right)}{m^*} = \frac{d\vee_k}{dt} = \frac{1}{\hbar}\frac{d^2E_k}{dkdt} = \frac{1}{\hbar^2}\frac{d^2E}{dk^2}\left(\hbar\frac{dk}{dt}\right) = \frac{1}{\hbar^2}\frac{d^2E}{dk^2}\left(F^*\right) \tag{3.124}$$

wherefrom the so-called effective mass m^* of electrons in solids results as:

$$\frac{1}{m^*} = \frac{1}{\hbar^2}\frac{d^2E_k}{dk^2} \tag{3.125}$$

The result (3.125) asserts, somehow surprising considering that the electrons mass is an universal constant (by classical means), that the electrons mass in crystal depends on the k- wave vector, while from Eq. (3.123), it may carry also the negative velocities (!) for the E-k dependence cases through the displayed negative slope.

How can these results be interpreted?

For clarify, these results will be effectively applied on the model of the tightly bound electrons, for the uni-dimensional version for which the eigen-energies E_k are provided by the relation (3.109), with the immediate statements (Putz, 2006):

$$\vee_k = -\frac{2a\beta}{\hbar}\sin(ka) \ \& \ m^* = \frac{\hbar^2}{2a^2\beta\cos(ka)} \tag{3.126}$$

From Eq. (3.126), the fact that the velocity, depending on the "sinus" function, an odd function, varies from positive values to those negative, reconfirms the electronic movement from a side to another of the first Brillouin zone, for any band of energy, Figure 3.31-left, and also how the effective mass becomes negative at the margin of the first Brillouin zone (with $k = \pm \pi/a$), see Figure 3.31-right (Putz, 2006).

But what physical meaning have these negative values?

If the negative values were associated to the velocity's reflection and respectively to the mass' absence, thus everything would make sense: the electrons motion with negative velocity means their reflection (also called as *Bragg scattering*) and corresponds to the electrons motion inside the first Brillouin zone, Figure 3.32-left, while the negative mass at the frontier of the Brillouin zone is associated with an absence of the electrons and the creation of some particles with identical mass, yet inversing the charge, the so-called holes that assume the motion of the unoccupied states by the electrons of the first Brillouin zone, Figure 3.32-right (Putz, 2006).

Thus, the two types of motion in solids, i.e., the electronic and the one of the holes, are naturally considered, with the holes being

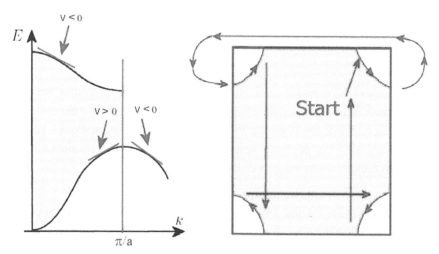

FIGURE 3.31 Phenomenological representation for the velocities sign (left) and the electronic motion (right) in the energetic bands reduced at the first Brillouin zone; after (Further Readings on Quantum Solid 1936–1967; Putz, 2006).

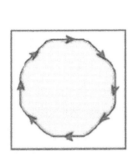 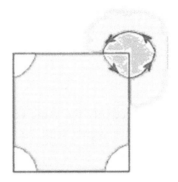

FIGURE 3.32   Representation of the electronic motion (left) and respectively for holes motion (right) in the first Brillouin zone; after (Further Readings on Quantum Solid 1936–1967; Putz, 2006)

associated with the absence of the electrons, but relative to the states that they would be "entitled" to occupy, according to Fermi distribution and Fermi level!

As long as Fermi distribution (3.122) indicates the probability with which an electron occupies the state of eigen-energy E, the probability that this state is *not* occupied by an electron, i.e., to be occupied by a hole, will be:

$$1 - f(E) = 1 - \frac{1}{1 + \exp\left[(E - E_F)/(k_B T)\right]} = \frac{1}{1 + \exp\left[(E_F - E)/(k_B T)\right]}$$

(3.127)

This behavior indicates the fact that the holes' energy E_h is corresponding to minus the electronic energy E_e:

$$E_h(k_h) = -E_e(k_e)$$

(3.128)

and, for which, also the associated wave vectors satisfy the same relation

$$k_h = 0 - k_e = -k_e$$

(3.129)

as based on the fact that the total wave vector in a band is zero (any external field if is absent), Figure 3.32.

From the energetic inversion relation (3.11) there results the opposite curves (3.125) for electrons and holes, i.e., with effective opposite masses

$$m_h = -m_e \qquad (3.130)$$

yet, with respectively the same variation for the associated wave vectors (3.129); in other words with the same slope, i.e., with the same group velocity (3.123):

$$\vee_h = \vee_e \qquad (3.131)$$

Thus, the holes appear out from released electronic states by the electronic passing of the valence band VB into the conduction band CB of a solid, Figure 3.27.

An analysis of the electrons of CB prescribes their number as being evaluated from the density of electronic states $g_C(E)$, the analogue of Eq. (3.68), however multiplied by the probability that such a state to be occupied (according with Fermi distribution) and integrated after the energetic spectrum starting from the first energy of the conduction band E_C (Putz, 2006):

$$N_e^C(T) = \int_{E_C}^{\infty} g_C(E) f(E) dE = \int_{E_C}^{\infty} \frac{g_C(E)}{1 + \exp\left[(E - E_F)/(k_B T)\right]} dE \qquad (3.132)$$

Similarly, for the remaining holes in VB they occupy the energetic states until the first energy of valence band E_V with states' density $g_V(E)$ and with the probability of occupancy of each such a state given by the holes distribution (3.127):

$$N_h^V(T) = \int_{-\infty}^{E_V} g_V(E)[1 - f(E)] dE = \int_{-\infty}^{E_V} \frac{g_V(E)}{1 + \exp\left[(E_F - E)/(k_B T)\right]} dE. \qquad (3.133)$$

These integrals can be further specialized if the occupancy probabilities for electrons in CB and for holes in VB are close to 1, so ignoring the exponentials; what results in the so-called degenerate case corresponding to the metals; for the other extreme situation where the exponential contributions for the probabilities case close to 0 prevails, the so-called no degenerate case of semiconductors is generated.

For the last case, the relations (3.132) and (3.133) are written such as:

$$N_e^C(T) \cong \int_{E_C}^{\infty} g_C(E) \exp\left[(E_F - E)/(k_B T)\right] dE$$

$$= \exp\left[(E_F - E_C)/(k_B T)\right] \left(\int_{E_C}^{\infty} g_C(E) \exp\left[-(E - E_C)/(k_B T)\right] dE \right)$$

$$\equiv \exp\left[(E_F - E_C)/(k_B T)\right] (N_C(T)),$$

$$N_h^V(T) \cong \int_{-\infty}^{E_V} g_V(E) \exp\left[(E - E_F)/(k_B T)\right] dE$$

$$= \exp\left[(E_V - E_F)/(k_B T)\right] \left(\int_{-\infty}^{E_V} g_V(E) \exp\left[-(E_V - E)/(k_B T)\right] dE \right)$$

$$\equiv \exp\left[(E_V - E_F)/(k_B T)\right] (N_V(T))$$

$$(3.134)$$

which allows the elimination of Fermi energy by multiplication and inserting of the width $E_g = E_C - E_V$ of the VB-CB gap, resulting in *the law of masses action*:

$$N_e^C(T) N_h^V(T) = \exp\left[-(E_C - E_V)/(k_B T)\right] N_C(T) N_V(T)$$

$$\equiv \exp\left[-E_g/(k_B T)\right] N_C(T) N_V(T) \qquad (3.135)$$

which further connects the information regarding the distribution of electrons and holes in a semiconductor.

Even the result of Eq. (3.135), based on Fermi distribution, does not involve the energy of Fermi level, its knowledge is essential for establish the way in which a semiconductor acts as a metal, i.e., facilitating the electric conduction through the conduction band.

The establishment of the Fermi level for two different ways to occupy the CB with electrons and thus producing holes in VB, with the consequence of the appearance of the electric current at the application of a potential difference between the solid ends is the subject of the following section.

3.5.2 SEMICONDUCTORS AND JUNCTIONS

Solids can be insulating, conductors (metals) but also semi-conductors, a classification that can be explained at quantum level by the width E_g of energetic gap that separates the valence band VB form the one of conduction CB, Figure 3.33; the present discussion follows (Putz, 2006).

The intermediate case, the semiconductors' one, is the most interesting as long as they allow the manipulation of the materials from this category, either as insulating or as conductors, depending on the practical needs.

Therefore, quantum the analysis of the semiconductors behavior and especially when they become conductors has a particular interest.

This analysis starts from the previously established law of masses (3.135) with the purpose to define the Fermi level that separates the maximum occupied from the free states.

Further, the attributing of the Fermi level depends on the way in which the semiconductors are "treated" aiming to transform it in a conductor,

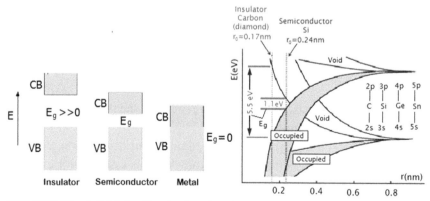

FIGURE 3.33   Solids' classification based on the width E_g between conduction band CB and the valence band VB, at quantum level (left), with Si and C-diamond exemplification (right); after (Further Readings on Quantum Solid 1936–1967; HyperPhysics, 2003; Putz, 2006).

or, in other words, on the way in which the electrons from VB are excited towards CB through passing the energy barrier E_g.

The first method in this concern is included in the so-called *intrinsic regime*, the relative semiconductors being also considered as *intrinsic semiconductors*, and is merely based on thermal excitations through considering the temperature T that drives the dependence in all the electronic distributions in solids by the Fermi distribution (3.122).

One assumes that with the intrinsic regime (almost) all the electrons in CB were excited from VB, i.e., do not come from impurities or defects in solid, therefore leaving behind an equal number of holes concentration $(n = N/v)$ in VB:

$$N_e^C(T) = N_h^V(T) := N_i(T) \qquad (3.136)$$

The Eq. (3.136) with the statements (3.134) and the law of masses (3.135) will form the system that will give the Fermi energy of intrinsic semiconductors.

For evaluating the number of electrons/holes of CB/VB according to Eq. (3.134) the evaluation of the density of states associated, $g_C(E)$ and respectively $g_V(E)$ are required. For this approach we will firstly use the general definition of the density of states (3.68) and (3.66) such as (Putz, 2006):

$$g(E) = 2g(k)\frac{dk}{dE} = 2\left(\frac{L}{2\pi}\right)^3 4\pi k^2 \frac{m^*}{\hbar^2 k} = \frac{v}{\pi^2}\frac{m^* k}{\hbar^2} \qquad (3.137)$$

where the total energy dependence on the effective mass has been considered

$$E = (\hbar k)^2 /(2m^*) \qquad (3.138)$$

Further, to customize holes and electrons, is taking into account the parabolic dependence of type (3.113) for BV (in the middle of the first Brillouin area) rewritten based on its last level E_V:

$$E(k_h) = E_V - \frac{\hbar^2 k_h^2}{2m_h^*} \Rightarrow k_h = \sqrt{\frac{2m_h^*}{\hbar^2}(E_V - E)} \qquad (3.139)$$

and respectively of type (3.116) for CB (at the border of first Brillouin zone) rewritten as based on its first level E_C:

$$E(k_e) = E_C + \frac{\hbar^2 k_e^2}{2m_o^*} \Rightarrow k_e = \sqrt{\frac{2m_e^*}{\hbar^2}(E - E_C)} \qquad (3.140)$$

With the density (3.137), for example for the electrons with the aid of the electronic wave vector (3.139), the integral $N_C(T)$ of Eq. (3.134) can be calculated through proceeding to the replacement $q=(E-E_C)/(k_B T)$, so obtaining the result:

$$N_C(T) = \frac{v\sqrt{2}\left(k_B T m_e^*\right)^{3/2}}{\pi^2 \hbar^3} \int_0^\infty \sqrt{q}\exp(-q)dq = \frac{1}{4}v\left(\frac{2k_B T m_e^*}{\pi \hbar^3}\right)^{3/2}$$

$$(3.141)$$

when replacing of the standard (of Poisson type) integral value (see the appropriate Appendix of the Volume I of the present five-volume set)

$$\int_0^\infty \sqrt{q}\exp(-q)dq = \sqrt{\pi}/2 \qquad (3.142)$$

For holes, the steps are identical, so re-gaining the result (3.141) but with the effective mass of the holes m_h^* instead the one of the electrons.

From now one, the Fermi level evaluation for intrinsic semiconductors can be performed. Firstly one combines (3.136) with the masses law (3.135) and with the statements (3.141), so obtaining:

$$N_i(T) = \exp[-E_g/(2k_B T)]\sqrt{N_C(T)N_V(T)}$$

$$= \exp[-E_g/(2k_B T)]\frac{1}{4}v\left(\frac{2k_B T}{\pi \hbar^3}\right)^{3/2}\left(m_e^* m_h^*\right)^{3/4} \qquad (3.143)$$

By the identity (3.136), the dependence (3.143) can be equalized with that one of electrons (3.134) of CB – for example, while through inserting (3.141) the Fermi energy equation is generated:

$$\exp[-E_g/(2k_BT)]\frac{1}{4}v\left(\frac{2k_BT}{\pi\hbar^3}\right)^{3/2}\left(m_e^*m_h^*\right)^{3/4}$$

$$=\frac{1}{4}v\left(\frac{2m_e^*k_BT}{\pi\hbar^3}\right)^{3/2}\exp[(E_F-E_C)/(k_BT)] \qquad (3.144)$$

wherefrom the solution of the Fermi level energy for the regime of intrinsic semiconductors results as (Putz, 2006):

$$E_F = E_C - \frac{1}{2}E_g + \frac{3}{4}k_BT\ln\left(\frac{m_h^*}{m_e^*}\right) \qquad (3.145)$$

There is immediate from Eq. (3.145) that for temperatures around the zero absolute ($T=0K$) the Fermi level is placed exactly at the half of energetic gap between CB and VB, see Figure 3.34 (Putz, 2006).

Nevertheless, is also true that in general, the statement (3.145) indicates the movement of Fermi level towards the level of lesser effective mass (for a less density of the states at the border of the respective band), during the increase of temperature.

However, because the effective masses of the electrons and holes have similar magnitudes this results in the non-displacement of Fermi level away of the center of the VB-CB gap.

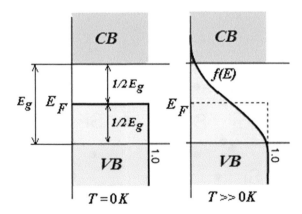

FIGURE 3.34   Fermi level representation for the intrinsic semiconductors regime; after (Further Readings on Quantum Solid 1936–1967; HyperPhysics, 2003; Putz, 2006).

The consequence shows in a relatively low efficacy for the semiconductors of the intrinsic regime to become conductors.

In order to increase this efficacy, other method is used, i.e., that one of semiconductors' doping, this way being introduced the category of doped semiconductors.

Taking into account that the basic elements of semiconductors are the metalloids of the group IVA, i.e., Si & Ge, i.e., the atoms of the corresponding solid structures have 4 electrons in the valence shell, the electric charges can be created by semiconductors doping, in two ways:

1. by pentavalent impurities with atoms of the group VA, thus automatically creating free electrons, the semiconductors being of *type n*
2. and by trivalent impurities with atoms of the group IIIA, thus automatically creating free holes of type *p*, see Figure 3.35 (Putz, 2006).

Also in this case we are interested in establishing the Fermi level for such doped semiconductors. For this aim, there is considered, for example, the situation of the combination of a semiconductor of Si or Ge with impurities such as the pentavalent atoms (P, As, Sb). Thus, a *n* type semiconductor by inter-atomic coordination is created, Figure 3.35-left.

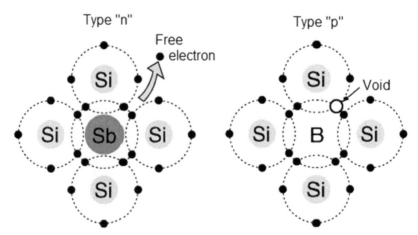

FIGURE 3.35 Semiconductors doping of type "n" (left) and type "p" (right); after (Further Readings on Quantum Solid 1936–1967; HyperPhysics, 2003; Putz, 2006).

In addition, there is assumed that (Putz, 2006):

(i)   the resulted free electron is very less attracted by the atom of impurity from which it comes;

(ii) the free electron becomes in the first approximation an electron of CB at lowest energy E_C;

(iii) the free electron evolves in crystalline potential thus having the effective mass m_e^*;

(iv) the electron "sees the rooting nucleus from crystal" by dielectric constant ε_r, i.e., it is shielded from the atom impurity nucleus that released the free electron.

Under these circumstances, the motion of this electron can be characterized using the adaptation of quantum model of hydrogen scaled as:

$$\varepsilon_0 \rightarrow \varepsilon_0 \varepsilon_r \ \& \ m_0 \rightarrow \frac{m_e^*}{m_0} m_0 \tag{3.146}$$

which will result in bonding energy (with principal quantum number $n=1$) and respectively the first Bohr radius accordingly modified:

$$E^* = -13.6 \frac{(m_e^* / m_e)}{\varepsilon_r^2} [eV] \ \& \ a_0^* = a_0 \frac{\varepsilon_r}{(m_e^* / m_e)} \tag{3.147}$$

For Si the values $m_e^* = 0.2 m_e$ and $\varepsilon_r = 11.7$ can be employed thus obtaining:

$$E^* = -0.02 [eV] \ \& \ a_0^* = 3 [nm] \tag{3.148}$$

i.e., especially corresponding to an effective radius of many inter-atomic distances order, according to initial assumptions, the above assumption (iii) in special, of and of an very small energetic correction relative to initial assumption, the above assumption (ii) condition,

$$E_{donating \atop electrons} = E_C + E^* = E_C - 0.02 \ [eV] \tag{3.149}$$

which situates the last level of electrons slightly under first level of CB, Figure 3.36-left. Similarly, for semiconductors of p-type, with impurities of acceptors type, the last level of the holes results as being slightly over

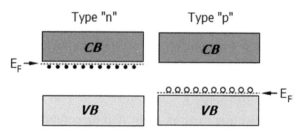

FIGURE 3.36   The Fermi level in a semiconductor of n-type (left) and respectively of p-type (right); after (Further Readings on Quantum Solid 1936–1967; HyperPhysics, 2003; Putz, 2006).

the last level of VB, Figure 3.36-right. Thus, the considerations of intrinsic regime can be re-applied, i.e., of thermal excitations of Figure 3.34, yet considering the new last levels of electrons and holes, wherefrom the Fermi levels associated to those doped semiconductors will be again at the half (for T=0K) of energetic bands remained between these last levels of electrons/holes and those of CB/VB, respectively, see Figure 3.36 (Putz, 2006).

There results a greater efficiency in transition of electrons from semiconductors of n-type in CB and in combination with neighboring VB holes in semiconductors of p-type (therefore creating the conducting holes in VB), at normal temperatures, these semiconductors becoming conductors when applying a potential difference at the ends of the solid.

Moreover, by the virtue of the type of transported charge (the electrons of n-case and the holes of p-case) the doped semiconductors can be coupled in the so-called hetero-junctions thus obtaining a new kind of material: the pn-junction, see Figure 3.37 (Putz, 2006), having a reliable role in the ordering of charge transport (of electric current) in a single direction (leading with the so-called *diode effect*).

At the formation of a pn-junction, firstly, the Fermi levels for each type of semiconductor in part are readjusted, until a common level is reached, at equilibrium, see Figure 3.37 right-down, as also illustrated in Figure 3.30.

Thus, some electrons of n-region migrate in p-region occupying the respective holes; at junction level positive ions are thus formed (in n-region wherefrom the electrons have migrated) and respectively the negative ions (in p-region where the holes have been occupied with electrons coming from n-region); thus an empty region appears, Figure 3.37-up, which, by the Coulomb barrier created further impedes any other further migration of free electrons remained in n-region to the remained holes of p-region.

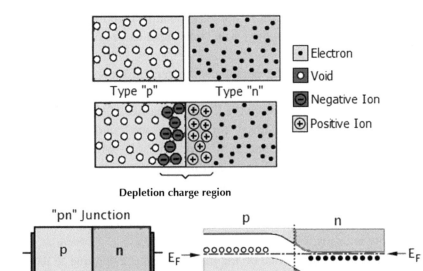

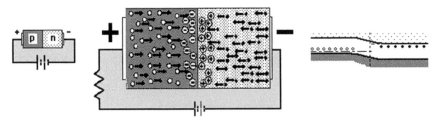

FIGURE 3.37 Illustration of a *pn*-junction formation at equilibrium; after (Further Readings on Quantum Solid 1936–1967; HyperPhysics, 2003; Putz, 2006).

However, the Coulomb barrier of the empty region can be exceeded by applying, in addition, from the outside, of a potential difference (with "+" on *p*-region and "–" on *n*-region), so that the difference of Fermi levels is again unbalanced with the Fermi level of *n*-region overlapping the one of *p*-region; the re-balance is made re-equalizing the Fermi levels, by driving the migration of electrons/holes through the emptying (depletion charge) area, Figure 3.38 (Putz, 2006).

The electrons and holes, combining at junction, allow a continuous current through junction, defining the direct regime on *pn*-junction operation.

FIGURE 3.38 The direct regime of *pn*-junction operating; after (Further Readings on Quantum Solid 1936–1967; HyperPhysics, 2003; Putz, 2006).

The created electric current varies as depending on potential difference applied at the ends, so that the *pn*-junction to act as a regulator (rectifier) of continuous current.

The reverse regime, the potential difference at the ends of junction is with "+" on the *n*-region and with "–" on *p*-region, Figure 3.39 (Putz, 2006), will increase even more the Coulomb barrier of empty (depletion charge) area; a transition current (in opposite way comparing to direct regime) will be formed as long as the electrons holes will be rejected by junction; when the applied negative difference of external potential will cause an increase or the Coulomb barrier until the equalization with it, also the transition current stops; only a thermal current (based on the thermal agitation that causes accidental combinations of electrons in holes) rests of a very low intensity, Figure 3.40-left (Putz, 2006).

Because diodes are based on the *pn*-junction behavior, due to the different behavior by generated current, depending on the polarities applied at the ends, Figure 3.40-left, they can be combined also with other components in order to serve various practical requirements.

As example, by including an intrinsic semiconductor region ("*i*") between the *p* and *n* regions the so-called PIN (*p&i&n*) diode may be formed, which for the inverse regime it acts almost as a condenser, while towards the direct regime it acts as a resistor with variable resistance, so that it can work for the modulation of the signals in alternative current (for which the energy is transited between the storage in resistor and respectively in condenser, at periodic change of polarity of the potential applied).

PIN diode has applications in the system of commutation of the microwaves.

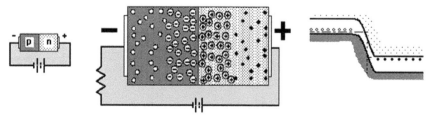

FIGURE 3.39 Reverse operating regime of *pn*-junction; after (Further Readings on Quantum Solid 1936–1967; HyperPhysics, 2003; Putz, 2006).

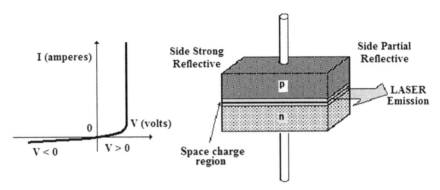

FIGURE 3.40 Direct and reverse current produced by a diode (left) and constitution of a laser diode (right); after (Further Readings on Quantum Solid 1936–1967; HyperPhysics, 2003; Putz, 2006).

A version of ordinary *pn*-diode assumes a reduction-in-steps of the doped level while approaching the junction, so the commutation time between the current of the reverse regime and the one of direct regime is reduced.

Remarkably, also, the direct current (or direct regime) is more quickly established (in terms of increase of the potential applied) than in common *pn*-junction.

The LASER effect (light amplification, monochromatic and coherent) can also be obtained by a *pn*-junction (by *p* and *n*-doping of semiconductors of GaAs, AlGaAs, GaInAsP type) for which the junction length (of contact surface between the *p*- and *n*-regions) is exactly correlated with the wavelength of the light that is in view to be emitted (GaAs: 840[nm], AlGaAs: 760[nm], GaInAsP: 1300[nm]): it features a strongly reflective end, while having the other only partially reflective – to allow the amplification and the final emission, respectively, Figure 3.40-right.

In the direct regime recombination (reoccupation of the holes of VB with electrons from a superior level of VB which emits photons) appears and an incoherent light as in light emitting diode (LED) is produced; over a certain limit of the current increase (Figure 3.40-left) are generated the photons which parallel move with initial junction, so contributing to the stimulated emission (excitation followed by decaying electrons) and of the LASER action.

The applications of these laser diodes are numerous in construction technology of "CD Player"(GaAs), of LASER printers (AlGaAs) and in communication by optical fibers (GaInAsP). Finally, the doped semiconductors can be combined also in three, so that a *bipolar junction* or a *transistor* can be formed, see Figure 3.41 (Putz, 2006), for a transistor in *npn* junction (for a *pnp* junction all polarities are inversed).

The resulted regions are re-named respectively as: *the collector* – the largest region being connected also with a heat reservoir because it dissipates the most heat in operation; *the base* – very thin (about 10 wavelength of light) region to ensure the passing of electric charge through it; and *the emitter* – the region less large than the collector, but the most doped in order to facilitate the conduction.

The collector-base polarization ensures the operation in a reverse regime, while the base-emitter polarization supports direct regime. Note that a small current in center of junction, *in base*, can be used to control a high intensity current directed from collector to emitter (electrons moves inverse!)

The collector current I_C is proportional with that one of the base I_B through the relation

$$I_C = \beta I_B \tag{3.150}$$

or, more exactly, is proportional to the base-emitter potential V_{EE}, with β being the amplification factor which, for a base current as small as possible, ensures the amplification of current from emitter, Figure 3.41-right (Putz, 2006).

Using an amplifier as transistor in an electric circuit it imposes constraints for the maximum values such that the applied potential can not

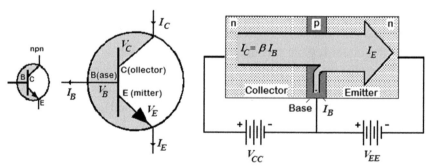

FIGURE 3.41 The simplified symbol (left), detailed (middle) and junction description for a npn-type transistor (right); after (Further Readings on Quantum Solid 1936–1967; HyperPhysics, 2003; Putz, 2006).

overcome. For example, a typical set of such values, for a silicon transistor (of 2N2222type) recommends maximum values: 60[V] for potential collector-base, 5[V] for potential base-emitter, 30[V] for potential collector-emitter, with a maximum 500[mW] power dissipation at temperature 125[°C].

The appearance of transistor junction and especially its quality as amplifier and switcher for the current, was in 1956 awarded with Nobel Price in Physics for the researches Shockley, Bardeen and Bratain; it fundamentally contributed for the replacement of vacuum triodes (vacuum tube with cathode electronic emission – the so-called *cathode rays* – with three terminals developed at the beginning of XX century following the studies of Thomas Edison and JJ Thompson) Figure 3.42-left (Putz, 2006), and had important applications for the construction of the first electronic calculators, yet with huge sizes, containing tens and hundreds of thousands of triodes in vacuum!

The advantage of first transistors, Figure 3.42-middle, do not consist in reduced sizes towards the classical triode, but especially by its low power consumption; however, the idea to concentrate the transistors as circuit element of small (viz. nano-) dimensions was very quickly expressed, and developed as an element of electric circuits, as resistors and condensers, Figure 3.42-right.

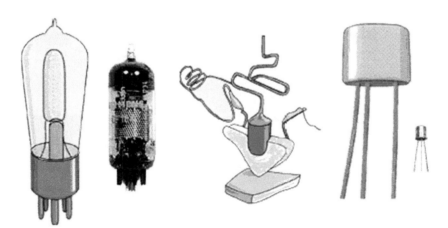

FIGURE 3.42   Scheme and sample for vacuum triode (left) and for transistor (middle and right); see (Further Readings on Quantum Electronics 1958-2000; HyperPhysics, 2003).

The next step was the compaction of transistors and electric circuits in the so-called *integrated circuits* (or *processors*), abbreviated CIP, Figure 3.43 (Putz, 2006).

The revolution of compacting and integrating (interconnection) of electric circuits in a single CIP, resulted from a single monolith block of semiconductor, by stepping technique – see Figure 3.43-left; it brings the huge advantage of conductor fibers abolition inside the components of a circuit (thus reducing also the time in which electrons pass from one element of circuit to another), so contributing at the increase of velocity of calculation in the new calculators and computing era, see (Further Readings on Quantum Electronics, 1973–2000).

Just at the dawn of the XXI, in 2000, this revolution in technology and communication was again recognized as being first degree in human-being evolution and knowledge, by further awarding the Nobel Prize in Physics, for the invention (to Jack Kilby), development (to Zhores I. Alferov) and applications (to Herbert Kroemer) of integrated circuits. But the story does not end here. In the same year, 2000, the Nobel Prize in Chemistry was awarded to the discovering and development of polymer conductors (Alan J. Heeger, Alan G. MacDiarmid and Hideki Shirakawa), i.e., for creating conductor materials from insulating (i.e., from plastic).

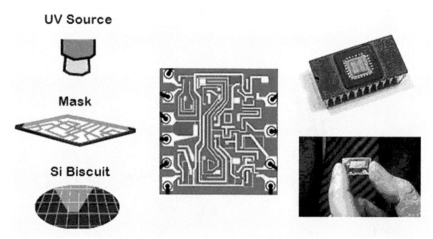

FIGURE 3.43 The scheme of a CIP creation by stepping (left) and the samples obtained (right); after (Further Readings on Quantum Electronics 1958-2000; HyperPhysics, 2003).

What is the consequence of this discovery? Molecular electronics, abbreviated as moletronics is seen as the next challenge of integrated circuits, until the level of atom and molecule. How this thing is possible along the implications in fundamental and applied knowledge are further discussed.

3.5.3 MOLETRONICS

Is there possible that a plastic to become a conductor? In particular conditions, it can, see (Further Readings on Quantum Polymers, 1973–1996).

Plastics are created on polymers, which are long strings and repetitive structures of carbon. Already the repetitive structure makes us think to potential periodicity, specific to; the present discussion follows (Putz, 2006).

Yet, for a polymer to be able to imitate a metal the free electrons should exist, with the possibility of motion along the chain and not being bonded by the atoms.

For this aim, first condition is that the polymer to consist in an alternation of simple and double bonds in chain, i.e., to contain conjugate double bonds. The simplest candidate is polyacetylenes, Figure 3.44 (Putz, 2006).

There remains the question about the long chains. Why are they necessary?

The approach to simple quantum model of free electrons for a string of periodic atoms, as in polyacetylenes, is enlighten.

Given N atoms bonded in a chain separated at d-distance one form another, so that the total length of the chain to become $(N–1)d$, and for $N>>1$ it can be further approximated as being $L=Nd$.

Now, taking into account the quantification energies of a free electron in a unidimensional potential well of L-width (with zero potential in inside

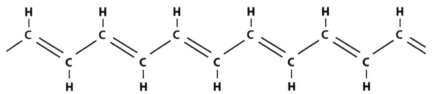

FIGURE 3.44 Polyacetylenes chain: a polymer with conjugate double bonds; after (Further Readings on Quantum Polymers 1973-1996; Putz, 2006)

and infinite outside), Figure 3.45-right (Putz, 2006), the eigen-energies such as, Eq. (3.58) with Eq. (3.62) are yielded with the present form:

$$E_n = \frac{2\pi^2 \hbar^2}{m_e} \frac{n^2}{L^2} = \frac{2\pi^2 \hbar^2}{m_e} \frac{n^2}{(Nd)^2} \quad , n = 1,2,3,... \tag{3.151}$$

Moreover, one should distinguish for a polymer description between σ(sigma) – bonds localized and immobile forming the covalent bonds between atoms of carbon inside the chain; and those π(pi)-conjugate double bonds of those electrons being also partly localized yet not so tight as the σ-electrons.

Therefore, the π electrons will be considered as being the N free electrons in potential well occupying energetic levels (according to Pauli principle: an electron of a certain spin in each orbital, Figure 3.45) until the highest occupied molecular orbital with $n=N/2$ in Eq. (3.151) so defining the actual HOMO, analogous of E_V level – the last level from VB in quantum solid, with the eigen-energy:

$$E_{HOMO} = \frac{2\pi^2 \hbar^2}{m_e} \left(\frac{N}{2}\right)^2 \frac{1}{(Nd)^2} \tag{3.152}$$

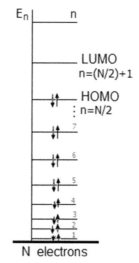

FIGURE 3.45 Energetic levels occupancy for free electrons within an infinite potential well; after (Further Readings on Quantum Polymers 1973-1996; Putz, 2006).

Similarly, for the eigen-energy of the lowest unoccupied molecular orbital level, the LUMO is analogous to E_C level – the first level of CB in quantum solid, one will write for it, with $n=(N/2)+1$ in Eq. (3.151), the expression:

$$E_{LUMO} = \frac{2\pi^2\hbar^2}{m_e}\left(\frac{N}{2}+1\right)^2\frac{1}{(Nd)^2} \qquad (3.153)$$

The differences between the LUMO (3.153) and HOMO (3.152) energies correspond to the E_g energy of VB to CB in quantum solid, thus generating an energetic gap with the dependence:

$$\Delta E = E_{LUMO} - E_{HOMO} = \frac{2\pi^2\hbar^2}{m_e}\frac{N+1}{(Nd)^2} \xrightarrow{N\gg1} \frac{2\pi^2\hbar^2}{m_e d^2}\frac{1}{N} \qquad (3.154)$$

The result (3.154) provides the reduction of energy gap LUMO-HOMO for increasing of the polymer chain (when one records also the increase of N) so predicting the gap cancelling for macroscopic dimensions, thus obtaining the answer to the necessity of long chains of polymers in order to ensure the conduction conditions, such for a metal.

Yet, the periodicity conditions being satisfied for double bonds and of the length in polymeric chain is not enough.

To become a conductor, a plastic material should be perturbed, either by released (oxidation) of electrons or by their acceptance (reduction), these processes being known as doping, in the obvious similarity with the existing processes at the level of semiconductors.

For polyacetylenes, Figure 3.44 (Putz, 2006), the two exposed typical processes of doping are: oxidation with halogen (p-doping)

$$[CH]_N + 3x/2\, I_2 \longrightarrow [CH]_N^{x+} + x\, I_3^- \qquad (3.155)$$

and by reduction with an alkaline metal (n-doping):

$$[CH]_N + x\, Na \longrightarrow [CH]_N^{x-} + x\, Na^+ \qquad (3.156)$$

In their researches, Heeger, MacDiarmid and Shirakawa (Nobel Prize for Chemistry on 2000 for "discovery and development of conducting polymers") have applied the oxidation of a thin film of polyacetylenes with iodine vapors, observing the increase of material conductivity of million times.

How can we explain this behavior?

Firstly, we have to note that a doped polymer is a salt where not the ions of iodine or sodium are those who create the current, but the electrons of conjugate bonds.

Yet, in the presence of a strong enough electric field as a potential difference (perpendicular to the film) applied to polymer, the ions of iodine/sodium can move back and forth along the chain, which result in the possibility to control the doping direction and also controlling the polymer conduction in switching on/off regime.

Passing on the reaction analysis in Eq. (3.155), firstly, at the level of conduction in polyacetylenes after oxidation, see Figure 3.46(a), the polyacetylenes molecule positively charged is obtained acting as a radical cation called *polaron*, Figure 3.46(b).

The remained electron in the double bond, where an electron was eliminated by oxidation, it can further move, by the virtue of conjugate double bond, along the polymeric chain, see Figure 3.46(c).

On the other side, due to electrostatic attraction, the positive charge of molecule will also move following the free electron, see Figure 3.46(d,e), until the pair of electric charges at long range distance is regain, see Figure 3.46 (b), so building a *soliton*.

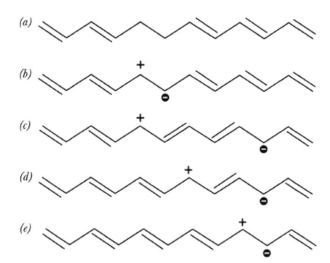

FIGURE 3.46 The scheme of electric charges conduction in a polymeric matter; after (Further Readings on Moletronics 1956-1997; Putz, 2006).

In a polyacetylenes chain, strongly oxidized, these solitons are responsible by the electric charges transport along the chain and between the polymeric chains at macroscopic scale, se also (Further Readings on Moletronics, 1956–1997).

The applications of the polymeric conduction were quickly proved to be brilliant. For example, polymeric semiconductors are going to replace the traditional inorganic photo-diodes used for generating light (for domestic and urban use), which, in turn, may replace the classical bulb, while saving even more energy and dissipating much less heat. The phenomenon to which this application is based is called *electro-luminescence* through the light is emitted by a thin layer of semiconductor polymer when its electrons are excited by an applied potential difference, Figure 3.47, see (Further Readings on Electroluminescence in Conjugated Polymers, 1990–2000).

The reduction mechanism of HOMO-LUMO gap combined with dopped polymers targeting the conductor properties, is currently explored also for more various chains of organic molecules, by direct doping with ad-atoms or/and by contact with metallic surface, Figure 3.48-left, and is going to be extended also for proteins interconnected by metallic clusters and nanotubes, conductors with bio-sensor role, able to react with particular agents that emit light, electrons, radicals, atoms, etc., see Figure 3.48-right.

Thus, from the conductor polymers one is passing to the electronics of conductor polymers, to molecular electronics and to bio-electronics, all these phenomena taking place at atomic and molecular scale, so at nano-scale, creating new molecular ensembles by modeling the materials atom by atom, at nano-metric level (10^{-9} [m]), i.e., at quantum level.

Remarkably is the fact that at this level, the electronic de Broglie associated wavelength is comparable with the size of the created element

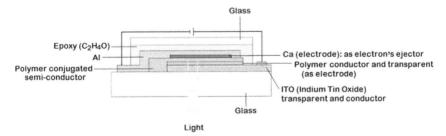

FIGURE 3.47 Emission of light by electro-luminescence by semiconductor's polymers; after Further Readings on Electroluminescence in Conjugated Polymers (1990–2000).

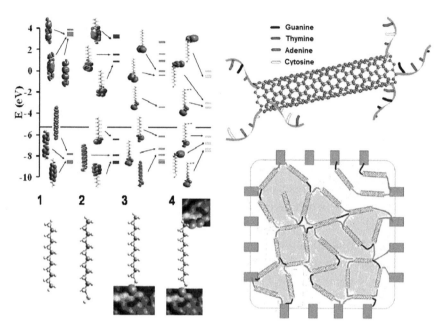

FIGURE 3.48   Left: successive movement of HOMO-LUMO gaps for organic molecule doped by contact with metallic surfaces. Right: endings of DNA biomolecules connected by clusters and metallic nanotubes (up) and a network of such type of connections (down); after Further Readings on Moletronics by Seminario's Group (1997–2002).

of molecular circuit; said reversely, the circuit, the CIP or the integrated molecular ensemble with conducting properties has a quantum size, generically renaming them as quantum dot(s).

In this picture, the transistor is considered a quantum dot, as are the molecular or biologic assemblies with conducting properties, so that the quantum dots are molecules and artificial solids that integrately operate at quantum level.

What means the electronic bond for such a new system?
It is a combination between Coulombian interaction (ionic bond) and quantum tunneling (covalent bond).

Multi-dot quantum systems are thus the new generation of super-molecules and quantum solids, for which the quantum-wave nature of electron prevails and commands the system's behavior and properties, see Figure 3.49 and (Further Readings on Moletronics by Seminario's Group, 1997–2002).

The versatility of eigen-functions and so of the quantum nature of matter, remember the "dream" of Wigner and Seitz exposed at the beginning

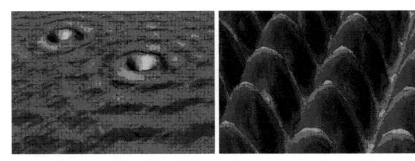

FIGURE 3.49 Two quantum dots (left) sunken in a quantum sea of electronic waves (right-detail); after Further Readings on Quantum Corals (1990–1995).

of this chapter (also adopted as the generic motto of the present volume), fulfills its role through combining of inorganic matter with the organic and biologic ones at atomic and molecular level, in solids and lattice/networks of quantum dots, with very interesting and important macroscopic properties.

The study and the applications of the coherent transport (by constructive interference of the waves associated to the electrons) of information charge (signals) between these quantum dots and the nodes of integrated molecular circuits to which they are taking part, stay as one of the biggest future challenge of nano-science and nano-technology.

3.6 CONCLUSION

The main lessons to be kept for the further theoretical and practical investigations of the quantum roots of crystals and solids that are approached in the present chapter pertain to the following:

- Identifying the basic quantum rules used to model the electronic quantum behavior in solid states as being related with averaging observables (operators) over the given (optimized) state functions, commutators and allied operations, along the eigen equation under the Schrodinger equation form;
- Employing the superposition principles with periodic constrains to formulate the wave function general forms specific to solid state – the Bloch theorem and orbitals;
- Writing the main relationships in the reciprocal space, at the unit cell level, being it associated with the space where electrons manifest their quantum behavior in the solid state through the de Broglie quantification of their wave vectors;

- Dealing with quantum models of crystals viewed as related yet distinct levels of approximations for electronic evolution/nature in the periodic lattice of a crystal structure;

- Characterizing crystal structures as ordered quantum wells with electronic wave function distributed within them by the orthogonalization feature respecting its valence and conducting or nuclear core and valence shells' realizations, respectively;

- Understanding the free electrons in crystals as a manifestation of valence electronic shells respecting the ordered nuclear + electronic core lattice nodes (structural points or dots);

- Describing the Brillouin zone in quantum crystal as the reciprocal space correspondence with the direct space unit cell;

- Learning the parabolic dependency of the quantified energy with the electronic wave vector ($E \sim k^2$), further correlated with density of the electronic number of electrons in the unit cell ($k \sim N_e^{1/3}$), in (Thomas-Fermi) accordance with parabolic energy-to-electronic number for valence behavior of electrons in atoms and molecules; Volumes II and III of the present five-volume set on Quantum Nanochemsitry (Putz, 2016a, 2016b);

- Treating the average of the total (valence) electronic energy with respect with the total number of the electrons in accordance with Thomas-Fermi theory $\langle E \rangle \sim N_e^{5/3}$ so better closing with the parabolic E-to-N dependency prescribed for atoms and molecules, as compared with the previous point;

- Solving the quasi-free electronic behavior in crystals by means of the quantum perturbation method in the first order so prescribing the one-energetic gap in electronic energies associated with two electronic wave functions: the valence and the (excited) conducting states, at the level of the first Brillouin zone, respectively;

- Formulating the extended perturbation model through infinite ordered quantum well in crystal (*Kronig-Penney model*) so generating the energetic multiple gaps in spectra of electrons in solids – the electronic structure of bands in crystals;

- Interpreting and therefore adapting the Hückel molecular model to the tight-binding model in crystals when only the closest neighboring of ordered/periodic crystals dots/structural points (attractive centers) act;

- Connecting the Fermi level by its quantum quantities as of Fermi energy, wave-vector, velocity, momentum, and wave-vector, along its extension to the quantum characterizations of solid;

- Developing the implication of Fermi quantum statistics to model the semiconductors and junctions, including the transistor phenomenology, in modeling the quantum solids at the conducting level.

Finding applications of the quantum behavior of electrons in crystals and solids to advance the molecular-electronics, leading with new conducting materials as coming from substances otherwise insulators (e.g., polymers), so that proving the quantum driving potency in controlling the concerned observable effects.

KEYWORDS

- **Bloch theorem**
- **bonding electrons in crystals**
- **Fermi distribution**
- **free electrons**
- **junctions**
- **moletronics**
- **quantum models of crystals**
- **quantum postulates of matter**
- **quantum solids**
- **quasi-free electrons**
- **Reciprocal Lattice**
- **semiconductors**
- **tight-binding electrons in crystals**

REFERENCES

AUTHOR'S MAIN REFERENCES

Putz, M. V. (2016a). *Quantum Nanochemistry. A Fully Integrated Approach: Vol. I. Quantum Theory and Observability*. Apple Academic Press & CRC Press, Toronto-New Jersey, Canada-USA.

Putz, M. V. (2016b). *Quantum Nanochemistry. A Fully Integrated Approach: Vol. II. Quantum Atoms and Periodicity*. Apple Academic Press & CRC Press, Toronto-New Jersey, Canada-USA.

Putz, M. V. (2016c). *Quantum Nanochemistry. A Fully Integrated Approach: Vol. III. Quantum Molecules and Reactivity.* Apple Academic Press & CRC Press, Toronto-New Jersey, Canada-USA.

Putz, M. V. (2010). *Environmental Physics and the Universe* (in Romanian), The West University of Timişoara Publishing House, Timişoara.

Putz, M. V. (2006). *The Structure of Quantum Nanosystems* (in Romanian), West University of Timişoara Publishing House, Timişoara.

Putz, M. V., Lacrămă, A. M. (2005). *Exploring the Complex Natural Systems* (in Romanian), Mirton Publishing House, Timişoara, Chapters 1–3.

SPECIFIC REFERENCES

Davisson, C., Germer, L. H. (1927). Diffraction of electrons by a crystal of nickel. Phys. Rev. 30, 705.

HyperPhysics (2003) Semiconductors [http://hyperphysics.phy-astr.gsu.edu/hbase/solids/semcn.html] Department of Physics and Astronomy Georgia State University. Atlanta, Georgia

Kittel, C. (1996). *Introduction to Solid State Physics*, Wiley, New York.

Lewis, G. N. (1916). The atom and the molecule. *J. Am. Chem. Soc.* 38(4), 762–785.

Pettifor, D. (1995). *Bonding and Structure of Molecules and Solids*, Oxford Science Publications, Clarendon Press, Oxford.

Rabi, I. (1936). On the process of space quantization. *Phys. Rev.* 49, 324–328.

Rupp, E. (1928). Über die Winkelverteilung langsamer Elektronen beim Durchgang durch Metallhäute. Ann. Phys. 390(8), 981–1012.

Slater, J. C. (1939). *Introduction to Chemical Physics*, McGraw-Hill, New York.

Thompson, G. P. (1928). Experiments on the diffraction of cathode rays. *Proc. Roy. Soc.* 117, 600–609; ibid. (1928). Experiments on the diffraction of cathode rays. II. 119, 651–663.

Verma, A. R., Srivastava, O. N. (1982). Crystallography for Solid State Physics, Wiley Eastern Limited, New Delhi.

FURTHER READINGS

[*on Quantum Crystal*]:
Ascroft, N. W., Mermin, N. D. (1976). *Solid State Physics*, Holt, Rinchart and Winston, New York.

Clark, H. (1968). *Solid State Physics*, Macmillan, London.

Ghatak, A. K., Kothari, L. S. (1971). *Introduction to Lattice Dynamics*, Addison-Wesley, Reading.

Madelung, O. (1978). *Introduction to Solid State Theory*, Springer-Verlag, New York.

Omar, M. A. (1975). *Elementary Solid State Physics*, Addison-Wesley, Reading.

Seitz, F. (1940). *Modern Theory of Solids*, McGraw-Hill, New York.

Wannier, G. H. (1959). *Elements of Solid State Theory*, Cambridge Univ. Press, Cambridge.

Ziman, J. M. (1964). *Principles of Theory of Solids*, Cambridge Univ. Press, Cambridge.

[*on Quantum Solid*]:
Altmann, S. L., Cracknell, A. P. (1965). Lattice harmonics, I. Cubic groups. *Rev. Mod. Phys.* 37, 19–32.

Anderson, P. W. (1958). Absence of diffusion in certain random lattices. *Phys. Rev.* 109, 1492–1505.

Asdente, M., Friedel, J. (1961). 3D Band Structure of Cr. *Phys. Rev.* 124, 384.

Bell, D. G. (1954). Group theory and crystal lattices. *Rev. Mod. Phys.* 26, 311–319.

Bohm, D., Pines, D. (1951). A collective description of electron interactions. I. magnetic interactions. *Phys. Rev.* 82, 625; ibid. (1953). A collective description of electron interactions: III. Coulomb interactions in a degenerate electron gas. 92, 609.

Bouckaert, L. P., Smoluchowski, R., Wigner, E. (1936). Theory of brillouin zones and symmetry properties of wave functions in crystals. *Phys. Rev.* 50, 58–67

Callaway, J. (1961). Energy bands in lithium. *Phys. Rev.* 124, 1824–1827.

Eckelt, P., Madelung, O., Treusch, J. (1967). Band structure of cubic ZnS (Korringa-Kohn-Rostoker Method). *Phys. Rev. Lett.* 18, 656–657.

Gell-Mann, M. (1957). Specific heat of a degenerate electron gas at high density. *Phys. Rev.* 106, 369–372.

Gell-Mann, M., Brueckner, K. A. (1957). Correlation energy of an electron gas at high density. *Phys. Rev.* 106, 364–368.

Ham, F. S. (1962). Energy bands of alkali metals. I. Calculated bands. *Phys. Rev.* 128, 82–97; ibid. (1962). Energy bands of alkali metals. II. Fermi surface. 128, 2524–2541.

Ham, F. S., Segall, B. (1961). Energy bands in periodic lattices—green's function method. *Phys. Rev.* 124, 1786–1796.

Kohn, W. (1952). Variational methods for periodic lattices. *Phys. Rev.* 87, 472–481.

Kohn, W., Rostoker, N. (1954). Solution of the Schrödinger equation in periodic lattices with an application to metallic lithium. *Phys. Rev.* 94, 1111–1120.

Kuhn, T. S., Van Vleck, J. H. (1950). A simplified method of computing the cohesive energies of monovalent metals. *Phys. Rev.* 79, 382–388.

Melvin, M. A. (1956). Simplification in finding symmetry-adapted Eigen functions. *Rev. Mod. Phys.* 28, 18–44.

Pines, D. (1953). A Collective description of electron interactions: IV. Electron interaction in metals. *Phys. Rev.* 92, 626–636.

Pines, D., Bohm, D. (1952). A collective description of electron interactions: II. Collective vs individual particle aspects of the interactions. *Phys. Rev.* 85, 338–353.

Segall, B. (1961). Energy bands of aluminum. *Phys. Rev.* 124, 1797–1806.

Shockley, W. (1937). The empty lattice test of the cellular method in solids. *Phys. Rev.* 52, 866–872.

Van Hove Léon (1953). The occurrence of singularities in the elastic frequency distribution of a crystal. *Phys. Rev.* 89, 1189–1193.

von der Lage, F. C., Bethe, H. A. (1947). A method for obtaining electronic Eigen functions and eigenvalues in solids with an application to sodium. *Phys. Rev.* 71, 612–622.

Wannier, G. H. (1943). Energy eigenvalues for the coulomb potential with cut-off. Part, I. *Phys. Rev.* 64, 358; ibid. (1937). The structure of electronic excitation levels in insulating crystals. 52, 191–197.

[*on Quantum Electronics*]:

Batlogg, B. (1991). Phisycal Properties of High-Tc Superconductors. *Physics Today* 44–50.

Blackwood, O. H., Kelly, W. C., Bell, R. M. (1973). *General Physics*, 4th Edition, Wiley.

Blatt, F. J. (1992). *Modern Physics*, McGraw-Hill.

Brown, W. F. (1958). *Magnetic Materials, Handbook of Chemistry and Physics*, Condon and Odishaw (Eds.) McGraw-Hill, Chapter 8.

Clarke, J. SQUIDs *Scientific American* 271(2), 46–53.

de Leeuw, D. (1999). Plastic Electronics. *Physics World* 31.

Diefenderfer, J., Holton, B. (1994). *Principles of Electronic Instrumentation*, 2nd Ed., Saunders College Publ.

Fishbane, P. M., Gasiorowicz, S., Thornton, S. (1996). *Physics for Scientists and Engineers*, 2nd Ed extended, Prentice Hall.

Floyd, T. L. (1991). *Electric Circuit Fundamentals*, 2nd Ed., Merrill.

Floyd, T. L. (1992). *Electronic Devices*, 3rd Ed., Merrill.

Gleason, R. E. (2000). How far will circuits shrink. *Sci. Spectra* 20, 32–40.

Horowitz, P., Hill, W. (1980). *The Art of Electronics*, 2nd Ed, (1989), Cambridge University Press.

Jones, E. R., Childers, R. L. (1990). *Contemporary College Physics*, Addison-Wesley.

Kittel, C. (1996). *Introduction to Solid State Physics*, 7th Ed., Wiley.

Kittel, C., Kroemer, H. (1980). *Thermal Physics*, 2nd Ed., W. H. Freeman.

Lubkin, G. B. (1996). Power applications of high-temperature superconductors *Physics Today* 49, 48.

Myers, H. P. (1997). *Introductory Solid State Physics*, 2nd. Ed., Taylor & Francis.

Ohanian, H. (1989). *Physics, 2nd Ed. Expanded*, WW Norton.

Schroeder, D. V. (2000). *An Introduction to Thermal Physics*, Addison-Wesley.

Simpson, R. E. (1987). *Introductory Electronics for Scientists and Engineers*, 2nd Ed., Allyn and Bacon.

Sproull, R. L., Phillips, W. A. (1980). *Modern Physics: The Quantum Physics of Atoms, Solids and Nuclei*, Wiley.

Thornton, S. T., Rex, A. (1993). *Modern Physics for Scientists and Engineers*, Saunders College Publishing.

[*on Quantum Polymers*]:

Blackwood, O. H., Kelly, W. C., Bell, R. M. (1973). *General Physics*, 4th Edition, Wiley.

Chiang, C. K., Druy, M. A., Gau, S. C., Heeger, A. J., Louis, E. J., MacDiarmid, A. G., Park, Y. W., Shirakawa, H. (1978). Synthesis of highly conducting films of derivatives of polyacetylene, (CH)x. *J. Am. Chem. Soc.* 100(3), 1013–1015.

Chiang, C. K., Fischer, C. R., Park, Y. W., Heeger, A. J., Shirakawa, H., Louis, E. J., Gau, S. C., MacDiarmid, A. G. (1977). Electrical conductivity in doped polyacetylene. *Phys. Rev. Lett.* 39, 1098–1101 .

Feast, W. J., Tsibouklis, J., Pouwer, K. L., Gronendaal, L., Meijer, E. W. (1996). Synthesis, processing and material properties of conjugated polymers. *Polymer* 37(22), 5017–5047.

Ito, T., Shirakawa, H., Ikeda, S. (1974). Simultaneous polymerization and formation of polyacetylene film on the surface of concentrated soluble Ziegler-type catalyst solution. *J. Polym. Sci.: Polym. Chem. Ed.* 12, 11–20.

Kanatzidis, M. G. (1990). Conductive polymers. *Chem. Eng. News* 38(49), 36–40.

Roth, S. (1995). *One-Dimensional Metals*, Weinheim VCH.

Salaneck, W. R., Lundström, I., Rånby, B. (Eds.) (1993). *Nobel Symposium in Chemistry: Conjugated Polymers and Related Materials: The Interconnection of Chemical and Electronic Structure*, Oxford.

[*on Moletronics*]:
Larsson, S., Rodriguez-Monge, L. (1997). Conductivity in polyacetylene. VI. Semiconductor—metal transition of alkali-doped polymer. *Int. J. Quant. Chem.* 63(3), 655–665.

Marcus, R. A. (1956). On the theory of oxidation reduction reactions involving electron transfer. I. *J. Chem. Phys.* 24, 966–978.

Marcus, R. A. (1964). Chemical and electrochemical electron-transfer theory. *Annual Rev. Phys. Chem.* 15, 155–196.

Winokur, M., Moon, Y. B., Heeger, A. J., Barker, J., Bott, D. C., Shirakawa, H. (1987). X-Ray Scattering from Sodium-Doped Polyacetylene: Incommensurate-Commensurate and Order-Disorder Transformations. *Phys. Rev. Lett.* 58, 2329–2332.

[*on Electroluminescence in Conjugated Polymers*]:
Berggren, M., Inganäs, O., Gustafsson, G., Gustafsson-Carlberg, J. C., Rasmusson, J., Andersson, M. R., Hjertberg, T., Wennerström, O. (1994). Light-emitting diodes with variable colors from polymer blends. *Nature* 372, 444–446.

Burroughes, J. H., Bradley, D. D. C., Brown, A. R., Marks, R. N., Mackay, K., Friend, R. H., Burns, P. L., Holmes, A. B. (1990). Light-emitting diodes based on conjugated polymers. *Nature* 347, 539–541.

Friend, R. H., Gymer, R. W., Holmes, A. B., Burroughes, J. H., Marks, R. N., Taliani, C., Bradley, D. D. C., Dos Santos, D. A., Bredas, J. L., Lögdlund, M., Salaneck, W. R. (1999). Electroluminescence in conjugated polymers. *Nature* 397, 121–128.

Groenendaal, L. B., Jonas, F., Freitag, D., Pielartzik, H., Reynolds, J. R. (2000). Poly(3,4-ethylenedioxythiophene) and its derivatives: past, present, and future. *Adv. Mater.* 12(7), 481–494.

[*on Moletronics by Seminario's Group*]:
Balbuena, P. B., Derosa, P. A., Seminario, J. M. (1999). Density functional theory study of copper clusters. *J. Phys. Chem. B* 103, 2830–2840.

Boudreaux, E. A., Seminario, J. M. (1998). SCMEH-MO calculations on lanthanide systems, V. A comparison of DFT and SCMEH-MO methods on Nd(CO)6. *J. Molec. Struct. (Theochem)* 425, 25–28.

Choi, D.-S., Huang, S., Huang, M., Barnard, T. S., Adams, R. D., Seminario, J. M., Tour, J. M. (1998). Revised structures of n-substituted dibrominated pyrrole derivatives and their polymeric products. termaleimide models with low optical band gaps. *J. Org. Chem.* 63, 2646–2655.

Derosa, P. A., Seminario, J. M. (2001). Electron transport through single molecules: scattering treatment using density functional and green function theories. *J. Phys. Chem. B* 105, 471–481.

Derosa, P. A., Seminario, J. M., Balbuena, P. B. (2001). Properties of Small Ni-Cu Bimetallic Clusters. *J. Phys. Chem. A* 105, 7917–7925.

Derosa, P. A., Zacarias, A. C., Seminario, J. M. (2002). *Application of density functional theory to the study and design of molecular electronic devices: the metal-molecule interface*, Sen, K. D. (Ed.) *Reviews in modern quantum chemistry: A celebration of the contributions of Robert G Parr*, World Scientific, Singapore, pp. 1537–1567.

Harrison, B. C., Seminario, J. M., Bunz, U. H. F., Myrick, M. L. (2000). Lowest electronic exited states of poly (para-cyclobutadienylenecyclopentadienylcobalt) butadienylene. *J. Phys. Chem. A* 104, 5937–5941.

Politzer, P., Murray, J. S., Seminario, J. M., Lane, P., Grice, M. E., Concha, M. C. (2001). Computational characterization of energetic materials. *J. Molec. Struct. (Theochem)* 573, 1–10.

Politzer, P., Seminario, J. M., Concha, M. (1998). Energetics of ammonium dinitramide decompositions Steps. *J. Molec. Struct. (Theochem)* 427, 123–129.

Seminario, J. M. (1997). *A Combined DFT/MD procedure for the study and Design of Materials*; In: Cisneros, G., J. A. Cogordan, M. Castro and, C. Wang *Computational Chemistry and Chemical Engineering*, World Scientific, Singapore, pp. 255–267.

Seminario, J. M., De La Cruz, C. E., Derosa, P. A. (2001). A theoretical analysis of metal-molecule contacts. *J. Am. Chem. Soc.* 123, 5616–5617.

Seminario, J. M., Derosa, P. A. (2001). Molecular gain in a thiotolane system. *J. Am. Chem. Soc.* 123, 12418–12419.

Seminario, J. M., Derosa, P. A., Bastos, J. L. (2002). Theoretical interpretation of switching in experiments with single molecules. *J. Am. Chem. Soc.* 124, 10266–10267.

Seminario, J. M., Tour, J. M. (1997). Systematic study of the lowest energy states of AUN(n=1–4) using DFT. *Int, J. Quantum Chem.* 65, 749–758.

Seminario, J. M., Zacarias, A. G., Derosa, P. A. (2001). Theoretical analysis of complementary molecular memory devices. *J. Phys. Chem. A* 105(5), 791–795

Seminario, J. M., Zacarias, A. G., Derosa, P. A., (2002). Analysis of a dinitro based molecular device, *J. Chem. Phys.* 116, 1671–1683.

Seminario, J. M., Zacarias, A. G., Tour, J. M. (1998). Theoretical interpretation of conductivity measurements of a thiotolane sandwich. a molecular scale electronic controller. *J. Am. Chem. Soc.* 120, 3970–3974.

Seminario, J. M., Zacarias, A. G., Tour, J. M. (1999). Molecular alligator clips for single molecule electronics. studies of group 16 and isonitriles interfaced with au contacts. *J. Am. Chem. Soc.* 121, 411–416.

Seminario, J. M., Zacarias, A. G., Tour, J. M. (1999). Molecular current-voltage characteristics. *J. Phys. Chem. A* 103, 7883–7887.

Seminario, J. M., Zacarias, A. G., Tour, J. M. (2000). Theoretical study of a molecular resonant tunneling diode. *J. Am. Chem. Soc.* 122, 3015–3020.

Tour, J. M., Kozaki, M., Seminario, J. M. (1998). Molecular scale electronics: synthetic and computational approaches to nanoscale digital computing. *J. Am. Chem. Soc.* 120, 8486–8493.

Zacarias, A. G., Castro, M., Tour, J. M., Seminario, J. M. (1999). Lowest energy states of small Pd clusters using density functional theory and standard ab initio methods. A route to understanding metallic nanoprobes. *J. Phys. Chem. A* 103, 7692–7700.

[on Quantum Corals]:
Crommie, M. F., Lutz, C. P., Eigler, D. M. (1993). Imaging standing waves in a two-dimensional electron gas. *Nature* 363, 524–527.

Crommie, M. F., Lutz, C. P., Eigler, D. M. (1993). Confinement of electrons to quantum corrals on a metal surface. *Science* 262, 218–220.

Crommie, M. F., Lutz, C. P., Eigler, D. M., Heller, E. J. (1995). Waves on a metal surface and quantum corrals. *Sur. Rev. Lett.* 2(1), 127–137.

Eigler, D. M., Schweizer, E. K. (1990). Positioning single atoms with a scanning tunneling microscope. *Nature* 344, 524–526.

CHAPTER 4

CHEMICAL CRYSTALLOGRAPHY

CONTENTS

ABSTRACT

Summarizing the experimental results accumulated, V.M. Goldschmidt stated *fundamental law of crystal chemistry* that meets certain factors which determine the forming of a certain crystalline structure from a set of particles. Under this law, a crystal structure is determined by the number and size of particles in the elementary cell, by their nature and by the nature of chemical bonds that are established between them. The present chapter is so devoted to systematic study of this law, thus developing the crystal chemistry.

4.1 INTRODUCTION

"When God established the heavens, I was there: when He set *a compass* on the surface of the deep."

Universal Law says so in the *Book of Proverbs*, 8–27. The universe is harmony. Harmony means proportion. The proportion determines the geometry. Geometry permits the construction, progression, evolution.

Each set of material particles defines a system. The homogeneous portions of the system, delimited by the other parts by surfaces in front of which occur a sudden variation, essentially for the physical properties' manifestation, are called *phases*.

A finite system can consist of a single gas phase and/or of one or more liquid phases and/or solid; the present discussion follows (Chiriac-Putz-Chiriac, 2005).

After the way in which the materials particles are distributed inside the phase, we distinguish (after Tamman): isotropic phases (gaseous, liquid and amorphous solids) and anisotropic phases (crystalline solids).

Inside an isotropic phase the material particles are chaotically distributed, disorderly, and its homogeneity is therefore being a statistical behavior.

As a result of the statistical homogeneity, the resultant of the cohesion forces between the particles is the same in all the directions, making that the physical properties to be the same in all the phase expansion, and independent of direction.

Instead, an anisotropic phase is characterized by a real homogeneity, the material particles being distributed orderly, with periodic repetitions, the minimum distance between two identical particles along the same direction being constant and different, on two neighboring directions.

In the anisotropic bodies' case, the cohesion forces between the constituent particles discontinuously vary with the direction, with maximum and minimum intensity values in the neighboring directions, but with the same value on the parallel directions. As a result, the vectorial physical properties of the anisotropic substance, especially those related to the cohesion, will depend on the direction.

The ordered internal distribution of the material particles is becoming observable by the substances property to individually crystallizing themselves, under the specific conditions (temperature, pressure, etc.) of a free growth, as polyhedra limited by the more or less plane faces.

Bringing together these features, the crystal is defined as a body, usually anisotropic with a real homogeneity, externally limited by plane faces which are intersecting after straight edges, providing the regular geometric shape as an expression of its internal structure (Weller, 1973).

Therefore, the Crystallography, the science which deals with the crystals' study, depending on the different aspect under focus, may be regarded with the following complementary approaches (Bunn, 1954; Evans, 1964; Knox & Gold, 1964; Ramachandran, 1964; Flint, 1968; Weller, 1973; Penkala, 1974; Urusov, 1987):

- crystallogenesis: studies the factors which cause the formation and the crystals growth.
- geometric crystallography: is dealing with the geometrical characteristics of the internal structure of the crystallized substances and with establishing the laws for the appearance of the crystalline forms (in this volume presented in Chapter 2);
- physical crystallography: research the physical properties of the crystals depending on the internal and symmetry structure (here studied in Chapter 3 but also reloaded as X-Ray crystallography in Chapter 5 of the present volume);
- crystallochemistry: crystal chemistry established as a science along with crucial experience of Max von Laue (1912) who shown by X- ray diffraction reticular structure of crystalline substances. Further research revealed great diversity of crystalline structures and created necessary theoretical groundwork to establish close connection between constitutive particles characteristics and

their crystalline structure resulting from their aggregation. One of the fundamental results obtained was the conclusion that, besides some molecular crystals, the crystals are atomic or ionic systems in which each is equally surrounded by other particles, the whole being viewed as a giant molecule. Therefore, the crystal represents a complex of atoms or ions which represents free bonds, so one crystal face is always unsaturated, capable of growth and adsorption in favorable conditions. Accordingly, studying the factors which determine the crystalline structure, the relationship between the internal structure, the chemical composition and the crystal's properties – will be in the present Chapter unfolded.

4.2   THE CRYSTAL MODEL OF RIGID SPHERES

4.2.1   SPATIAL ORDERING OF IDENTICAL SPHERES

Passing from mathematical network of points to the actual arrangement of atoms, the maximum simplification we can approach is to consider the motif-atom as a rigid sphere of R-radius. Thus, in real network is no infinite free space (mathematical point scale with zero) but only the one which results by filling the nodes with spherical particles of R radius. Neglecting atoms interactions and approximating them with rigid spheres of R radius, the ordered distribution in the plane of these spheres can take only one of the options presented in Figure 4.1; the present discussion follows (Chiriac-Putz-Chiriac, 2005)

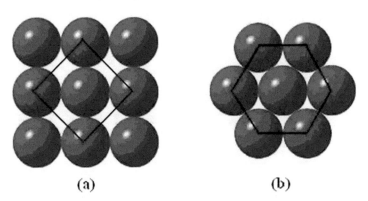

(a)   (b)

FIGURE 4.1   Ordering in plane of identical spheres' R radius.

In case (a) of Figure 4.1 any reference sphere (R_p) has four coplanar neighbors situated at the same distance (Chiriac-Putz-Chiriac, 2005).

The distance between the center of the sphere and the neighbor is d=2R. Number of proximity neighbors of a reference sphere is called *the coordination code or number* (CN) of the center sphere (being the lattice mark).

Therefore, this planar arrangement is provided for a particle CN=4, i.e., in which the parallel strings have the centers of spheres placed on a perpendicular line on the reticular strings direction.

However, free spaces remain between plane network particles, called *interstices* or *voids*. In (a)-type arrangement, the void shape is a square, defined by joining the centers of the four spheres from successive parallel strings, with the surface:

$$S=(2R)^2 - \pi R^2 = R^2(4 - \pi) \qquad (4.1)$$

In such a void a sphere with radius (r) may easily fall; it depends however on the radius of the spheres (R) that generates plane network as the deduction in Figure 4.2.

In the b) case of Figure 4.1 the reticular strings are moving one to the other by a R distance such that only the centers of spheres of odd rows (1, 3, 5,…), respectively, even (2, 4, 6, …) are placed on the same vertical lines. String sequence is of ABAB…. Type; each sphere has proximity neighbors with coordination number CN=6. Trigonal voids are formed between spheres of two reticular successive rows on which the spheres of radius r can be inserted, Figure 4.3.

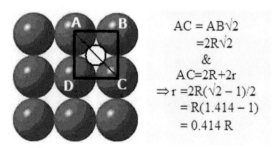

$$AC = AB\sqrt{2}$$
$$=2R\sqrt{2}$$
$$\&$$
$$AC=2R+2r$$
$$\Rightarrow r =2R(\sqrt{2} - 1)/2$$
$$= R(1.414 - 1)$$
$$= 0.414\,R$$

FIGURE 4.2   Deduction of the sphere radius of quadratic void from AA type (quadratic) plane network.

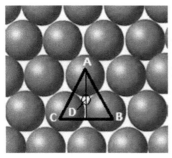

Since AC=2R,
And the centre of the small
circle is placed in the centre of
symmetry of the triangle ABC,
$\Rightarrow AD = (2/3) \, AC \sin60$
$= (2/3) \, 2R \, (\sqrt{3})/2 = 2R/\sqrt{3}$
$\& \; AD = R + r$
$\Rightarrow r = (2 - \sqrt{3})R/\sqrt{3}$
$\cong 0.155 \, R$

FIGURE 4.3   Deduction of sphere's radius for triangular void from flat AB network type (or planar hexagonal).

The (b) type arrangement with CN=6 provides a much higher "filling"–compaction of the space.

Passing to spatial arrangements we can highlight two possibilities for each flat arrangement.

Accordingly, if the plane lattice of AA type translates in space so that the center of spheres from two successive planes overlap to the 0z axis we get a simple cubic type network, with primitive cubic cell (PC), as in Figure 4.4 (Chiriac-Putz-Chiriac, 2005).

Each part of this arrangement has six proximity neighbors (so CN=6) placed in the vertices of an octahedron (as a polyhedron coordination).

In this network cubic interstitials (CN=8) are forming that can fit spheres of radius r whose value can be calculated, as exposed in the Figure 4.5.

If reticular AA type planes are horizontally moving toward each other so that the spheres of a plan to be placed in the next plan's cubic voids, so the resulting symmetry to be maximum, we have a sequence

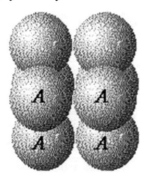

FIGURE 4.4   Identical spheres in the simple primitive cubic (PC) type arrangement, after Chiriac-Putz-Chiriac (2005).

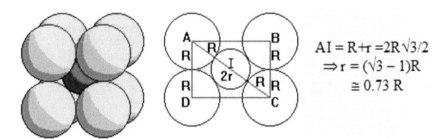

$$AI = R + r = 2R\sqrt{3}/2$$
$$\Rightarrow r = (\sqrt{3} - 1)R$$
$$\cong 0.73 R$$

FIGURE 4.5   Deduction of the interstitial sphere radius with CN=8 from AA cubic spatial network.

of ABAB type in vertical alignment, overlapping only the center of the spheres from odd and even vertical layers. The resulted arrangement is described by the body centered cubic (BCC) cell, Figure 4.6 (Chiriac-Putz-Chiriac, 2005).

For any referential sphere, of course, the coordination number is CN=8 and the corresponding polyhedron is a cube. Such situation is obvious if we take inland particle as reference yet, the same is inferred if the reference particle is from corner: since elementary cell corner is common to eight cubes, the reference particle is surrounded by the eight inner spheres of the neighboring cubes.

If in the primitive cubic network (PC) the sequenced particles from corners of the elementary cells are tangent, they came in contact in compact body centered cubic (BCC) network with the inner particle placed on cube A^3 axis, see Figure 4.7 (Chiriac-Putz-Chiriac, 2005):

$$AG = a\sqrt{3} = 4R \Rightarrow a = 4 R \sqrt{3} / 3 \cong 2.31R > 2R \qquad (4.2)$$

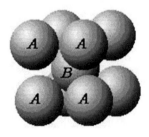

FIGURE 4.6   Identical spheres in compact body centered cubic (BCC) type arrangement.

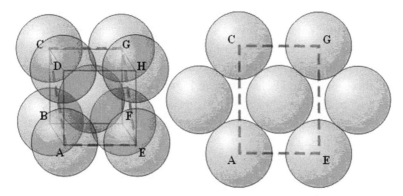

FIGURE 4.7   Contacts of identical spheres from BCC packing type.

In such an arrangement interstices with bi pyramidal squared shaped are formed between the common cubic faces, Figure 4.8 (Chiriac-Putz-Chiriac, 2005).

For a sphere to occupy such a gap the radius should have the value:

$$r \leq (a - 2R)/2 = (2\sqrt{3}/3 - 1)R = 0.155R \qquad (4.3)$$

In fact, such a sphere has only two neighbors at minimum distance (the two centers of the neighboring cubes, d=a/2). The others four neighbors, from face corners are placed at distances $d_1 = a\sqrt{2}/2 > d$. So, the real coordination number of such a particle is CN=2, and *coordination polyhedron is a line (II'/2)*.

For AB type flat network, the vertical overlapping of reticular planes can be achieved in two ways.

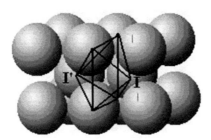

FIGURE 4.8   The square bi pyramid of the formed interstice by the BCC packing type of the identical spheres (II' = 2R +2r = a).

In first way the spheres centers of the reticular planes vertical are over-lapping as from two in two layers so making the vertical sequence 1212.... In this case, a reference sphere has six planar neighbors, placed in the vertices of regular hexagon and six others neighbors outside the reference plane, three above and below it, as in Figure 4.9.

The set of three neighbors placed on each of the hexagonal sides which includes the reference sphere are coplanar and are placed in the vertices of an equilateral triangle.

The two equilateral triangles, respecting to the reference, features a reflection plane. Coordination number in this arrangement has CN=12 value and the coordination is like cub-octahedron anti prism, see Figure 4.10.

In terms of syngony, the arrangement is hexagonal, since its elementary cell being the compact hexagonal prism (HC); note that the compact hexagonal prism is derived from primitive hexagonal prism (as a Bravais polyhedron) by occupying the sextants in center of the symmetry, within half of prism height, see Figure 4.11 (Chiriac-Putz-Chiriac, 2005).

The second way of overlapping of AB type of plane networks makes only the fourth layer to overlap the centers of the spheres over the first

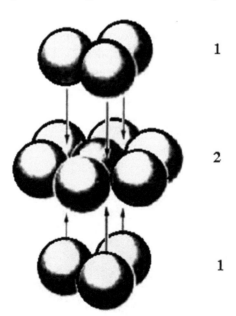

FIGURE 4.9 The 1212... sequencing of planar ABAB... ordering type of identical spheres; after Borg and Diens (1992).

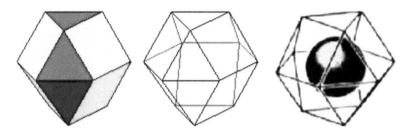

FIGURE 4.10 The coordination polyhedron as the cubic-octahedral prism for the reference sphere from the vertical 1212 arrangement of ABAB plane string in Figure 4.9; after Borg and Diens (1992).

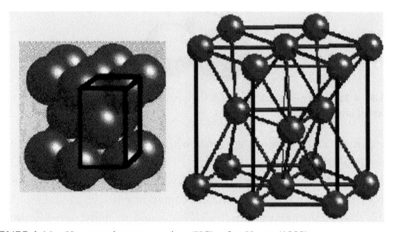

FIGURE 4.11 Hexagonal compact prism (HC); after Heyes (1999).

layer. Consequently, reticular planes sequence is of 123123… sequence type, see Figure 4.12.

The coordination number of a reference particle is as in the 1212… sequence equal so CN=12. As in previous case, the reference sphere is coordinated in its own plane by six neighbors hexagonally placed. Again, like in the previous case, there are six other neighbors placed three-by-three in the vertices of the equilateral triangles above and below hexagonal plane.

The difference between these two cases is represented by the relative position of the two triangles to the referential hexagonal plane. In the late case, the two triangles are rotated one to each other with $\pi/3$ angle. Therefore, the resulted structure features an A_3^6 axis, i.e., an inversion center; the coordination polyhedron of the particle with CN=12 in this

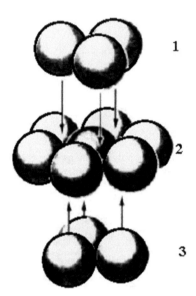

FIGURE 4.12 The plane ordering ABAB… type of identical spheres in vertical 123123… sequence; after Borg and Diens (1992).

arrangement is the tetra-decahedron, while the polyhedron is called as the cubic-octahedron, see Figure 4.13.

In a compact arrangement (CN=12) whether is hexagonal or cubic, the ratio of the radius reference (the coordination particle, r) and neighbors radius (the coordination particles, R) is equal to unity, see Figure 4.14 (Chiriac-Putz-Chiriac, 2005).

Elementary cell which describes the arrangement network from Figures 4.13 and 4.14 is the face- centered cubic compact polyhedron, Bravais-FCC, see Figure 4.15 (Chiriac-Putz-Chiriac, 2005). The analysis

FIGURE 4.13 The cubic-octahedral polyhedron of coordination for the reference sphere of the vertical 123123 arrangement of ABAB plane strings from Figure 4.12; after Borg and Diens (1992).

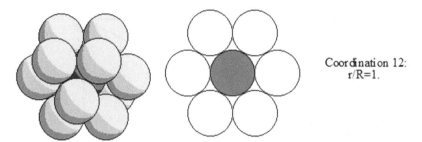

Coordination 12:
r/R=1.

FIGURE 4.14 Coordination with CN=12 for the arrangements of Figures 4.9 and 4.12; after Heyes (1999).

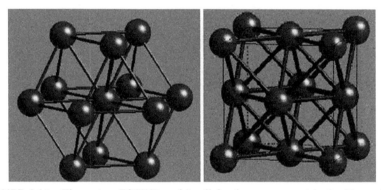

FIGURE 4.15 Elementary FCCF(Bravais) cell for the arrangement in the Figure 4.12; after Heyes (1999).

of this elementary cell allows us for the direct identification of the type (geometry) and number of interstices created by the compact arrangement. As an observation, the superposition of the AB type plane networks, any of them, leads to a three- dimensional compact arrangement.

Therefore, a vertex along the three centers of nearly faces that generate it by intersection associate with an interstice of tetrahedral geometry (T), see Figure 4.16 (Chiriac-Putz-Chiriac, 2005).

A particle placed inside such void will have CN=4 and a radius r with the value deduced from the construction shown in Figure 4.17 (Chiriac-Putz-Chiriac, 2005).

In a face-centered cubic cell tetrahedral interstices (T) are formed to each corner. Therefore, the cell has 8 T-interstices. However, the distribution of the particles in FCC cell causes the appearance to an additional type of voids: the octahedral interstices.

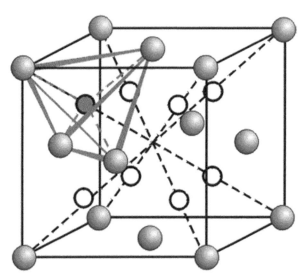

FIGURE 4.16   Tetrahedral intersitices in a compact cell FCC type; after Heyes (1999).

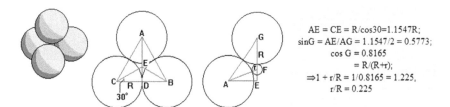

$AE = CE = R/\cos 30 = 1.1547R;$
$\sin G = AE/AG = 1.1547/2 = 0.5773;$
$\cos G = 0.8165$
$= R/(R+r);$
$\Rightarrow 1 + r/R = 1/0.8165 = 1.225,$
$r/R = 0.225$

FIGURE 4.17   The geometrical deduction the radius of sphere trapped into a tetrahedral gap/interstice, after Chiriac-Putz-Chiriac (2005).

This way, inside such a cell, the particles from the centered-faces form a "whole" octahedron, see Figure 4.18, see Chiriac-Putz-Chiriac (2005).

The radius of a sphere that may be included in such a gap is r=0.414R and the coordination number of the particle is CN=6, with the detailed design in the Figure 4.19.

Yet, in same cell, the ending particles of an edge, along with those from faces, centers which generate the edge, border the two faces of octahedron (1/4 from octahedron volume).

Then, since the edge is common to four cubes, in space, there is a completely octahedral interstice respecting an edge.

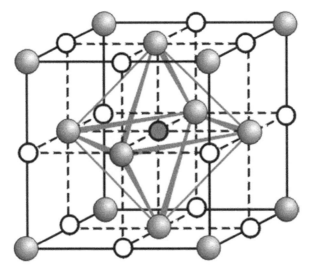

FIGURE 4.18 Octahedral interstices in a FCC compact cell; after Heyes (1999).

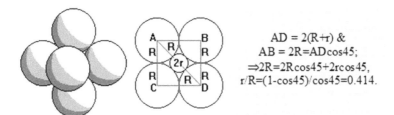

$$AD = 2(R+r) \ \&$$
$$AB = 2R = AD\cos45;$$
$$\Rightarrow 2R = 2R\cos45 + 2r\cos45,$$
$$r/R = (1-\cos45)/\cos45 = 0.414.$$

FIGURE 4.19 The geometrical deduction for the radius of sphere trapped into an octahedral gap/interstice; after Chiriac-Putz-Chiriac (2005).

However, effectively summing for the 12 edges of the elementary cell, follows that only 12 (1/4) = 3 complete octahedral interstices belonging per edge; thus, in total, the FCC cell has four complete octahedral interstitials.

There is important to remember that in FCC cell the number of tetrahedral interstices (8T) is double respecting those of the octahedral form (4O) and that the dimension of a tetrahedral spherical interstice is much smaller that the octahedral one, see comparatively the deduced values from Figures 4.17 and 4.19, respectively.

In compact hexagonal cell, Figure 4.11, there are also tetrahedral and octahedral interstices with a ratio identical with that one of face centered cubic cell, as previous discussed.

Tetrahedral interstices, as Figure 4.20, are identified by (Chiriac-Putz-Chiriac, 2005):

- combining the particle from the center of a base and the three indoor particles; the two tetrahedral cavities thus creating 2T interstices having a common face, being thus separated by a symmetry plane;
- Between bases' centers, the two vertices and an inner particle; this way six tetrahedral interstices (6T) appear which, two by two, having common vertices, are center- symmetric;
- Between one particle from one the bases' vertices and the closest three inner particles from alternative sextants from three hexagonal cells with common edge on which there is the particle from the vertex of the considered base; the common edge, that passes through the particles from the vertices of adjacent hexagonal bases, serves as the A^3 axis; being two bases, (6×2)/3=4T tetrahedral interstices are formed and placed on the edges, thus belonging to a compact hexagonal cell.

In total there there are 2T+6T+4T=12T tetrahedral interstices.

Octahedral interstices (holes or gaps) are formed in between the center of the base, two vertices and two inner particles, in the unfilled sextants with inner particles, see Figure 4.21 (Chiriac-Putz-Chiriac, 2005).

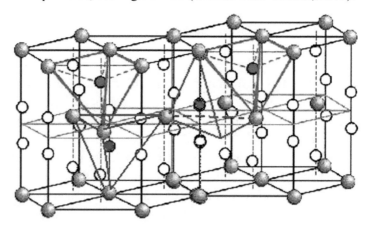

FIGURE 4.20 Tetrahedral holes in a compact hexagonal cell; after Chiriac-Putz-Chiriac (2005).

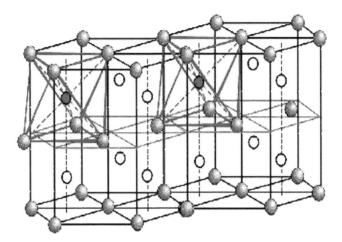

FIGURE 4.21   Octahedral holes in a compact hexagonal cell; after Chiriac-Putz-Chiriac (2005).

Worth remarking how, although none of octahedral holes formed do not completely belong to compact cell, since closing outside of the elementary cell, yet, due to the fact that the same effect appears to all elementary cells, in the global network, what a cell "lose" from a hole to a neighboring cell "wins" from another on the distance of the opposite sextant.

Thus, effectively, to one compact elementary cell 6O octahedral interstices belong, i.e., at a rate of ½ respecting the tetrahedral interstices allowed (i.e., 12T, as previously discussed).

4.2.2   SPATIAL CRYSTAL STABILITY BY THE MODEL OF IDENTICAL SPHERES

Establishing an identical spheres arrangement' stability depends of the degree of compaction, so on its potential energy; the present discussion follows (Chiriac-Putz-Chiriac, 2005).

The degree of compaction is defined as the ratio between spheres volume belonging to an elementary cell (number of spheres "n" × sphere's volume "v") and its volume (V).

The higher the degree of compaction is the more stabile is the arrangement.

Let's consider the possibilities of compactness for basic crystal structures.

The most affined structure corresponds to the simple cubic arrangement, Figure 4.4.

In this case, the spheres come into contact according with a cubic structure so that with:

- the cube edge: a=2R;
- the cube volume: V=a³=8R³;
- the cube content: one single particle⇒n=1;
- the particle volume: v=(4/3)πR³;
- the degree of occupancy will be:

$$\frac{1 \cdot v}{V} = \frac{\frac{4}{3}\pi R^3}{8R^3} = \frac{\pi}{6} = 0.5236 \tag{4.4}$$

In body centered cubic structure, see Figure 4.6, the spheres come into contact by A³ axis (the space diagonal) of the cube.

Thus there is:

- the cube edge: a=4R/√3;
- the cube volume: V=a³=4³R³ /(3√3);
- the cube content: two particles⇒n=2;
- and the resulting degree of occupancy (compactness) will be:

$$\frac{2 \cdot v}{V} = \frac{2\frac{4}{3}\pi R^3}{\frac{4^3}{3\sqrt{3}}R^3} = \frac{\sqrt{3}\pi}{8} = 0.68017 \tag{4.5}$$

In primitive-simple hexagonal structure, the spheres come into contact on each reticular string so that:

- the hexagon edge has the length: a=2R;
- the hexagonal base surface: S=6 [a (a √3/2) (1/2)]=6√3R²;
- the prism height: h=2R;
- the elementary prism volume: V=S h=12√3R³
- the elementary cell content: n=3=12 (1/6)+2(1/2);
- the degree of space occupancy will be:

$$\frac{3 \cdot v}{V} = \frac{3\frac{4}{3}\pi R^3}{12\sqrt{3}R^3} = \frac{\pi}{3\sqrt{3}} = 0.6046 \tag{4.6}$$

In the compact cubic arrangement, the FCC cell of Figure 4.15, the particles are in contact on the face's diagonal so that:

- the diagonal of cube faces has: $d=4R=a\sqrt{2}$;
- the cube edge has: $a=2\sqrt{2}R$
- the cube volume is: $V=a^3=2^4\sqrt{2}\ R^3$;
- the cell contain four particles: $n=4=8(1/8)+6(1/2)$;
- the degree of space occupancy CCF ordering will be:

$$\frac{4\cdot v}{V}=\frac{4\frac{4}{3}\pi R^3}{2^4\sqrt{2}R^3}=\frac{\pi}{3\sqrt{2}}=0.7405 \tag{4.7}$$

In compact cubic arrangement (HC), of Figure 4.11, the spheres are in contact on each plane while the spheres from two adjacent layers are placed in intermediary layer interstices.

Thus, the bases' centers rest on the three interior spheres with which a tetrahedron is built, as in Figure 4.22 (Chiriac-Putz-Chiriac, 2005).

Noting that the elementary cell height is twice tetrahedron height, we can summarize the following geometrical information:

- the base edge is: $a=2R$;
- the hexagonal base surface is: $S=6\ [a\ (a\ \sqrt{3}/2)\ (1/2)]=6\sqrt{3}R^2$;
- the prism height is: $h=AI\ \sin(\angle AIC)=2\ (a\sqrt{3}/2)\ (2\sqrt{2}/3)=4R\sqrt{2}/(\sqrt{3})$;

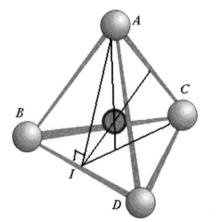

$AI \perp BD$;
$BD=2R=a \Rightarrow AI=a\sqrt{3}/2$;
$\triangle AIC$ is isosceles $\Rightarrow AI=IC= a\sqrt{3}/2$
& $AC=a$;

$\cos(\angle ACI)=(a/2)/IC=\sqrt{3}/3$
$\Rightarrow \sin(\angle ACI)=\sqrt{2}/\sqrt{3}$;

From sine theorem:
$\sin(\angle AIC)/a= \sin(\angle ACI)/AI$
$\Rightarrow \sin(\angle AIC)= 2\sqrt{2}/3$

FIGURE 4.22   The formed tetrahedron by the interior particles with base's particles from the compact hexagonal cell.

- the elementary prism volume is: $V=S\,h=24\sqrt{2}\,R^3$
- the cell contains six effective particles: $n=6=12\,(1/6)+2(1/2)+3\,(1)$;
- the degree of occupancy in HC ordering will be:

$$\frac{6\cdot v}{V}=\frac{6\frac{4}{3}\pi R^3}{24\sqrt{2}R^3}=\frac{\pi}{3\sqrt{2}}=0.7405 \tag{4.8}$$

Thus the two compact arrangements (FCC and HC) have the same compaction degree.

4.2.3   THE CRYSTALLINE CONSTITUTIVE PARTICLES' DIMENSION

In crystal chemistry, particles sizes are represented by ionic and atomic radii. Admission of such size is related to the assumption that the ions and the atoms have spherical shapes and a determined volume, impenetrable to others ions or atoms. Thus, the inter-ionic or inter- atomic distance that is established between two particles in a network, i.e., the distance between the mass centers of these particles is equal to the sum of radius of the two particles. Based on this assumption certain determinations to establish the ionic and atomic radii are therefore possible. However, aiming to determine the size of the constitutive particles of crystalline lattice/networks numerous experimental or calculation methods have been proposed; the present discussion follows (Chiriac-Putz-Chiriac, 2005).

For instance, according to Pauling, the ionic radius can be calculated using the formula:

$$r=\frac{C_n}{Z-S} \tag{4.9}$$

where Z is the atomic number of the element, S corresponds to the shielding effect of the nucleus towards the electronic shell, i.e., $Z-S$ represents the effective charge of the nucleus, and C_n is a constant determined as a function of principal atomic number n of valence shell of electrons.

Thus, applying this formula for NaF, where both ions have the electronic configuration of neon, and shielding coefficient deduced from theoretical considerations, for example, Slater rules, is $S=4.52$, see Volume II of the present five-volume set, and Borg and Diens (1992); so we obtain:

- for Na^+ $Z = 11$: $Z\text{-}S = 6.48 \rightarrow r_{Na^+} = \dfrac{C_n}{6.48}$

- for F^- $Z = 9$: $Z\text{-}S = 4.48 \rightarrow r_{F^-} = \dfrac{C_n}{4.48}$ $\qquad$ (4.10)

Knowing the internodal distance (accessible to direct determinations using röntgeno-structural/X-ray analysis), $d_{Na(+) - F(-)} = 2.31\text{Å}$, there results:

$$d_{Na(+) - F(-)} = r_{Na(+)} + r_{F(-)} = C_n \left(\frac{1}{6.48} + \frac{1}{4.48} \right) = 2.31 \text{ Å} \qquad (4.11)$$

where: $C_n = 6.13$, while $r_{Na(+)} = 0.95\text{Å}$ and $r_{F(-)} = 1.36\text{Å}$.

However, the obtained values by different methods departs each other due to the complexity of factors considered in modeling the influence of the inter-node distance. Regardless this difference, data analysis on nodal/structural particle size highlights some very important regularity. They are:

- within one group, the ionic and atomic radii increase with increasing of order number;
- within the small periods, the atomic radii decreases, generally, between groups IA-VIIA with increasing of the atomic number, and then increasing for the inert gases group;
- if certain element presents several oxidation states, the radii are increasing with negative charge;
- the radii of positive ions decrease for elements of a period with increasing of atomic number;
- on the diagonals of the Periodic Table there are atoms and ions with identical or very similar radii;
- the cations' radii (0.4–1.7Å) are generally smaller than those of anions (1.3–2.7Å).

4.2.4 THE CRYSTALLINE PARADIGM FOR THE GENETIC CODE

Although the atomic Lewis cubic model was abolished, see the Introduction of the present Chapter, crystalline paradigm continued its applicability journey to higher level of organization of matter, now for the bio-molecules.

In particular, the language of biological systems teaches us that: the forming, transmission and decryption of genetic code arise in nature by trial and error through structural combination, so evolving step by step towards accumulation of small changes (structural and functional), thus (paradoxically) avoiding a definitive optimal solution of development, named the *optimal choice*, such as was the case for atomic combinations in molecules, within the quantum paradigm approach; the present discussion follows (Chiriac-Putz-Chiriac, 2005).

For this reason, explaining and quantifying the biologic language is vital to understanding the life and evolution of species (Putz, 2010a-b), while proves to be a very tricky enterprise, yet not in a stage of a complete theory.

Biologists prefer to see in language of organismal genes and proteins the *frozen accidents* for which every small change affects their "life" (structure, function and metabolism).

However, something is already known accurately, namely that the genetic language is *universal*, i.e., for all living organism, and it shows an efficient *casing* (packing, condensation, coding after all) of genetic information (preserving minimum correlation, maximum entropy), so suggesting that a mathematical model for choosing an optimal form of genetic code does exist (Putz, 2010b).

In other words, by questions like: *Why carbon? Why peptide chain?, and Why 20 nucleic acids- no more no less?* one seeks for paradigmatic answers, abstracted from a general value, possible universal.

Fortunately, to these questions a mathematical- chemical- physical answer can already be formulated, yet posing a high degree of formalization, which we prefer to call as the *crystalline paradigm of genetic code*.

Thus, the question about proteins description can be reformulated by finding a *discreet language*, able to codify the arbitrary form of a one-dimensional chain in a three- dimensional chain, analogously to the fact that proteins can appear in different structures, from primary (linear) to quaternary one (representing sets of linear chains deformed and compacted in three- dimensional/tertiary structures) (Putz, 2010b).

Then, of all ways to mathematically quantifying a space, the simplest unity of lattice (!) which can successfully describe the space in any dimension (D) is the so-called *simplex* structure, representing the volume element formed from a set of ($D+1$) points in D-dimensions.

TABLE 4.1   The Face-To-Face Numerical and Structural Information Encoded in Simplex One- and Three-Dimensional Systems, Respectively

Properties	Simplex 1-dimensional	Simplex 3-dimensional
Information:	*Numeric*	*Structural*
Discrete Space:	integers	Network/ lattice
Basic variables:	$\{0,1\}$	$l=\{0,1\}$
Implementation:	Input/ Output	Tetrahedron
Operations:	Addition/ multiplication	Translation/ rotation

The way the digital information is characterizing the discreet 1-dimensional space becomes information in a simplex three-dimensional through the correspondences summarized in Table 4.1.

The key for protein description relays in the properties of 3-dimensional simplex.

The three-dimensional simplex is a network/lattice whose structural element is a tetrahedron. Thus, arbitrary constructions can be created by successive joining of tetrahedrons with common vertices and edges.

For example, a tetrahedron corresponding to harmonic representation associated to s and p orbital types, carrying the orbital quantum number taking the values $l=\{0,1\}$, is a regular tetrahedron, describing four equivalent and equidistant states. In quantum theory, it is said that these 4 states, described by so- called hybridization sp^3, are forming an *orthogonal basis*.

The ideal element that can meet these requirements is Carbon, under its diamond type structure, Figure 4.23-left (Chiriac-Putz-Chiriac, 2005).

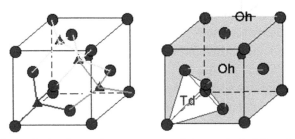

FIGURE 4.23   Left: the diamond type structure; right: tetrahedral T_d and octahedral O_h interstices highlighted in face-centered cubic structure; after Heyes (1999).

The carbon atom situated in center of a tetrahedron is able to achieve four covalent bonds on the directions oriented to the 4 vertices of the tetrahedron (Purves et al., 2001).

On the other side, the proteins carry information in 3-dimensional form, appropriately packed, but also having by their linear (1-dimensional) forms an important role, for example in their crossing process through membranes and cellular walls. Such process implies unpacking of the tertiary structure to the primary 1-dimensional one that can penetrate membranes or cell walls without damaging the information held (i.e., the sequences of amino acids, respectively of RNS- DNA), being afterwards re-packed into the 3-dimenisonal tertiary structure level, so sending (through its unique spatial form) the information for specific function for that protein in the body (Purves et al., 2001).

The tetrahedral structure allows this possibility of unbinding, inter-weaving and packing by forming a *tetrahedral chain* inside a cubic type structure with faces-centered, in which the permitted T_d and O_h interstices are occupied by carbon atoms, as in Figure 4.23-right.

Thus, one can already conclude that the tetrahedral structure around a carbon atom is the simplest (and therefore the most likely) structure that allows the proteins' encoding and transport of properties and function.

Tetrahedral geometry plays a crucial role in describing the shapes of proteins and even their packed structure. On the other hand, the simplest chained structure carbon based is that of polyethylene, $(-CH_2-)_n$.

This carbon atoms' skeleton is adequate to rotations, but is too flexible to maintain its own space form. Rigidity of this skeleton can be increased when the carbon atoms form, in addition to simple bonds, also the double bonds that would resist to rotations. In this way, the carbon atoms chain will have flexible segments, favoring the rotations, as well as the fixed segments that specify the chain form.

The next element after carbon that can be inserted in such structure is the nitrogen. Therefore, the $C_\alpha - N$ bond froming the chain can contribute to different rotations around the alpha (central) Carbon atom.

Furthermore, the next element after nitrogen capable of being inserted in such structure is that of oxygen, while contributing by its double bond with carbon to fixing the carbon chain form.

All these structural requirements can be achieved if, around the alpha carbon atom an acid grouping (–COOH) and a basic grouping (–NH$_2$) are

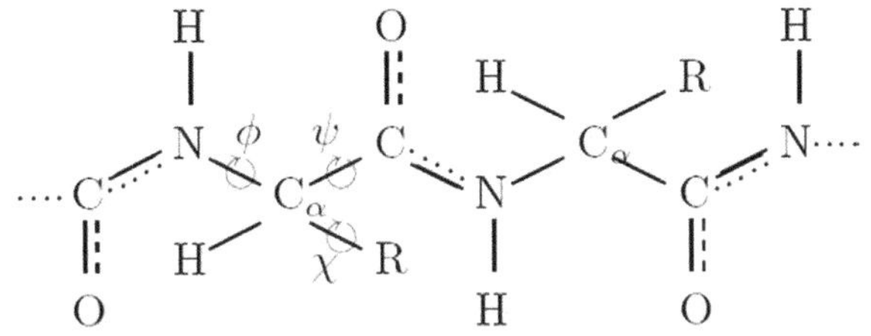

FIGURE 4.24 The chemical structure of a polypeptide chain.

attached too, so allowing the peptide bond formation by their neutralization. In this way, the polypeptide chain is developed, see the Figure 4.24 (Chiriac-Putz-Chiriac, 2005).

Now, each peptide unit in a chain contains two rotating bonds of C_α atom, denoted by ϕ and ψ angles in Figure 4.24. When projecting these angular revolutions to tetrahedral geometry only three possibilities (any of, 120°) rotations are fixed for each of these angles, covering 9 combinations in total (Purves et al., 2001).

There should be noted at this point, for correctness in analogies, that, along a polypeptide chain, not all the formed angles are rigorously equal to angles in a regular tetrahedron, because not all formed bonds are of the same length (as would all edges be equal in a regular tetrahedron), based on the fact that the properties of N atom are not the same with those of C atom (as would all vertices of a regular tetrahedron be equivalent). However, variations in bonds and angles length in a polypeptide chain does not exceed 10%, and therefore, one can rationalize properties of a crystalline diamond structure for a polypeptide chain wrapped in tertiary structure of proteins.

Put differently, the "Nature" is not able to organize itself in a perfect geometric structure as that of the cubic structure with faces-centered, but it has "evolved" so as to approach it.

The *trans-cis* structural options can be added to the nine ideal guidelines, easily to accommodate in diamond structure because of the local exchange symmetry between the faces centered cubic and the hexagonal lattices, see Figures 4.11 and 4.15, or between tetrahedral T_d and octahedral O_h groups of symmetry associated to interstices in Figure 4.23-right. The nine possible angular orientations plus the *trans-cis* option cover all elementary operations necessary to a polypeptide chain to pack into a compact tertiary structure (similar to diamond).

If we multiply these operations (10 in total) by the factor 2 (which would index the polar/no polar properties of R-groups from the polypeptide chain) 20 occupancy possibilities for groups R are obtained, which is precisely the total number of amino acids present in the standard genetic code.

Remarkably, the resulted total number (20) is exactly the same as the available number of T_d and O_h interstices (excluding central octahedral interstice) in face-centered cubic structure, see the right side in Figures 4.23 and 4.25; these holes, once occupied by the R groups of amino acids will compact the tertiary structure of proteins.

At this point one should be noted that, indeed, the *packing fraction* or *occupancy of space* (the ratio between occupied and free space) for proteins is very close to that of packing of identical spheres (with two types of cavities in interstices), as corresponding to the T_d and O_h symmetries of Figure 4.23-right in face centered cubic structure, see the relation (4.7).

This feature provides another argument for crystalline paradigm for proteins structure rationalization.

Moreover, the fact that the R groups of amino acids can be classified in two groups according to their weight is paralleling the two types of cavities available in a face centered cubic structure, with interstices precisely occupied by these R groups (Purves et al., 2001).

Ending this section, one remains with the fact that the crystalline paradigm of genetic code, here presented, provides a geometric, chemical and physical insight on how the Nature "chooses" its ways of evolution, tending towards ideal (compact) structures for chemical-to-biological life.

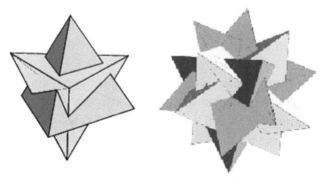

FIGURE 4.25   Left: the intercrossing way of two tetrahedral structures; right: the compact intercrossing of tetrahedral elements centered on T_d and O_h interstices of a face-centered cubic type structure; after Wolfram Research (2003).

4.3 CRYSTALLINE SUBSTANCES: TYPICAL NETWORKS

4.3.1 SYSTEMATICS OF CRYSTALLINE STRUCTURE

Under the conditions of temperature and pressure on the Earth, except the rare gases, all the other elements are found in bonding state; the present discussion follows (Chiriac-Putz-Chiriac, 2005).

If the bonding atoms belong to the same element, the substance is called *simple substance = elementary substance = unary substance* (A_x).

If the bonding atoms belong to different elements we have a *composed substance*, a combination. Based on their qualitative composition, the compound substances can be binary (A_xB_y), ternary $(A_xB_yC_z)$, etc.

A way of classification is based on the stoichiometry of the relative substances, while noting each class of network/lattice with a certain letter. Thus, we distinguish (Chiriac-Putz-Chiriac, 2005):

- networks of type A for the elements,
- networks of type B – the binary compounds AB,
- networks of type C – AB_2,
- networks of type D – A_nB_m,
- networks of type E composed by more types of particles than two, without forming of atomic groups $(CaTiO_3)$,
- networks of type F – showcase groups with 2 or 3 atoms $(NaClO_2)$,
- networks of type G – showcase groups with 4 atoms $(CaCO_3)$,
- networks of type H – showcase groups with 5 atoms $(BaSO_4)$,
- networks of type I – showcase groups with 7 atoms (K_2PtCl_6),
- networks of type K – showcase complex groups of atoms $([Cu(NH_3)_4]I_2)$,
- networks of type L – alloys,
- networks of type O – organic compounds and,
- networks of type S – silicates.

The Table 4.2 presents a systematics of certain (representative) crystalline structures after the main crystalline types. There are indicated: the structural type, the representative combination, the spatial group, the number of the structural units in the elementary cell (Z) and the main iso-type combinations (Chiriac-Putz-Chiriac, 2005).

Reviewing the main classes of the crystalline combinations once more underline the wide structural variety of the crystals, with factors that can influence the type of network.

TABLE 4.2 Systematics of Crystalline Structures; after (Chiriac-Putz-Chiriac, 2005)

Structural type	Representative compounds	Spatial groups	No. of structural units per unit cell (Z)	Isotype structures
A – Elements				
A1	Cu	Fm3m	4	Ca, Sr, Al, Ce, La, Pb, Th, γ-Fe, β-Co, Ni, Pt, Ag, Au. Ne, Ar, Kr, Xe $AuCu_3$, $CaSn_3$
A2	W	Im3m	2	Li, Na, K, Rb, Cs, Ba, V, Nb, Ta, Cr, Mo, a-Fe, d-Mn β-Ag_3Al, β-$BeCu_2$, β-Pd_2Cd_3, β-CuZn
A3	Mg	$P6_3/mmc$	2	Be, Mg, Se, Y, La, Ti, Zr, Hf, Tc, Re, Ru, Os, Co, Zn, Cd Ag_3Ga, Au_2Ti, Au_4In, V_2Rh_3
A4	C – diamond	Fd3m	4	C, Si, Ge, Sn
A5	β – Sn	I4/amd	4	Si, Ge at very high pressure
A6	In	F4/mmm	2	γ-Mn Mn_2Cu
A7	As	$R\bar{3}m$	2	Sb, Bi
A8	Se	$C3_12$	3	Te

TABLE 4.2 Continued

Structural type	Representative compounds	Spatial groups	No. of structural units per unit cell (Z)	Isotype structures
B – AB Compounds				
B1	NaCl	Fm3M	4	Hydrides of: Li, Na, K, Rb, Cs
				Halides of: Li, Na, K, Rb
				CsF, AgCl, AgBr, MgO, CaO, SrO, BaO, MnO, FeO, CoO, NiO, CdO
				Sulfides and selenides of:
				Mg, Ca, Sr, Ba, Pb, Mn, La, Th, U
				Tellurides of:
				Ca, Sr, Ba, Pb, La, U
				Nitrides and phosphides of:
				La, Ce, Pr, Nd, TiN, ZrN
				Carbides of: Ti, Zr, Hf, Th
				V, Nb, Ta, U, ZrB, HfB, SnAs, SnSb, LaSb, LaBi, ThAs, ThSb, UAs, USb
B2	CsCl	Pm3m	1	NH$_4$Cl, NH$_4$Br, NH$_4$I, CsBr, CsI, TlCl, TlBr
				SrMg, LaMg, CeMg, FeTi, CoBe, NiBe, CuBe, AgLi, AgLa, AuMg, ZnLa, ZnAg, CdLa, HgLi, HgMg, GaNi, InNi, TlLi, TlCa, TlBi, TlLa

TABLE 4.2 Continued

Structural type	Representative compounds	Spatial groups	No. of structural units per unit cell (Z)	Isotype structures
B3	ZnS – blende	F43m	4	CuF, CuCl, CuBr, CuI, α-AgI
				Sulphides, selenides and tellurides of:
				Be, Zn, Cd, Hg
				β-MnS, β-MnSe
				BN, BP, BAs, AlP, AlAs, AlSb,
				Interstice, GaAs, GaSb,
				InP, InAs, InSb, SiC
B4	ZnS – wurtzit	P6₃mc	2	β-AgI; BeO, ZnO
				MgTe, β-MnS, γ-MnSe, CdS, CdSe
				CuH, AlN, GaN, InN, SiC
B5	SiC – politypes	R3m	4	Politypes of ZnS
B6		P6₃mc	6	
B7		P6₃mc	15	
B8	NiAs	P6₃/mmc	2	Sulphides, arsenide and tellurides of:
				V, Ti, Cr, Fe, Co, Ni, Mn
				TiSb, CoSb, MnAs, MnSb, CrSb,
				NiSb, NiBi, CuSn, PtSb, PtSn, AuSn,
				PtPb, MnBi, FeSb, IrPb, RhBi
B10	PbO	P4/mmm	2	SnO, LiOH
B32	NaTl	Fd3m	8	ZnLi, CdLi, LiAl, GaLi, InLi, InNa

TABLE 4.2 Continued

Structural type	Representative compounds	Spatial groups	No. of structural units per unit cell (Z)	Isotype structures
MoC	MoC	P6/mmm	1	WC, MoN
C–AB$_2$ Compounds				
C1	CaF$_2$	Fm3m	4	Ir$_2$P, Be$_2$C, Be$_2$B, PtAl$_2$, AuAl$_2$, PbMg$_2$
				SrF$_2$, BaF$_2$, PbF$_2$, CdF$_2$, HgF$_2$, SrCl$_2$, PaO$_2$, CeO$_2$, ThO$_2$, UO$_2$
Anti C1				Li$_2$O, Na$_2$O, K$_2$O, Rb$_2$O
				Sulphides, selenides and tellurides of Li, Na, K, Rb$_2$S
C2	FeS$_2$ – pyrite	Pa3	4	MgO$_2$, ZnO$_2$, CdO$_2$
				Disulphides, diselenides of
				Mn, Ni, Co, Rh, Ru, MnTe$_2$, RhTe$_2$
				PtP$_2$, PdAs$_2$, PdSb$_2$, AuSb$_2$, PtBi$_2$
C3	Cu$_2$O	Pn3m	2	Ag$_2$O, Pb$_2$O
anti-C3				Zn(CN)$_2$, Hg(CN)$_2$
C4	TiO$_2$ – rutil	P4/mmm	2	MgF$_2$, MnF$_2$, FeF$_2$, CoF$_2$, NiF$_2$, PdF$_2$, ZnF$_2$
				CoO$_2$, MnO$_2$, GeO$_2$, OsO$_2$

TABLE 4.2 Continued

Structural type	Representative compounds	Spatial groups	No. of structural units per unit cell (Z)	Isotype structures
C6	CdI_2	$P\bar{6}c2$	1	$TiCl_2$, $MgBr_2$, $TiBr_2$, VBr_2, $MnBr_2$, $FeBr_2$, $CoBr_2$, MgI_2, CaI_2, TiI_2, ZnI_2, PbI_2
				$Mg(OH)_2$, $Ca(OH)_2$, $Mn(OH)_2$, $Fe(OH)_2$, $Co(OH)_2$, $Ni(OH)_2$, $Cd(OH)_2$
				Sulphides and selenides of:
				Ti, Zr, Pt, VS_2, SnS_2, $CoTe_2$, $NiTe_2$
anti-C6				Ag_2F
C7	MoS_2	$P6_3/mmc$	2	WS_2, WSe_2
C8	SiO_2 – quartz a	$P6_2 22$	3	BeF_2, GeO_2, $AlPO_4$, $FePO_4$
C9	SiO_2 – cristobalite	Fd3m	8	BeF_2, GeO_2, $AlPO_4$, $FePO_4$
C11	CaC_2	I4/mmm	2	CaC_2, SrC_2, BaC_2, LaC_2, CeC_2, UC_2
				SrO_2, BaO_2
				$MoSi_2$
C14	$MgZn_2$	$P6_3/mmc$	4	VBe_2, $CrBe_2$, $MnBe_2$, $FeBe_2$, $MoBe_2$, WBe_2, $CaMg_2$, $LaLi_2$, $TiMn_2$
C15	$MgCu_2$	Fd3m	8	$NaAu_2$, KBi_2, $AgBe_2$, $CuBe_2$, $LaMg_2$, $CaAl_2$, UAl_2, $ZrCo_2$, Au_2Bi
C19	$CdCl_2$	$R\bar{3}m$	1	$MgCl_2$, $MnCl_2$, $FeCl_2$, $CoCl_2$
				$NiBr_2$, $ZnBr_2$, ZnI_2, NiI_2, $CdBr_2$

TABLE 4.2 Continued

Structural type	Representative compounds	Spatial groups	No. of structural units per unit cell (Z)	Isotype structures
D –A_n B_m Compounds				
$DO5$	BiI_3	$R3$	2	$ScCl_3$, $TiCl_3$, VCl_3, $FeCl_3$, $CrBr_3$, $FeBr_3$, AsI_3, SbI_3
$DO1_2$	FeF_3	$R\bar{3}C$	2	TiF_3, VF_3, CoF_3, RuF_3, RhF_3, PdF_3, IrF_3
$DO1_9$	Mg_3Cd	$P6_3/MMC$	2	Ni_3Sn, Fe_3Sn, Ni_3In, Co_3W, Cd_3Mg
$D1_1$	SnI_4	Pa_3	8	GeI_4, $TiBr_4$, TiI_4, $ZrCl_4$
$D2_1$	CaB_6	$Pm3m$	1	CaB_6, SrB_6, BaB_6, LaB_6, CeB_6, ThB_6
$D5_1$	Al_2O_3 - corundum	$R\bar{3}c$	2	Cr_2O_3, a-Fe_2O_3, V_2O_3, Ti_2O_3, Ga_2O_3 $MgTiO_3$, $FeTiO_3$: homeotype ($E2_1$)
$D5_2$	La_2O_3	$C3m$	1	Ce_2O_3, Pr_2O_3, Nd_2O_3, Mg_3Sb_2, Mg_3Bi_2, Th_2N_3, U_2N_3
$D5_3$	Mn_2O_3 - bixby x	$Ia3$	16	Sc_2O_3, Y_2O_3, In_2O_3, Tl_2O_3, Sm_2O_3, Gd_2O_3, La_2O_3, Be_3N_2, Be_3P_2, Mg_3N_2, Mg_3P_2, Mg_3As_2, Zn_3N_2, Cd_3N_2, U_2N_3
$D5_9$	Zn_3P_2	$P4/nmc$	8	Cd_3P_2, Zn_3As_2, Cd_3As_2

TABLE 4.2 Continued

Structural type	Representative compounds	Spatial groups	No. of structural units per unit cell (Z)	Isotype structures
E –A$_x$B$_y$C$_z$ Compounds				
EO$_1$	PbFCl	P4/nmm	2	BaFCl, BaFI, PbFBr, CaFCl CaHCl, CaHBr, CaHI, SrHCl, SrHBr, SrHI, BaHCl, BaHBr, BaHI ThOS, UOS, YOF, YOCl, LaOF, LaOCl, LaOBr, LaOI, CeOCl, BiOF, BiOCl, BiOBr, BiOI
E2$_1$	CaTiO$_3$ – perovskit	Pm3m	1	SrTiO$_3$, BaTiO$_3$, PbTiO$_3$, SrZrO$_3$, BaZrO$_3$, PbZrO$_3$, SrSnO$_3$, BaSnO$_3$, CaMnO$_3$, LaWO$_3$, LaFeO$_3$ RbCaF$_3$, CsCaF$_3$, BaLiF$_3$, KMnF$_3$, KFeF$_3$, KCoF$_3$, KNiF$_3$, CsMgF$_3$, NaZnF$_3$, KCrF$_3$, CsCdCl$_3$, CsHgCl$_3$
F – Compounds of complex ions bi- and tri-atomic				
F1$_1$	CoAsS	P2$_1$3	4	NiAsS, NiSbS Hg(CN)$_2$
F5$_1$	NaHF$_2$	R$\bar{3}$m	1	LiVO$_2$, LiCrO$_2$, LiNiO$_2$, NaVO$_2$, NaCoO$_2$, NaFeO$_2$, NaNiO$_2$, CuFeO$_2$ NaCNO, CsICl$_2$, LiHF$_2$
F5$_2$	KHF$_2$	I4/mcm	2	KCNO, KN$_3$, RbN$_3$

TABLE 4.2 Continued

Structural type	Representative compounds	Spatial groups	No. of structural units per unit cell (Z)	Isotype structures
G – Compounds of complex ions tetra-atomic				
GO_1	$CaCO_3$ – calcite	$R\bar{3}c$	2	$MgCo_3$, $MnCO_3$, $CoCO_3$, $ZnCO_3$, $CdCO_3$,
				$LiNO_3$, $NaNO_3$,
				$ScBO_3$, YBO_3, $InBO_3$
GO_3	$NaClO_3$	$P2_13$	4	$NaBrO_3$
H – Compounds with complex ions penta-atomic				
HO_3	$ZrSiO_4$ – zircon	I4/amd	4	$ScPO_4$
				YPO_4, $YAsO_4$, $YNbO_4$, $YTaO_4$
				$CaCrO_4$
HO_4	$CaWO_4$ – scheelite	$I4_1/a$	4	$NaIO_4$, KIO_4, NH_4IO_4, $AgIO_4$
				$CaMoO_4$, $SrMoO_4$, $BaWO_4$, $CdMoO_4$, $LiLa(MoO_4)_2$, $KLa(MoO_4)_2$
$H1_1$	$MgAl_2O_4$ – spinel	Fd3m	8	MgV_2O_4, $MgCr_2O_4$, $MgFe_2O_4$, $TiMg_2O_4$, VMg_2O_4, $FeAl_2O_4$, $FeCr_2O_4$, Fe_3O_4, Co_3O_4, $CuCo_2O_4$
				$BeLi_2F_4$, Li_2SO_4, $ZnSn_2O_4$
				$K_2Zn(CN)_4$, $K_2Hg(CN)_4$
				$MnCr_2S_4$, $ZnCr_2S_4$, Co_3S_4

TABLE 4.2 Continued

Structural type	Representative compounds	Spatial groups	No. of structural units per unit cell (Z)	Isotype structures
$H1_2$	$(Mg, Fe)_2SiO_4$ – olivine	Pmcn	4	Mg_2SiO_4, Fe_2SiO_4, Mn_2SiO_4, $CoMgSiO_4$, $CaMgSiO_4$, $Y-Ca_2SiO_4$, Co_2SiO_4
$H1_5$	$K_2[PtCl_4]$	P4/mmm	2	$(NH_4)_2[PtCl_4]$, $K_2[PdCl_4]$, $(NH_4)_2[PdCl_4]$
J – Compounds with complex ions hexa-atomic				
$J1_1$	$K_2[PtCl_6]$	Fm3m	4	$K_2[SiF_6]$, $(NH_4)_2[SiF_6]$, $Rb_2[SiF_6]$ $Cs_2[TiCl_6]$, $Cs_2[GeF_6]$, $K_2[SnCl_6]$, $(NH_4)_2[PbCl_6]$, $K_2[NiF_6]$ $(NH_4)_2[SeCl_6]$, $K_2[SeBr_6]$, $Rb_2[TeBr_6]$
S – Silicate				
SO_1	Al_2SiO_5 – distena	P$\bar{1}$	4	
SO_2	Al_2SiO_5 – andalusite	Pnnm	4	$Cu_2[PO_4(OH)]$, $Cu_2[AsO_4(OH)]$, $Zn_2[AsO_4(OH)]$
SO_3	Al_2SiO_5 – sillimanite	Pbnm	4	
$S1_1$	$ZrSiO_4 \equiv HO_3$	I4/amd	4	$ThSiO_4$, $USiO_4$

TABLE 4.2 Continued

Structural type	Representative compounds	Spatial groups	No. of structural units per unit cell (Z)	Isotype structures
$S1_2$	$Mg_2SiO_4 \equiv H1_2$	$C2/n$	2	$Mg_2P_2O_7$
$S2_1$	$Sc_2Si_2O_{7-}$ tortveitite	$C2/n$	2	$Sc_2Ge_2O_7$
$S4_1$	$Mg_2Si_2O_6$ – clinoenstatite	$C2/c$	4	$CaMgSi_2O_6$, $CaFeSi_2O_6$, $CaMnSi_2O_6$, $CaNiSi_2O_6$, $LiAlSi_2O_6$, $NaAlSi_2O_6$
$S4_2$	$Ca_2Mg_5(OH)_2$ $(Si_4O_{11})_2$ –tremolite	$C2/m$	2	Amfiboli

The amplification of the number of the structures by polymorphic modifications, more and more numerous as a result of the investigations of the extreme thermodynamic domains' parameters (e.g. very high or very low temperatures, coupled with pressures variation from the very deep vacuum to tens and hundreds of thousands of atmospheres) requires finding certain criteria of systematization that allow the coverage of the entire existing experimental data as well as the current and future research results.

In this context, the development of the compact packing theory and of the politype seems to lead in a not far away future to the realization of a new systematization and crystalline structures.

4.3.2   CRYSTALS' STOICHIOMETRY

Taking into consideration the membership of each particle to a certain elementary cell, the number of particles, respectively of structural units, can be easily found, see the values of Z in Table 4.2, and contained therein.

The total number of particles thus counted for an elementary cell, or a multiple integer and a small of it, represents the rough formula of the substance; the present discussion follows (Chiriac-Putz-Chiriac, 2005).

A crystalline substance can be described by its elementary cell just like a gaseous substance is defined by the knowledge of its molecular structure.

The crystalline structures can be also defined by the shape of the coordination polyhedra, respectively by the type of packing with the indications of the interstices which are occupied, if the forming particles of the crystal are different.

For example, the structure of the most of AB compounds can be rendered by the packing of some rigid spheres, as will be exposed in the forthcoming sections.

The characterization of a structure by the elementary cell and coordination polyhedra underline the mutual relations of coordination between the component particles of a given substance, the symmetry of the network (crystallographic system) and the symmetry of the motif (the symmetry class).

This complete picture, yet lengthily for the crystalline structures notation, can be simplified e by the crystallochemical formulas as will be next exposed.

In this regard, the crystallochemical formula of a substance underlines its essential structural features.

To obtain such a formula, one starts form the common stoichiometric formula that will be completed with information regarding the mono-, bi- or tri-dimensional character of the structure, regarding the mutual coordination of the particles and of the crystallographic system. The mono-, bi- or tri-dimensional character of the structure is indicated by the symbols 1, 2, 3, on left side of the formula, respectively. These kinds of symbols have the following significances (Chiriac-Putz-Chiriac, 2005):

∞ – indicates a macromolecule (polymer),

1 – indicates a linear structure (chain structure),

2 – a planar/flat structure (layer structure)

and 3 – a tridimensional structure.

The crystallographic system is indicated at the end of the formula, abbreviated as: t_c – triclinic, m_c – monoclinic, r_b – rhombic, t_t – tetragonal, t_g – trigonal, h- hexagonal, c – cubic. The coordination number of a particle is indicated as an exponent in square brackets in the right side of the symbol of the considered particle. If the nature of the particle that coordinate a given particle is not obvious, or there are variable distances for the identical particles of the same spheres (within the maximum limit of 20%), the exact situation is specified considering as exponent not the coordination number, but the number of each kind of coordination particle, respectively the number of the particle corresponding to each substance, as a sum. Here follows some examples (Chiriac-Putz-Chiriac, 2005):

$2C^{[3]}h$ – Carbon hexagonal system, the structure represents the infinite layer where each atom of carbon is surrendered by other three.

$1Se^{[2]}h$ – Selenium hexagonal system, the structure represents the infinite layers where each atom of selenium has two neighbors.

$3C^{[4]}c$ – Carbon cubic system, infinite tridimensional structure where each atom has four neighbors.

$3W^{[8+6]}c$ – Wolfram cubic system, infinite tridimensional structure where each atom has eight neighbors to a minimum distance and other 6 to a slightly higher distance (at the under 15% difference).

$3Na^{[6]}Cl^{[6]}c$ – Cubic sodium chloride, infinite tridimensional structure, where each particle has six neighbors (Cl⁻: 6 sodium ions, and Na⁺: chloride ions).

$3Ba^{[12]}S^{[4]}O^{[1S+3Ba]}$ r_b – barium sulphate rhombic system, infinite tridimensional structure where each barium ion is surrendered by 12 oxygen ions, each sulfur ion with 4 oxygen ions and each oxygen ion with 4 ions (1 of sulfur and 3 of barium). Usually, this formulas is abbreviated excluding the first part and writing together the acid radicals, respectively the complex ions. Thus, the abbreviated crystallochemical formula of the barium sulphate will be: $Ba^{[12]}SO_4 r_b$.

4.3.3 UNARY CRYSTALLINE SUBSTANCES

In the unary substances case, the atoms, being identical, have equal radii and electronegativity, so their organization in solid state depends only by the distribution of the external electrons; the present discussion follows (Chiriac-Putz-Chiriac, 2005).

Form this point of view, the atoms can be classified in four categories, as follows:

(a) Atoms with deficit of electrons: the number of electrons outside the atomic core is smaller than the number of the external atomic orbitals of close energy.

This is the situation of the elements which, in the solid state, form a metallic bonding. This kind of bonding is not oriented in the space, so the ordering of the atoms represent the principle of the maximum compactness for most of the situations.

The most common stable metallic networks/lattices under standard thermodynamic conditions are of cubic compact type (CC) respectively the hexagonal compact (HC), both with the atomic cores in the nodes of the metallic networks. Yet, there are also metals with a different type of crystalline lattice, as shown in Figure 4.26 (Chiriac-Putz-Chiriac, 2005).

The deficit of electrons is obvious for the elements of groups IA-IIIA, of transitions and of the internal transition elements (with 4f and 5f orbitals). For the metals of the p-block (Sn, Pb, Sb, Bi,

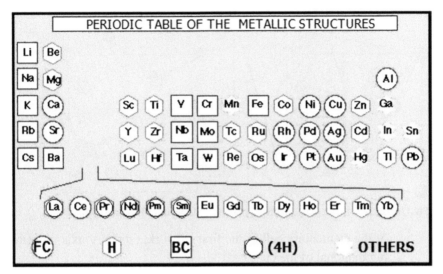

FIGURE 4.26   The metals of the periodic system with the specific types of packing: (face) CC-cubic compact (or FCC), HC-hexagonal compact, body centered cubic (BCC), 4H-hexagonal alternate; after Heyes (1999).

At) the deficit of the electrons results from the establishment of the shell of the valence layer.

(b) Atoms with equal number of external electrons with external atomic orbitals of close energy.

This is the case of the light elements of the group IVA: C, Si, Ge.

For these atoms, the accomplishment of the electronic configuration of maximum stability (octet) in the conditions of minimum inter-electronic repulsion, the hybridization state sp^3 is imposed by the formation of four covalent σ-bonds oriented on the hybrids lobs, i.e., tetrahedral, so that each atom has CN=4. The crystalline network formed by this kind of identical atoms is a covalent network having in the structural nodes/points of the lattice the atomic cores with the directed external electronic clouds. The preeminent example is the diamond network, Figure 4.27 (Chiriac-Putz-Chiriac, 2005).

The elementary cell of the diamond network is a cubic cell with centered faces where half of the tetrahedral interstices are alternatively occupied.

It can be considered that the diamond network results by the overlap of two compact cubic networks, so that the corner of the elementary cell of the second network to be placed at 1/3 of the height

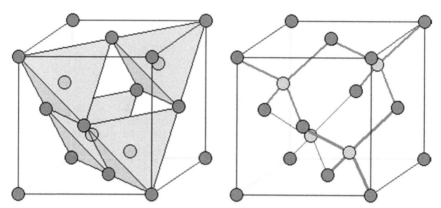

FIGURE 4.27   Elementary cell from diamond lattice; after Heyes (1999).

of the elementary cell of the first network, on the A^3 axis of it (the big diagonal of the cube).

(c) The atoms with excess of electrons: are those in which the number of electrons from the last shell (nsnp) is higher than the number of atomic orbitals (AO) of these shells: $4 < nr. e^- \leq 7$.

The unary substances of these elements have in the nodes of the network small molecules. The number of the neighboring atoms in a such stable molecule is determined by the Humme-Rothery rule = the crystallochemical rule = 8-N rule, where N is the number of the group where the element is placed.

Therefore, the elements of the group VIIA have CN=8–7=1 and the molecule is diatomic (A_2); the elements of the group VIA have CN=8–6=2 and thus the molecule should be infinite (A_∞)

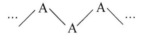

or cyclic (e.g., A_8).

For the group VA, an atom should have CN=3. Three neighbors equivalent in space can be provided only by a tetrahedron with a vertex occupied by the reference one.

The elements of the group IVA should have CN=4 which impose that the polyhedron of coordination to be the tetrahedron with the center occupied by the (A_4) reference.

This arrangement was deduced in other way for no. e^-/no. AO ratio as previously exposed; the unary substance is, in this case, the infinite molecule, and thus having another reason to consider the diamond network as being covalent in nature.

Instead, the boron element of the IIIA group, although is an element with deficit of electrons should form a metallic network, by following the crystallochemical rule and forming a finite structure where CN=8–3=5.

The cause is the kino (changing) – symmetric feature of the orbital where its distinguishing (valence) electron is placed. The finite B_{12} structure is a polyhedron with 18 icosahedron-faces, see Figure 4.28 (Chiriac-Putz-Chiriac, 2005).

The unary substances of the elements of this category form the lattice that have molecular nodes while the van der Waals interactions are exercised between them.

The lattices of this type are molecular networks on which the symmetry essentially depends on the symmetry of the nodal particle.

(d) A last possible situation for the electronic configuration of an atom is that when all the external sub-shells are completely occupied by electrons.

The atom is electronically saturated having the octet (Ne, Ar, Kr, Xe, Rn) or doublet (He) configuration.

The crystallochemical rule asserts that such an atom in a bonding state has CN=8–8=0, i.e., the atom can not be bounded to any other identical atom.

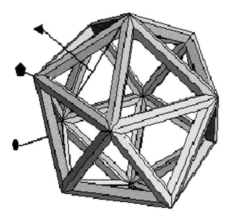
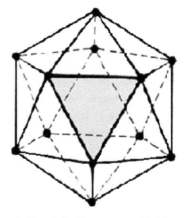

FIGURE 4.28   The Icosahedron: the symmetry axes (left) and the B_{12} structure (right).

The atoms of the rare gases form molecular (saturated) structure with the mono-atomic molecule; Only van der Waals interactions occur between the identical atoms, un-oriented, so that their ordering in the network respects the maximum compactness rule. The resulted lattices can be cubically or hexagonally compacted. However, since atoms are placed in the nodes of the network, the rare gases lattice can be called as atomic network or particular case of a molecular network.

4.3.4 CRYSTALLINE COMPOUNDS WITH AB STOICHIOMETRY

Two different atoms, A and B, have different radii, different electronic configurations and different electronegativity. All these three properties "have their say" regarding the ordering of the two partners in the solid state; the present discussion follows (Chiriac-Putz-Chiriac, 2005).

The electronic configurations of partners are those that impose the mixing ratio x/y; it should provide greater stability for each produced by a reaction species as well as the electroneutrality of the system.

For the main group elements, the (AB) compounds with x: y = 1 stoichiometry correspond to the isoelectronic horizontal series, with the interacting elements symmetrically placed in relation to group IVA, see Table 4.3 (Chiriac-Putz-Chiriac, 2005).

In an isoelectronic horizontal series, one is remarking that with the increasing ΔZ of the elements, increases also the atoms' electronegativity difference ($\Delta \chi$) so emphasizing on the ionic character of the interaction.

This increase is manifested also in the increasing tendency of compaction of the arrangement, i.e., the increasing of the coordination number of partners. Starting from a compound of a horizontal isoelectronic series, vertical isoelectronic series are obtained by substitution of one of

TABLE 4.3 Crystalline Substances of AB Type

Group A Period	IV	III+V	II+VI	II+VII
2	C (diamond)	BN	BeO	LiF
3	Si (diamond)	AlP	MgS	NaCl

TABLE 4.4   Isoelectronic Series From AB Type Crystalline Substances

NaCl Series		AlP Series	
LiCl	NaF	BP	AlN
KCl ↓	NaBr ↓	Interstice ↓	AlAs ↓
RbCl	NaI	InP	AlSb
CsCl		TlP	AlBi

its element with one of its analogues form the same group; such are those of Table 4.4 (Chiriac-Putz-Chiriac, 2005).

In the vertical isoelectronic series the replacement of the less electronegative element (positively polarized) with high polarizing capacity, with a heavier one ($\Delta Z > 0$) has two consequences: increasing the electronegativity difference between partners and the increasing of the cation's radius.

The two phenomena are acting in the same way toward the degree of ordering in compactness and result in the increase of the number of coordination of the partners. Indeed, the CsCl, where there is the highest percentage of ionic character of the bond, $CN = 8$ while all other halogens of the isoelectronic series have $CN = 6$.

Replacing the more electronegative element in the vertical isoelectronic series leads to the decrease the difference of electronegativity between partners. This means that in some cases the increase of the covalent bonding character (e.g., for NaCl series) or of its increase in metallic character (the series of AlP). In the last case the "metallization" of bonding is manifested by narrowing the forbidden bands of the solid structure.

To describe the structure of a heteroatomic network the following steps are required:

- To specify the reference particle (the particle that generates the network of whom the elementary cell describes the structure of the ensemble, of the basis' particle).

  In the stoichiometric AB compounds, the reference particle is indifferent and for this reason the choice of "anion" is preferred as reference.
- To establish the "motif" of the lattice, i.e., the position of the non-reference particles in the elementary cell.

  This is imposed by the coordination constant law which, for the stoichiometric AB compounds, it leads to the simple solution:

$$x \cdot N \cdot C_A = y \cdot N \cdot C_B \Rightarrow N \cdot C_A = N \cdot C_B \qquad (4.12)$$

- To establish the number of the formula unit $(AB)_n$ belonging to the elementary cell.

The "multiplicity" of the formula unit should be a natural number: $n=1, 2, \ldots$

The structural formula should illustrate all these information in the resumed way:

$$\text{Crystallographic system } \left(A_x^{[NC_A]}B_y^{[NC_B]}\right)_n \qquad (4.13)$$

The most important structures of AB stoichiometry, with coordinative network, will be further exposed.

4.3.4.1   CsCl Crystalline Structure

Such an arrangement, Figure 4.29, is adopted by the compounds where the interaction between elements has a predominantly ionic character, with the ratio:

$$\frac{r_{CATION}}{r_{ANION}} > 0.7 \qquad (4.14)$$

The ordering is performed by overlapping two simple cubic networks, of each species, so that the corner of the elementary cell of a network to reach the center of the elementary cell of the other one.

Therefore, the elementary cell of the structure is the primitive cube $(0, 0, 0)$ of which center is occupied by the atom of the other species $(\frac{1}{2}, \frac{1}{2}, \frac{1}{2})$.

Taking into account the participation quota of the particle for the elementary cell, we obtain:

- for the reference particle ● Cl^-: $8\,(1/8) = 1\ Cl^-$
- for the other particle ○ Cs^+: $1\,(1) = 1\ Cs^+$

so that the multiplicity has the value $n=1$.

The internal particle position obviously shows its location in a interstice of cubic symmetry so that its polyedron of coordination is the cube,

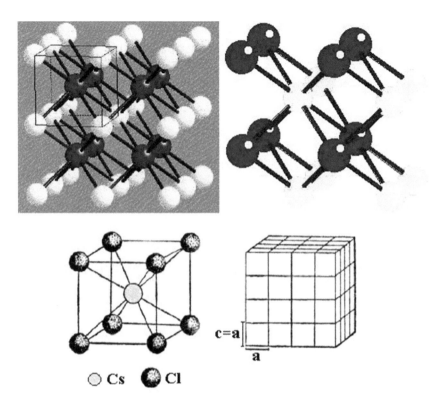

○ Cs ◉ Cl

FIGURE 4.29 The CsCl elementary cell in various shapes of representation; after (Apostolescu 1982; Heyes, 1999).

in our example, $CN_{Cs^+} = 8$. According with the coordination conservation law (4.12) we obtain $CN_{Cl^-} = 8$. Indeed, each corner of the cube has as proximate neighbors the 8 centers of the common (shared) cubes that are disposed as the vertices of a cube.

Thus, the formula of the CsCl structure is: *cubic simple* (*primitive*) $\left(Cs^{[8]}Cl^{[8]}\right)_1$. Some binary substances that adopt the CsCl structure and their network parameter are exposed in Table 4.5.

4.3.4.2 NaCl Crystalline Structure

This type of structure is adopted by the compounds where the bonding has a considerable share of ionicity, with the ratio:

TABLE 4.5   Cristalline Substances of CsCl Type; after (Chiriac-Putz-Chiriac, 2005).

Structure of CsCl type	Cell' Parameter a[Å]
CsBr	4.30
CsCl	4.12
CsI	4.57
TlBr	3.98
TlCl	3.86
TlI	4.21

$$0.7 > \frac{r_{CATION}}{r_{ANION}} > 0.4 \qquad (4.15)$$

There can be asserted that the network is formed by the overlap of two cubic compact networks so that, by translation, the face of the elementary cell of a network to reach the middle of the edge of the elementary cell of the other network.

Therefore, the elementary cell of the lattice is the cube with the centered faces of whose edges and center occupied by non-reference particles.

By the participation quota to the particular positions of the cube, we obtain:

- for the reference particle • Cl^-: 8 (1/8) + 6 (1/2) = 4 Cl^-
- for the non-reference particle • Na^+: 12 (1/4) + 1 (1) = 4 Na^+

so that the multiplicity is n = 4.

The position occupied by the non-reference particles (center and edges) shows, according to the previous demonstration, that they occupy all the octahedral interstices of the compact cubic network constituted by the bases. Therefore, $CN_{Na^+} = 6$, as shown also in Figure 4.30 (c). The constancy of the coordination imposes the same number of coordination also for the reference particle (base), so $CN_{Cl^-} = 6$. Thus, the structural formula of NaCl type is:

Face centered cubic $(Na^{[6]}Cl^{[6]})_4$

Some substances with structure of NaCl type and their lattice parameter are listed in Table 4.6.

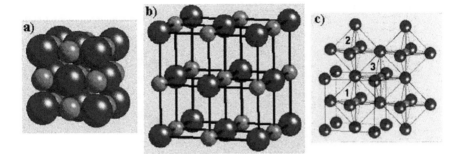

FIGURE 4.30 The NaCl packing structure ($\bullet$ Cl⁻; • Na⁺); (a) the ordering of the atoms in NaCl arrangement; (b) elementary cell of the NaCl lattice; (c) coordination polyhedron of Na⁺ sub-lattice; after Heyes (1999).

TABLE 4.6 Crystalline Substances of NaCl Type; after (Chiriac-Putz-Chiriac, 2005)

Structure of NaCl type	Cell's Parameter a[Å]
AgBr	5.78
BaO	5.53
CaTe	6.35
CoO	4.26
CsF	6.02
CsH	6.39
FeO	4.29
KCl	6.29
KF	5.34
KH	5.71
LiCl	5.14
RbCl	6.55

4.3.4.3 ZnS (Sphalerite/Blende) Cubic Structure Type

Sphalerite type structure, Figure 4.31, is adopted by the compounds where the covalent character of the bonding prevail and has the radii ratio:

$$0.4 > \frac{r_{CATION}}{r_{ANION}} > 0.2 \qquad (4.16)$$

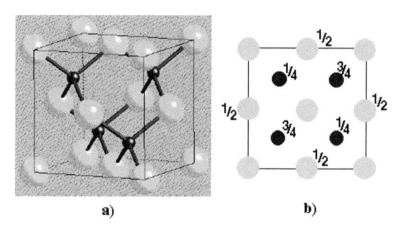

FIGURE 4.31 The Structure of ZnS type (● S^{2-}; ● Zn^{2+}); (a) elementary cell; (b) basal projections of particles from the elementary cell; after Heyes (1999).

The network results by the overlap of two cubic networks with centered faces, by gliding along the large diagonal of the elementary cell of one of them, so that the corner of the other one to be place at ¼ of its length. There a result that the elementary cell of the structure is the cube with the centered faces where half of the tetrahedral interstices are alternatively occupied. The content of the cell and the multiplicity are obtained from the participation quota of the particles:

- the reference particle (O) S^{2-}: 8 (1/8) + 6 (1/2) = 4 S^{2-}
- the non-reference particle (•) Zn^{2+}: 4 (1) = 4 Zn^{2+}

thus, the multiplicity of the structure having the value n=4.

Form the structure formation description appears that the ions Zn^{2+} occupy the tetrahedral interstices, then $CN_{Zn^{2+}} = 4$. Compulsory, also the ions S^{2-} should be tetrahedrally surrounded, then $CN_{S^{2-}} = 4$.

Indeed, the corner of the cube, commonly shared to eight neighbor cubes where the tetrahedral interstices are alternatively occupied, is tetrahedrally surrounded. Thus, the formula of the sphalerite type structure looks like:

$$\text{cubic with the faces centered } (Zn^{[4]}S^{[4]})_4$$

Some compounds which adopt the sphalerite structure are present in Table 4.7.

The stoichiometric compounds AB cannot form structures where NC ≤ 3.

TABLE 4.7   Crystalline Substances of ZnS Type; after (Chiriac-Putz-Chiriac, 2005)

Cubic Structures of ZnS Type	Cell's Parameter a[Å]
AgI	6.48
AlAs	5.63
AlP	5.43
AlSb	6.11
BeS	4.87
BeSe	5.14
BeTe	5.62
CSi	4.37
CdS	5.82
CdSe	6.05
CdTe	6.47
CuBr	5.69
CuCl	5.42
CuF	4.26
CuI	6.06
HgS	5.85
HgSe	6.08
HgTe	6.45
ZnS	5.43
ZnSe	5.67
ZnTe	6.10

4.3.5   TYPE STRUCTURES OF THE AB$_2$ STOICHIOMETRIC COMPOUNDS

The present modification of the stoichiometric ratio, respecting the previous AB case, has as consequence the action of the coordination constant law, such as; the present discussion follows (Chiriac-Putz-Chiriac, 2005):

$$N \cdot C_A = 2N \cdot C_B \qquad (4.17)$$

Due to this fact, the choice of the reference particle is not indifferent anymore; it is chosen, despite the fact that the anions have higher radii, so that the description of the structure to be as simplest as possible. Taking into account the possible coordination polyhedra in the crystalline structures: the line (CN=2), the triangle (CN=3), the square or the tetrahedron (CN=4), octahedron (CN=6), cube (CN=8) and cubo-octahedron (CN=12) in coordinative networks, along the ratio constraint for the numbers of coordination, but also taking into account that the atomic radii ratio is different from unity, all these, are assured only on three situations: 4:8; 3:6, and respectively 2:4. Indeed, there are only three networks of type AB_2 as in following described.

4.3.5.1 CaF_2 (Fluorite) Type Structure

The fluorite structure, Figure 4.32, is assumed by the compounds with still prevalent ionic character of the bonding.

There can be considered that the present network may result from an overlap of a compact cubic network of the reference ions (the base is compulsory represented by the Ca^{2+} ions) with a simple cubic network of the counter-ions (F^-) through the gliding of the last one along the A^3 axis so that the corner of its elementary cell to be placed to a quarter of the length of the large diagonal of the elementary cell of the base. This means that in the cubic cell with centered faces of the mandatory reference (Ca^{2+}) lattice, all the tetrahedral interstices are occupied ion anions F^-. Thus the elementary cell contains:

- the basic particles • Ca^{2+}: 8 (1/8) + 6 (1/2) = 4 Ca^{2+}
- the counter-ions v F^- 8 (1) = 8 F^-

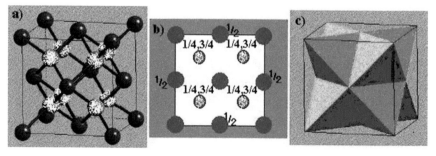

FIGURE 4.32   Structure of type CaF_2 (• Ca^{2+}; ⊛ F^-); (a) elementary cell of fluorite type; (b) atomic positions projected on the base of the elementary cell; (c) union of the coordination polyhedra of Ca^{2+}; after Heyes (1999).

and the multiplicity is n=4. From the description of the position of the counter-ions one yields $CN_{F^-} = 4$, while for the reference ions one has $CN_{Ca^{2+}} = 8$ (the eight neighbors are provided by the eight tetrahedral interstices of each shared-common cube to their corners). Therefore, the structural formula of the CaF_2 type is:

$$\text{face centered cubic } \left(Ca^{[8]}F_2^{[4]}\right)_4$$

Very important is the fact that a similar arrangement is obtained if the positions of the ions are changed so that the reference ion becomes the natural one: an anion.

In this case the structure of anti-fluorite is obtained, as characteristic to the oxides of the alkaline elements. For example: face centered cubic $\left(Na_2^{[4]}O^{[8]}\right)_4$; the compounds with fluorite and anti-fluorite structure are in Table 4.8 exposed.

4.3.5.2   TiO$_2$ (Rutile) Type Structure

The rutile structure, Figure 4.33, is assumed by the compounds where the ionic character of the bonding is comparable with its covalent character, a fact that is manifested in the decrease of the coordination number of the anion.

The elementary structure of this type has as references the Ti^{4+} ions and a tetragonal body centered symmetry. To the cell belong 8 (1/8) +1 (1) = 2 ions of titanium.

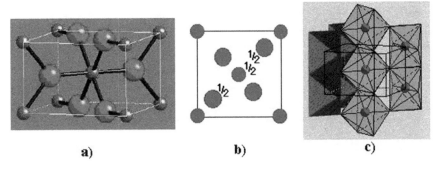

a)   b)   c)

FIGURE 4.33   The Structure of TiO$_2$ type ($\bullet$ Ti^{4+}; $\bullet$ O^{2-}); (a) elementary cell of TiO$_2$ type; (b) basal projections of the atoms of the elementary cell; (c) the biding polyhedra of the Ti^{4+}coordination; after Heyes (1999).

TABLE 4.8   Fluorite (CaF_2) and Anti-Fluorite (Na_2O) Types of Crystalline Substances; after (Chiriac-Putz-Chiriac, 2005)

Structure of (anti) Fluorite Type	Cell's Parameter a[Å]
$AuAl_2$	6.01
BaF_2	6.20
Be_2C	4.34
CaF_2	5.46
CdF_2	5.41
CeO_2	5.41
CuF_2	5.42
HgF_2	5.55
K_2O	6.45
K_2S	7.40
K_2Se	7.695
K_2Te	8.17
Li_2O	4.63
Li_2S	5.72
Li_2Se	6.01
Li_2Te	6.51
Mg_2Pb	6.85
Mg_2Si	6.40
Mg_2Sn	6.78
Na_2O	5.56
Na_2Se	6.82
Na_2Te	7.32
PbF_2	5.94
Rb_2S	7.66
$SrCl_2$	6.99
SrF_2	5.79
ThO_2	5.58
ZrO_2	5.08

The anions are placed on the symmetric bases toward their centers, as two on a diagonal of the face, and the other two being internal and symmetrically placed toward the center of the prism in its other vertical diagonal plan, see Figure 4.33(b).

The elementary cell contains 4 (1/2) + 2 (1) = 4 ions of O^{2-} so that the multiplicity has the value as n=2.

Further, the anion coordination number can be easily revealed as following the occupancy of a diagonal plane of the elementary prism.

The O^{2-} anion is placed in a triangle composed by the particles of an edge together with the inner particle of the elementary prism.

Therefore $CN_{O^{2-}} = 3$ and we obtain for Ti^{4+} the coordination number

$$CN_{Ti^{4+}} = 2 \times CN_{O^{2-}} = 6$$

The coordination polyhedron of titanium is a deformed octahedron. The combination of these polyhedra, corresponding to this ordering, is shown in Figure 4.33(c).

Thus, the structural form of the arrangement of rutile type will be:

$$\text{Tetragonal prism internal centered} \left(Ti^{[6]}O_2^{[3]}\right)_2$$

The compounds which crystallize as rutile networks are shown in Table 4.9.

4.3.5.3   Cu$_2$O (Cuprite) Type Structure

For the cuprite type arrangement there are two possibilities in choosing of the elementary cell, Figure 4.34. Accordingly, the compact cubic network (CCF) of Cu^+ ions can be considered where only two octahedral interstices, diagonally opposite, are coupled with O^{2-} anions, see Figure 4.34(a).

From a different perspective, we can consider a body centered cubic network constituted by O^{2-} anions (the more natural choice) where the cations are located on large diagonals of the cube, in a symmetrically alternatively up and down occupation, see Figure 4.34(b).

In this framework of the elementary cell the linear coordination of the cation: $CN_{Cu^+} = 2$ is obvious along the tetrahedral coordination of the anion $CN_{O^{2-}} = 4$ (the eight shared-common cubes to their corners, alternatively

TABLE 4.9   Crystalline Substances of Rutile (TiO_2) Type; after (Chiriac-Putz-Chiriac, 2005)

Structures of TiO2 (Rutile) Type	Cell's Parameter a[Å]	Cell's Parameter c[Å]	Ratio c/a
CoF_2	4.70	3.20	0.68
CrO_2	4.72	2.88	0.61
FeF_2	4.68	3.28	0.70
IrO_2	4.50	3.15	0.70
MgF_2	4.67	3.08	0.66
MnF_2	4.88	3.32	0.68
MnO_2	4.45	2.89	0.65
MoO_2	4.87	2.78	0.57
NiF_2	4.72	3.12	0.66
OsO_2	4.52	3.26	0.70
PbO_2	4.94	3.36	0.68
PdF_2	4.94	3.36	0.68
RuO_2	4.52	3.12	0.69
SnO_2	4.73	3.17	0.67
TeO_2	4.80	3.79	0.79
TiO_2	4.60	2.94	0.64
VO_2	4.55	2.87	0.63
WO_2	4.87	2.78	0.57
ZnF_2	4.73	3.12	0.66

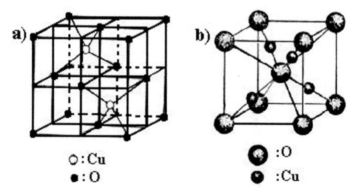

a)   O : Cu
       ● : O

b)   ◉ : O
       ◐ : Cu

FIGURE 4.34   The cuprite two types of elementary cell arrangements; after Apostolescu (1982).

offer each one a diagonal cation). Therefore, the cuprite structure formula can be written as:

$$CI \left(Cu_2^{[2]}O^{[4]}\right)_2$$

Some of the substances which crystallize in this type of crystalline structure are in Table 4.10 listed.

4.3.5.4 SiO$_2$ (β-Cristo-balite) Structural Type

The β-cristo-balite network, Figure 4.35, can be described by analogy with that of diamond, Figure 4.27. The elementary cell is the cube with centered faced where half of the tetrahedral interstices are alternatively occupied with the reference atoms of the silicon. Therefore, a number of reference particle belongs to the elementary cell as follows: Si^{4+}: 8 (1/8) + 6 (1/2) + 4 (1) = 8 Si^{4+}.

The counter-ions of oxygen are located inside the elementary cell tetrahedrally coordinating each internal atom of silicium.

The SiO$_4$ tetrahedra through its O-vertices coordinate each with four atom of silicium: one of the vertices and three from the faces of the elementary cell. The atoms of oxygen are collinearly located, at the half of the distance between two atoms of silicium. Therefore, the elementary cell contains a number of counter-ions O^{2-}: 4 (4) = 16 O^{2-}.

The multiplicity of the structure has the $n = 8$ value and the structural formula of β-cristo-balite looks like:

$$CCF \left(Si^{[4]}O_2^{[2]}\right)_8$$

TABLE 4.10   Crystalline Substances of Cuprite (Cu$_2$O) Type; after (Chiriac-Putz-Chiriac, 2005)

Structures of Cu2O (Cuprite) Type	Cell's Parameter a[Å]
Ag$_2$O	4.73
Ag$_2$S (>180°)	4.91
Pb$_2$O	5.39
Cu$_2$O	4.26

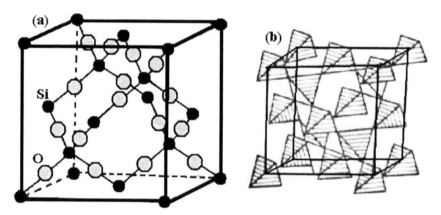

FIGURE 4.35 β-cristo-balite arrangement; (a) the elementary cell; (b) the Si^{4+} coordination polyhedra for the elementary cell; after Apostolescu (1982).

4.3.6 CRYSTALLINE STRUCTURES WITH AB_3 STOICHIOMETRY

Knowing that in coordinative structures the coordination number can not be less than 2, the application of the constant coordination law give us for the coordination numbers of the two species of a compound with AB_3 stoichiometry, the following pairs of values; the present discussion follows (Chiriac-Putz-Chiriac, 2005):

(a) $CN_B=2$; $CN_A=6$
(b) $CN_B=3$; $CN_A=9$
(c) $CN_B=4$; $CN_A=12$.

Because the coordination polyhedra in solid can be just CN=2, 3, 4, 6, 8 and 12 the solution (b) is excluded. Then, because CN=12 is achieved only for some particles of equal radii the situation (c) is also excluded.

Thus only the solution (a) possible remains. However, such an ordering of particles, where the anion has the minimum number of coordination, corresponds to the compounds with a high degree of covalence in bonding and it is illustrated by the network of type ReO_3 in next described.

4.3.6.1 ReO_3 Type Structure

The elementary cell of rhenium trioxide has as reference the ions of Re^{6+} placed in a simple cube, Figure 4.36.

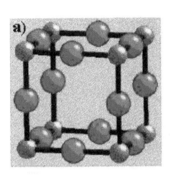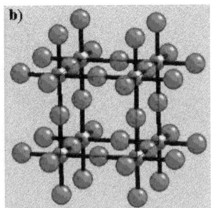

FIGURE 4.36   The structure of ReO_3 type ($\bullet$ Re^{6+}; $\bullet$ O^{2-}); (a) the elementary cell; (b) the mutual coordination of the particles from the rhenium trioxide network; after Heyes (1999).

To the elementary cell belongs a single ion of the base:

- Base particles: $\bullet$ Re^{6+}: 8 (1/8) = 1 Re^{6+}
  While the counter-ions, O^{2-}, are placed on the edges of the cube, in the middle of its side. To the elementary cell so belongs:
- Counter-ions: $\bullet$ O^{2-}: 12 (1/4) = 3 O^{2-}

Multiplicity of the structure is then $n=1$.

Coordination numbers, revealed by representation of Figure 4.36b are: CN=2 for oxygen (linear coordination) and CN=6 for Re^{6+} (octahedral coordination).

The structure formula for rhenium trioxide is:

$$\text{Cubic simple } \left(Re^{[6]} O_3^{[2]}\right)_1$$

The increase of the number of atoms in the formula unit allows another ordering of the particles of AB_3 stoichiometry compounds. Such instance is possible due to the fact that the atom of B-type can have different coordination. This arrangement is illustrated by the $BiLi_3$ structure.

4.3.6.2 $BiLi_3$ Type Crystalline Structure

Elementary cell of the $BiLi_3$ network, Figure 4.37, is composed by Bi atoms and it is a cube with centered faces, while Lithium atoms occupy all the tetrahedral and octahedral interstices of the cell.

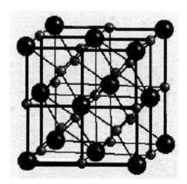

FIGURE 4.37   The BiLi$_3$ type lattice structure ($\bullet$ Bi; $\bullet$ Li); after Heyes (1999).

Therefore, an elementary cell contains a number of reference atoms:

- the base particles $\bullet$ Bi: 8 (1/8) + 6 (1/2) = 4 Bi
  and a number of counter-ions:
- the non-reference particles $\bullet$ Li: 12 (1/4) + 1 (1) + 8 (1) = 12 Li.

Multiplicity has the value $n = 4$.

Depending on various location, Lithium atoms have different coordinative valences: those placed in (8) tetrahedral interstices have CN=4, and those placed in (4) octahedral interstices have CN=6. Yet, the constancy coordination law imposes that the Bismuth atoms to have $CN_{Bi} = (8.4 + 4.6)/4 = 14$. Indeed, each atom of Bi is surrounded by eight tetrahedral Lithium atoms at a distance $a\sqrt{3}/4 \cong 0.41$ and by 6 octahedral Lithium atoms at a distance $a/2$ (0.5 a). Coordination polyhedron of the Bismuth atom is the pyramidal cube.

The structure formula of the Li$_3$Bi network will therefore look like:

$$FCC \left(Li^{[6]} Li_2^{[4]} Bi^{[14]} \right)_4$$

4.3.7   POLYATOMIC COMPLEX IONS' STRUCTURES

From the E and H types of compounds in Table 4.2, the most significant ones will be here detailed, i.e., the perovskit and spinel types; the present discussion follows (Chiriac-Putz-Chiriac, 2005).

4.3.7.1   The Perovskit Type

The Perovskit structure had been discovered in the 19th century by the Russian mineralogist C. L. A. Perovski, wherefrom its name, originally under the form of the calcium titanate, $CaTiO_3$, see Figure 4.38.

The ideal structure is a cubic primitive with an effective unit of $CaTiO_3$ per unit cell, with the motifs: Ti at (0, 0, 0); Ca la ($^1/_2$, $^1/_2$, $^1/_2$) and 3O at ($^1/_2$, 0, 0), (0, $^1/_2$, 0), (0, 0, $^1/_2$) .

The Ca atoms are 12-coordinated by the O-atoms in cuboctahedra CaO_{12} that share their faces, while the Ti atoms are 6-coordinated by the O-atoms in octahedra TiO_6 that share only the vertices, see Figure 4.38. In turn, the O-atoms are coordinated in distorted octahedra with 4 atoms of Ca and 2 atoms of Ti. Moreover, the perovskite structure can be derived from that one of the rhenium oxide ReO_3, from Figure 4.36, by completing the central location in the unit cell by the Ti atom. Compounds as (Mg, Fe) SiO_3

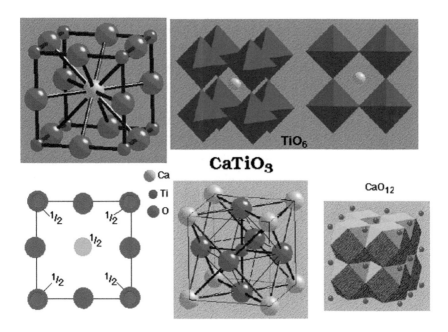

FIGURE 4.38 The geometry, coordinates, unit cell and the coordination polyhedra representations, respectively, for the calcium titanite as perovskite type; after Heyes (1999).

and $CaSiO_3$ are transformed in the perovskite structure at very high pressures, while being major components of the mantle of the Earth.

Over 50 metallic ions can be combined in perovskite structure, the only condition being that the positive and negative charges to be equal in number, such as $M^{II}M^{IV}O_3$ or $M^{I}M^{II}X_3$, so being characterized by the big radius of the cation, see Table 4.11.

The most popular perovskites are the oxides (in majority) followed by fluorides, sulphides, halides and some selenides.

A special property of perovskites is the supercondictibility, the appearance of the electronic conduction (electronic current) at superior temperatures (also called critical temperatures T_c) respecting the temperature of (absolute zero) 0K.

In Figure 4.39, such an superconductor (oxide) is presenting as having the La_2CuO_4 {or K_2NiF_4} structure, also known as the perovskite of type A:B due to the fact that it can not be made by an ABAB... arrangement of perovskite cells.

The first superconductor oxide at a high critical temperature (~ 40 K), was discovered in 1986, and was that one based on the La_2CuO_4 structure, but with a doped $La_{2-x}Sr_xCuO_4$ form, received the Nobel Prize for Physics on 1997 awarded to Georg J. Bednorz and Alexander K. Müller *"for their important break-through in the discovery of superconductivity in ceramic materials"*.

Instead, the first superconductor material at the temperature of the liquid nitrogen ($T_c > 77$ K) is the structure $YBa_2Cu_3O_7$ (also called 1:2:3 superconductor) also derived from three cells of perovskites by removing the oxygen between the adjacent cells, as shown in Figure 4.40.

However, worth to specify the fact that, by increasing the concentration of Cu also the critical temperature T_c increases as it is the case for the

TABLE 4.11 The Compounds of Perovskite Type

A^{+n}	B^{+m}	X_3	n:m
Na	W	O_3	1:5
Ca	Sn	O_3	2:4
Y	Al	O_3	3:3
Cs	Cd	Br_3	1:2
$K_xLa_{(2/3-x/3)}$	Ti	O_3	2:4

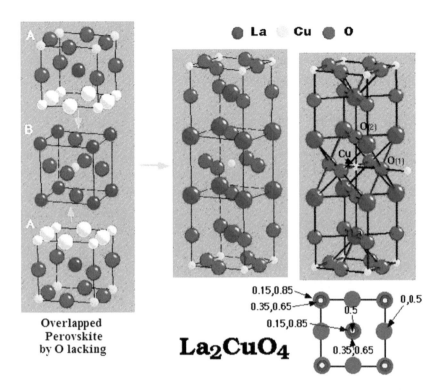

FIGURE 4.39   The Geometry, coordinates and the unit cell of the perovskite type La$_2$CuO$_4$ (A: B) superconductor A:B; after Heyes (1999).

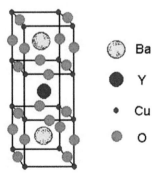

FIGURE 4.40   The unit cell of perovskite type (1:2:3) YBa$_2$Cu$_3$O$_7$ superconductor.

compounds: YBa$_2$Cu$_3$O$_{6+x}$, Bi$_2$Sr$_{3-x}$Y$_x$Cu$_2$O$_8$ and La$_{2-x}$A$_x$CuO$_4$, with A=Ba, Sr, Ca or Na.

This behavior is since the substitution of the ions with small radii (e.g., Ba, Sr, Ca, Na) is causing a correspondent decreasing in the Cu–O distance and, in consequence, leads to the increase of T_c. This effect can be further amplified by the successively increase of the pressure.

4.3.7.2 The Spinel Type

Regarding the spinel type $-\,^3_\infty Mg^{[4]}Al_2^{[6]}O_4^{[1+3]}$ c. – this structure is found to the compounds of $M^{II}M_2^{III}O_4$ type or, generally, to AB_2O_4 crystals that fulfill the condition according which the positive sum of the ions A and B to be 8 or 4.

The spinel AB_2O_4 structure, being a simple or primitive compact cubic PCC structure for the O^{2-} the ions, has A^{2+} and B^{3+} cations as occupying 1/8 of tetrahedral and octahedral generated interstices. So, the structure is characterized by the compact cubical arrangement of the oxygen ions, the cations being placed in tetrahedral interstices (bivalent ions) and octahedral (trivalent ions), as shown in Figure 4.41.

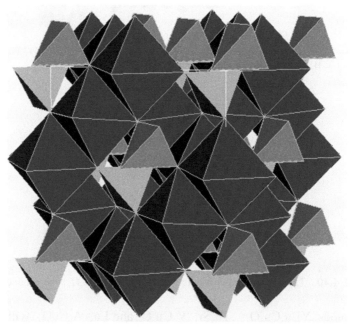

FIGURE 4.41 The Spinel Network; after The Chemistry of Silicates (2003).

The above description corresponds to the "normal" type of spinel where the octahedral interstices are occupied in proportion of 1/2 with the more polarized B^{3+} cations, while the tetrahedral interstices are filled only in 1/8 proportion by the A^{2+} cations with lower polarization ability.

The representation of the direct spinel with the oxygen in three (blue) alternate layers is drawn in the Figure 4.42, while the octahedral interstices are by the green and yellow colors symbolized, with the tetrahedral one depicted in purple color, respectively.

When, instead, the tetrahedral interstices are occupied by the B^{3+} cations, and the octahedral ones are occupied by a cationic A^{2+} and B^{3+} combination the resulting structure is said of "reverse" spinel.

Thus, a "reverse" spinel structure appear with octahedral interstices occupied by both the cation species, in equal proportions, and statistically

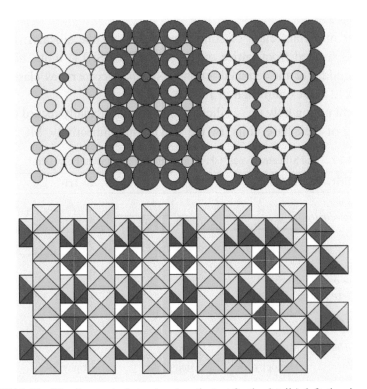

FIGURE 4.42 The direct spinel structure (see the text for the details); left: the view along the 4th order axes of symmetry; right: the view of packed structure; after Heyes (1999).

assigned on 1/2 of these interstices, while being the rest of the ions with high ability of polarization in the tetrahedral interstices.

In the "normal" structure of spinel one has the crystalline systems: $MgAl_2O_4$ (spinel), $FeAl_2O_4$ (*hercynite*), $FeCr_2O_4$ (*cromite*), with the "reverse" structure of spinel being represented by: $MgFe_2O_4$, $TiMg_2O_4$, $MgGa_2O_4$, etc.

Actually, the most spinels are a combination between the direct and inverse extreme structures.

4.3.8   SILICATES CRYSTALLINE STRUCTURES

Further the significant silicates compounds of Table 4.2. will be analyzed; the present discussion follows (Chiriac-Putz-Chiriac, 2005).

4.3.8.1   Silicates' Classification

Another important class of solid compounds are the silicates: they are minerals formed from lattice of the tetrahedral $[SiO_4]^{4-}$ units, having Si in the middle of the tetrahedron and oxygen in corners, as classified in Table 4.12 and exemplified in Table 4.13.

Worth noting how these silicates' compounds actually record covalent bonding of their base units, by the maximum connectivity, MC=8-N, of

TABLE 4.12   The Silicates Types' Classification by the Adjacent Polyhedra*

Multiplicity→	1: Mono-	2: Di-	3: Tri-	...
Maximum Interconnectivity ↓				
0: Oligo- or Ortho-(terminal)	Nezo-	Soro-		
1: Ciclo- (annular)	Monociclo-	Diciclo-	Triciclo-	...
2: Poly-(Ino-)	Monopoly-	Dipoly-	Tripoly-	...
3: Phyllo-	Monophyllo-	Diphyllo-	Triphyllo-	...
4: Tecto-	Tecto-			

*After The Chemistry of Silicates (2003).

TABLE 4.13   Examples of Silicate*

NEZOSILICATES:	SOROSILICATES:	CYCLOSILICATES:
$[SiO_4]^{4-}$ Olivine (Mg, Fe)$_2[SiO_4]$	$[Si_2O_7]^{6-}$ Hemimorphite: $Zn_4(OH)_2[Si_2O_7].H_2O$	$[Si_pO_{3p}]^{2p-}$ Beryllium (p=6): $[Si_6O_{18}]^{12-}$

INOSILICATES (on a single row): $[SiO_3]^{2-}$ Pyroxene (Mg,Fe)SiO_3	INOSILICATES (on two rows): $[Si_4O_{11}]^{6-}$ Tremolite $Ca_2Mg_5[Si_8O_{22}](OH)_2$	PHYLLOSILICATES $[Si_2O_5]^{2-}$ Okenite Ca$[Si_2O_5]$2 H$_2$O

  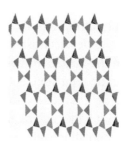

TECTOSILICATES $[SiO_2]_{2p}$ Cristo-balite SiO_2	TECTOSILICATES $[SiO_2]_{2p}$ Quartz Si_2O_4	ZEOLITES ZSM-5: $M^{n+}_{x/n}[(AlO_2)_x(SiO_2)_y]\times mH_2O$ Thomsonite NaCa$_2[Al_5Si_5O_{20}]$ $\times$ 6 H$_2$O

*After The Chemistry of Silicates (2003).

oxygen anions, with the total of N number of the participant valence electrons, as a generalization of the octet rule of Lewis for valence electrons (see Introduction to Chapter 3 of the present Volume).

The structural unit of all silicates is the SiO_4 group, where Si^{4+} ion has ionic radius 0.39Å, while the oxygen reaches 1.32Å in ionic radius. Corresponding to the ratio $0.39/1.32 = 0.29$, the ensemble SiO_4 is assigned to a tetrahedral structure with Si in the middle and the oxygen ions placed in the vertices of the tetrahedron.

The distance Si-O inside the fundamental tetrahedron is 1.6 Å, and the distance O–O=2.6Å.

The mesodesmic character of silicate network is manifested by the ability of these groups to join together, i.e., to form structural types composed by 2–3 or more tetrahedra.

But there are also other structural types where the SiO_4 tetrahedra are independent.

The bonding *between* the tetrahedra of SiO_4 is made through their vertices and not through their edges or faces, such behavior resulting from certain simple electrostatic considerations. Accordingly, if the distance between the centers of two tetrahedra, which by necessity establishes an electrostatic repulsion, is considered equal to the unity for the union case by vertices, it is reduced to 0.58 when the union is made through edges and to 0.33 if it will made by faces.

Correspondingly, the electrostatic repulsion will be four times higher in bonding case by edges and as nine times higher if the union will be by faces.

Thus, the ability of chemical bonding rationalization appears, in general, by geometrical models, an idea which will be detailed in the next sections.

However, as based on the union types of the SiO_4 tetrahedra, according with the Table 4.13, we distinguish the following structural types.

4.3.8.2   Nesosilicates and Sorosilicates

Silicates with island structure, also called nezosilicates, correspond to the orthosilicates with the O:Si=4:1 ratio.

The network is composed by isolated groups of SiO_4 bonded between them by ions of Mg, Fe, etc.; the most popular examples are *forsterite* (Mg_2SiO_4) and *zirconium* ($ZrSiO_4$) of Figure 4.43. The silicates having

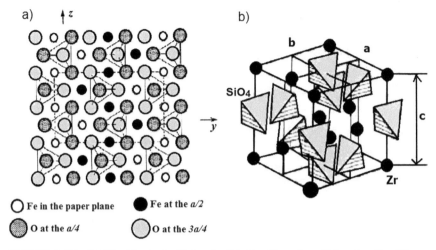

a)

O Fe in the paper plane   ● Fe at the $a/2$
◉ O at the $a/4$ ◐ O at the $3a/4$

FIGURE 4.43   (a) The Forsterite Network; (b) The Zirconium Network; after (Bunn, 1954; Evans, 1964; Knox and Gold, 1964; Ramachandran, 1964).

groups of two tetrahedra SiO_4 with the O:Si = 7:2 ratio are called *sorosilicates*. Their structure is similar to the anisodemic network (as for the islands silicates), with the complex $[Si_2O_7]^{6-}$ being the anion. This group has as typical representant the *thortveitite* $Sc[Si_2O_7]$.

4.3.8.3   Cyclosilicates

The silicates with cyclic groups of tetrahedra SiO_4 are called *cyclosilicates*.

In the structure of cyclosilicates, tetrahedra are united between them by two corners, forming cycles of 3,4 or 6 structural units.

The O:Si ratio in cyclosilicates is 3:1, while the charge of the anion being the double of the number of tetrahedra inside the cycles. A typical representative of this is *benitoite*, $BaTi[Si_3O_9]$, of Figure 4.44, having cycles formed by 3 tetrahedra.

Other important cycle silicates are: beryl, tourmaline, dioptase, (6), axenite (4), etc.

4.3.8.4   Inosilicates

The silicates with uni-dimensional finite groups of SiO_4 tetrahedra are called inosilicates.

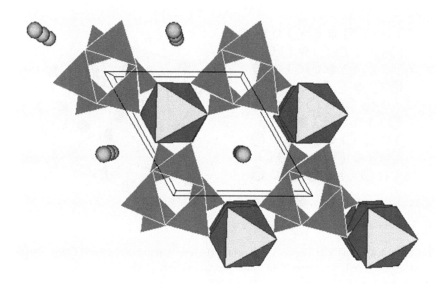

FIGURE 4.44   The Benitioite network; after The Chemistry of Silicates (2003).

The uni-dimensional organization of the tetrahedra can be made in simple or double chains (the "band" structure) and the periodicity along the strings can vary from 1 to 7. The O:Si ratio is 3:1, while the rough formula corresponds to the meta-silicates for simple chain and 11:4 for double chains.

The chains or bands are united in the respective structures by the electrostatic interaction with cations placed between strings.

This way, as an example, for the thermolite structure $Ca_2Mg_5[(OH)(Si_4O_{11})]_2$ the double chains with periodicity 2 are united by nextpolyhedra: $[MgO_6]$, $[Mg(OH)_2O_4]$ and $[CaO_8]$, see Table 4.13 and Figure 4.45.

The representative structures of this group are pyroxenes (clinoenstatite, diopside, egyrine) formed by simple chains of tetrahedra, Figure 4.46 (a); with clinoenstatite and amphiboles (antophyllite, tremolite, hornblende) formed by bands of tetrahedra, as in Figure 4.46 (b).

4.3.8.5   Phyllosilicates

The silicates with bidimensional finite groups of tetrahedra form the *phyllosilicates*.

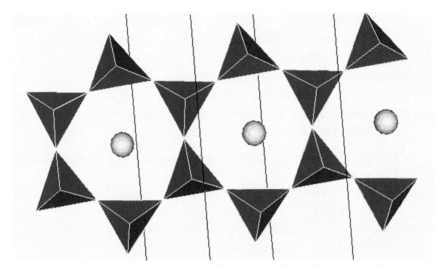

FIGURE 4.45   The *thermolite* structure; after The Chemistry of Silicates (2003).

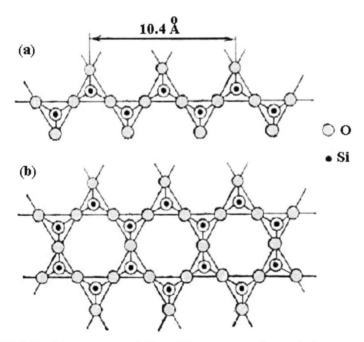

FIGURE 4.46   (a) pyroxenes, and (b) amphiboles structures by tetrahedra concatenation; after Apostolescu (1982).

The stratified structure results by the unlimited association of the bands. In these structures the tetrahedra are united through three common corners, the O:Si ratio being equal to 2.5. The structural types of phyllo-silicates can be subdivided after the symmetry of the lattice's closures for the resulted planes and after the number of the layers that are framed by the cations which ensure the cohesion between the layers by ionic bonds.

After the lattice's closure, the layers can be:

- tetragonal, when 4 or 8 tetrahedra are united in a lattice's closure, Figure 4.47(a);
- hexagonal, when the lattice' closure is composed by 6 tetrahedra, Figure 4.47(b).

Considering the number of the bi-dimensional units united through electrostatic bonding ("packages"), we distinguish:

- phyllosilicates with two layers package (e.g., *kaolinite* $Al_2[(OH)_2 - (Si_2O_5)_2]$, Figure 4.48(a),
- and phyllosilicates with three layers package (e.g., *talcum* $Mg_3[(OH)_2 - (Si_2O_5)_2]$, Figure 4.48(b).

As seen in Figure 4.48, in the first case the package is realized through electrostatically united bonds, while for talcum the bonding between packages are of van der Waals type, which allows the easy cleavage on it.

An important structure is that of *micas*. The micas' group has the formula: $R^IR_2^{III}[(OH)_2AlSi_3O_{10}]$ where R^I is K^+, and R^{III} stands for Al^{3+}, Fe^{3+},

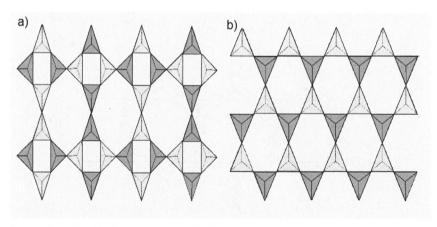

FIGURE 4.47   (a) Tetragonal; and (b) hexagonal types of stratified phyllo-silicates' structures; after The Chemistry of Silicates (2003).

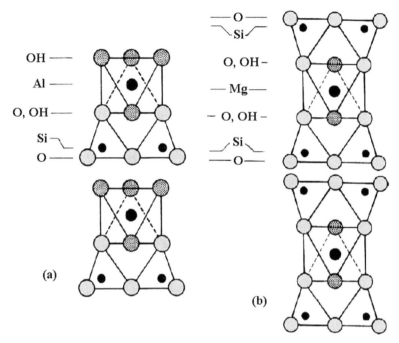

FIGURE 4.48 The structure of (a) kaolinite; (b) talcum phyllo-silicates; after (Bunn 1954; Evans 1964; Knox and Gold, 1964; Ramachandran, 1964).

Mn^{3+}, etc.. Their structure, illustrated in Figure 4.49, is composed by two external hexagonal layers formed by tetrahedra of SiO_4 and AlO_4 (overlapped in 3:1 ratio) and a middle layer formed by $Al(OH)_2O_4$ octahedra. The packages of layers are united through the potassium ions.

4.3.8.6 Tectosilicates

The silicates with tridimensional skeleton of tetrahedra form the *tectosilicates* class. In the tectosilicates structure, the tetrahedra are united through all the four vertices, forming a tridimensional skeleton.

The O:Si = 2:1 ratio, required by this organization, is found only for the SiO_2 varieties. Other representatives of this group contain ions of aluminum in their composition which also form $[AlO_4]$ tetrahedra. The tridimensional skeleton will be formed by tetrahedral units of oxygen ions of whom centers are occupied by ordered (or static) ions of silicium and

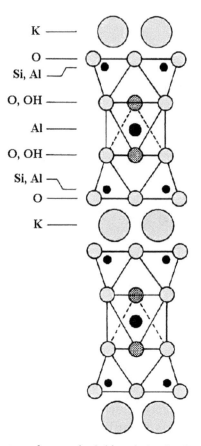

K

O

Si, Al

O, OH

Al

O, OH

Si, Al

O

K

FIGURE 4.49   The structure of muscovite (white mica); after (Bunn, 1954; Evans, 1964; Knox and Gold, 1964; Ramachandran, 1964).

aluminum, so that two ions of aluminum to be separated by at least one tetrahedron of SiO_4.

Representative structures of tecto-silicates are found at crystalline modifications of silicon dioxide (quartz, cristo-balite, trydimite), feldspars, feldspatoids and zeolites. The modifications of silicon dioxide vary upon the mutual arrangement of SiO_4 tetrahedron as shown in Figure 4.50. Such an arrangement determines the network with cubic symmetry of cristo-balite, rhombic symmetry for trydimite, and hexagonal or rhombohedral for quartz, see Figure 4.51.

The feldspars are aluminum-silicates of composition like $Na[AlSi_3O_8]$ – *albite*; $Ca[Al_2Si_2O_8]$ – *anorthite*; $K[AlSi_3O_8]$ – *orthose*, etc.

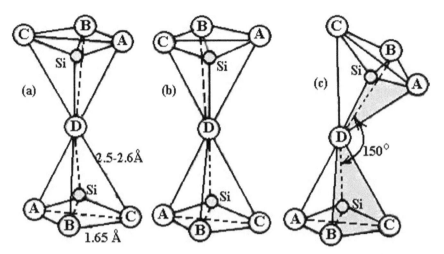

FIGURE 4.50   The arrangement of SiO₄ tetrahedra in: (a) cristo-balite; (b) tridymite; (c) quartz tectosilicates; after Sirotin and Şaskolskaia (1981).

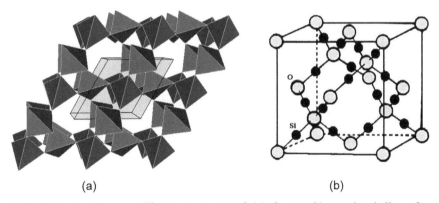

FIGURE 4.51   The tectosilicates structures of (a) Quartz (b) α cristo-balite; after (Apostolescu 1982; The Chemistry of Silicates, 2003).

The structure is of tridimensional skeleton, Figure 4.51(b), formed by the concatenation of [AlO₄] and [SiO₄] tetrahedra, while requiring the compensation of negative charge brought by aluminum for the formation of the O:(Si+Al) = 2:1 ratio, a fact which is realized by the presence of the other cations which are placed in the interstices of the skeleton.

The additional cation should have a large enough dimension to occupy these interstices (Na⁺, K⁺, Ca²⁺, Ba²⁺) and from this reason the small cations such as Fe²⁺, Mg²⁺ do not appear in the structure of feldspars.

The feldspatoids have the same structural type; however, they differ from the feldspars only by their higher content of cations in the interstices of tridimensional skeleton.

For example, in structure as $K[AlSi_2O_6]$ or in *nepheline* $Na[AlSiO_4]$, the Me^I:(Al + Si) ratio is 1:3 and respectively as 1:2, instead of 1:4 characteristic ratio for the feldspars.

4.3.8.7   Borates

For orthobotares are known by the planar groups of isolated BO_3 structures.

These are bounded between them by foreign cations; it is the cause for which the orthoborates crystallize as the rhomboedric carbonates type, for example, *Nordenskjold*, $CaSn(BO_3)_2$.

The cyclic groups occur at metaborates, as the rings of $(B_3O_6)^{3-}$, for example, $K_3B_3O_6$, or as groups in infinite chains of B-O bonding, with the infinite repetition of the $(BO_2)^-$ formula, for example, CaB_2O_4. Note that the stratified structures are less spread to borates than to silicates.

Thus, for B_2O_3, which should crystallize in a network of graphite type as a result of a hybridization sp^2 of valence orbitals of boron, one has, according with some facts, a molecular B_4O_6 structure, while according to other facts, a structure with tridimensional skeleton with a particular form. The molecules B_4O_6 would be united in the crystalline network by bonding with intermediate role between the electrostatic and van der Waals bonding. The second assumption allows the tridimensional skeleton form where the B^{3+} ion is tetra-coordinated by the vertices of an irregular tetrahedron. Yet, a corner of tetrahedron is common for two tetrahedra, while the other corners simultaneously belong to three tetrahedra.

The borates' layer structure is certainly found in *datolite* $Ca[BSiO_4(OH)]$; the network is formed by planes with tetragonal and hexagonal closure, as for phyllosilicates, Figure 4.47(a), united by Ca^{2+} and OH^- ions. The planes are formed by tetrahedra of BO_4 and SiO_4 which mutually surrounding.

The structures with tridimensional skeleton are formed by BO_4 tetrahedra and BO_3 triangles which are united in cycles, usually hexatomic. The cycles are assembled through the atoms of oxygen in spirals and toward a tridimensional skeleton. Such structures have been found as CsB_3O_5 and KB_5O_8.

4.4 THE SOLID AND CRYSTAL STRUCTURE BY CHEMICAL BONDING

4.4.1 THE CHEMICAL BOND IN SOLIDS

The solids' cohesion is determined by the electrostatic interaction of attraction between the negative charge of the electrons and the positive charge of the nuclei that make the particle system. The magnetic and gravitational forces have a small contribution upon the stability of a crystalline structure. The natures of the connecting particles condition the interactions that settle between them. The type of bonding which is establishing in atom ensemble depends on the ionization energy and on the affinity it has for the electron; the present discussion follows (Chiriac-Putz-Chiriac, 2005).

If we refer to a single atom specie, for high values of the ionization energy and the electron affinity, the formation of a *covalent bonding;* for low values of the two sizes, the formation of a *metallic bonding is favored*, and if the ionization energy is very high and the affinity for the electrons very low between the atoms-there may appear only weak interactions – *the van der Waals bonding is favored*. If the ensemble is made of two species having the ionization energy and the affinity for the electrons very different, the system will be made of ions, the interaction between the particles being of an electrostatic nature, favoring the formation of a new *ionic bonding*.

The periodicity of the physical and chemical characteristics offers a clear orientation upon the relative values of the ionization energy and also of the affinity for the electron – and, in this way, it gives, in general, the possibility to forecast the type of bonding which settles in a certain atom ensemble.

The crystalline lattices that present a single type of bonding between the particles are called *homodesmics*, and those that present multiple types of bonding are called *heterodesmics*.

4.4.1.1 The Ionic Compounds. The Formal Ions' Paradigm

The ions formation in a particle ensemble is predictable, when, in the first place, stable electronic configurations of the particles are made (electronic configurations with null or maximum spin). In case of the solid substances, the ions formation is much more encountered; on the diversity of the obtaining experimental condition is corresponding a relatively wide variety of ions.

To make an example, we specify that the ions Cr^{2+}, V^{2+}, O^{2-}, $(XeO_6)^{4-}$ are encountered almost exclusively in solid state. The interaction of the ions in the lattice is of an electrostatic nature. It consists of the attraction between cations and anions and the rejection between ions of the same sign. The force field which is created by a mono-atomic ion is uniformly distributed in the surrounding space.

The lack of ionic interaction localization brings to the tendency of the anions and cations to surround them reciprocally as many as possible.

The mono-atomic ions (K^+, O^{2-}, etc.) can be considered with a good approximation as being of spherical form. In many crystals there can be encountered polyatomic ions, which bring to a more complicated geometry of the structure of the electrostatic field.

By polyatomic ions we understand only those groups that can be passed unmodified in another combination.

Thus, the group that has a tetrahedral shape $(SO_4)^{2-}$ represents a polyatomic ion, different than the groups, that are tetrahedral as well, e.g. MgO_4, from the structure of the spin Al_2MgO_4 and that are not alike in the chemical interactions.

The bonds which unify between them the particles of the polyatomic ion have a predominantly covalent character, and thus are stronger than those between the polyatomic ion and the other ions of the lattice.

In the polyatomic ions there can be often distinguished a central (nucleus) atom (ion) surrounded by a determined number of other atoms (ions).

The great majority of the polyatomic ions represent anions in which the central particle has the oxidation number higher than the ones that surround it.

Using as systemization criteria the *complexity degree* the following categories of ions in the solids structures there can be distinguished:

(a) elementary ions (monatomic):

- *Cations*: Na^+ (in NaCl), Ca^{2+} (in CaO), Th^{4+} (in ThO), etc.;
- *Anions*: Cl^- (n NaCl), O^{2-} (n MgO), S^{2-} (n Na_2S), etc.

(b) polyatomic ions with a finite number of particles

- polyatomic ions without central nucleus: O_2^- (in KO_2), O_2^+ (in O_2PtF_6), I_3^- (in KI_3), N_3^- (in NaN_3), SCN^- (in KSCN);

- *mononuclear ions*: NO_2^- (in KNO_2) UO_2^+ (in $UO_2(NO_3)_2$), CO_3^{2-} (in $CaCO_3$), SO_4^{2-} (in $BaSO_4$), $AuCl_4^-$ 9in $KAuCl_4$), PtF_6^- (in O_2PtF_6), ZrF_7^{3-} (in K_3ZrF_7O, XeF_8^{2-} (In Cs_2XeF_8), etc.;
- *polynuclear ions*: $Si_2O_7^{6-}$ (in $Sc_2Si_2O_7$), $Tl_2Cl_9^{3-}$ (in Cs_3-Tl_2Cl_9), $Si_3O_9^{6-}$ (in $BaTi(Si_3O_9)$), $B_3O_6^{3-}$ (in KBO_2), $P_4O_{12}^{4-}$ (in $(NH_4)_4P_4O_{12}$, etc.

(c) polyatomic ions with indefinite number of particles:

- unidimensional (linear) macroanions: $(SiO_3)_2^{2n-}$ (in pyroxenes), $(PO_3)_n^{n-}$ (in $NaPO_3$), $(AlF_5)_n^{2n-}$ (in Tl_2AlF_5), etc.;
- *bidimensional macroanions*: $(AlSi_4O_{10})_n^{n-}$ (în micas), $(NiF_4)_n^{2n-}$ (in K_2NiF_4), etc.;
- *tridimensional macroanions*: $(AlSi_3O_8)_n^{n-}$ (in feldspars), $(B_3O_7)_n^{5n-}$ (in CsB_3O_5O, etc.

The variety of ions in the solids is amplified by the anions and cations presented in the acids salts, respectively the organic basis. These ions represent the diversity of shapes that are characteristic to those certain molecules.

In what concerns the ionic lattices, even though there doesn't exist any particle system in which is possible to completely lack the change interaction, the lattices that are formed by ions can be considered as being built exclusively through "ionic bonding".

For these it is considered the reciprocal polarization of the particles and it shouldn't be omitted the confrontation of the conclusions with the experimental data.

In representing the structures there is operated currently with "*the formal ions paradigm*", monatomic ions E^{n+}, even though their existence is impossible. Maintaining the approximation, according to which, any particle of the lattice is a monatomic ion with the electric charge that is resulted from the stoichiometric formula, in which it is attributed to the anions' full charges, the ionic lattices classify according to the size of the *electrostatic valence of the cations*:

$$p = z/n \qquad (4.18)$$

respecting the half of the anion valence ($y/2$) in the isodesmic lattices ($p < y/2$), the mesodesmic lattices ($p = y/2$) and the anisodesmic lattices ($p > y/2$).

(i) In case of the isodismic lattices, the inequality:

$$p < y/2 \qquad (4.19)$$

shows that the anions, on a certain given direction, are not attracted not even on the half of their capacity to connect. Thus, there also exist other directions to which the anion behaves identically; in other words, the force field around the anions and the cations is uniformly distributed in the space.

For example in the perovskite structure of the calcium titanium $CaTiO_3$ (Figure 4.38), for the two coordination polyhedra CaO_{12} and TiO_6, p has smaller values (1/6 respectively 2/3) than 1 ($y=2$). This, because Ca^{2+} is 12-coordinated by $O^{2-} \Rightarrow p(Ca^{2+}) = {}^2/_{12} = {}^1/_6$, and Ti^{4+} is 6-coordinated by $O^{2-} \Rightarrow p(Ti^{4+}) = {}^4/_6 = {}^2/_3$.

For the isodesmic structures which have the AX formula there are two important types: the type of the cesium chloride ${}^3_\infty Cs^{[8]}Cl^{[8]}$ c , Figure 4.29, and the type of the sodium chloride ${}^3_\infty Na^{[6]}Cl^{[6]}$ c , Figure 4.30, while for the isodesmic structures which have the AX_2 formula the representative are the types of the fluorine ${}^3_\infty Ca^{[8]}F_2^{[4]}$ c , Figure 4.32, and of the rutile ${}^3_\infty Ti^{[6]}O_2^{[3]}$ tt, Figure 4.33.

In what concerns the isodesmic complex structures, there are the perovskite types, $Me^{II}Me^{IV}O_3$, respectively $Me^IMe^{II}X_3$, Figures 4.38–4.40, of the spin, $Me^{II}Me_2^{III}O_4$ – generally AB_2O_4 or particularly as ${}^3_\infty Mg^{[4]}Al_2^{[6]}O_4^{[1+3]}$ c , Figures 4.41 and 4.42, as the type of the corundum ${}^3_\infty Al_2^{[6]}O_3^{[4]}$, Figure 4.52.

The structure comes from a hexagonal –compact packing of the oxygen ions, in which 2/3 from the octahedral holes are occupied by the aluminum ions. The coordination polyhedron of the oxygen is a deformed tetrahedron.

In this type of lattice there crystallize Cr_2O_3, Fe_2O_3, Y_2O_3, etc. A similar structure presents the ilimenite $FeTiO_3$ in which the trivalent cation from the structure Al_2O_3 is alternatively replaced by Fe and Ti.

(ii) In the mesodesmic lattices the ions paradigm imposes:

$$p = y/2. \qquad (4.20)$$

In this case the anion bonds are equal in two directions and thus are able to connect equally only two cations, resulting lattices which have chain shapes, rings with a tridimensional skeleton.

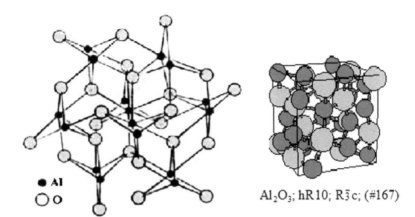

FIGURE 4.52 Left: the lattice of the corundum; after Apostolescu (1982). Right: the structural cell of Al_2O_3 under which the Pearson spatial notation is specified, along the international notation, and the number of the spatial group, according to the Tables 2.17 and 2.16, respectively; after U.S. Naval Research Laboratory/Center for Computational Materials Science (2003).

So, for the tetrahedron $[SiO_4]$ we have the cation charge $z=4$, coordinated to a number of anions equal to $n=4$, which results in an electrostatic valence of the cation equal to $p=4/4=1$, while the valence of the anion is $y=2$, concluding the equality $1=2/2$ $(p=y/2)$.

The mesodosmic structures form every time the number of coordination of the cation equaling to its charge, which happens when the electrostatic valence of a cation froming a lattice is unity (1).

This condition is accomplished only by three classes of compounds: the silicates and the germanites in tetrahedral coordination and borates in trigonal coordination (see the Section 4.3.8).

The equality of the electrostatic valence of the cation with half of the anion valence allows unifying the structural units among them, so resulting into a structural diversity which is impossible to encounter at other classes of inorganic mixtures.

(iii) In the anisodesmic lattices, the formal ions paradigm implies:

$$p > y/2 \tag{4.21}$$

The inequality that has been specified previously means the formation in the inside structure of some groups in which the lattice anion is connected to a certain cation in a stronger way than with the other cations.

The common characteristic of these lattices is the presence of the complex anions in spins, formed as a result of the polarization interaction between the cations that have a greater polarization capacity (high charge and small radius) and the monoatomic anions of the lattice.

The forms of the complex anions are extremely diverse depending on the number of particles and the electronic structure of the cation that generates these groups.

There can be noticed the following types of complex anions (Chiriac-Putz-Chiriac, 2005):

- *bi-atomic* XY: CN^-, C_2^{2-}, S_2^{2-}, N_2^- etc.
- *tri-atomic* X_3, XY_2 or XYZ:
  - linear: N_3^-, Cl_3^-, ICl_2^-
  - angular: ClO_2^-, NO_2^-
- *tetra-atomic* BX_3:
  - plane trigonal CO_3^{2-}, NO_3^-
  - trigonal pyramidal: PO_3^{3-}, IO_3^-
- *penta-atomic* BX_4:
  - tetrahedral SiO_4^{4-}, SO_4^{2-}, MnO_4^-
  - plane squared: $Ni(CN)_4^{2-}$, $PtCl_4^{2-}$
- *hepta-atomic* BX_6 (octahedral): SiF_6^{2-}, $PtCl_6^{2-}$

The lattice symmetry is directly influenced by the complex anion symmetry. Generally, the structure can derive from simple structures considering the deformation of the symmetry as a result of the orientation of the complex anion in relation to the symmetry elements of the lattice.

Therefore, the type of the pyrite, $_\infty^3 Fe^{[6]}S_2^{[6]}$ c, and the type of the calcium carbide $_\infty^3 Ca^{[6]}C_2^{[6]}$ t, derives from the structure of the NaCl type, that is, they are made of two compact interpenetrated cubic lattices.

The complex anion lattice though, hasn't got the symmetry of the cation lattice anymore but, and for this reason, the overall symmetry of the lattice will be the *cubical meriedry* at the pyrite, Figure 4.53(a), and tetragonal CaC_2, Figure 4.53(b).

Identically, the $CaCO_3$ lattice (the calcite variety) can be deducted from NaCl as this structure is also made of two compact interpenetrated cubic lattices.

The trigonal symmetry of the CO_3^{2-} ion, deforms the symmetry of the whole lattice, which will thus present a single ternary axis, perpendicular to the plane of the CO_3^{2-} ions.

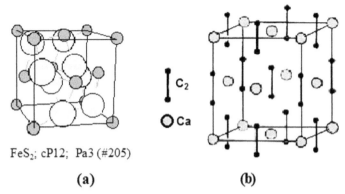

FeS₂; cP12; Pa3 (#205)

(a)   **(b)**

FIGURE 4.53 (a) The lattice of the pyrite where its structural type cell it is specified by the Pearson spatial notation, by the International notation, and also by the number of the spatial group, according to the Tables 2.17 and 2.16, respectively; after U.S. Naval Research Laboratory/Center for Computational Materials Science (2003). (b) the lattice of the carbide compound; after Apostolescu (1982).

The calcite lattice will have trigonal symmetry with the main axis in the direction of the ternary axis of the cube, after the deformation has happened.

In the Figure 4.54, the structure of the calcite perpendicular to the ternary axis has been presented. One observes that the Ca^{2+} ions are arranged in compact structures, with the ABCA succession, which is characteristic to the cubic packing.

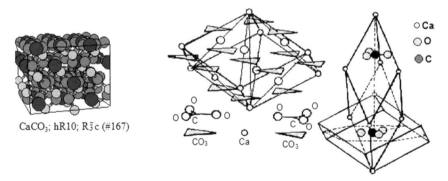

CaCO₃; hR10; R3̄c (#167)

FIGURE 4.54 Left: the calcite lattice; after U.S. Naval Research Laboratory/Center for Computational Materials Science (2003). Right: the structural cell of $CaCO_3$ with specifying of its Pearson spatial notation, along the International notation, and also by the number of the spatial group, according to the Tables 2.17 and 2.16, respectively; after Penkala (1974).

Another variety of $CaCO_3$ – the aragonite – presents a compact hexagonal settlement of the calcium ions, the placement of the anions between the layers being made in the way that the calcium is coordinated with 6 ions of oxygen, which determines the deformation of the 120°C angle between the calcium ions and the rhombic symmetry of the aragonite.

Other two characteristic structural types are offered by the platinum's complexes with the chlorine, Figure 4.55.

In the potassium tetrachloroplatinite $^3_\infty K_2^{[4]} [PtCl_4]^{[8]} t_t$, Figure 4.55(a), every ion of K is coordinated by 4 complex ions, and around a complex ion, flat-square, the potassium ions are put after the edges of a cube.

Instead, the tetrachloroplatinite of K, Figure 4.55(b), presents a lattice that is similar to the one of the fluorine, in which the place of the Ca ions is occupied by the octahedral complex anion, and the place of the F ions, by the K cation of the lattice.

The lattices of the same stoichiometry in which the anions and the cations change reciprocally their places, are being defined through the antistructure object.

Thus, the hexachloroplatinite of K, presents a structure of an antifluorine type: $^3_\infty K_2^{[4]} [PtCl_6]^{[8]} c$.

Some structures, even though are made of polyatomic anions, having their own symmetry smaller than the one of the lattice, the symmetry of

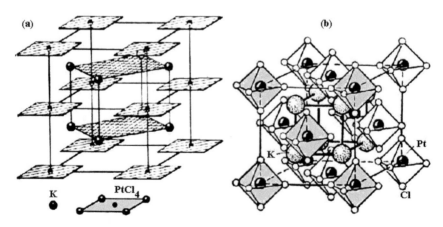

FIGURE 4.55   (a) The Lattice of $K_2[PtCl_4]$ complex; (b) The lattice of $K_2[PtCl_6]$ complex; after (Flint 1968; Penkala 1974; Urusov, 1987).

the lattice does not modify. Thereby, NaCN presents a cubic lattice, identical to the one of the NaCl, without the deformation that can be observed at FeS_2 or CaC_2. The explanation consists in the possibility of the free rotation of the diatomic particle in the structure of the crystal, which offers it a spherical pseudo- symmetry. The free rotation can be also observed in crystals made of polyatomic cations, for example in the cubic variety, with a lattice of CsCl, a NH_4NO_3 type.

As a conclusion to this sub-section, there should be specified that (Chiriac-Putz-Chiriac, 2005):

- The number of the ionic combinations studied upto the present largely exceeds the number of the other types of lattices.
- The ionic crystals structure results from a pure geometrical settlement of ions with radii that are more or less different made in a way to obtain a structure that is as stable as possible from an energetic point of view. This condition makes that, for a certain type of lattice should correspond a very large number of combinations, if the ratio of the radii varies in small limits, and if the charges are identical.
- For the ionic lattices, the lattice type is conditioned by the geometric factor even more than the chemical nature of the lattice ions.
- For the classification of the many ions' lattices one should be taken into account two factors: the character of the bond and the charge of the ions.

4.4.1.2   The Atomic Solid Compounds

There are two kinds of combinations where nodal particles can be considered as atoms: the crystals of inherent gases and covalent crystals of diamond type.

In the crystals of inherent gases the cohesion between atoms is given by the van der Waals interaction. The origin of this interaction consists in the fact that the electrons are in movement around the nucleus, even in the lowest energetic state, so even for 0K. The electrons movement produces a dipole moment instantaneous non-null, which will induce a dipole moment instantaneous in the neighboring atom and so one. The *van der Waals interaction is un-oriented,* thereby, in these structures will appear the tendency that an atom will be surrounded with a maximum number from other atoms.

The van der Waals lattices – homodesmic – are found, as shown, only at inert gases and correspond to the compact structures. Indeed Ne, Ar, Kr, Xe present a compact cubical lattice, see Figure 4.15.

Also in this class the crystalline structures of the gases can be integrated with diatomic molecules where it is accomplished a compact packaging due to a free rotation of molecule in crystal. Thus CO, N_2, HCl and HBr present the compact cubical structure, Figure 4.15, while H_2 crystallizes under compact hexagonal structure, Figure 4.11.

There are a small number of compounds in which the atoms from the lattice nodes are united by covalent bonds (diamond, silicon crystals, germanium, etc.). The union of the atoms through strong covalent bonds confers to the structure the character of infinite tridimensional macromolecule. The electronic configuration of the atoms in covalent crystals corresponds to an excited state (in diamond, Figure 4.27, for example, the atoms of carbon have the configuration $2s2p^3$ towards the fundamental state $2s^22p^2$), the energy necessary to the excitation being compensated by the formation of the bonds; the corresponding atomic orbital are hybrids of sp^3 type for the considered example.

In the valence bond paradigm (V.B.) is considered that the covalent bond results through the overlapping of atomic orbital of two atoms having electrons with not compensated spin, see Volume III of the present five-volume set (Putz, 2016) and (Sirotin & Şaskolskaia 1981; Becherescu et al., 1983).

The formation of a covalent bond suppresses the spherical symmetry of electrons cloud around an atom and, therefore, the bonds resulted will have various directions, exactly oriented in space or not, in other words, particles, covalently bonded by a given particle, can occupy only certain positions, strictly determined around it.

Inter-atomic distances which are established as a result of covalent bond of two atoms are constant measures and independently form the aggregation state of the relative substance.

This means that the atoms participant to a covalent bond have a fix position one form other and can not be removed from these positions without destroying the molecule.

Thus, the covalent bond is characterized by the rigidity and by its direction. The angle between two covalent bonds is a constant measure and characteristics for a certain electronic structure of an atom (ion).

By covalent bonds are united with a given particle a limited number of particles (the number of neighbors being conditioned by the electronic structure of the central particle).

Regarding the covalent lattices, this group of structures is the last numerous because in this category only homodesmic structures are included, which is reduced as number.

We distinguish some types of homeopolar lattices (Chiriac-Putz-Chiriac, 2005):

1. *Diamond type* – $3C^{[4]}c$, described in Figure 4.27.

2. *Blende type* – $^{3}_{\infty}Zn^{[4]}S^{[4]}c$, Figure 4.56(a), shows a pronounced similarity with the diamond lattice, unless the nodes are occupied by different atoms. The elementary parallelepiped is a cube with centered faces of the atoms of Zn, in octants founding the atoms of sulfur. Each atom of Zn or S is tetrahedrally surrounded by 4 atoms of the other element. The blende lattice can be considered as being composed of two cubical lattices with centered faces of Zn and S, displaced one to other with 1/4 from the cube diagonal.

3. *Wurtzite type* – $^{3}_{\infty}Zn^{[4]}S^{[4]}h$, Figure 4.56(b), hexagonal variety of ZnS, is very similar with the blende structure, the coordination polyhedra being also tetrahedra. But unlike the blende, the structure of type

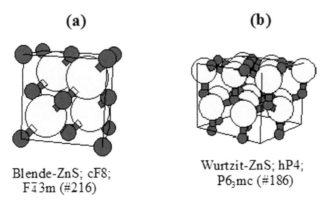

(a)

Blende-ZnS; cF8;
F$\bar{4}$3m (#216)

(b)

Wurtzit-ZnS; hP4;
P6$_3$mc (#186)

FIGURE 4.56 (a) The Blende structure; (b) The Wurtzite structure. Under the cell of each structural type there is specified: the Pearson spatial notation, the International notation, as well as the number of spatial group, according to the Table 2.17 and respectively Table 2.16; after U.S. Naval Research Laboratory/Center for Computational Materials Science (2003).

wurtzite comes from the overlapping along A^3 of two compact hexagonal lattices of S and Zn.

Both structures appear to binary combinations of the elements where the sum of valence electrons is 8 (BeO, CaS, AlN, SiC, AgI, MgTe, CdSe, CuCl, InSb).

4.4.1.3 The Metallic Compounds

Specific metals properties: electric conductibility, glossiness, malleability and ductility, suggest that their internal structure should be different to the other solids.

The simplest structure model of the metals corresponds to an ensemble of positive ions, surrounded by mobile electrons (the so-called electronic gases), the total electrons number being equal with the sum of the positive charges of the ions placed in the nodes of the lattice.

The metallic lattices ions are not identical to the common cations from the combinations of relative elements.

For electrons to be able to freely move in crystal is necessary that the nodal particles ("metallic atoms") to present free orbital where the electrons from neighboring particles can pass through.

The representation of nodal particles by ions actually is related with this possibility of electrons migration from an atom to other.

Free electrons movement is limited to some directions (the directions where are placed the particles in neighbors), on this preferential directions locating the maximum density of electrons.

But, metallic bond is not directed in space and thus metallic structures will be characterized by the tendency (already found at inherent crystalline gases and at compounds formed by monatomic ions) that a particle to be surrender, with a maximum number possible, by other particle. The strength of metallic bond is determined, in first approximation, by the number of peripheral electrons of the atoms through which occur the interaction between particles. There are also metallic compounds where the nodes of the lattice are occupied by complex groups of atoms, as the Cs_3O. In such compounds, the nodes are represented by the groups Cs_3O united between them by the electrons common to entire lattice.

Metallic lattices are found at metals and also at some binary combinations, i.e., at alloys.

Most metals crystallize in one or more of the following three structures (Chiriac-Putz-Chiriac, 2005):

1. *Copper type* – Figure 4.57(a) – is a cubical lattice with centered faces $3Me^{[12]}c$. Elementary cell is a cube having the side of 3.86[Å] and which has in the corner and the center of each side one each atom. This lattice is found at: Cu, Ag, Au, Ca, Sr, Al, Pb, Fe, Ni, Pt.

2. *Wolfram type* – Figure 4.57(b) – is a cubical lattice internally centered $3Me^{[8]}c$. Elementary cell is a cube having the atoms in the corners and in the cube center. This structure is found at alkaline metals Ba, Cr, Mo, Fe^{α}, Fe^{β}.

3. *Magnesium type* – Figure 4.57(c) – corresponds to a compact hexagonal lattice $3Me^{[12]}h$. Elementary cell is composed by an hexagonal prism with the bases centered, having in the middle of three trigonal prisms which alternate another three particles. Is found at Be, Yn, Cd, Te, Ti.

Interstitial structures are composed by atoms of transitional metals with compact lattice and metalloids of small dimensions as H, B, C, N, which occupy the holes of a compact lattice.

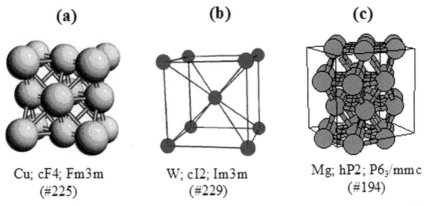

(a)   **(b)**   **(c)**

Cu; cF4; Fm3m   W; cI2; Im3m   Mg; hP2; P6₃/mmc
(#225)   (#229)   (#194)

FIGURE 4.57 (a) The metallic Cu-lattice; (b) The metallic W-lattice; (c) The metallic Mg-lattice. Under each structural elementary cell there are specified: the Pearson spatial notation, the International notation, as well as the number of spatial group, according to Tables 2.17 and 2.16, respectively; after U.S. Naval Research Laboratory/Center for Computational Materials Science (2003).

Crystalline phases with interstitial structures are characterized by some relations of combination Me:X, but the composition can present deviations in large limits toward the theoretical one.

The phases of MeX type are characterized by the placement of the metalloid in octahedral holes of compact cubical lattice, as for TiC, ZrC, NbC, TiN, ZrN, YN, NbN, NbO – Figure 4.58(a).

The structures of *Me$_2$X* type rarely appear in compact cubical lattice when X occupies 1/2 of octahedral holes (W_2N, Mo_2N). They are often found in compact-hexagonal lattices, the metalloids being found both in octahedral coordination (Mn_2N, Co_2N, Fe_2N, Fe_2P – Figure 4.58(b), W_2C, Mo_2C, V_2C) as also in tetrahedral coordination (Zr_2H, Ta_2H, Ti_2H).

Hydrures *1:1* are characteristic for the compact cubical lattices where H is tetrahedrally coordinated (ZrH, TiH). There are also known the cubical hydrures *2:1* (Pd_2H) and even 4:1 (Zr_4H). Even rarely, the interstitial structures can occur also in simple hexagonal lattice (MoN, WC). The alloys formed by various metals have the lattice similar with those previously presented. They are the solid solutions composed by a single type of lattice i.e., in each elementary cell enter atoms of all the elements of the alloy.

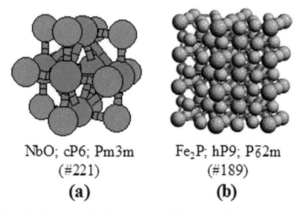

NbO; cP6; Pm3m   Fe$_2$P; hP9; P$\bar{6}$2m
(#221)   (#189)
(a)   **(b)**

FIGURE 4.58 (a) the interstitial lattice of NbO type; (b) The interstitial lattice of Fe$_2$P type . Under each structural elementary cell there are specified: the Pearson spatial notation, the International notation, as well as the number of spatial group, according to the Tables 2.17 and 2.16, respectively; after U.S. Naval Research Laboratory/Center for Computational Materials Science (2003).

4.4.1.4 Molecular Solids' Compounds

In the three categories of compounds previously described, the nodes are represented by individual particles (simple or complex) united between them by the same strength, i.e., the relative solid can be considered a giant molecule. Yet there is a manifold class of compounds where the lattice is composed by molecule completely isolated between them. Inside the lattice, the atoms of a molecule are found at much shorter one from other, comparing to the closest atoms of the neighboring molecules. The corresponding structures are of heterodesmic type, the intra-nodal bonds being in all the cases stronger than the inter-nodal ones. The molecular solids category is quite varied, Figure 4.59. In this category the crystals of the substances are found in gaseous state in common conditions (H_2, O_2, N_2, Br_2, Cl_2, CO_2, SH_2), for a set of inorganic compounds (SnI_4, $HgCl_2$, FeS_2, As_4O_6), and all the organic compounds are included, Figure 4.59 (Chiriac-Putz-Chiriac, 2005).

Intra-nodal bonds are covalent or ionic-covalent (prime covalent) and the nodal particles of molecular lattice are united by van der Waals bond of by hydrogen bonds. Van der Waals interaction is determined by the attraction of opposite poles of polar molecules, between the dipoles induced

(a)   **(b)**

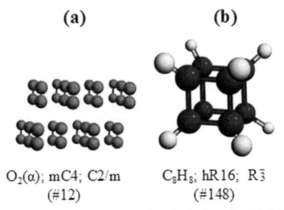

$O_2(\alpha)$; mC4; C2/m C_8H_8; hR16; R$\bar{3}$
(#12)   (#148)

FIGURE 4.59 (a) The lattice of O_2 type; (b) The lattice of solid Cuban C_8H_8 type. Under each structural elementary cell there are specified: the Pearson spatial notation, the International notation, as well as the number of spatial group, according to the Tables 2.17 and 2.16, respectively; after U.S. Naval Research Laboratory/Center for Computational Materials Science (2003).

by the dipole of some molecules in the non-polar neighboring molecules or the interaction between the instantaneous dipoles that are formed due to the vibrations of the electron shell towards the atomic nuclei. Van der Waals interactions are weaker than the other types of bonds, which are manifested in the physical properties of the compounds where only they are presented: high volatility, low hardness, etc..

The hydrogen bond is the inter-nodal bond inside the solids which contain atoms of hydrogen bonded by the very electronegative elements: F, O, N, of those atoms have non-participant electrons. The best known example of solid compound with intermolecular hydrogen bonds is the ice, where, each atom of oxygen of a molecule is conjunct with two atoms of hydrogen of two different molecules of water, by hydrogen bonding, Figure 4.60.

The hydrogen bond is weak comparing to the ionic bond or covalent, but stronger that the van der Waals bond. As an example, we remind that the ice melts at 0°C while the crystal of H_2S, where the molecules are stabilized by van der Waals bonds, melts at (– 85.6°C).

Being generated by the ends of some polar covalent molecules, the hydrogen bonds are weaker enough, being based on the electrostatic attraction between a positively end polarized of a molecule and another negatively end polarized from another molecule. Although these hydrogen bonding are weak at individual level, when are manifested between a big number of polar molecules, become strongly enough to imprint a tridimensional form to the ensemble of molecules.

The H bond in ice

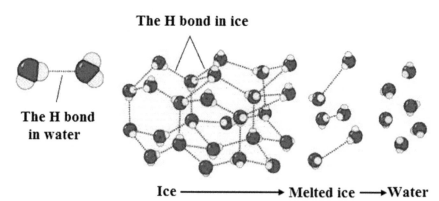

The H bond in water

Ice ⟶ Melted ice ⟶ Water

FIGURE 4.60   The Hydrogen bonding in ice and in water; after Purves et al. (2001).

An example particularly important in this sense represents the polar regions form the nucleic acids (Thymine T, Adenine A, Cytosine C, Guanine G), basic components in DNA structures (dezoxiribonucleic acids), which are bounded by hydrogen bonding (also called *hydrogen bridges*), Figure 4.61, forming chains of macromolecules (or *biopolymers*), Figure 4.62.

The DNA molecules thus resulted are the molecules that form the "living world", so encapsulating the information necessary for the procreation, the *genetic code*.

The diversity of molecular forms is properly reflected also in the organization of the structure of molecular compounds in solid state.

A classification of the molecules as nodal particles is possible taking into account the development of the molecule after the three directions of the space.

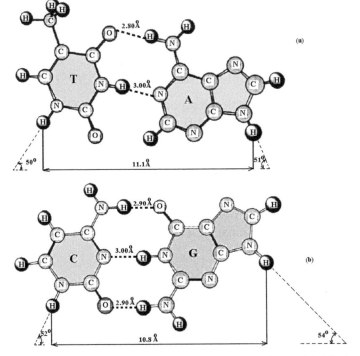

FIGURE 4.61 The nucleotide DNA bases coupling: (a) Thymine T with Adenine A; (b) Cytosine C with Guanine G, by means of hydrogen bonding; after Crick (1966, 1968).

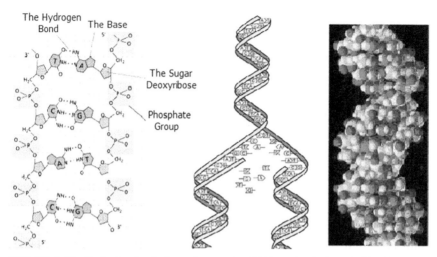

FIGURE 4.62 Left: the detailed two layers of a DNA molecule bounded by hydrogen bonding; middle: the chains of nucleic acids bounded by hydrogen bridges form a double spiral (one progressing in up, the other in down) of a DNA molecule; right: the spatial representation of DNA molecule where, towards the outside the light shades the sugar-phosphate substrate are underlined, and towards the middle of the spiral the dark shades, are representing the nucleotide bases (T, A, C, G); after (Crick, 1966, 1968; Kimball's Biology Pages, 2003).

Thus, we can distinguish (Chiriac-Putz-Chiriac, 2005):

- simple symmetric molecules (point-like): H_2, N_2, O_2, CH_4, NH_3, CCl_4;
- one-dimensional molecules (linear):
  finite: aliphatic hydrocarbons with elongate chains, fatty acids;
  indefinite: Se, Te, cellulose;
- bi-dimensional molecules (flat);
  finite: flavored hydrocarbons; sugar;
  indefinite: graphite, BN, black phosphor;
- complex tridimensional molecule:
  finite: protein;
  indefinite: glyptics.

In conclusion, there is resumed that:

- structural types of molecular compounds are extremely varied;
- corresponding lattices – called molecular- are characterized by covalent bonds inside the structural unit (molecule) and by van der Waals bonding or of hydrogen between the structural unites;

- structural unit of these lattices can be: the atom – for the rare gases, the proper molecule, i.e., the finite molecule for elementary substances of type S_8, P_4, X_2 ($X=$ I, Cl, F, Br), or of binary compounds from class CH_4, CI_4, etc., infinite associations of atoms – linear (Se, Te), flat (graphite) or tridimensional (globular proteins).

4.4.1.5 Relations between Chemical Bonds: The Heterodesmic Lattices

The four types of bonds represent two classes of interactions between the particles of a lattice: *electrostatic interaction of Coulombian type* and *exchange interaction*, the last one having a quantum nature.

Quantum theory allows the explication also for ionic bond, though the above division is useful because limits the domain where classical physical representations can be used for the study of crystalline structures. Moreover, there is demonstrated that Coulombian interactions and of exchange interactions always coexist in a sort of bond, or more exactly, there are no structure completely absent of exchange interactions.

On the other side, the physical meaning of van der Waals bond (residual) indicates the fact that also in this kind of interaction should be present almost one of its components in any system of particle.

Obviously, when coexist more kinds of interactions of very different intensities in the same direction, those of low intensity (e.g., the residual one) may be neglected. Aiming to characterize the way of distribution of electric charges between two particles, i.e., for the interaction modeling, two paradigms are used: the delocalization of the valence electrons (or the electrons collectivization) and the bonds' delocalization.

In Figure 4.63, the distribution of electronic distribution between two nodes of the four kinds of lattices are illustrated.

As shown in Figure 4.63, electronic density can be almost null at a certain distance between two nuclei for the ionic interaction, effectively being equal to zero for van der Waals homodesmic lattice.

For covalent bond and metallic ones, the electronic density differs from zero for all the intermediate r-distances.

Yet, diagrams of Figure 4.63 suggest the type of interaction, but do not indicate the bond delocalization or of the valence electrons.

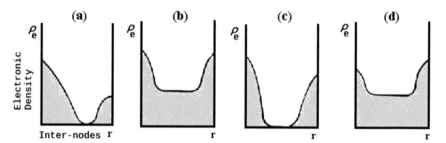

FIGURE 4.63   The electronic density variation diagrams in various types of lattices: (a) for ionic bond; (b) for covalent bond; (c) for van der Waals bond (residual); (d) for metallic bond; after Drăgulescu and Petrovici (1973).

The existence of a null electronic density or almost null leads to the conclusion that *the electrons are exclusively located around nodal particles.* Indeed, Fourier diagram of electronic density in NaCl illustrated in Figure 4.64(a) confirms this assumption and in the same time it presents the un-oriented character, delocalized in space, of electrostatic interaction.

Unlike of NaCl, Fourier diagram of electronic density of diamond in Figure 4.64(b) shows that for covalent bond, *the two bonding electrons are strictly collectivized between the two bound atoms*, thus the bonding being oriented.

Metallic bond is characterized by the electrons collectivization of the atoms of entire lattice, and therefore it is an oriented bond.

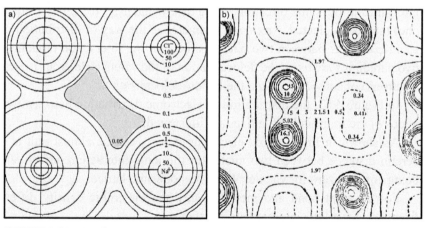

FIGURE 4.64   (a) The Fourier diagram of NaCl; and (b) the Fourier diagram of diamond; after Drăgulescu and Petrovici (1973).

The ordering of constitutive particles of a crystalline lattice facilitates the methods of quantum mechanics application to determine the energetic structure of the crystal. One of the most used methods in crystal study was firstly developed by Bloch to elucidate metallic bond. Subsequently applied to the study of other types of bonds, the method allowed the delineation of a unified theory of crystalline solids (applicable, with some approximations, also for amorphous solids).

At this stage, the paradigm known as the *energy band theory* requires assumptions on the type of dominating bond (see Chapter 3 of the present Volume). This is deduced from the nature of the particles to come together in crystalline lattice (metallic or nonmetallic atoms, ions or molecules). The energy band paradigm advantage consists in the fact that allows the provision of the properties connected by a certain reticular organization. Thus, the crystalline lattice is replaced by an electromagnetic field of periodic potential where occurs the migration of valence electrons. In this treatment, the energetic levels of the valence electrons belong to the crystal and not to the particles of it. The possible energetic states of those electrons were for a set of bands, separated by the forbidden zones, Figure 4.65(a).

Accordingly, by the atoms' union and respectively by the ions in a crystalline lattice, there can be considered that the molecular orbital merges itself in bands separated by the forbidden zones, analogous to the case of quantified allowed levels of the electrons in isolated quantum particles. Usually, the width of the allowed bands increase with the increase of corresponding energetic level, of the electron in nodal particle, paralleling the narrowing of the forbidden zones. In Figure 4.65(b) to Figure 4.65(d) there are illustrated the relations between the configuration of energetic zones and levels of the atoms (respectively ions) based on the r-distance between particles.

It can be noted that, regardless the bond character, the reduction of the distance between particles leads to the enlargement and overlapping of the energy levels. In fact, there can be considered that the proximity of particles determines the stepping increase of the degeneration of the levels of the same kind of the isolated particles; i.e., from a level with N-identical atoms, and thus with a multiplicity of N order, resulting a band with N nondegenerated levels.

Estimatedly, taking into account the fact that 1 cm^3 of solid contains, in a coarse approximation, 10^{22} particles, there results that a band will be

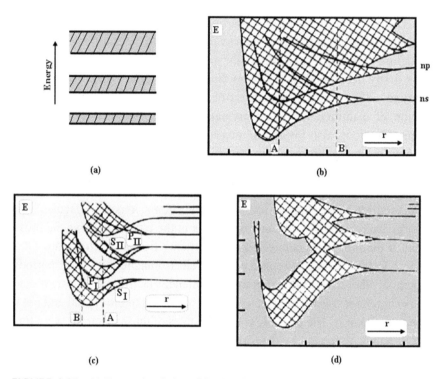

FIGURE 4.65 (a) Energy bands in solids; (b) The relation between the configuration of the zones and energetic levels of the atoms in metallic Na; (c) The relation between the configuration of the zones and the energy levels of ions in LiF; (d) The relation between the configuration of the zones and the energy levels of atoms in diamond; after Drăgulescu and Petrovici (1973).

composed from 10^{22} discreet levels of energy. Due to this huge number of allowed levels, the distribution of energy inside the band can be considered as continuous. The real energetic state of the solid corresponds to a certain distance r at equilibrium, i.e., the parameter of the reticular string upon which the compactness is made.

"The overlapping" of the bands determines a set of properties of the solid. As this overlap is more pronounced, as it is easier the passing of electrons from the inferior occupied bands in the unoccupied superior ones.

If the width of a forbidden zone which separates the occupied from those unoccupied is large, the solid will be an *insulator*, otherwise will be a *semiconductor*; this because the conductibility through electrons is associated with their passing from the last occupied band – valence band – to the first free band – *conduction band*, without energy consumption.

One has illustrated already that the bond between two polyatomic particles is the result of more types of interactions, the pure bond being an approximation or a particular case.

For example, the inert gases crystals are formed only by van der Waals bond.

So on, neglecting the van der Waals interaction, there can be admitted the existence of a pure covalence in diamond, or of a pure metallic bond in compact lattices of metals.

In any case, the possibility of the coexistence of various interactions for the same structures suggests a continuous passing between the four fundamental types of chemical bonds as for the corresponding homodesmic lattices, by the heterodesmic lattices.

In the Figure 4.66, the tetragonal paradigm of chemical bonding is illustrated.

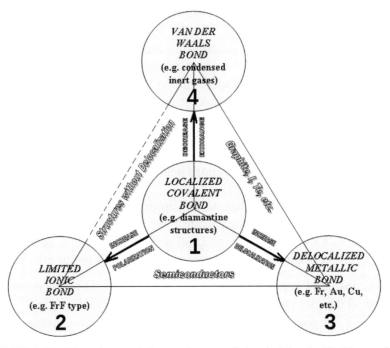

FIGURE 4.66 The scheme of the coexistence of chemical bonds (1)–(4): covalent, ionic, metallic and van der Waals chemical bonds throughout the quantum concepts of localization, polarization, delocalization, and exchange interactions, respectively; after Putz and Chiriac (2008).

The *vertices of tetrahedron* correspond to the homodesmic structures (e.g., 1: diamond, 2: CsF, 3: Na, 4: He).

The tetrahedron edges represent the continuous passing from a kind of bond to another, respectively the coexistence inside the same bond (homodesmic lattices) or inside the same structure (heterodesmic lattices) of two types of interactions (Chiriac-Putz-Chiriac, 2005):

- The edge 1–2 corresponds to some structures of oxides (CaO, Al_2O_3), silicates of type Mg_2SiO_4, etc.
- The edge 1–3 represents the stepping delocalization (collectivization) of valence electrons as for the case of graphite lattice, etc.
- Along the segment 1–4 the intra-nodal covalent bonds and the inter-nodal van der Waals bonds are grouped as like are the crystalline lattices of the combinations with the diatomic molecules found in gaseous state in common conditions: H_2, O_2, N_2, Br_2, etc.
- The edges 2–3 and 3–4 correspond to the alloys from various metals and metallic structures (ZnHg), which present deviations from the compact arrangement.
- The edge 2–4 represents only the presence of the van der Waals interaction in any ionic structure (usually in negligible proportion).
- The face 1-2-4 corresponds of the vast majority of crystalline combinations such as: salts of oxygenated acids, organic combinations with simple bonds, etc., so generally the structures where the electrons delocalization is excluded.
- The face 1-3-4 interests the crystallochemistry of organic combinations with double combinations delocalized (benzene, some macromolecules of rubber type).
- The face 1-2-3 describes the structures of interstitial compounds of type TiC, of semiconductors, etc.
- The face 1-3-4 has the same significance as the edge 2–4, i.e., the (negligible) van der Waals interaction is recognized as for the alloys case.

The inside of tetrahedron includes the structures where there are presented all the four types of fundamental bonds.

In this category are included the structures with hydrogen bonding as the compounds of: MoS_2, CdI_2, proteins, etc.

As resulted from Figure 4.66, the lattices having the particles reunited in a single type of bond represent particular cases. Yet, since we consider only the predominant interaction, the framing of the types previously discussed for the corresponding homodesmic lattices is completely justified. The group of heterodesmic lattices corresponding to the faces 1–2–3 and 1–3–4 of tetrahedron of Figure 4.66, after the bond ensuring the lattice integrity, encounters difficulties caused by the continuous transition between the fundamental types of bonding. Indeed, excepting some combinations such as: organic compounds or compounds of CH_4 type, the molecules Cl_2, Br_2, etc., that can be located on the edge 1–4, and of the ionic lattices (edge 1–2), the other studied structures present properties which impose the consideration of at least three types of bonds; thus, the compounds are located on a face of paradigmatic tetrahedron of Figure 4.66.

A typical example is given by the graphite $^2_\infty C^{[3]}$ h, Figure 4.67. In graphite structure the atoms of C form flat hexagonal lattices, parallel, each atom placed in the node of such a lattice having coordination 3 with a hybridization sp^2 of covalent bonds. Inter-nodal distance is 1.24 Å, and the distance between planes has the value 3.41 Å. The electrons π are delocalized in the carbon atoms plane, conferring to the graphite properties characteristic to the metals: metal glossiness, high optical absorption, electric conductibility. The bond between layers is of van der Waals type and explains the low hardness of the graphite and the tendency to cleavage after (0001) planes. Thus, in the graphite structures three types of bonds coexist: covalent (between two atoms of flat hexagonal lattices), van der Waals (between two flat lattices) and metallic (between all the atoms of a reticular plane).

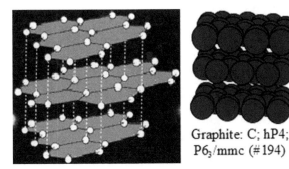

Graphite: C; hP4;
$P6_3/mmc$ (#194)

FIGURE 4.67   Left: the graphite structure: with density 2.26 g/cm³ is the most common form of the Carbon, where the atoms of Carbon are placed in the corners of some united hexagons and disposed in parallel layers; after Heyes (1999).Right: structural cell of graphite, under which there is mentioned: the Pearson spatial notation, the International notation, as well as the number of spatial group, according to the Tables 2.17 and 2.16, respectively; after U.S. Naval Research Laboratory/Center for Computational Materials Science (2003).

The more important structures of heterodesmic type are: cuprite type, galene type, nickel arsenide and the structure types of the molecular compounds. The molecular compounds were previously discussed in subsection 4.4.1.4.; the other compounds will be shortly presented in the following.

Cuprite type, $^3_\infty Cu_2^{[2]}O^{[4]}c$, has covalent features, ionic and metallic, Figure 4.34.

Galena structure $^3_\infty Pb^{[6]} S^{[6]} c$ is similar to the one of Na chloride (see Figure 4.30); a species is found with the particles in the octahedral holes of compact cubical lattice of the other species. After the type of bonds, the combination is located inside the tetrahedron of Figure 4.66, in the proximity of the face 1–2–3, which explains its semiconductor character. The structures of galena type also characterize the sulphides of Mg, Ca, Sr, Ba, etc.

The structures of nickel arsenides is similar, $^3_\infty Ni^{[6]}As^{[6]}h$, with the distinction that the shape of coordination polyhedra is different. The nickel is octahedrally coordinated by the arsenide atoms and the coordination polyhedron of arsenide is a trigonal prism. The structure is formed by combining a simple hexagonal lattice of atoms of Ni with a compact hexagonal lattice of atoms of arsenide. This type of structure, Figure 4.68, is found

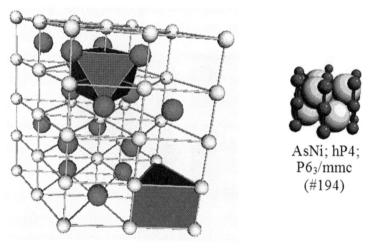

AsNi; hP4;
P6$_3$/mmc
(#194)

FIGURE 4.68   Left: the NiAs lattice [32]. Right: the structural cell of NiAs, under which there is mentioned: the Pearson spatial notation, the International notation, as well as the number of spatial group, according to the Tables 2.17 and 2.16, respectively; after U.S. Naval Research Laboratory/Center for Computational Materials Science (2003).

to a various range of binary compounds such as: sulphides, selenides and tellurides of Fe, Co, Ni; tellurides of Mn, Pd, Pt; arsenides of Mn and Ni; stibia of Cr, Mn, Fe, Co, Ni, Pd ecc.

The center of tetrahedron of Figure 4.66 corresponds to the compounds where the weight of the 4 fundamental types of bonds is almost the same.

An example of such a structure of this class is the molibdenite MoS_2 type along the CdI_2.

The molibdenite lattice, $^2_\infty Mo^{[6]}S_2^{[3]}h$, is formed by planes of molibdenite atoms, overlapped between each two planes of the atoms of sulfur. Thus, the lattice appears to be formed by layers inside of which the atoms are covalently bond.

Coordination polyhedra are: trigonal prism for Mo and trigonal pyramid for S, Figure 4.69. This type of structure is characteristic to the sulphides, selenides of Mo and W tellurides.

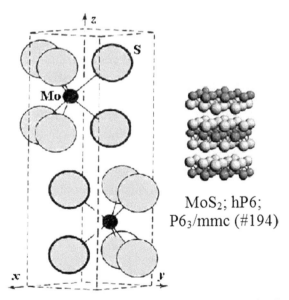

MoS₂; hP6;
P6₃/mmc (#194)

FIGURE 4.69   Left: the Molibdenite lattice [9]. Right: the structural cell of MoS₂, under which there is mentioned: the Pearson spatial notation, the International notation, as well as the number of spatial group, according to the Tables 2.17 and 2.16, respectively; after U.S. Naval Research Laboratory/Center for Computational Materials Science (2003).

4.4.2   PAULING RATIONALIZATION FOR THE COMPLEX IONS' BONDING

Until now the geometric conditions and solid structures possible for systematization in crystalline classes have been analyzed, from a combined view of the symmetric arguments and those of compaction; the present discussion follows (Chiriac-Putz-Chiriac, 2005).

Already, by introduction the compaction notion, by the consequence of the interstices, there is underlined the importance to considering the interactions, the repulsion-attraction forces, suggesting the possibility of geometric rationalization of the paradigm of formal ions.

Thus, to the question: can the chemical bonds (in solids) being geometrically rationalized? – the answer is affirmative, but the criteria based on which these rationalizations are made should be analyzed, and moreover, used very carefully, not being yet established a geometric model with an universal degree of validity for the chemical bonds and compounds, therefore valid in any conditions.

In other words, also in terms of general philosophy, this geometric limitation resides in the impossibility of geometrization of quantum nature of the matter. In this sense, even more generalizing the paradigm of the formal ions, Linus Pauling was the one who firstly formulated a set of coherent rules that rationalize in geometric terms the energetic stability of solid structures, and therefore these are generically called "Pauling's Rules" (Evans, 1964; Borg & Diens, 1992).

Rule 1 (also called the principle of ionic radii) defines the ionic crystals as a set of "co-bonding" polyhedra, with the anions around a central cation forming the so-called coordination polyhedron; moreover, the distance cation-anion is simply calculated as the ions' radii sum (also called the *Pauling or Goldschmidt* radius)

$$r_0 = r_+ + r_-$$
(4.22)

Instead, their ratio defines the maximum- number of coordination in contact cation-anion framework.

As long as the cations and the anions commonly occupy compact locations and respectively the interstitials' structures, Rule *1* indicates, shortly, how, from the cations radii ratio of (r_+) with the one of anions (r_-) there can be identified the crystalline type, by the corroboration with the coordinative number, according to the determinations of Section 4.2, and re-summarized in Table 4.14 (Chiriac-Putz-Chiriac, 2005).

The example which works best in agreement to the Rule 1 Pauling referees to the coordination of cations in minerals of silicates type.

Thus, if one rewrites the general silicates formula: $AX_mY_n(Z_pO_q)W_r$ with A, X, Y and Z referring to the cations, therefore, based on the cation species and its location in relation to the oxygen, there results, as in Table 4.15 the cationic radii, the ratio towards the anion radius of oxygen and, at the end, the coordination in agreement to the Rule 1 of Pauling, compare with Table 4.14.

Now arises the question: does the 1 Pauling rule also an energetic correspondent? Yes, it has and consists, by the aid of Coulombian electrostatic potential,

$$V_{ij} = \frac{z_+ z_- e_0^2}{r_{ij}} \quad , e_0 = \frac{e}{4\pi\varepsilon_0} \tag{4.23}$$

in the so-called *Madelung constant* (M), that mediates until the convergence to all the repulsion-attraction bonding energies (E_{ij}), in equal number with the number N of molecules in crystal, between a reference cation and the N_i anions and cations from the successive layers, allowing, at the end, the energy lattice rewritten (E_L) in an unitary way (Borg & Diens, 1992):

$$-E_L = N \sum_{ij} N_i V_{ij} \cong M \frac{z_+ z_- e_0^2}{r_0} \tag{4.24}$$

TABLE 4.14   The Limiting Values of the Spherical Dimensions Ratio for a Certain Number of Coordination; after (Chiriac-Putz-Chiriac, 2005)

Coordination	Ration of Radii r_+/r_- (Relative Dimension)
2	$0 - 0.15$
3	$0.15 - 0.22$
4	$0.225 - 0.414$
6	$0.414 - 0.732$
8	$0.732 - 1.37$
12	1

TABLE 4.15 The Pauling Rule No. 1 As Applied to the Silicates Compounds*

Location	Ion	Radius (Å)	Radius ratio	Coordination
Z	Si^{4+}	0.42	0.300	4
	Al^{3+}	0.39	0.279	
Y	Fe^{3+}	0.64	0.457	6
	Mg^{2+}	0.66	0.471	
	Ti^{4+}	0.68	0.486	
	Fe^{2+}	0.74	0.529	
	Mn^{2+}	0.80	0.571	
X	Na^+	0.97	0.693	6–8
	Ca^{2+}	0.99	0.707	
A	K^+	1.33	0.950	8–12
	Ba^{2+}	1.34	0.957	
	Rb^+	1.47	1.050	
	Sr^{2+}	1.47	1.050	

*After The Chemistry of Silicates (2003).

Equation (4.24) is reduced on the distance r_0 of the anion-cation bonding distance.

Thus, the Madelung energetic rationalization reduces the lattice energy to the anion-cation bond energy, in the formal ions paradigm context.

For example, for the crystalline state NaCl in Figure 4.70 each cation Na^+ (of smaller dimensions – even if here isn't represented – than the anion with which coordinates) is surrender by anions of Cl^- and by other cations at distances expressible by Miller indices, so adapting the relation (4.29) at the inter-nodal distances inside the cubical system (Chiriac-Putz-Chiriac, 2005):

$$r_{[[hkl]]} = r_{[[100]]}\sqrt{h^2 + l^2 + k^2} \tag{4.25}$$

Combining the relations (4.23)–(4.25) while taking into account of the correspondence:

$$r_{[[100]]} = r_0, \ r_{[[hkl]]} = r_{ij} \tag{4.26}$$

the energy of the lattice (4.24), inside the cubical system, can be expressed such as:

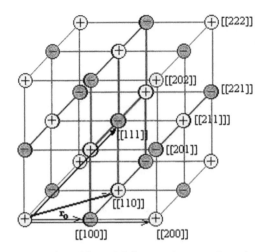

FIGURE 4.70   The Illustration of the nodal distances in successive spheres of neighboring for a reference cation Na^+ in the NaCl structure; after (Chiriac-Putz-Chiriac, 2005).

$$-E_L = \left[\sum_{i,h,k,l}(-1)^{h+k+l} p_{hkl} N_i \left(h^2 + l^2 + k^2\right)^{-1/2}\right] N \frac{z_+ z_- e_0^2}{r_0} \qquad (4.27)$$

Here remains to explicate the measures $(-1)^{h+k+l}$ and p_{hkl}, related to the way in which one makes the summing after the set of Miller indices. Firstly, the sign resulted from the series evaluation of (4.27) will decide the stability of the lattice: if it is negative, there is expected a lattice to be stabile due to the dominant attractive feature, while the positive sign will indicate the dominance of repulsive forces, prescribing the crystalline non-existence.

In order to explain the parameter p_{hkl} of (4.27) there will be analyzed the case of NaCl of Figure 4.70, for which the Table 4.16 is considered (Chiriac-Putz-Chiriac, 2005).

One notes that in Table 4.16 the so-called *weight of charge* has been introduced, in the second column, under fractional form p_{hkl} since based on the observations according which the anion-cation interaction can be classified in layers of influence, upon the maximum value found along the indices Miller indexing the lattice ions. Thus, there is noted in Table 4.16 the existence of two layers of influence: the first layer where the ions are indexed by Miller indices and where the value "1" is the highest, along the second layer where the value "2" is the highest existed.

TABLE 4.16   The Specific Parameters to the NaCl Lattice for the Calculation of Madelung Constant; (Chiriac-Putz-Chiriac, 2005)

Miller Indices	$phkl=$ 2^{-c}, $c:=Card\{max(h,k,l)\}$	N_i	$\left(h^2 + l^2 + k^2\right)^{1/2}$	$(-1)^{h+k+l} N_i \left(h^2 + l^2 + k^2\right)^{-1/2}$
100	1/2	6	$\sqrt{1}$	-6
110	1/4	12	$\sqrt{2}$	+8.485
111	1/8	8	$\sqrt{3}$	-4.620
200	1/2	6	$\sqrt{4}$	+3
201	1/2	24	$\sqrt{5}$	-10.730
211	1/2	24	$\sqrt{6}$	+9.800
202	1/4	12	$\sqrt{8}$	+4.244
221	1/4	24	$\sqrt{9}$	-8.00
222	1/8	8	$\sqrt{12}$	+2.310

Further, taking into account that for each structural point there can be made the combinations as ± for the maximum value of Miller indices associated, there results that for the N_i surrounding ion of class [[hkl]], placed at the same distance (4.25) from the central cation, each of them contributes to the interaction with the weight:

$$p_{hkl} = 2^{-Card\{max(h,k,l)\}} \tag{4.28}$$

Moreover, there is noted that the values of the weight (4.28) can be only 1/2, 1/4 and 1/8 (as the Miller indices are just one, two or all three equal to a maximum value, which however indicates the layer of interaction).

The last observation also gives the weight significance for (4.28) and of the necessity of its inclusion in the Madelung constant calculation: if one imagined the reference cation as being located in the cube's center of Figure 4.70, resulting in the Figure 4.71, then, by restricting the analysis only at the first layer of influence, there is noted that the structural dots of this layer, of Table 4.16, correspond to the center of the faces, edges and vertices of the cube included in the first layer of interaction, respectively.

Hence, immediately results how, between the first layer of interaction and the next one, the dots on the faces, edges and vertices belong to the weights 1/2, 1/4 and 1/8 of the cube associated to the first layer of interaction, so finding again the values p_{hkl} of Table 4.16.

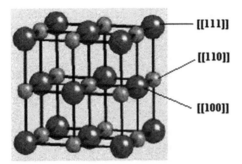

FIGURE 4.71 Representation of the associated cube for the first layer of cation-anion interactions in NaCl structure; after Heyes (1999).

Consequently, the lattice energy of Eq. (4.27) can be reduced to the Madelung constant, which, in turn, can be calculated in various orders of approximation, based on the layers of influence taken into account, following the formula:

$$M = \sum_{i,h,k,l}(-1)^{h+k+l}\, 2^{-Card\{max(h,k,l)\}}\, N_i\left(h^2 + l^2 + k^2\right)^{-1/2} \qquad (4.29)$$

Thus, if only the first layer of influence is considered, by using the data of Table 4.16, the Madelung constant is obtained from Eq. (4.29):

$$M^{I}_{NaCl} \cong -\frac{6}{2} + \frac{8.485}{4} - \frac{4.620}{8} = -1.457 \qquad (4.30)$$

Instead, if also the second layer of influence is considered, this one integrally containing the first one, thus in the formula (4.29) the weight will be customized by (4.28) only for the rest of the interactions in Table 4.16, so resulting in a better Madelung constant:

$$M^{II}_{NaCl} \cong (-6 + 8.485 - 4.620)$$
$$+ \left(\frac{3}{2} - \frac{10.73}{2} + \frac{9.8}{2} + \frac{4.244}{4} - \frac{8}{4} + \frac{2.31}{8}\right) = -1.750 \qquad (4.31)$$

Continuing this procedure for the influence layers increasingly higher, one gets the exact value $M_{NaCl} = -1.74755$, very close of Eq. (4.31).

Similarly, there are calculated the Madelung constants also for other structures, with the results mentioned in Table 4.17.

TABLE 4.17   The Pauling Rule No. 1 Rationalized by the Madelung Constant*

Structure type	Madelung constant	Coordination
CsCl	-1.763	8:8
NaCl	-1.748	6:6
ZnS (Wurtzite)	-1.641	4:4
ZnS (Blende)	-1.63806	4:4
Cu_2O (Cuprite)	-4.3224	2:4
TiO_2 (Rutile)	-4.816	6:3
CaF_2 (Fluorite)	-5.03878	8:4
Al_2O_3 (Corundum)	-25.0312	6:4

*After Borg and Diens (1992).

Table 4.17 establishes the fact that increasing of coordination corresponds to the increase of the Madelung constant (potential).

Yet, one should mention that the Rule 1 of Pauling approximately works, being valid only for 50% of the cases, so being rather an instrument for prediction than an effective rationalization.

Therefore, the Madelung constant analysis is recommended throughout the energetic calculation, for the confirmation of the structural coordination.

The Rule 2 of Pauling (also called the principle of electrostatic valence) refers to the local electro-neutrality and assets that in the stable ionic structures, the charge of an anion X^{x-} surrounded by n cations M^{m+} is balanced by the sum of *strengths of electrostatic bonds* (abbreviated SEB = m/n) with which polyhedrally coordinates:

$$\Sigma(m/n) = x \qquad (4.32)$$

Therefore, this second principle, or rule, correlates the number of coordination (which establishes the geometry of the polyhedron of coordination) not with the dimension of the radii, but with the valences of the ions found in the chemical bond.

For example, for the class of perovskite, analyzing the structure $CaTiO_3$ of Figure 4.38, there results that Ca^{2+} is 12-coordinated by O^{2-}, wherefrom there results that the bond Ca-O has SEB = $^2/_{12} = ^1/_6$; on the other side, Ti^{4+} is 6-coordinated by O^{2-}, wherefrom there results that the bond Ti-O has SEB = $^4/_6 = ^2/_3$.

As long as O^{2-} has a total valence 2 there results that the unique combination that can be satisfied in agreement with the Rule 2 of Pauling recommends $2 = \{4\ ^1/_6\ (Ca^{2+})\} + \{2\ ^2/_3\ (Ti^{4+})\}$, i.e., certifying the fact that the oxygen should be common at 4-cube-octahedra CaO_{12} and at 2 octahedra TiO_6.

Thus, this rule is closer by the rationalized character of the structures that it describes, especially helping for the verifications of the structures, previously determined.

Rule 3 of Pauling (also called the principle of polyhedral connection) prescribes how the decreasing in the structures establishment is ordered paralleling to the union of the polyhedra in the order: vertices, edges, faces.

In Figure 4.72, various polyhedra coordination are exemplified by the various polyhedra connections.

The rule no. 3 of Pauling, actually, geometrically rationalizes the principle of minimizing the electrostatic energy (of repulsion) inside a lattice.

The rule recommends a low number of coordination and the coordination through the vertices for the high cationic charges (e.g., Al^{3+}, Si^{4+}), thus applying also for the low ionic rations r_+/r_-, see Table 4.14, until the stability limit in coordination.

The rule is satisfied by the compounds with high polarity (that record an acute separation between the centers of positive and negative charges) as is the case of fluorines and oxides, and is less applied for the compounds with low polarity (for example: SiO_2 is coupled by the vertices of the tetrahedra, SiS_2 is coupled by edges).

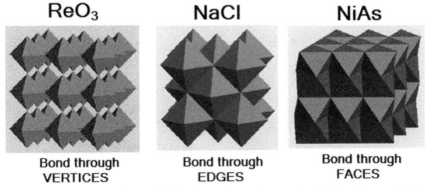

ReO₃	NaCl	NiAs
Bond through VERTICES	Bond through EDGES	Bond through FACES

FIGURE 4.72 Types of coordination by vertices, edges and polyhedra faces, with the distance between the coordination cations ranging in 1:0.58:0.38 proportion, thus properly increasing the repulsive potential between them; after Heyes (1999).

Rule no. 4 of Pauling (also called the principle of cation evasion) comes as a continuation of the Rule 3, covering the case of much more types of cations found in the structures, for which one recommends that those cations with high valence and low number of coordination tend to couple as little their polyhedra of coordination.

In Figure 4.73, the rule no. 4 is detailed for the perovskite structure $CaTiO_3$, for which the cube-octahedron CaO_{12} (Ca^{II} being 12-coordinated) couples its faces, respecting the octahedron TiO_6 that confines only with the vertices (Ti^{IV} being 6-coordinated).

Another series of compounds which fully satisfy this rule are the minerals of silicates type for which, the tetrahedra, as coordinated polygons for cations of silicon tend to couple in a minimal way, according to the Rules 3 and 4, see for example the series of Table 4.13.

Rule 5 of Pauling (also called the principle of crystallographic locations) recommends the minimum contributing species for a structure in crystalline form, i.e., there is prescribed an environment as similar as possible with the one of the atoms or the species base-motif for a crystal, or a minimum number of interstitial cations.

For a good application of this last rule, it will be used in conjunction with the previous ones, especially with Rule 2 of local electro-neutrality.

For example, treating the garnet mineral ($Ca_3Al_2Si_3O_{12}$) as an ionic crystal, in the formal ions paradigm context to which the Pauling's rules are subscribed, there is made the Table 4.18, as resulted from the calculation of the strength of electrostatic bond for the coordination of the present cations with oxygen.

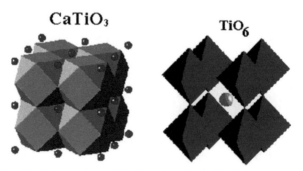

FIGURE 4.73   The Cubo-octahedral (left) and octahedral (right) couplings in perovskite structures $CaTiO_3$; after Heyes (1999).

TABLE 4.18   The Pauling Rule No.5 Applied for the Garnet Mineral Here Treated As Ionic Crystal in the Context of the Formal Ions Paradigm, With the Associate *Strengths of Electrons in Bonding* (SEB)

	Ca^{2+}	Al^{3+}	Si^{4+}
Coordination	8	6	4
SEB	$^1/_4$	$^1/_2$	1

Thus, the anion O^{2-} has the valence 2 to be satisfied, from the cations combinations of Table 4.18, according to the Rule 2. Yet, the possible combinations are reduced to a single one, for the requirement of the satisfaction of the Rule 5, so that each anion O^{2-} to be surrendered by the same cations combination in Table 4.18; there results the coordination illustrated in Figure 4.74.

Nevertheless, there should be mentioned that, in general, none of Pauling's rule is, by itself, the decisive criterion for the assessment of a structure, or a coordination of geometry for a crystalline compound.

For this reason, these rules should be combined between them and, moreover, confronted with other criteria, energetic or geometric-topological. Thus, the various energetic approaches for the rationalization of the structures and chemical bonds will be further presented.

4.4.3   THE POTENTIAL FUNCTIONS' PARADIGM OF THE CHEMICAL BONDING

Assessing of a certain bond between the particles composing a crystalline structure influences in a direct way the coordination ratios between

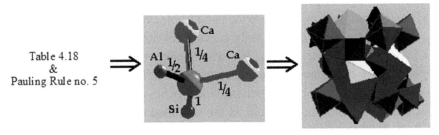

FIGURE 4.74   Assessing the structure of Garnet crystal based on the Pauling rule no. 5; after (Borg & Diens, 1992; Heyes, 1999).

particles, their dimensions and the assessment of the substance's edifice (Putz, 2006).

For the pure covalent bonds, the coordination is exclusively determined by the electronic atomic configurations which interact. The strength of covalent bond (i.e., bonding energy) depends on the overlap measure of valence orbital of the partners; the present discussion follows (Chiriac-Putz-Chiriac, 2005).

Thus, the bonding energy measures the exchange effect, i.e., the repartition of electric charges based on the mutual spin orientation, according to the Pauli principle, being a quantum measure of Coulombian interaction.

The un-oriented bonds tend to group the particles in as much compact structures, based on their relative dimensions.

But, there should be noted that there is no relation between the bond strength and the structure compactness, a fact demonstrated by the compactness of affine structure of diamond with the one compact of the inert gases' crystals.

For the metallic structures, the delocalization of peripheral electrons directly determines the bonds strength and the physical properties of the crystals.

For the ionic structures, the coordination in lattice suppresses the spherical symmetry of electrostatic field, the charge in density having the tendency of localization towards particles' directions on that coordinate.

The tendency of electrons localization is more pronounced as much the bond has a pronounced covalent character, i.e., the more the ratio between (integral) influence of exchange towards the (integral) Coulombian influence (classic-electrostatic) is bigger. After Pauling, the ratio between covalent and ionic character of a bond can be appreciated based on the electronegativity of the atoms participant inside the bond (Putz, 2012). More the difference of electronegativity is bigger more the bond is ionic pronounced.

Recently, the electronegativity concept has reached its complete quantitative and qualitative form, by identification with minus of chemical potential of the system (atom, molecule, radical) (Putz, 2011). Thus, for an isolated atom or involved in bonds, with (hybrid) orbital "i" that competes at chemical bond, there is defined the so-called *orbital electronegativity* (Putz, 2008):

$$\chi_i = -\frac{\partial E}{\partial N_i}\bigg|_{N_a} \tag{4.33}$$

where E is the atom total energy, N_i is the charge number in the considered orbital i, and N_0 has the meaning that all the other occupied orbital are kept with the fixed number of occupation.

For the orbital of valence shell, i.e., the region of chemical interaction, the electronegativity (4.33), written in the approximation of finite differences, successively takes the forms:

$$\chi = -\frac{\partial E_N}{\partial N}\bigg|_{N_0} \cong -\frac{E_{N_0+1} - E_{N_0-1}}{2}$$

$$= \frac{(E_{N_0-1} - E_{N_0}) + (E_{N_0} - E_{N_0+1})}{2} = \frac{IP + EA}{2} \equiv \chi_M \qquad (4.34)$$

The expression (4.34) defines the so-called Mulliken electronegativity, in terms of ionization potential IP and of electronic affinity EA, so determining by their semi-sum the so-called scale of electronegativity, see Table 4.19.

Following the electronegativity variation inside the periodic system of elements there is noted that the general tendency, with some exceptions, is to decrease down groups (vertical columns) and to increase in periods, once with the increase of the atomic number, see Figure 4.75.

Moreover, from Figure 4.75, there is noted that for the binary compounds AB, the difference in electronegativity, $\chi(A)$-$\chi(B)$, naturally quantifies the ionic character, covalent and metallic of chemical bonds.

The simplest way to view this tendency is given by the paradigmatic construction of the triangle (Figure 4.76), where the sides correspond to the chemical bonds and the corners of the extreme limits: covalent (C), ionic (I) and metallic (M) (Ketelaar, 1958).

Thus, immediately appears that between the metallic and the covalent character of chemical bond, on the side MC of Figure 4.76, there is recorded the same electronegativity, $\chi(A)=\chi(B)$, in accordance with the quantum fact according which the covalent bond and the metallic one are associated with "the same nature of maximum overlapping of exchange forces" (Slater, 1939).

Thus, along the side C-M in Figure 4.76, we actually move from the right to the left in Periodical System, see Table 4.19. Complex modern approaches of electronegativity (4.33), as those relating with Mulliken

TABLE 4.19 The Mulliken Scale of Electronegativity (χ_M) Calculated for the First Four Periods of Periodical System, Based On the Relation (4.34), As the Semi-Sum of Ionization Potentials (IP) and of Corresponding Electronic Affinity (EA)*

Legend:

Element	χ_M	IP	EA
H	7.18	13.62	0.73
He	12.27	24.65	-0.21

Element	χ_M	IP	EA
Li	3.02	5.41	0.62
Be	3.43	9.36	-2.5
Na	2.80	5.02	0.52
Mg	2.6	7.7	-2.39
K	2.39	4.37	0.52
Ca	2.29	6.14	-1.66
Sc	3.43	6.55	0.21
Ti	3.64	6.86	0.42
V	3.85	6.76	0.94
Cr	3.74	6.76	0.62
Mn	3.85	7.49	0.31
Fe	4.26	7.90	0.62
Co	4.37	7.90	0.94
Ni	4.37	7.7	1.14
Cu	4.47	7.8	1.25
Zn	4.26	9.46	-0.94
Rb	2.29	4.16	0.52
Sr	1.98	5.72	-1.77
Y	3.43	6.45	0.52
Zr	3.85	6.86	0.83
Nb	4.06	6.86	1.14
Mo	4.06	7.18	1.04
Tc	3.64	7.28	0.00
Ru	4.06	7.38	0.62
Rh	4.26	7.49	1.04
Pd	4.78	8.42	1.25
Ag	4.47	7.59	1.35
Cd	4.16	9.05	-0.62

Element	χ_M	IP	EA
B	4.26	8.32	0.21
C	6.24	11.34	1.25
N	6.97	14.56	-0.62
O	7.59	13.62	1.46
F	10.4	17.47	3.33
Ne	10.71	21.63	-0.31
Al	3.22	6.03	0.42
Si	4.68	8.22	1.25
P	5.62	10.5	0.73
S	6.24	10.4	2.08
Cl	8.32	13	3.64
Ar	7.7	15.81	-0.42
Ga	3.22	6.03	0.42
Ge	4.58	7.90	1.25
As	5.3	9.78	0.83
Se	5.93	9.78	2.08
Br	7.59	11.86	3.43
Kr	6.86	14.04	-0.42
In	3.12	5.82	0.31
Sn	4.26	7.38	1.25
Sb	4.89	8.63	1.04
Te	5.51	9.05	1.98
I	6.76	10.5	3.12
Xe	5.82	12.17	-0.42

*All the units are in eV (electron-volts). Conversion at units J/mol is made by following the formula 1MJ/mol=10.4 eV (Putz, 2006); conversion to Pauling scale (χ_P) is made according to the $\chi_P \approx 3.5\chi_M - 0.2$ transformation (Putz, 2006).

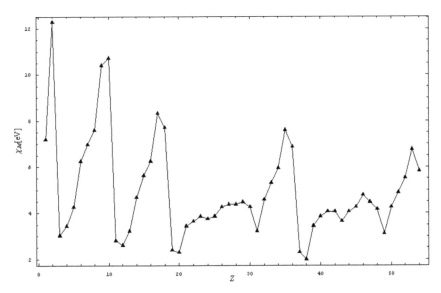

FIGURE 4.75   The Mulliken electronegativity variation, upon the Table 4.19; after Putz (2006).

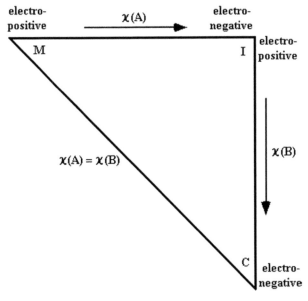

FIGURE 4.76   The schematic representation of the triangle of ionic (I), covalent (C) and metallic (M) tendencies of chemical bond, for the binary AB compounds, in terms of the differences of electronegativity paradigm; after (Ketelaar, 1958; Putz, 2006).

version (4.34), in a systematic manner, make the object of the latest research in the field (Putz, 2006, 2008, 2011, 2012).

Remarkable, if the electronegativity definition relation (4.33) is formally integrated by taking into account the nature of the electronegativity potential as minus chemical potential for the system in focus, the total energy would result as an integral that would correspond to the summing (4.24).

This way, the electronegativity, *as potential for chemical bond formation*, quantitatively and qualitatively contains the competition between the repulsive character (through IP) and the attractive one (by EA) of the multi-electronic system at equilibrium (or in interaction).

Hence, the idea that the formal ions paradigm at the level of the bonding energies in chemical combinations can be analytically rationalized by considering in the bonding energy (or latter in the lattice) two components, one corresponding to electrostatic attraction of the ions with opposed sign, and the other determined by the electrostatic repulsion of the ions with the same sign.

Further, the quantitative analytical unfold of the introduction of the so-called potential functions that modeling the bonding energies will be approached, at the level of the molecular combinations, wherefrom the energies of cohesion in the "lattice" that actually "freezes" the relative chemical bond will be then abstracted.

Considering as starting point the general form of the inter-atomic potential, Figure 4.77, there is noted that, essentially, the potential function, as the electronegativity, in a qualitative meaning, contains the competition of two tendencies: repulsive and attractive,

$$V(r) = V_r(r) + V_a(r) \qquad (4.35)$$

wherefrom, in a quantitative meaning (with sign), it takes the general form:

$$V(r) = \frac{A}{r^n} - \frac{B}{r^m} \ , \ n > m \qquad (4.36)$$

where n, m, A and B depend by the bond type, and with the values that can be experimentally determined.

From the competition of the attractive term with the repulsive one, in Eq. (4.36), the chemical bond, characterized by the equilibrium conditions applied to the total potential, is modelled. Thus, there can be written:

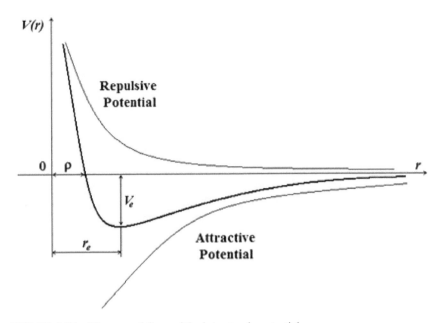

FIGURE 4.77 The general form of the interatomic potential.

$$\left(\frac{\partial V(r)}{\partial r}\right)_{r=r_e} = 0 = -n\frac{A}{r^{n+1}} + m\frac{B}{r^{m+1}} \Rightarrow \frac{nA}{mB} = r_e^{n-m} \qquad (4.37)$$

wherefrom, at equilibrium, the potential (4.36), with (4.37), will take the forms:

$$V_e \equiv V(r_e) = \frac{B}{r_e^m}\left(\frac{m}{n} - 1\right) = \frac{A}{r_e^n}\left(1 - \frac{n}{m}\right) \qquad (4.38)$$

which allows the abstraction of the parameters A and B as functions of V_e:

$$A = \frac{m}{m-n}V_e r_e^n \;,\quad B = \frac{n}{m-n}V_e r_e^m \qquad (4.39)$$

Finally, with (4.39) back in the original potential (4.36) there results the general form of the prototype of the potential function of a diatomic bond:

$$V(r) = \frac{V_e}{m-n}\left[m\left(\frac{r_e}{r}\right)^n - n\left(\frac{r_e}{r}\right)^m\right] \qquad (4.40)$$

known as the *Mie equation*, after the name of its inventor, see (Borg & Diens, 1992).

Often, the Eq. (4.40) is rewritten as based on the dissociation energy (V_d), experimentally accessible, grounded on the relation with the equilibrium energy V_e:

$$V_d \equiv V(r \to \infty) - V(r = r_e) = 0 - V_e = -V_e \qquad (4.41)$$

Here should be mentioned that, even if there are known the potential type which overlap the phenomenology of diatomic bonds, among the traditional functions being those of Lennard-Jones, Morse and Born, these are, in fact, special cases of Mie equation (4.40), as will be further seen.

For example, for the weak interaction of van der Waals type, specific to the inert gases, a form of Mie equation (4.40) is employed as based on the minimum distance ρ of Figure 4.77, for which the interaction potential between the two atoms is zero. Thus results the condition:

$$V(r = \rho) = 0 \Rightarrow m\left(\frac{r_e}{\rho}\right)^n = n\left(\frac{r_e}{\rho}\right)^m \Rightarrow r_e = \rho\left(\frac{n}{m}\right)^{1/(n-m)} \qquad (4.42)$$

as the potential function characteristic to the weak chemical bond; by combining the relation (4.42) with the form of the Mie equation (4.40) written with the energy of dissociation (4.41) one gets:

$$V(r) = \frac{V_d}{n-m}\left(\frac{n^n}{m^m}\right)^{1/(n-m)}\left[\left(\frac{\rho}{r}\right)^n - \left(\frac{\rho}{r}\right)^m\right] \qquad (4.43)$$

The values of m and n, as notified also previously, are experimentally determined, through a calculation based on the relation (4.43), and then compared with the measured values, e.g. the thermodynamic parameters, as the isotherm compressibility and the free Helmholtz energy. Thus, empirically, there was established that, most often, there are obtained good results based on the choice:

$$n=2m=12 \qquad (4.44)$$

which, by the substitution in Eq. (4.43), generates the so celebrated *Lennard-Jones potential* (LJ), or "6–12" potential, with the form:

$$V_{6-12}^{Lennard-Jones}(r) = 4V_d \left[\left(\frac{\rho}{r} \right)^{12} - \left(\frac{\rho}{r} \right)^6 \right] \qquad (4.45)$$

Turning now to the "freezing" bond in solid state, neglecting the kinetic energy of the inert gases atoms, the associated lattice energy is obtained, as already prescribed in relation (4.24), as the sum of the Lennard-Jones potential towards the all (N) pairs of atoms in crystal:

$$-E_L = N \sum_{ij} V(r_{ij}) \qquad (4.46)$$

Because in writing of the Lennard-Jones potential (4.45) the potential sign was taking into account, further, by recalling the initial discussion of the potential function (4.36), for performing the sum (4.46) with (4.45) - the sign is implicitly, and the cohesion energy of the lattice will be:

$$E_L^{Lennard-Jones}(r) = 4NV_d \sum_{ij} \left[\left(\frac{\rho}{d_{ij}r} \right)^{12} - \left(\frac{\rho}{d_{ij}r} \right)^6 \right] \qquad (4.47)$$

The sums in equation (4.47) are evaluated for the (d_{ij}) distances between a reference atom (j) and the atoms (i) which coordinate it, resulting, for the cases of compact cubical and hexagonal structures of Figures 4.15 and 4.11, respectively, the values (Kittel, 1996):

$$(FCC) \begin{cases} \sum_{ij} d_{ij}^{-12} = 12.13188 \\ \sum_{ij} d_{ij}^{-6} = 14.45392 \end{cases} ; (HC) \begin{cases} \sum_{ij} d_{ij}^{-12} = 12.13229 \\ \sum_{ij} d_{ij}^{-6} = 14.45489 \end{cases} \qquad (4.48)$$

as it can be seen, very close to 12, i.e., the number of the neighbors of first order (the closest) of FCC, see Figure 4.15-left.

Because the structure FCC is the one specific to the NaCl salt, the prototype of the paradigm of the formal ionic bond, here it is generalized, by the potential functions also for other bonds and structures, see the set FCC of Eq. (4.48) for the further energetic evaluations. Thus, the energy (4.47) firstly takes the parametric shape:

$$E_L^{Lennard-Jones}(r) = 4NV_d \left[12.13\left(\frac{\rho}{r}\right)^{12} - 14.45\left(\frac{\rho}{r}\right)^6 \right] \tag{4.49}$$

which can be simplified by imposing the equilibrium condition

$$\left(\frac{\partial E_L^{Lennard-Jones}(r)}{\partial r}\right)_{r_0} = 0 = 4NV_d \left[(12)(12.13)\frac{\rho^{12}}{r_0^{13}} - (6)(14.45)\frac{\rho^6}{r_0^7} \right]$$

$$\Rightarrow \frac{r_0}{\rho} = 1.09$$

$$\tag{4.50}$$

resulting in the general form of the rare gases lattices' energy:

$$E_L^{Lennard-Jones}\left(\frac{r_0}{\rho} = 1.09\right) = -(4.30)(4NV_d) \tag{4.51}$$

with the values of the Table 4.20.

Now, there is remarked how the universal ratio (4.50) is found with a very good accuracy for all the rare gases thus confirming the reliability of the potential form, as Lennard-Jones potential, for the treatment of the weak chemical bond.

TABLE 4.20   The Lennard-Jones Energy (4.51) and the Properties of the Crystals of the Inert Gases (At 0K and Zero Atmospheric Pressure)*

Element Properties	*He (liquid)*	*Ne*	*Ar*	*Kr*	*Xe*
$V_d (10^{-16}$ erg)	14	50	167	225	320
$- E_L^{theoretical}$(KJ/mol)	1.45	5.178	17.297	23.305	33.145
$- E_L^{experimental}$(KJ/mol)	-	1.88	7.74	11.2	16.0
r_0 (Å)	-	3.13	3.76	4.01	4.35
ρ (Å)	2.56	2.74	3.40	3.65	3.98
r_0/ρ	-	1.14	1.11	1.10	1.09

*For the calculation of $E_L^{theoretical}$ a mole with crystalline substance was considered in (4.51), N = 6.02217×10^{23} mol^{-1}, while using the conversion factor 1 erg = 10^{-7}J (Kittel, 1996).

On the other side, from Table 4.20, there is noted a discrepancy between the theoretical values and the experimental ones for the lattices cohesion energy.

The explication consists in the full neglecting of the quantum effects in lattice.

For example, only the consideration of the quantum kinetic energy:

$$T = \frac{p^2}{2M_X} = \frac{h^2}{2M_X \lambda^2} \tag{4.52}$$

written as based on the direct use of the Broglie relation ($p=h/\lambda$, with h Planck constant and λ the wavelength associated to the particle momentum p), induces corrections that reduce the bond and the cohesion energy in a percent about 28, 10, 6 and 4% according to the ratio of the atomic masses, $M_X = 20.2, 39.9, 83.8, 131.3$, for $X = $ Ne, Ar, Kr and Xe, respectively.

In any case, the quantum treatments describe more accurate the cohesion energies for the crystals of the rare gases as the increasing of the complexity of approach and corresponds to a special direction of research in the current physical-chemistry.

Turning to the case of diatomic interaction around the equilibrium potential V_e of Figure 4.76, the chemical bond will be quantitatively modeled by the difference of the local potential:

$$\Delta V(r) = V(r) - V_e = \frac{V_e}{m-n}\left[m\left(\frac{r_e}{r}\right)^n - n\left(\frac{r_e}{r}\right)^m \right] - V_e$$

$$= \frac{V_e}{m-n}\left[m\left(\frac{r_e}{r}\right)^n - n\left(\frac{r_e}{r}\right)^m - (m-n) \right]$$

$$= \frac{V_d}{n-m}\left\{ n\left[1 - \left(\frac{r_e}{r}\right)^m \right] - m\left[1 - \left(\frac{r_e}{r}\right)^n \right] \right\} \tag{4.53}$$

If the potential (4.53) based on the relative displacement position respecting the equilibrium:

$$x = \frac{r - r_e}{r_e} \Rightarrow \frac{r_e}{r} = \frac{1}{1 + x} \tag{4.54}$$

is rewritten under simplified form:

$$\Delta V(x) = \frac{V_d}{n - m} \left\{ n\left[1 - (1 + x)^{-m}\right] - m\left[1 - (1 + x)^{-n}\right] \right\} \tag{4.55}$$

then, based on the equivalence of the involved binominal series with the exponential series, when restricting their expansion at the approximation of second order, $(1+x)^{-m} \cong e^{-mx}$, the equivalent potential form for Eq. (4.55) is obtained:

$$\Delta V(x) = \frac{V_d}{n - m} \left\{ n\left[1 - e^{-mx}\right] - m\left[1 - e^{-nx}\right] \right\} \tag{4.56}$$

Wherefrom, for the case in which $n=2m$, as in Eq. (4.44) but without the numerical specification from there, the celebrated *Morse potential* also results:

$$\Delta V_{n=2m}^{Morse}(x) = \frac{V_d}{m} \left\{ 2m\left[1 - e^{-mx}\right] - m\left[1 - e^{-2mx}\right] \right\} = V_d \left[1 - e^{-mx}\right]^2 \tag{4.57}$$

rewritten, most often, as based on the relation (4.54), such as:

$$V^{Morse}(r) = V_d \left[1 - \exp\left(-m\frac{r - r_e}{r_e}\right)\right]^2 \tag{4.58}$$

Worth underlining that the expression (4.58) incorporates bond quantum effects, by the exponential dependence virtue of the same nature with the function of molecular wave associated to the system.

Under these conditions, in solid state, when the chemical bond "crystallizes", the considered potential will be written as a combination between its Coulombian electrostatic nature (4.23) and the quantum one, suggested by the Morse potential (4.58), generating the so-called *Born-Mayer potential*:

$$V_{Born-Mayer}(r_{ij}) = V_{ij} + \Delta V = \frac{z_+ z_- e_0^2}{r_{ij}} + a\exp(-mr_{ij}) \qquad (4.59)$$

The associated cohesion energy, as long as for the potential (4.59) wasn't quantitatively considered the interactions sign, will be written as for the general case (4.24):

$$-E_L^{Born-Mayer} = N\sum_{ij} V_{Born-Mayer}(r_{ij})$$

$$= NM\frac{z_+ z_- e_0^2}{r} + 6Na\exp(-mr) \qquad (4.60)$$

the summation being restricted at the level of the neighbors of first (I) order for the NaCl lattice, i.e. the prototype of the paradigm of the formal ions.

The last assertion is supported by the values very close to the Madelung constants associated to the various types of coordination, around M_{NaCl} as resulted from Table 4.17.

Then, for the cohesion energy at equilibrium, on the relation (4.60) the minimum condition will be applied,

$$\left(\frac{\partial E_L^{Born-Mayer}}{\partial r}\right)_{r_0} = 0 = -M\frac{z_+ z_- e_0^2}{r_0^2} - 6am\exp(-mr_0)$$

$$\Rightarrow 6a\exp(-mr_0) = -M\frac{z_+ z_- e_0^2}{mr_0^2} \qquad (4.61)$$

wherefrom, by replacing the result (4.61) in the general form (4.60) the lattice energy expression for a binary formal ionic combination results, according to Born and Mayer, such as:

$$-E_L^{Born-Mayer}(r_0) = NM\frac{z_+ z_- e_0^2}{r_0}\left(1 - \frac{1}{mr_0}\right) \qquad (4.62)$$

For the practical application of the relation (4.62) one should perform a set of numerical specifications for the parameters that intervene.

Thus, for a mole of crystalline substance will be customized N with the Avogadro's number, $N = 6.02217 \times 10^{23}$ mol^{-1}, which allows the calculation of the product $Ne_0^2 = Ne^2/(4\pi\varepsilon_0)$, based on the notation from Eq. (4.23) and of the equivalence $1/(4\pi\varepsilon_0) = 10^{-7} c^2$, where $c = 2,997,925$ [m/s] is the speed of light in vacuum, and $e = 1.60219 \times 10^{-19}$ [C] the elementary electric charge; there is obtained: $Ne_0^2 = 332.388$[Kcal $\overset{o}{A}$/mol], where the conversion factor 1 cal $= 4.18$ [J] has been taken into account.

Moreover, one fixes $1/m$ of Eq. (4.62) to an empirical value, $1/m = 0.345\text{Å}$, and considered approximately the same for all crystals.

The Madelung constant will be written based on its value for NaCl, of Table 4.17, but modified so that to contain the semi-sum of the number of the ions of the gross formula, $(1/2)\Sigma_i n_i$ (a kind of formal average bond, for example equal to 2 for NaCl, or 3 for TiO_2).

Thus, in Eq. (4.62) we will have:

$$M = -1.7476 \,(1/2)\, \Sigma_i n_i \tag{4.63}$$

With all this considerations, the so-called *Kaputinski lattice energy* it will be semi-empirically written:

$$E_L^{Kaputinski} = 290.441 \cdot \sum_i n_i \frac{z_+ z_-}{r_+ + r_-} \left(1 - \frac{0.345}{r_+ + r_-}\right) [\text{Kcal/mole}] \tag{4.64}$$

The energy of Kaputinski type (4.64) is calculated in Table 4.21 for some crystalline substances of NaCl type, see also Table 4.6, and compared to the correspondent experimental values calculated as based on the Haber-Born cycle.

This calculations can be extended also to other crystalline combinations, initially applying the formal ions paradigm, with the prototype of the NaCl case generalized as above, where can be made corrections for the ionic radii of coordination 6:6 (specific for the NaCl case) as indicated in Table 4.22.

Finally, there can be concluded that, regardless the relation used, the lattice energy increases once with the strength of the bond, i.e., once with the increasing of the charge of the ions and the reduction of the distance between them, until a certain limit when the repulsive interaction of the ions of the same sign can annihilate the attraction effect.

TABLE 4.21   The Lattice Energies, Theoretically Calculated Based On the Eq. (4.64) of Kaputinski Type and Compared with Those Experimentally Deduced from the Born-Haber cycle, for Crystalline Substances of NaCl Type*

Structure of Na^z+Cl^z- type	r+[Å]	r-[Å]	$-EL^{\text{Kaputinski}}$ [Kcal/mol]	$-E_L^{\text{Born-Haber}}$ [Kcal/mol]
Li$^+$F$^-$	0.68	1.33	239.392	241.2
Li$^+$I$^-$	0.68	2.20	177.534	175.4
Li$^+$Cl$^-$	0.68	1.81	200.963	198.2
Li$^+$Br$^-$	0.68	1.95	191.895	188.5
Na$^+$Cl$^-$	0.97	1.81	183.019	183.8
K$^+$Cl$^-$	1.33	1.81	164.668	166.8
K$^+$F$^-$	1.33	1.33	190.053	191.5
Rb$^+$Cl$^-$	1.49	1.81	157.622	162.0
Cs$^+$F$^-$	1.70	1.33	169.882	171.0
Ag$^+$Br$^-$	1.15	1.95	166.528	-
Co^{2+}O^{2-}	0.65	1.40	942.681	954
Fe^{2+}O^{2-}	0.61	1.40	957.569	937

*After Borg and Diens (1992).

TABLE 4.22   Correction Factors, As Based on the Current Coordination Number, for the Radii of the Ions of 6:6 (Standard) Coordination; after (Chiriac-Putz-Chiriac, 2005)

Coordination Number	Correction Factor
12	1.12
8	1.03
6	1.00
4	0.94

The reduction of inter-nodal distances, once with increasing the ions charge (which equalizes with the reduction of the radii in the hypothesis of rigid spheres), leads therefore to increasing of the exchange interaction, i.e., is the expression of the changing character of the ionic bond.

In conclusion, one can say that, regardless the bond type, if the bond strength is strong enough, the particle (atom or ion) from a lattice has a lower symmetry of its electronic shell comparing with the identical free particle. A consequence of this phenomenon is the dependence of inter-nodal distances of coordination. More the coordination number if bigger, more is higher the symmetry of charge distribution around a particle, while its approximation with a rigid sphere being closer to the reality.

Thus, to the higher coordination numbers the large inter-nodal distances will correspond, as illustrated in Table 4.22 and re-synthesized in Table 4.23 for ionic and metallic lattices.

4.4.4 POLARIZATION IN CRYSTALS

The inter-nodal distances dependence by the coordination shows that the assumptions according to which the particles of a lattice do not exercise any influence on toward other have a very limited validity; the present discussion follows (Chiriac-Putz-Chiriac, 2005).

The particles interaction in ionic lattices is manifested by increasing the bond covalence degree. This deformation can be intuitively represented by the modification of the symmetry of the electronic cloud of a particle under the action of another particle, i.e., by polarization notion.

We have to mention that the representation of the interactions between the ions with the aid of this notion has a more qualitative value. But it is very helpful for the ionic combinations systematization, since representing the most important category of inorganic compounds.

By polarization one refers to the force field symmetry modification around a particle under the influence of another particle field.

As a result of polarization, the gravity center of the positive and negative charges from inside of a ion distances itself, the concerning ion becoming a dipole. Any particle can be characterized by the measure of the influence that exercises towards the symmetry of the electronic cloud of other

TABLE 4.23 The Relation Between the Coordination Number and the Inter-Nodal Distances; after (Chiriac-Putz-Chiriac, 2005)

Number of Coordination	Inter-Nodal Distance (%)	
	Ionic lattices	**Metallic lattices**
12	100	100
8	92	98
6	89.3	96
4	83.8	88

particles, i.e., by its action of polarization but also through the measure of the influence that supports from other particles, i.e., by its polarization.

The increase of positive charge of a particle and the reduction of its measures lead to the increase of its polarization ability, and the increase of the particle dimensions and its negative charge lead to the increase of its polarization.

Polarizability of a particle depends also its electronic structure. Thus, the ions with external shell with 8 electrons are less polarizable and less polarizing than the ions with external shell with 18 electrons, and the cations with incomplete external shell having a very high ability of polarization.

The polarization phenomena have an important role for establish a certain kind of lattice. Thus, the alkaline earth oxides: CaO, SrO, and BaO have the ratio $r_+/r_- = 0.80$, 1.08, and 0.96, respectively. These oxides crystallize in the lattices of NaCl type with the coordination number equal to 6, and not 8 or 12, according to the value of the ration, see Table 4.14; this fact confirms the paradigmatic approach of formal ions developed on the NaCl prototype in the discussion of previous section.

Very interesting, the polarization phenomena may lead to the impossibility of the existence in solid state of some combinations, such as CuI_2. Due to the electron shell of the ions with iodine, the combination CuI_2 is decomposed forming CuI where, by the reduction of the cation charge, the effects of polarization are reduced and the combination becomes stable. This fact can be possible by increasing the cation radius – not by reduction, but by its complexation (e.g., with ammonia). The combination resulted in such a process (e.g. tetra-aminocopper iodine – $[Cu(NH_3)_4]I_2$) is fully stable in common conditions.

For identical structures of electrons shell, we can establish ionic series where the polarizability variation is parallel to the ionic charges' variation, regardless of the position in periodic system and the atomic number. Thus, the polarization decreases in the series

$$O^{2-} > F^- > Ne > Na^+ > Mg^{2+} > Al^{3+} > Si^{4+}$$

paralleling with the increasing ability of polarizing.

Another consequence of the polarization interaction is the reduction of the distances between particles, due to the movement of electron shell of anions towards the cations. The bond C-O of CO_3^{2-} is of 1.26 Å, while the radius of the ion O^{2-} is of 1.32 Å.

From the presented examples the existence of some different situations of polarization is indicated, as based on the type of combination studied. Thus, in the alkaline earth oxides the polarization is symmetrical around each particle, and therefore, the effect is the decrease of coordination number simultaneously with the increase of covalent degree of bond.

Instead, for $CaCO_3$, Figure 4.54, the polarization in the lattice is asymmetrical; a ion of carbon C^{4+}, due to the big positive charge and the small radius, has a polarizing action much larger than the ion of Ca^{2+}. As a result, the electron shell of O^2 will be deformed, meaning the movement towards C^{4+}; in other words, the electronic density is mainly located along the bond C-O. The overlapping of valence orbital of the two ions leads to the reduction of the distance between them with the formation of a covalent bond.

This reasoning shows that, taking into account the mutual particles polarization, one can interpret the possible structures formed by different particles, even if the relative ions do not have a real existence in lattices.

The representation of all the particles in a lattice by monatomic ions, formal or real, in a first approximation of the formal ions paradigm, and through the appreciation of the interaction between them as based on the density of charge, allows obtaining some useful conclusion (even of pure qualitative nature) for the understanding of the crystallochemical relations.

For a better specification of these notions, the combination $MgCO_3$ will be further analyzed. From the three elements present in the compound, the oxygen is the most electronegative, see Table 4.19. Thus, we can allow that the crystalline lattice is formed by simple ions (formal or real) O^{2-}, Mg^{2+} and C^{4+}. Considering the corresponding ionic radii: $r_{Mg}^{2+} = 0.78$ Å, $r_C^{4+} = 0.26$ Å and $r_O^{2-} = 1.32$ Å, the ratios r_+/r_- are calculated since both Mg^{2+} and C^{4+} will be coordinated by the only single ion with opposite sign O^{2-}, so obtaining:

$$\frac{r_{Mg^{2+}}}{r_{C^{2-}}} = 0.59 \overset{o}{A}, \quad \frac{r_{C^{4+}}}{r_{O^{2-}}} = 0.19 \overset{o}{A} \tag{4.65}$$

For these values of the radii ratio the coordination number will be 6 for the cation of Mg^{2+} and 3 for C^{4+}, according to Table 4.14.

Thus, the form of polyhedra CO_3 and MgO_6 are obtained by taking into account that both cations have the configuration of a rare gas, with the directions of coordination corresponding to a hybridization sp^2 at C^{4+} and sp^3d^2 for Mg^{2+}.

Hence, the structure $MgCO_3$ will be formed by the concatenation of the polyhedra of MgO_6 and CO_3, with octahedral respectively triangular shape.

Moreover, by applying the Pauling no. 2 rule, Eq. (4.32), of electrostatic valence, along the bond C-O results for $p = z/n$ of Eq. (4.18) the value $p = 4/3$ while for the bond Mg-O it corresponds to the $p = 2/6 = 1/3$ value.

Oxygen having the "–2" charge and compulsory getting involved by both cations, the compensation of its valence requires the coordination through C^{4+} and $2Mg^{2+}$.

Thus, the lattice $MgCO_3$ will be composed by triangular units CO_3 where the distances C-O are much shorter than the distance Mg-O as a result of the electrostatic attraction stronger between C^{4+} and O^{2-}. In other words, the polyhedron of oxygen coordination will be asymmetrical, the anion O^{2-} being closer to the corner with C^{4+} than towards the corners with Mg^{2+}.

Taking into account by the bigger charge and the smaller radius of the ion of C^{4+} in relation to Mg^{2+}, a unilateral polarization of the ion of O^{2-} results along the symmetry deformation of the distribution charge towards C^{4+}. The C-O bond will have a pronounced covalent character which gives to the group CO_3 an independent existence in the lattice. Thus, the lattice is actually made from the CO_3^{2-} ionic complex and the simple Mg^{2+} ion.

The conclusion is obviously real. But, the bond length C-O, as shown, is just 1.26Å respecting 1.58Å as would result in the hypothesis of the ordering of maximum compactness. This difference once again shows the formal character of the representation of any particle by simple ions and, on the other side, underlines the intensity of polarization phenomena.

Anyway, the interpretation of the interactions as a polarization consequence allows the intuitive explanation of a set of experimental observations. Thus, $CaCO_3$ – variety of calcite with the same structure such as $MgCO_3$ – decomposes to an higher temperature (the decomposition starts about 500°C while $MgCO_3$ decomposes at 300°C).

The difference in thermal stability of the two compounds is due to the stronger polarization action of Mg^{2+} than that of Ca^{2+} upon the O^{2-} ion.

The O^{2-} ion polarization by a cation weakens the bond C-O by the partial redistribution of de charge density.

Symmetrical polarization appears as a particular case of asymmetrical polarization (for the case previously discussed, toward an anion is exercised equal polarizing actions, from opposite directions).

If the particles are different, the polarization of opposite sign to the one induced by a reference cation, is called *counter-polarization*.

The counter-polarizing action of bivalent cations of small radius is noted also in the lattices of sulphates, nitrates, orthosilicates, etc. in terms of reduction of thermal stability (the decrease of decomposition or melting temperature).

In the AB_2 type lattices or more complex compounds, the polarization interactions can lead to the appearance of stratified structures (Chiriac-Putz-Chiriac, 2005).

Thus, in CdI_2, which has an ion strongly polarizable (I^-) and another strongly polarizing (Cd^{2+}), the relative dimension of the particle indicates the existence of the polyhedra of coordination CdI_6 and ICd_3.

The strong polarization of I^- respecting Cd^{2+} determines the shortenes of the Cd – I distances while keeping the symmetry of polyhedron CdI_6. The counter-polarizing action of another cadmium ion should ensure the symmetry for the polyhedron ICd_3 since a single kind of cations of the lattice.

However, through existing a double number of anions toward the cations, the shortening of the Cd-I distances causes the intercalation between 2 ions of Cd by 2 ions of I and the "asymmetrization" of the CdI_3 polyhedron. There results a structure in layers, Figure 4.78, where the strength of the bonds is stronger inside a layer than that between two layers. The lattice of cadmium iodine can be so represented as being formed by infinite bi-dimensional macromolecules, connected by van der Waals bonds.

Such lattice stratification is not characteristic only for the ionic compounds. Any collectivization of bonding electrons upon the two directions of a plane causes the appearance of layered structures.

Still, there is the possibility that the electrons collectivize along a single direction, i.e., the orientation of the bonds to be made upon a single direction. In this case, a structure of infinite chains results, since the lattice

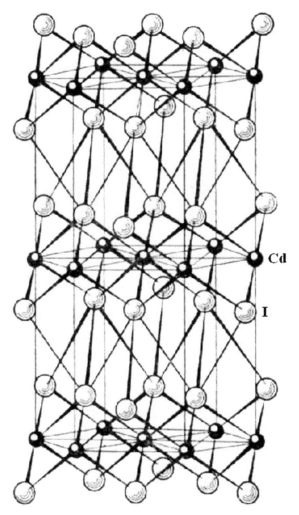

FIGURE 4.78   The CdI$_2$ Lattice; after Flint (1968); Penkala (1974); and Urusov (1987).

being composed by linear macromolecules united by residual bonds or by other types of bond, weaker than those inside the chains, Figure 4.79.

The Figure 4.79 illustrates the structure with linear organization of the clinoenstatite MgSiO$_3$. There is noted that the chains are formed by SiO$_4$ tetrahedral, with strong bonds as a result of the polarization interaction between Si^{4+} and O^{2-}, and the layers of tetrahedra being connected by the octahedra MgO$_6$.

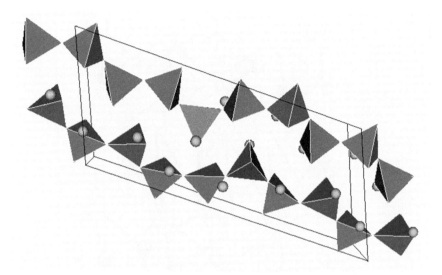

FIGURE 4.79   The Clinoenstatite structure; after The Chemistry of Silicates (2003).

The ionic $Mg - O$ bonds, as the bonds between the layers too, are more weaker comparing to the $Si - O$ bond, as resulted from their length or from the application of the Rule no. 2 of Pauling, of electrostatic valence (4.32), taking into account that applying (4.18) we obtain that $p_{Si^{4+}} = 4/4 = 1$ and $p_{Mg^{2+}} = 2/6 = 1/3$.

Homodesmic lattices have an equal tridimensional development because the bonds strength after the three directions is the same. For the heterodesmic lattices there are two possibilities of realization of the strength to be equal upon the three dimensions. The most often situation corresponds to the analogous organization, as for the homodesmic lattices, i.e., the groups resulted after the interaction of polarization have independent particles' role in the crystalline lattice (e.g., diamond, NaCl-homodesmic; $CaCO_3$, $ZrSiO_4$-heterodesmic).

The interaction between SiO_4 tetrahedral and ions of Zr^{4+} in zirconium structure, $ZrSiO_4$, illustrated in Figure 4.68(b), is of ionic nature, the formed tetrahedron thus having a nodal role of the complex lattice. The coordination number of Zr^{4+} towards the oxygen is CN=8, while the shape of the polyhedron of coordination being the doubled bisphenoid.

The arrangement is more or less compact as based on the degree of covalence of the bond between cation and anion of the complex lattice.

Another type of tridimensional organization of the particles is differentiated by the existence of a distribution of a more intense bond (analogous to the layered or in chains lattices) only between certain particles of the lattice. The bonds' distribution thus remains tridimensional; a "skeleton" of the particles united by much stronger bonds is formed, with the rest of the particles placed inside the skeleton according to the interstices' symmetry and of their radius measure. An example for this organization is offered by feldspar – alkaline aluminum-silicates and alkaline earth, $Me^I[AlSi_3O_8]$ and $Me^{II}[Al_2Si_2O_8]$, respectively, in which a skeleton structure is recognized as formed by SiO_4 and AlO_4 tetrahedra (Figure 4.80).

The bonds that form the skeleton have a pronounced covalent character due to the polarization interactions, while the alkaline ions are connected by the weaker ionic bond, upon the vertices of the polyhedron of KO_{10} coordination. The formation of the tridimensional skeleton is found to a reduced number of combinations and only for compounds able to accomplish the tridimensional concatenation of the basic structural (C, Si) units.

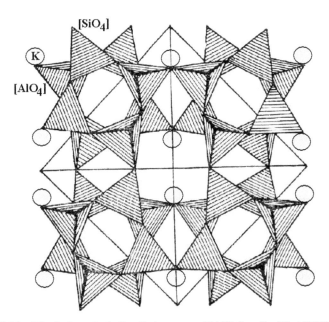

FIGURE 4.80   The Orthose (or hollandite) structure $KAlSi_3O_8$; after Flint (1968); Penkala (1974); and Urusov (1987).

The interstitial lattices are approaching this type of organization as long as they are formed by a typical metallic lattice in which interstices are populated by particles of different nature without an essential modification of the charge distribution characteristic to the metallic lattice.

In conclusion, the Table 4.24 includes the examples of the most important combinations with the predominant bond indication along the strongest bonds distribution in crystalline structures.

TABLE 4.24   The Relation Between the Type of Bond and the Way of Structural Organization; after (Chiriac-Putz-Chiriac, 2005)

Way of organization bond		Structures more or less compact	Structures constituted by infinite chains	Structures constituted by infinite layers	Structures with tridimensional "skeleton"
Ionic	Homo-desmic	NaCl, MgO, CaF_2, $CaTlO_3$, $MgAl_2O_4$	–	–	–
	Hetro-desmic	Na_2SO_4, $CaSiO_4$, $CaCO_3$	Pyroxenes, Amphiboles	Micas, Gypsum	Feldspars, Zeolites
Covalent	Homo-desmic	Diamond, Si, Ge, ZnS	–	–	–
	Hetero-desmic	FeS_2, NiAs	–	–	–
Metallic	Homo-desmic	Metals, Alloys	–	–	–
	Hetero-desmic	Zn, Cd, Hg, Sn	Se, Te, Sb_2S_3	As, Sb, Bi, MoS_2	Interstitial compounds (TiC, ZrB_2)
van der Waals	Homo-desmic	Inert gases	–	–	–
	Hetero-desmic	O_2, H_2, HCl, Sulfur, Organic compounds	Cellulose, Rubber, Fibrous Proteins	Graphite, CdI_2, Flat macromolecules	"Globular" proteins

4.5 CONCLUSION

The main lessons to be kept for the further theoretical and practical investigations of the chemical crystallography that are approached in the present chapter pertain to the following:

- Identifying the structural nodes in a lattice by rigid spheres of various radii (for reference and interstices) towards modeling the hexagonal and cubic packing both in plane and in space, respectively;
- Employing the model of rigid spheres for modeling the crystal packing and compactness and stability degree of a given Bravais lattice;
- Writing the total number of aminoacids (20) as the exact available number of tetrahedral (T_d) and octahedral (O_h) interstices (excluding central octahedral interstice) in face-centered cubic structure this way providing the crystalline paradigm of the genetic code;
- Dealing with the substances systematics under their crystalline form by combining the crystalline stoichiometry with the spatial group information they belong;
- Characterizing the unary compounds by the cases of electrons' deficit, equal and excess respecting the available atomic orbitals in valence shells of elementals atoms the unary crystals are rooting;
- Understanding the AB stoichiometry and crystal structure as the equivalence between the choice of referential particle among the two unit cell's constituents;
- Describing the AB_2 stoichiometry of crystal structure through the coordination number equation with 4:8; 3:6, and 2:4 solution for the paradigmatic CaF_2, TiO_2, and Cu_2O structural types, respectively, while having cation as the referential (lattice's basis) atoms;
- Learning that only coordination numbers as $NC_B=2$; $NC_A=6$ can be acceptable for crystals stoichiometry as AB_3 having the ReO_3 and $BiLi_3$ as the representative structures so far;
- Treating the complex crystals structures through multiple references' coordinates (e.g., the perovskites and spinel structures) and fractional stoichiometry associated with special manifested properties (as is the supraconductibility in the perovskite type cases);
- Solving the crystallographic space covering by tetrahedral $[SiO_4]^{4-}$ units, having Si in the middle of the tetrahedron and oxygen in corners by the celebrated silicates structures and of their combinations and various bindings by adjacent polyhedra;
- Formulating the chemical bonding in solids by the formal ions' paradigm opening the way of understanding the atomic solids, along the metallic and molecular solids compounds, in qualitative manner however;
- Interpreting the chemical bonding in solids by the mutual relationships between the various chemical bonds (ionic, covalent, van der Waals, and metallic) eventually further characterized by associated

(quantum) electronic density distribution at the unit cell levels and for crystal bands' energies;

- Connecting the chemical bond in solids and crystals by atomic/ionic radii involved in bonding so consecrating the Pauling rules of bonding, with relevance in computing the crystal potential involving the collective Madelung constant averaging the anion-cation interaction over the crystal;

- Developing the general potential paradigm of potential function accounting for attractive and repulsive influence alike towards producing the various working models of chemical bonding and lattice interactions: Mie equation, Lennard-Jones potential, Morse potential, Born-Mayer potential, till the Kaputinski lattice energy as the semi-empirical approach fro total energy of a given lattice, eventually involving the electronegativity as a potential of the chemical bonding formation itself, in molecules and solids;

- Finding applications of crystal structures by employing the polarization potential through the ionic constituents which eventually drives the intensity of chemical bond, thus the type of bonding and the allied properties, toward predicting the manifested (observed) features.

KEYWORDS

- borates
- chemical bond in solids
- Cristo-balite
- crystal's stoichiometry
- Fluorite
- Heterodesmic lattices
- ionic compounds
- metallic compounds
- Pauling rationalization of chemical bonding in solids
- Perovskite type
- polarization in crystals
- potential function model of chemical bonding
- rigid sphere model of crystals

- **Rutile**
- **silicates**
- **Sphalerite/Blende**
- **Spinel type**
- **typical crystalline structures**
- **unitary substances**

REFERENCES

AUTHOR'S MAIN REFERENCES

Chiriac, V., Putz, M. V., Chiriac, A. (2005). *Crystalography* (in Romanian), West University of Timişoara Publishing House, Timişoara.

Putz, M. V. (2016). *Quantum Nanochemistry. A Fully Integrated Approach: Vol. III. Quantum Molecules and Reactivity*. Apple Academic Press & CRC Press, Toronto-New Jersey, Canada-USA.

Putz, M. V. (2012). *Quantum Theory: Density, Condensation, and Bonding*, Apple Academics, Toronto.

Putz, M. V. (2011). Electronegativity and chemical hardness: different patterns in quantum chemistry. *Current Physical Chemistry* 1(2), 111–139 (DOI: 10.2174/1877946811101020111).

Putz, M. V. (2010a). *Environmental Physics and the Universe* (in Romanian), The West University of Timişoara Publishing House, Timişoara.

Putz, M. V. (2010b). Cosmos, order and obligations: the big CO_2. *Int. J. Environ. Sci.* 1(January-June), 1–8.

Putz, M. V. (2008). *Absolute and Chemical Electronegativity and Hardness*, Nova Publishers Inc., New York.

Putz, M. V., Chiriac, A. (2008). Quantum Perspectives on the Nature of the Chemical Bond. In: *Advances in Quantum Chemical Bonding Structures*, Putz, M. V. (Ed.), Transworld Research Network, Kerala, Chapter 1, pp. 1–43.

Putz, M. V. (2006). Systematic formulation for electronegativity and hardness and their atomic scales within density functional softness theory. *Int. J. Quantum Chem.* 106, 361–389 (DOI: 10.1002/qua.20787).

SPECIFIC REFERENCES

Apostolescu, R. E. (1982). *Crystallography-Mineralogy* (original in Romanian: Cristalografie-Mineralogie), Didactic and Pedagogic Publishing House (Editura Didactică and Pedagogică) Bucharest.

Becherescu, D., Cristea, V., Marx, F., Menessy, I., Winter, F. (1983). The Chemistry of Solid State (original in Romanian: *Chimia Stării Solide vol. I*) Scientific and Encyclopedic Publishing House (Editura Ştiinţifică and Enciclopedică), Bucharest, Vol. I..

Borg, R. J., Diens, G. J. (1992). *The Physical Chemistry of Solids*, Academic Press, Boston.

Bunn, C. W. (1954). *Chemical Crystallography. An Introduction to Optical and X-Ray Methods*, Oxford at the Clarendon Press.

Crick, F. H. C. (1968). The origin of the genetic code. *J. Mol. Biol.* 38(3), 367–379.

Crick, F. H. C. (1966). Codon-anticodon pairing: the wobble hypothesis. *J. Mol. Biol.* 19(2), 548–555.

Drăgulescu, C., Petrovici, E. (1973). *Introducere în Chimia Anorganică Modernă* (original in Romanian: Introduction to Modern Anorganic Chemistry), Facla Publishing House, Timişoara.

Evans, R. C. (1964). *An Introduction to Crystal Chemistry*, University Press, Cambridge.

Flint, E. (1968). *Principes de Cristallographie* (translation in French from Russian), Mir, Moscow.

Heyes, S. J. (1999). *Four Lectures in the 1st Year Inorganic Chemistry Course*, Oxford University.

Ketelaar, J. A. A. (1958). *Chemical Constitution*, Elsevier, New York, Chapter 1.

Kimball's Biology Pages (2003, 2013), http://users.rcn.com/jkimball.ma.ultranet/Biology-Pages/

Kittel, C. (1996). *Introduction to Solid State Physics*, Seventh Edition, John Wiley & Sons, New York, Chichester.

Knox, R. S., Gold, A. (1964). *Symmetry in the Solid State*, W. A. Benjamin Inc. New York, Amsterdam.

Penkala, T. (1974). *Za Ochersky Kristalokhimii* (translation in Russian from Polish), Izd. Khymia, Leningrad.

Purves, W. K., Orians, G. H., Heller, H. C. (2001). *Life: The Science of Biology*, 6th Edition, by Sinauer Associates, Sunderland (MA), (www.sinauer.com) and WH Freeman, New York (www.whfreeman.com).

Ramachandran, G. N. (1964). *Advanced Methods of Crystallography*, Academic Press, London-New York.

Sirotin, I. I., Şaskolskaia, M. P. (1981). *Crystal Physics* (translation from Russian in Romanian: Fizica Cristalelor), Scientific and Encyclopedic Publishing House (Editura Ştiinţifică and Enciclopedică), Bucharest.

Slater, J. C. (1939). *Introduction to Chemical Physics*, McGraw-Hill, New York, Chapter 22.

The Chemistry of Silicates (2003, 2013), http://ruby.chemie.uni-freiburg.de/Vorlesung/silicate_0.html

U.S. Naval Research Laboratory/Center for Computational Materials Science (2003, 2013), http://cst-www.nrl.navy.mil/.

Urusov, V. S. (1987). *Teoreticheskaya Kristalokhimia* (original in Russian), Izd. Moskovskaia Univ.

Weller, P. F. (1973). *Solid State Chemistry and Physics—An Introduction*, Marcel Dekker, New York.

Wolfram Research (2003), http://mathworld.wolfram.com.

CHAPTER 5

X-RAY CRYSTALLOGRAPHY

CONTENTS

ABSTRACT

The journey toward the depths of the structure of condensed state will eventually proceed through an *"external non-destructive intervention"*, with the aid of the X-radiations on the crystals; we should note that X-rays interact with the electrons of the atoms or the groups of atoms from the network and not with their nuclei. Thus, the picture of diffraction (reflection or scattering) of the X-ray will generate an "electronic map" of the bodies investigated, so characterizing their structure. Moreover, this electronic map is a manifestation of the so-called structure factor of the crystal analyzed a quantity in direct (Fourier) relationship with the associated electronic density. This way, the union between the experimental methods with X-ray diffraction on crystals and their quantum characterization is so retrieved by means of the electronic densities characteristic to the structures. Practical ways of refining of this merging between crystals and X-ray diffraction are the main focus of the present chapter.

5.1 INTRODUCTION

Why the X-ray? Why exactly the X-rays are used for the electronic structures investigation? Due to their high penetration in the structure of quantum levels of atoms, having wave characteristics of the atoms sizes order (wavelength about 10 nm and associated frequencies of about 3×10^{16} Hz) as carrying of quantum energies comparable with the energies of the deepest atomic quantum levels (about 124 eV) (Putz, 2014).

This chapter is dedicated to fundament the crystallographic approach of the solid state structure and properties by presenting the main features the geometric as well the dynamical theory of X-ray diffractions may reveal for a perfect crystal. Here will be studied their fields and intensities, the equations that connect them as also the solutions of propagation for the non-absorbent or respectively absorbent crystals; the present discussion follows Putz and Lacrămă (2005).

However, the study will be limited to the approximation of two waves: the incident wave ($\vec{K}_0$) and the diffracted wave ($\vec{K}_h$), coupled to the first through the Bragg's law:

$$\vec{K}_0 + \vec{h} = \vec{K}_h \tag{5.1}$$

i.e., through a single node of the reciprocal network, outside its origin O, on the Ewald's sphere, Figure 5.1. In other words, for the given incident wave, a single set of planes (*hkl*) participates to the diffraction process. This is the most frequent case met in the conditions of some experimental setup. By taking into account the absorption process, but also the processes that occur at the propagation of the two types of waves (transmission and diffraction) through the crystal, the dynamical theory offers a

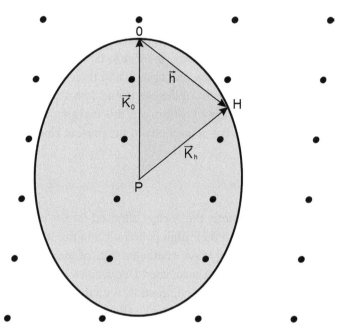

FIGURE 5.1   The Ewald's sphere and the representation of the Bragg's equation in the approximation of the two (transmitted-diffracted) waves.

precise description of their propagation processes through the crystal places at the Bragg's angle.

Therefore, the dynamical diffraction has the specificity of the coupling of the two wave fields, transmitted and diffracted, in a reciprocal interaction also with the crystal. The purpose of this chapter is the assessment of the general form of the fields in the crystal based on the shape of the external fields as well as the intensities related to the transmitted and diffracted directions.

In this framework the decomposition and re-composition in the dynamical propagation of X-ray in the crystals is to be emphasized. In this case, the decomposition on the alpha and beta branches, on the transmitted and diffracted directions, have been noted then followed by the re-compositions on common direction from different branches (Pendellösung), on common branches from different directions (Standing Waves), with the possibility of the effect of the direct anomalous absorption (i.e., the alpha branch free of absorption, with the beta branch anomalous absorbing) and respectively the reverse anomaly (with the alpha branch anomalously absorbing, while the beta branch is free of absorption), eventually followed by the decomposition of these standing waves of low and high absorption in the effect of fluorescence (Putz, 2014).

Since all these effects arise as a result of the phases differences which occur at the dynamical propagation, they can undergo the unified phenomenological through giving a global approach of the self-consistency in terms of the phases composition, so taking into account all the other phases that can occur in the inter-branches or inter-directions effects, i.e., in an interrelated way, with the above listed phenomena being identified in the final (natural) quantum mechanical description.

5.2   X-RAY DIFFRACTION APPLIED TO CRYSTALLINE SYSTEMS

5.2.1   X-RAY AND DIFFRACTION ON CRYSTALS

Which is the relation between X-rays and electrons? The electrons, as electric charge carries a weight of about 2000 times lower than the protons, so can be relatively easy to be thermal-ionized and then can be accelerated to a potential difference (of dozens of kV) toward a metal target, where are suddenly decelerated in the field of the electrons of the metallic atoms of the target, Figure 5.2-left, sending out the so-called braking radiation ("bremsstrahlung"-in German), of *continuous X-radiation*; the present discussion follows Putz and Lacrămă (2005).

When energy of the electrons which bombard the target increases also the intensity of the emitted continuous X-ray raises and a displacement of the maximum toward the electromagnetic spectrum with the highest frequencies and shortest wavelengths is noted, respectively, Figure 5.2-middle.

But there is also a second type of X-ray, the so-called characteristic X-radiations, appearing when the electrons bombing the target have the energies corresponding to the quantum energies of the first electronic levels in the atoms of the target, so that the electrons of these levels are injected, in their place immediately "falling" the electrons of the (immediate) higher levels, so resulting the transition that determines the emission of the characteristic X-radiation (depending on the type of the atoms of the target).

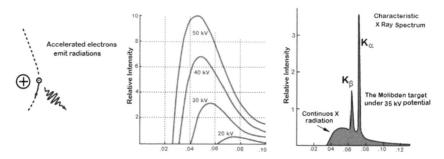

FIGURE 5.2 Left: deceleration and emission of X-radiation by electrons' scattering; middle: the spectrum of continuous X-radiation of a tungsten target (according to the information of Ulrey, 1918); right: spectrum of characteristic X-radiation, on continuous background; after HyperPhysics (2010).

For example, the transitions (to be called characteristic) occurring to the level with n=1 (K spectral level) will generate the X-radiations of K-type, and those will be called of Kα type if n=2 → n=1 and respectively of Kβ type if resulting from the transitions n=3 → n=1, Figure 5.2-right. Similarly, the X-ray of L type (with n=2) with the Lα subtypes (n = 3 → n = 2) and Lβ (n=4 → n=2) are defined. The typical X-radiations produced in order to be further used for determining the structural electronic maps of the other species are those produced by the target of Cu (producing the Kα line with λ=1.54Å) and of Mo (producing the Kα with λ = 0.71Å).

The two types of radiations previously described are produced in the so-called X-ray tube, Figure 5.3 (left-up) together appearing in the spectrum, Figure 5.3-right. In order to eliminate as much as possible the (background) continuous radiation the X-ray tube is supplemented with specific filters, Figure 5.3 (left-bottom).

Thus the quasi-monochromatic X-radiation can be obtained at the edge of absorption immediately before and after the Kα line of characteristic spectrum. The X monochromatic radiation may be further directed (collimated) by successive apertures of a strong absorbent material (commonly Pb-lead, from safety reasons) and finally "projected" into the analyzed sample (crystal), while the reflected (diffracted) sample radiation will be recorded by a detector, found at the double of the X-ray incidence angle on the crystal. Both the crystal and the detector can be involved in reciprocal rotations, keeping the condition of the double of incidence angle,

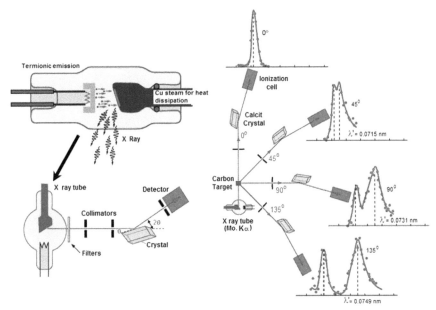

FIGURE 5.3 The X-ray tube (left-up), X-ray spectrometer (left-bottom) used for the Compton experiment (right); after HyperPhysics (2010).

the so-called *Bragg condition*. The described complex is called X-ray spectrometer, and primarily served to Arthur H. Compton for the cutting experimental evidence of electromagnetic radiation corpuscular nature (here in the X-ray region), originally by recording the effects of the collision between the X-photons and the electrons of the crystal of Calcite (in the original experience), Figure 5.3-right.

Due to the fact that the characteristic X-radiations are determined by the electronic transitions between the internal levels of atomic structure, on 1920' years, Henry G. Moseley (1887–1915), described by Rutherford as "the most talented of his student", measured and represented the frequencies in characteristic X-ray for about 40 elements of Periodic Table, forming the so-called Moseley representation of the elemental periodicity, Figure 5.4.

By combining the Planck postulate of the electromagnetic radiation quantification with that of Bohr regarding the quantification of electronic transitions between the stationary levels in atom, Moseley introduced the so-called shielding factor, $Z_{effective}$, related to the way in which the electrons

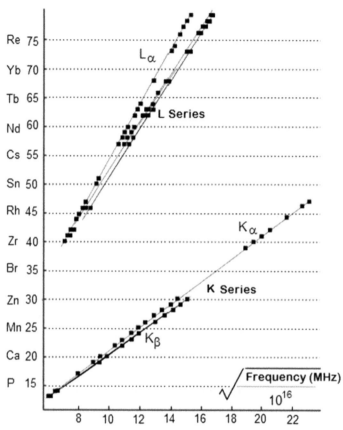

FIGURE 5.4 The Moseley graphical representation of periodic elements; after HyperPhysics (2010).

of K and L levels shield the nucleus for the electrons for the immediate higher level, found equal to Z-1 for Kα lines, and respectively with Z-7.4 for the Lα lines, obtaining the relation between energy, or the frequency of the lines characteristic for the X-ray, and the atomic numbers Z, see also Chapter 1 of the Volume I of the present five-volume work (Putz, 2016a):

$$hv_{K\alpha} = 13.6[eV](Z-1)^2\left[\frac{1}{1^2} - \frac{1}{2^2}\right] \qquad (5.2)$$

$$hv_{L\alpha} = 13.6[eV](Z-7.4)^2\left[\frac{1}{2^2} - \frac{1}{3^2}\right] \qquad (5.3)$$

If the second order radical of the characteristic frequencies for the X-ray is represented, the linear dependencies are obtained from the relations (5.2) and (5.3), as indicated in the Moseley's graphic of Figure 5.4. Moseley's contribution was remarkable for the decisive meaning in ordering the Periodic Table elements upon the atomic quantum number and not based on the atomic mass: thus, Co (Z=27, A=58.9) was earlier misplaced after Ni (Z=28, A=58.7), as like K (Z=19, A=39.10) was wrongly reversed with Ar (Z=18, A=39.95) till Moseley's landmark.

In the same way, Moseley predicted the existence of the Hafnium element (Z=72), further to be discovered in Bohr's laboratory in Copenhagen. Prematurely, Moseley, voluntarily enrolled in the World War I, and was eventually killed during the Battle of Gallipoli, at only 27 years.

Returning to the X-ray spectrometer: why crystals? If the X-ray are so penetrating atomically structure, and implicitly onto molecular systems too, why the analysis sample in X-ray spectrometer is preferred to be under crystalline form? There is a practical answer: the X-ray scattering by a single molecule produces a barely detectable small signal in detector, with very small possibilities to be detected over the "background noise" that includes the scattering from ubiquitous air and water. The atoms, but especially the groups of atoms and molecules, instead, once "multiplied" and ordered in condensed matter (Figure 5.5-left) can be summed under the effect of phase interference, to increasing the signal received in detector. Therefore, the crystalline state has an effective amplifier role!

At this point the X-radiation scattering aspect on crystals intervenes. The X-radiation can be scattered or re-emitted by the electrons of the

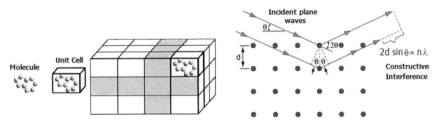

FIGURE 5.5   Left: a molecular crystal; right: the geometry of X-ray diffraction on the reticular strings of a crystal; after HyperPhysics (2010), Matter Diffraction (2003), X-Rays (2003).

crystalline structure in any direction. But, the electronic structure can be viewed only if simultaneously two conditions are satisfied (James, 1969):

1. (D1) in case in which this scattering is equivalent with a reflection, i.e., the resulted X-ray due to the scattering on the scattering centers inside the crystal, has a direction equal to the double of the angle of incidence θ; and

2. (D2) when the differences occurring between the scattered rays' paths is a multiple of the wavelength λ *of incident radiation,* Figure 5.5-right. The last condition corresponds to the condition of constructive interference, which ensures a maximum of intensity in the spectrum recorded by detector, and sets, along with the first condition, the famous relation known as the Bragg law of X-ray diffraction on crystals:

$$2d \sin\theta = n\lambda, n \in Z \qquad (5.4)$$

with d the inter-planar distance between the crystal reticular planes, Figure 5.5-right.

Before justifying the necessity of the two conditions above for determining the electronic structure of crystalline systems, we will note that the variation of the diffraction angle θ drives the fulfillment of the condition of constructive interference expressed by the relation (5.4), which at its turn corresponds to the summing in phase of the scattered waves, with the effect in recording of a more or less intense diffraction signal, Figure 5.6-left.

On the other side, there can be noted how the image of diffraction is not the direct image of the object producing the diffraction or the scattering of the X-ray, Figure 5.6-right. This observation will be explained after it will be justified the necessity of D1 condition in X-ray diffraction on crystals.

So, why there is necessary that the X-radiation to be reflected on the crystalline planes in order to be able to record its structure? The reason is simple: when the reflection angle is equal with the incidence one the X-rays that are in phase at "input" will be as such also at the "output" of the system, regardless the place where the diffraction-reflection occurs.

In Figure 5.7-left a graphical representation of this argument is shown: if the *ad* and *bc* segments differ in length, the waves of the relative rays would

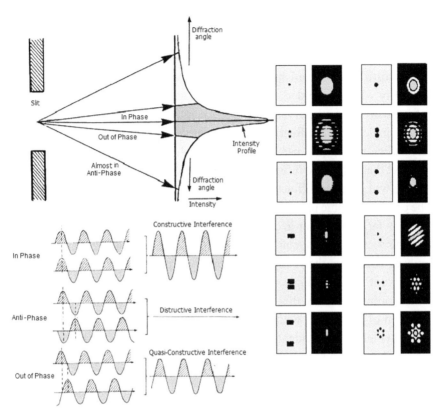

FIGURE 5.6   Left: the connection between diffraction, interference and recorded intensity for the various phase relations between the scattered rays; right: diffraction images for various types of slits-object; after HyperPhysics (2010), Matter Diffraction (2003), X-Rays (2003).

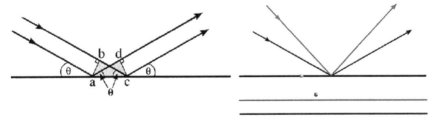

FIGURE 5.7   "Left" the graphical explanation for reflection condition of X-ray diffraction on crystals; right: the graphical explanation for crystalline structure resolution through X-ray diffraction on crystallographic planes; after (Matter Diffraction, 2003; X-Rays, 2003).

not sum themselves in phase at the level of the front dc; thus, the segments ad and bc are equal only if equal angles toward the segment ac are assumed. As a consequence, when the condition of constructive interference (5.4) of Bragg diffraction is satisfied, the "refection or diffractions spots" of high intensity will be recorded, i.e., the bonded electrons are located at the level of unit cell inside the crystal (Putz & Lacrămă, 2005).

A single spot of diffraction will give information about an entire family of crystalline planes (the parallel crystallographic planes). There is immediate the fact that, while the diffraction on reticular scattering centers of crystalline planes will generate diffraction waves in phase, the scattering centers between the crystallographic planes will produce dephased scattered waves (depending on how far is the center in cause toward a diffraction plane, Figure 5.7-right) and respectively producing spots of very weak intensity in the diffraction image. Therefore, the X-ray diffraction gives information about the relative position of the scattering centers (the electrons around the nuclei).

In order to clarify this concept even more, take the maximum of diffraction of first order, with $n=1$ in (5.4). Thus, the direct relation between the wavelength of incident X-radiation and the inter-planar distance of the network is obtained as:

$$d = \frac{\lambda}{2\sin\theta} \tag{5.5}$$

What does this connection say? If $\theta = 0$, there is no direction changing of incident X-radiation, i.e., the differences of the paths traversed by the X-ray in crystalline structure have the same value, regardless the location of the scattering centers, so actually there is no diffraction, and the crystal acts as it would have an infinite inter-planar distance, not supplying any spatial information! Instead, for the other extreme case when $\theta=90°$, the X-radiation is at a normal incidence on a family of crystalline planes, so the difference of path corresponding to the double of interplanar distance, $\lambda=2d$, i.e., the resolution of the information, is limited at half of the used wavelength: $d=\lambda/2$. Consequently, in order to increase the resolution, one should decrease the wavelength, which again justifies the utilization of X-radiations instead of the visible ones, for example, for determining electronic structures.

In any case, information is obtained about the objects' relative positions (electrons, and implicitly the nuclei that "bond" them) on the directions perpendicularly on the crystalline planes, but not also about positions parallel with the planes in focus. Therefore, in practice, the methods of X-ray diffraction are used when or the crystal or the detector assumes a rotational movement, for a complete "scan" of the spatial positions of the scattering centers.

There remains to elaborate on the issue of "diffraction image": the fact that indicates the localization of the electrons was argued, in turn bonded to the position of the nuclei inside the structure, but, what exactly indicates the diffraction image?

Take the incident X-radiation described by a plane wave characterized by the wave vector $\vec{K}_0$ with the versor s_0 and, respectively, with the diffracted plane wave characterized by the wave vector $\vec{K}$ with the versor s, Figure 5.8-left. The two vectors can combine so that to be reduced to a single one, *the scattering vector $\vec{s}$* (Putz & Lacrămă, 2005):

$$\vec{S} = \vec{K} - \vec{K}_0 = \frac{\vec{s}}{\lambda} - \frac{\vec{s}_0}{\lambda} \tag{5.6}$$

with a modulus that immediately results, firstly taking its scalar product with itself:

$$\vec{S} \cdot \vec{S} = \frac{1}{\lambda^2}\left(s^2 + s_0^2 - 2\,\vec{s} \cdot \vec{s}_0\right) = \frac{2(1 - \cos 2\theta)}{\lambda^2} = \frac{4\sin^2\theta}{\lambda^2} \tag{5.7}$$

and then the with the square root:

$$S = \frac{2\sin\theta}{\lambda} \tag{5.8}$$

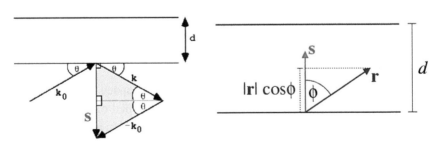

FIGURE 5.8 Description of X-ray diffraction in terms of wave vectors; after (Matter Diffraction, 2003; X-Rays, 2003).

The relation (5.8) of the scattering vector can be rewritten as based on the phase of the diffracted wave towards the incident one,

$$2\pi\left[\left(\vec{K}-\vec{K}_0\right)\cdot\vec{r}\right]=2\pi\left(\vec{S}\cdot\vec{r}\right) \tag{5.9}$$

for the adjacent planes (at the distance $r=d$) in relation with the plane that passes through the origin, i.e., for a phase corresponding to 2π, wherefrom the condition

$$\vec{S}\cdot\vec{r}=1 \tag{5.10}$$

is equivalent to the resolution:

$$S=\frac{1}{d} \tag{5.11}$$

From the identification of the relations (5.8) and (5.11) again results the Bragg's law, for the first order of diffraction, the relation (5.5). In this way, by introducing the scattering vector $\vec{S}$ as difference of the wave vector the directions of diffracted and incident radiations (5.6), in the space of the wave vectors, the Bragg's law of diffraction is naturally found (in the reciprocal space). This result further indicates how the reciprocal space (the space of the wave vectors) corresponds to the space of diffraction! Moreover, the relation (5.11) can be generalized, taking the scalar product between the scattering vector and an arbitrary position of the scattering center, Figure 5.8-right:

$$\vec{S}\cdot\vec{r}=Sr\cos\phi=\frac{r}{d}\cos\phi \tag{5.12}$$

which indicates the fraction from the inter-planar distance to which the position vector is located $\vec{r}$ and properly influencing the phase $2\pi(\vec{S}\cdot\vec{r})$ of the diffracted wave towards the incident one. The constructive interference appears when $\phi=0$, i.e., when the r distance is minimum, i.e., equal to the inter-planar distance d.

There is concluded that – even if in a reciprocal space – the X-ray diffraction is more sensitive while the objects overlapped on the diffraction (scattering centers) become closer in the real space! This intimate connection between the real space investigated and the reciprocal space in modeling, finally allows the expression of the real measures of localization, especially the electronic density, based on the diffraction characteristics, as will be further explained.

5.2.2   X-RAY AND THE ELECTRONIC DENSITY

There had been seen how the X-ray diffraction on crystals can be rewritten both in terms of wave vectors, of reciprocal space, and also through the path differences of the real space at the level of the crystal unit cell. Thus, the crystal and its diffraction picture belong to two different spaces, yet in reciprocal relation, Figure 5.9, in the same way in which the real space and the reciprocal ones transform each another; the present discussion follows (Putz & Lacrămă, 2005). Yet, as mentioned also in Figure 5.9, a new measure appears at the level of diffraction space which correspond to the scattering at the unit cell level inside a crystal: the structure factor F (Fanchon & Hendrickson, 1991).

The structure factor is the wave resulted by diffraction at the level of unit cell, thus being expressible through a complex number, with an amplitude and a phase, without dimension, but having the number of electrons as meaning units. This aspect can be even better illustrated by considering the structure factor associated to the diffraction produced by a single electron, therefore having the amplitude $1e$, with the waveform:

$$F\left(\vec{S}\right) = (1e)\exp\left(2\pi i \vec{S}\cdot\vec{r}\right) \tag{5.13}$$

When the diffraction picture is recorded from an object containing more electrons located in various points (e.g., a crystal), the waves scattered by each electron contribute to diffraction by summing the individual structure factors (5.13) in the total structure factor:

$$F\left(\vec{S}\right) = \sum_j \exp\left(2\pi i \vec{S}\cdot\vec{r}_j\right) \tag{5.14}$$

and which, for the case of the electronic continuous distribution, with density $\rho(\vec{r})$, takes the integral form:

$$F\left(\vec{S}\right) = \int_{\substack{CRYSTAL \\ SPACE}} dV \rho\left(\vec{r}\right)\exp\left(2\pi i \vec{S}\cdot\vec{r}\right) \tag{5.15}$$

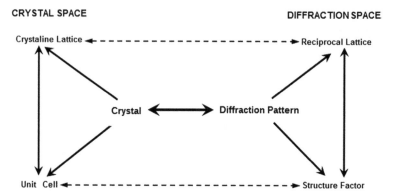

FIGURE 5.9 Schematic connections between the real and the diffraction spaces' characteristics.

The Eq. (5.15) is as interesting as important, since making the connection between the characteristic of the electronic distribution in the real space, through $\rho(\vec{r})$, and the picture of its distribution by scattering in the reciprocal space, through $F(\vec{S})$. Moreover, this connection defines the so-called *Fourier transformation*: one of its most importance features is that it can be reversed! Thus, the way of electronic density determination is opened from the measurement of the structure factor. Before proceeding to such inversion, one should be noted, again by appealing to the correspondences of Figure 5.9, that the structure factor corresponds to the unit cell from a crystalline arrangement, thus indicating the possibility of reducing the integral of Eq. (5.15) from the entire crystalline space to the space of the unit cell. This fact is also analytical doable if the scalar product (5.12) is rewritten by combining the direct and the reciprocal space, as based on the properties between the fundamental vectors which define them, namely:

$$\vec{S}\cdot\vec{r} = \left(h\,\vec{a_1^*} + k\,\vec{a_2^*} + l\,\vec{a_3^*} \right)\cdot\left(x\,\vec{a_1} + y\,\vec{a_2} + z\,\vec{a_3} \right) = hx + ky + lz \equiv \vec{h}\cdot\vec{x} \quad (5.16)$$

In these conditions, the Eq. (5.15) will be rewritten as the equation for the structure factor:

$$F\left(\vec{h}\right) = \int\limits_{\substack{CRYSTAL \\ UNIT\ CELL}} dV\,\rho\left(\vec{r}\right)\exp\left(2\pi i\,\vec{h}\cdot\vec{x}\right) \quad (5.17)$$

allowing from now the expression of the structure factor in terms of Miller indices (h,k,l) and of fractional coordinates (x, y, z), at the unit cell level.

Based on the Fourier inverse transformations properties, the inverse Fourier transformation of Eq. (5.17) involves only the inversion of the sign in the appeared exponential. Generally, the Fourier transformation of integral form has a inverse transformation still as an integral form, yet, in this case being about the electronic density in a crystal, i.e., presenting a periodicity. Thus, the inverse transformation of the Eq. (5.17) will be written as a sum, generating the electronic density equation:

$$\rho\left(\vec{x}\right) = \frac{1}{V}\sum_{\vec{h}} F\left(\vec{h}\right) \exp\left(-2\pi i \vec{h}\cdot\vec{x}\right) \tag{5.18}$$

There is remarkable how $F(\vec{h}=\vec{0})$ inside the equation of structure factor (5.17) has the significance of the number of electrons from the unit cell, while, the same measure in the equation of electronic density (5.18) should be divided to the volume of unit cell V, in order to contribute to the electronic density from the unit cell level.

The proof of electronic density equation (5.18) can be reproduced by using the elegant method of the successive Fourier transformations. Thus, if the Eq. (5.18) is real, then by its replacement in the equation of the structure factor (5.17) one should obtain the identity. The direct substitution firstly generates the equation:

$$G\left(\vec{h}\right) = \int_{\substack{CRYSTAL \\ UNIT\ CELL}} dV_{\vec{x}} \left[\frac{1}{V}\sum_{\vec{h'}} F\left(\vec{h'}\right)\exp\left(-2\pi i \vec{h'}\cdot\vec{x}\right)\right]\exp\left(2\pi i \vec{h}\cdot\vec{x}\right) \tag{5.19}$$

that can be rearranged by the exponentials' grouping

$$G\left(\vec{h}\right) = \int_{\substack{CRYSTAL \\ UNIT\ CELL}} dV_{\vec{x}}\sum_{\vec{h'}} \frac{1}{V} F\left(\vec{h'}\right)\exp\left[2\pi i \left(\vec{h}-\vec{h'}\right)\cdot\vec{x}\right] \tag{5.20}$$

and, finally, by inversing the integral with the summing operation's order:

$$G\left(\vec{h}\right) = \sum_{\vec{h'}} \frac{1}{V} F\left(\vec{h'}\right)\int_{\substack{CRYSTAL \\ UNIT\ CELL}} dV_{\vec{x}} \exp\left[2\pi i \left(\vec{h}-\vec{h'}\right)\cdot\vec{x}\right] \tag{5.21}$$

The last integral is identical null for all the non-null differences $(\vec{h} - \vec{h}')$ by the virtue of the fact that at the level of the unit cell the integral accounts for the variation of the fractional coordinates x, y, z between -1 and 1, and taking into account that the periodic trigonometric functions (cosines for the real part and sinus for the imaginary part of the exponentials with non-null complex arguments) cancel on average over a period. Thus, one further obtains the identity of the proof:

$$G\left(\vec{h}\right) = \sum_{\vec{h}'} \frac{1}{V_{\vec{h}'}} F\left(\vec{h}'\right) V_{\vec{h}} \delta_{\vec{h},\vec{h}'} = F\left(\vec{h}\right) \qquad (5.22)$$

The proof just exposed reveals another important fact, i.e., the orthogonality of the term $\exp(2\pi i \vec{h} \cdot \vec{x})$ on any other one with a different "vector" $\vec{h}'$ (respectively for a set of Miller indices):

$$\int_{\substack{CRYSTAL \\ UNIT\ CELL}} dV_{\vec{h}} \exp\left[2\pi i \left(\vec{h} - \vec{h}'\right) \cdot \vec{x}\right] = \delta_{\vec{h},\vec{h}'} \qquad (5.23)$$

This property allows the "absence" in the knowledge of a structure factor, without the need to compensate its lack by modifying the others, being thus of an extremely utility for the theoretical approaches of the interpretations of the X-ray spectra.

The next striking question relates to the electronic density inside the Eq. (5.18): how the electronic density may be a real measure since it comes from a sum of complex numbers (structure factor)? This behavior can be demonstrated in the context of *Friedel's law*, which assumes that all the electrons diffract the incident X-ray with the same relative phase, i.e., the reflections on Bragg planes fixed by the Miller sets $\vec{h}$ and $-\vec{h}$ having the same amplitude, but opposite phases, Figure 5.10.

In terms of the structure factors, the Friedel's law will be written such as:

$$F(\vec{h}) = F^*(-\vec{h}) \qquad (5.24)$$

with the asterisk "*" indicating the operation of "complex conjugate"; it analytically transposes into versorial forms inside the complex space, see Figure 5.10 (Putz & Lacrămă, 2005):

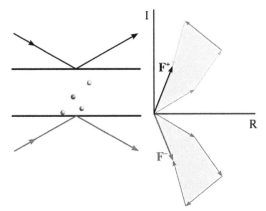

FIGURE 5.10 Left: the diffraction by the scattering centers fixed by the scattering vectors (of the planes characterized by the Miller indices) $\vec{h}$ and $-\vec{h}$; right: Friedel's law at the level of structure's factors; after (Matter Diffraction, 2003; X-Rays, 2003).

$$F\left(\vec{h}\right) = \left|F\left(\vec{h}\right)\right| \exp\left(i\alpha_{\vec{h}}\right); \quad F\left(-\vec{h}\right) = \left|F\left(\vec{h}\right)\right| \exp\left(-i\alpha_{\vec{h}}\right) \qquad (5.25)$$

Under the Friedel's law conditions, the equation of the electronic density (5.18), in the contributions of the scatterings from the hemispheres defined by the vectors $\vec{h}$ and $-\vec{h}$, can firstly be written with the summing restricted to a single hemisphere:

$$\rho\left(\vec{x}\right) = \frac{F(0,0,0)}{V}$$

$$+ \frac{1}{V} \sum_{\substack{+h \\ HEMISPHERE}} \left[F\left(\vec{h}\right)\exp\left(-2\pi i\,\vec{h}\cdot\vec{x}\right) + F\left(-\vec{h}\right)\exp\left(2\pi i\,\vec{h}\cdot\vec{x}\right)\right] \qquad (5.26)$$

Replacing the relations (5.25) we also have:

$$\rho\left(\vec{x}\right) = \frac{F(0,0,0)}{V}$$

$$+ \frac{1}{V} \sum_{\substack{+h \\ HEMISPHERE}} \left|F\left(\vec{h}\right)\right|\left\{\exp\left[i\left(\alpha_{\vec{h}} - 2\pi\,\vec{h}\cdot\vec{x}\right)\right] + \exp\left[-i\left(\alpha_{\vec{h}} - 2\pi\,\vec{h}\cdot\vec{x}\right)\right]\right\} \qquad (5.27)$$

and by the Euler's expansion of the exponentials, the electronic density is immediately obtained such as:

$$\rho\left(\vec{x}\right) = \frac{F(0,0,0)}{V} + \frac{1}{V} \sum_{\substack{+\vec{h} \\ HEMISPHERE}} \left|F\left(\vec{h}\right)\right| \cos\left(\alpha_{\vec{h}} - 2\pi \ \vec{h}\cdot\vec{x}\right) \qquad (5.28)$$

consecrating the real character of electronic density.

If the scattered electrons have transition energies close to those of the incident X photons, a phase departure in the associated waves to the diffracted X-ray, and respectively also inside the structure factors associated, will be introduced, so no longer fulfilling the relations of Eq. (5.25) type, and consequently the Friedel's law, this being the case of *anomalous scattering*. Under these conditions, the electronic density will have an imaginary component.

In any case, as based on Fourier's transformations that connect the electronic density (as real measure) with the structure factor (as complex measure), the *Parseval's theorem* is formulated which stipulates that the mean square value "on each side" of the Fourier's transformations are proportional, i.e., it can be written like (Cantor & Schimmel, 1980):

$$\int_{\substack{UNIT \\ CELL}} dV \rho\left(\vec{x}\right)^2 = \frac{1}{V}\sum_{\vec{h}}\left|F\left(\vec{h}\right)\right|^2 \qquad (5.29)$$

Based on the Parseval's theorem under the form (5.29) the *root mean-square* (rms) of electronic density it can be formulated,

$$rms(\rho) = \frac{1}{V}\sqrt{\sum_{\vec{h}}\left|F\left(\vec{h}\right)\right|^2} \qquad (5.30)$$

thus ensuring in any circumstances the real character of electronic localization through the electronic density. Even simpler, the *rms* of electronic density is proportional with the *rms* of structure factor,

$$rms(\rho) \propto rms(|F|) \qquad (5.31)$$

with the consequence, based on the additive of Fourier's transformations, into a similar relation also at the differences level between two densities and the structure factors in focus:

$$rms(\rho_1 - \rho_2) \propto rms(|F_1 - F_2|). \qquad (5.32)$$

The differences relation of (5.32) type indicates how, by exactly knowing the amplitude and randomly the phase of the structure factor, an error in *rms* can occur bigger even comparing to the structure factor, the case being different when the phase is exactly known and the structure factor amplitude is arbitrary chosen, Figure 5.11-left-up. Therefore, appears that the knowledge of the structure factor's phase is even more important than the knowledge of the amplitudes.

And yet, although very important, the phases are the most inaccessible experimental measures, forming the so-called *"Phase Problem"* (with uppercase "P")! Why this inaccessibility? Very simple: if the structure factor is written as the amplitude and the associated phase, for example as:

$$F = |F| \exp(i\alpha) \qquad (5.33)$$

while by expressing the diffraction (scattering) waves, the diffraction intensity will be assessed as induced by the scattering vector $\vec{S}$, through the product:

$$I = FF^* = |F|^2 \qquad (5.34)$$

so noting the disappearance of the phase information. Such absence from the diffraction pattern and consequently the inaccessibility of the direct knowledge of the scattering phase is the major problem of the X-ray diffraction studies: it should be "guessed", or theoretically calculated inside the so-called *"bias models"* (models that work)-Figure 5.11-left down, or indirectly estimated by minimizing the differences between the employed models-Figure 5.11-right (Putz & Lacrămă, 2005).

For example, in Figure 5.11-left-down a *bias* method is graphically illustrated: the real structure factor is F, the one calculated from the model is F_C (the sub-index "*C*" indicates the "calculated" nature)- which, despite indexed, provides the calculated phase $\exp(i\alpha_C)$ which by combining with the observed (measured) amplitude $|F|$ generates the bias model (calculation + experiment) of the *bias structure factor* $|F|\exp(i\alpha_C)$: this should be closer to the real one than the one F_C provided by applying of theoretical model alone. Thus, properties of the structure will be identified, as close to the real ones, namely: if inside the calculated model there were not certain atoms, they will appear in the *bias electronic map*, built on the bias structure factor.

But, in the same time, the bias structure factor will always be close also to the calculated structure factor F_C so that both the characteristics of the considered phase model, that are not specific for the real electronic map, will appear on the bias electronic map!

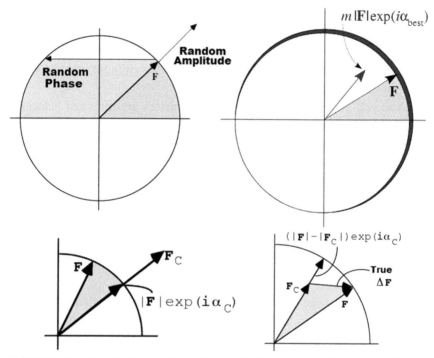

FIGURE 5.11   Amplitude and phase relations for the structure's factors; after (Matter Diffraction, 2003; X-Rays, 2003).

A simple way to identify this information, "in addition" respecting the real map, can be revealed by the subtracting the model map (theoretical, calculated) from the bias map, based on the additive property of the structure factors. Thus the bias difference appears

$$\Delta F_{bias} = \left(|F| - |F_c|\right)\exp(i\alpha_c) \qquad (5.35)$$

versus the real difference

$$\Delta F = F - F_c \qquad (5.36)$$

see Figure 5.11-right-down, for which the difference only on the direction parallel with F_c can be further performed, while the difference on the perpendicular direction being lost because there is no way to a priori know the accepted phase error.

A solution of this problem has been formulated by Blow and Crick as based on the Parseval's theory and of the relation (5.32): through minimizing the *rms* error towards the real electronic density between the structure factors within the complex plane. However, in order to apply this observation one should known the probabilities of the various possible choices of the phases, as indicated by the thickness of the line of the phase circle of Figure 5.11-right-up.; such weighted possibilities are then averaged; yet, the average of a complex number around a circle has as result another complex number, but inside the circle, with an amplitude indicated by the so-called *figure of merit* smaller than the circle radius, and with the phase called as the *best phase*. When the phase information is perfect, the figure of merit corresponds to $m=1$; nevertheless, it is reduced with the increasing of the phase attributing ambiguity, till the fixed equi-probability of $m=0$.

In any case, in order the molecular structures (through the associated molecular crystals) be assessed, the experiment for recording the diffraction pattern is indispensable, together with an optimal rationalization of the measured structure factors. The following sections present experimental diffraction techniques and the rationalization of their interpretation, in order to determine the maps of electronic localization.

5.2.3 THE EWALD X-RAYS DIFFRACTION RATIONALIZATION

A crystal arbitrary oriented in the way of an X-ray beam does not necessary produce a scattering beam and respectively its diffraction picture. Instead, diffraction can be rationalized through a construction in the reciprocal space, where the diffraction occurs, through the so-called *Ewald reflection sphere* that allows the prediction of the necessary orientation for satisfying the condition of Bragg's diffraction (Ewald, 1965, 1969).

The Ewald construction is based on the consideration of the scattering vector of Eqs. (5.6) or of (5.11) type, as being the vector which "transforms" the incident wave vector into the diffracted one, on a sphere, Figure 5.12 of $1/\lambda$ radius; such construction is possible according to the condition of reflection (D1 of preceding section) of the diffracted X-ray towards the incident one (wave vectors with equal $1/\lambda$ modulus; the present discussion follows Putz and Lacrămă (2005).

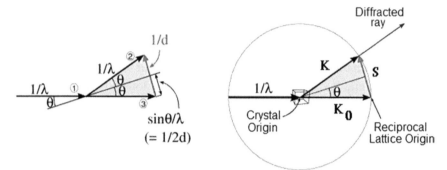

FIGURE 5.12   The construction (left) of the Ewald's sphere of diffraction (right), after (Matter Diffraction, 2003; X-Rays, 2003; Putz & Lacrămă, 2005).

Thus, the Ewald's reflection sphere corresponds to the geometrical place of the elastic scattering of incident X photons on the electrons of a crystalline lattice; if the momentum conservation law is applied to the elastic scattering the relation will next be written at the level of the associated wave vectors:

$$\vec{K_0} + \vec{S} = \vec{K} \tag{5.37}$$

and which, in the reflection condition virtue

$$K_0^2 = K^2 \tag{5.38}$$

can be rewritten such as:

$$\left(\vec{K_0} + \vec{S}\right)^2 = \vec{K}^2 \Leftrightarrow 2\vec{K_0} \cdot \vec{S} + \vec{S}^2 = 0 \tag{5.39}$$

the last version of the relation (5.39) corresponding to Bragg's law at the level of reciprocal space. This fact can be easily re-proofed if the Eq. (5.39) is transposed at a scalar level, as based on Figure 5.12 (Putz & Lacrămă, 2005):

$$2\frac{1}{\lambda}\frac{1}{d}\cos\left(\frac{\pi}{2}+\theta\right) + \frac{1}{d^2} = 0 \Leftrightarrow 2d\sin\theta = \lambda \tag{5.40}$$

Therefore, when the Ewald's sphere contains, passes by, or "touches" points of the reciprocal lattice the possibility of Bragg's reflection is predicted; the point from the reciprocal space, on the Ewald sphere, touched or fixed on the wave vector direction of the incident X-radiation on the crystal will also fix the origin of the reciprocal space for diffraction, Figure 5.13-left-up (the segment QO).

By rotating the crystal also the associated reciprocal space is rotated, around its origin, and respectively also the Ewald sphere is corresponding for the incident radiation (of ray $1/\lambda$), defining the so-called *limited sphere of diffraction*, containing all the points of reciprocal space that satisfy the condition of Bragg's diffraction; it further corresponds to all the families of planes of the network that give spots in the diffraction pattern; other planes or families of planes will never produce diffraction, regardless of how the crystal will be rotated, for a given incident radiation, Figure 5.13-down and right-up.

Therefore, one notes how the Ewald's sphere rationalizes the X-ray diffraction, by the reflection on the crystallographic planes (*hkl*), along the Bragg's law in terms of wavelengths of incident X-radiation on crystal, through the $1/\lambda$ radius of reflection's sphere.

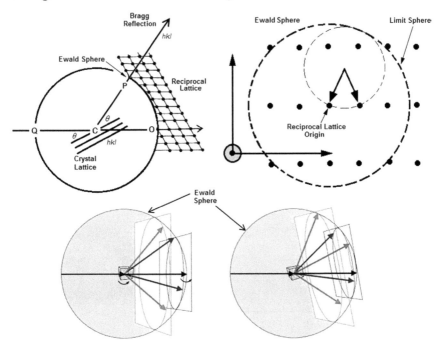

FIGURE 5.13 The Bragg's reflection rationalization through the Ewald's sphere of diffraction, after (Matter Diffraction, 2003; X-Rays, 2003; Putz & Lacrămă, 2005; HyperPhysics, 2010).

From the analysis of the Bragg's law of X-ray diffraction, the relation (5.40), there results how, excepting the interplanar d distance as a structure characteristic (internal to the crystal), the wavelength of incident radiation λ and the angle of reflection θ rest as variable (or external) parameters.

The wavelength of the X photons sets the Ewald's sphere, while the Bragg's θ angle characterizes the diffraction planes that will give the spots in the diffraction picture and correspond to the points of the reciprocal space of Ewald's sphere.

If the Bragg's angle is fixed, but not also the incident wavelength, the famous *Laue method* appears (Laue, 1952, 1953, 1960); in practice it is used for assessing the orientation of mono-crystals and consists in the crystal irradiation with a "white" beam of X-ray (i.e., with many monochromatic wavelengths, generally with $\lambda < 2.0$Å), Figure 5.14-left, commonly produced by a synchrotron source, thus allowing a very fast collection of the diffraction data, this way being the ideal method for generating the so-called *time resolved crystallography.*

The advantage of Laue's method is that the entire set of data is collected in a few films, the multitude of spots being caused by the polychromaticity of incident X- radiations, Figure 5.14-left, thus a significant area of reciprocal space being covered (Putz & Lacrămă, 2005).

The Laue's method supports two versions: *in transmission* and in *back-reflection*, Figure 5.15, depending on place where the film relative to the X-ray source and crystal is placed. For *transmission Laue diffraction* the film is placed behind the crystal and the spots will be recorded as the bases of ellipses' shape of some imaginary cones of scattering beams, Figure 5.15-left-up. Instead, for the back-reflection Laue's diffraction, the

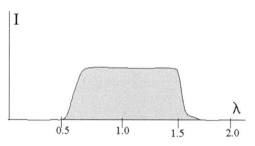
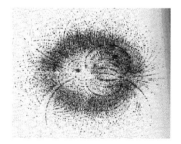

FIGURE 5.14 Polychromatic X-radiation intensity spectra (left) and the sample of a Laue's diffraction record (right); after (Matter Diffraction, 2003; X-Rays, 2003; Putz & Lacrămă, 2005; HyperPhysics, 2010).

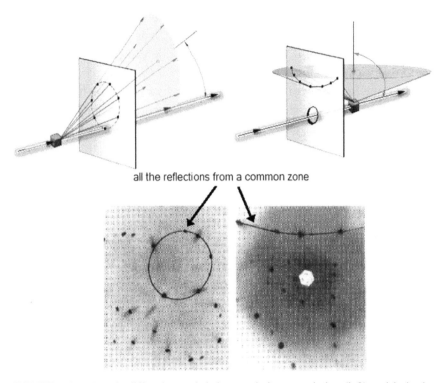

all the reflections from a common zone

FIGURE 5.15 Laue's diffraction and their records in transmission (left) and in back-reflection (right); after Matter Diffraction (2003).

film is placed between the X-ray source and the investigated crystal, and the recorded spots indicates the hyperboles resulted at the intersection between the cones of the X diffracted beams and the film, Figure 5.15-right (Putz & Lacrămă, 2005).

In Laue's diffraction, when the Bragg's angle is fixed for a set of crystallographic planes, from the incident radiation the corresponding diffracted wavelength is selected according to the Bragg's law, and for each such wavelength of the radiation a curve (with ellipse or hyperbole shape) will be generated inside the diffraction picture. The spots in a curve inside the diffraction picture belong to the reflections of the planes that, in turn, belong to the so-called *crystallographic zone*; thus, the common axis to the cones formed by the diffraction rays represents the so-called zone axis. The recorded spots are indexed, wherefrom the crystallographic planes involved in diffraction are identified. The Laue's method allows, by the spots indexing, also the assignment of the crystals perfection state along of its size and shape; if the crystal is bent or deformed, the spots are not recorded as points, but as stains.

Yet, the limitation of Laue's method consists in establishing the multiple reflections, the reflections generated by the families of (*nh, nk, nl*) planes, so generating the so-called *Cruikshank's problem;* this arises by rewriting of the Bragg's law under a multiple form:

$$2\sin\theta = \lambda d = (2\lambda)d/2 = (3\lambda)d/3 = ... \qquad (5.41)$$

involving therefore the existence of the incident radiation polychromatic spectrum.

The problem is graphically illustrated in Figure 5.16-up, where, as based on the construction of the Ewald's spheres for the wavelengths between λ_{max} and λ_{min} the dilemma of setting appears in choosing, form the multiplicities of reflections, for the equivalent reflections - the unique ones. The multiple reflections, are found "strung" on the directions of possible diffraction, as predicted by the directions that connect the center of diffraction with the allowed Ewald spheres. For example, in Figure 5.16-right-up 73 reflections are indicated, where 44 are multiple and 29 are unique. The construction of Figure 5.16-down will decide which of the multiple reflections are unique. Considering the difference λ_{max}-λ_{min} the radius $1/(\lambda_{max}$-$\lambda_{min})$ will be formed and a corresponding sector will be mapped, centered on the direction (of the wave vector) of the incident beam, and similarly the sector corresponding to the $1/\lambda_{max}$ sphere (Putz & Lacrămă, 2005).

The points of the reciprocal space placed inside the area delimited by the two sectors and the extreme Ewald's spheres for the extreme (min-max) wavelengths of the incident radiation will establish the number of unique or equivalent spots (respectively of crystallographic planes) that contribute to the picture/pattern of diffraction. Thus, form Figure 5.16-down, a total number of 19 reflections are found, where 2 are multiple (equivalent) and 17 (in proportion of 89%) singular. This type of problems of multiple covering of the reciprocal space, make the Laue's diffractions diagram to be less practicable for attributing the crystallographic classes, but being

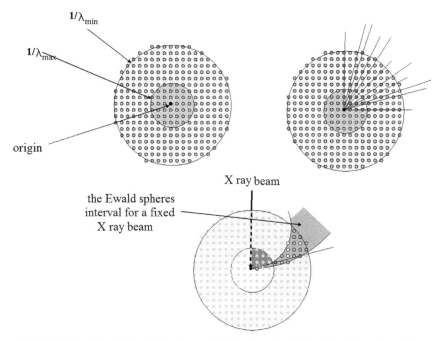

FIGURE 5.16   Up: the Ewald's spheres of diffraction for the Cruikshank problem; down: determination of unique spots from multiple Laue diffractions, by Ewald's construction; after (Matter Diffraction, 2003).

very useful for fixing the orientations of the zone axis of the crystalline samples, for further investigations.

At this point the second possibility of reproduction of the diffraction picture comes into attention, by fixing the wavelength of the incident X-radiation, but rotating the crystal, i.e., varying the angle of Bragg's reflection, Figure 5.17.

The diffraction methods with rotating crystal are used for the accurately determination of the unit cell parameters of the crystal subject to the X-ray scattering. In this framework the method of crystalline powder methods was consecrated.

Why the crystalline powder does it necessary? Being the monochromatic incident radiation (of the fixed wavelength), for a mono-crystal only one or two diffraction beams would be obtain and therefore the diffraction figure would be "poorer" in the structural information.

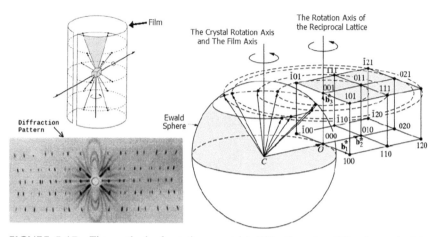

FIGURE 5.17 The method of rotating crystals' diffraction: the diffraction principle (left-up) and diffraction pattern (left-down); along the Ewald's rationalization on the reflection sphere of diffraction (right); after (Matter Diffraction, 2003; X-Rays, 2003; Putz & Lacrămă, 2005; HyperPhysics, 2010).

Instead, if one is working with a sample from a crystalline powder (small, scraps of crystals), each micro-crystal is oriented differently towards the incident beam, respectively exposing different crystallographic planes at diffraction; the diffracted ray's cones associated emerge in all directions, as imaginary cones, each of them intersecting the circular film in diffraction lines-arcs, as shown in Figure 5.18, enriching the structural information.

The simplicity and the efficiency of the Debye-Scherrer procedure will be further exposed by the algorithm of the assessing of unit cell sizes of a cubical structure. After exposure, the film with the diffraction of crystalline

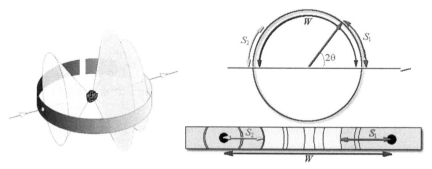

FIGURE 5.18 The Debye-Scherrer diffraction principle (left) and the rationalization of the obtained diffraction film (right); after Matter Diffraction (2003).

powder is flatly stretched and the S_1 and S_2 distances from the center of the diffraction arcs till the center (the hollow) of the film are measured in direct transmission and back-reflection, respectively, see Figure 5.18. The S_1 distances correspond to the diffraction 2θ angle, while the S_2 distances correspond to the π-2θ angle; in the same time, the W distance on the film between the transmission hollow and the one of back-reflection corresponds to the diffraction $\theta = \pi$ angle. Therefore, by the proportion rule the corresponding relation is found, for each case:

$$\theta = \frac{\pi S_1}{2W}; \ \theta = \frac{\pi}{2}\left(1 - \frac{S_2}{W}\right) \tag{5.42}$$

Once measured, the distances S_1 or S_2 the Bragg's θ angle can be established according to the (5.42) relations thus determining also the $\sin\theta$ measure that intervenes in the Bragg's law (5.40). This way, $\sin\theta$ can be related to the interplanar d distance, $\sin\theta = \lambda/(2d)$, which, for the cubical lattice it can be in turn expressed by the Miller indices and the network a-parameter (Verma & Srivastava, 1982):

$$d_{hkl} = \frac{a}{\sqrt{h^2 + k^2 + l^2}} \tag{5.43}$$

So results the connection:

$$\sin^2\theta = \frac{\lambda^2\left(h^2 + k^2 + l^2\right)}{4a^2} \tag{5.44}$$

wherefrom the lattice a-parameter can be established. In fact, one should consider that the sum of the squares of the Miller indices for a Bragg reflection is an integer, see Table 5.1, and, therefore, the multiplication "q" factor should be found as satisfying the relation

$$q\sin^2\theta = h^2 + k^2 + l^2 \tag{5.45}$$

based on which, by comparing with Eq. (5.44), the a dimension results in each case.

For example, in Table 5.2 the measurements and the determination of the cubical lattice parameter are exposed for Debye-Scherrer diffraction of a X-ray beam of wavelength λ=1.54Å, and with the diffraction picture having the W=180 [mm] dimension (Putz & Lacrămă, 2005).

TABLE 5.1   The Sum of Miller Indices' Squares for Various Families of Crystalline Planes

$h^2+k^2+l^2$	1	2	3	4	5	6	8	9	10	11	12	13
hkl	100	110	111	200	210	211	220	221	310	311	222	320

TABLE 5.2   Determination of the Lattice Constant/Parameter of a Cubical Lattice from the Debye-Scherrer Diffraction Data

$S_i(mm)$	θ	$sin^2\theta$	$q\ sin^2\theta$	$h^2+k^2+l^2$	*hkl*	$a(\mathring{A})$
38	19.0	0.11	3.0	3	111	4.05
45	22.5	0.15	4.1	4	200	4.02
66	33.0	0.30	8.2	8	220	4.02
78	39.0	0.40	10.9	11	311	4.04
83	41.5	0.45	12.3	12	222	4.02
97	49.5	0.58	15.8	16	400	4.04
113	56.5	0.70	19.1	19	331	4.03
118	59.0	0.73	19.9	20	420	4.04
139	69.5	0.88	24.0	24	422	4.01
168	84.0	0.99	27.0	27	511	4.03

From the families of Miller indices types corresponding to Bragg's diffraction planes also the investigated type of cubical structure can be established; in this case the face centered cubic (FCC) structure is identified, because the involved crystallographic planes have either odd or only even Miller indices.

For a grater accuracy in determining the lattice parameters the average of these lattice parameters can be considered from the last column of Table 5.2.

Finally, worth emphasizing the idea by which the Ewald rationalization of diffraction with crystalline powder will provide an Ewald's sphere (a single wavelength generates a single sphere in the reciprocal space of $1/\lambda$ radius) yet including and intersecting the various sets of points in the reciprocal space, for the various interplanar distances d_i: they correspond to the crystallographic planes' varieties that become Bragg's diffraction planes that, when projected into the reciprocal space, generate points placed at the $1/d_i$ distances, called *resolution distances*, because equal the modulus of the scattering vectors, see the relation (5.11) and Figure 5.19.

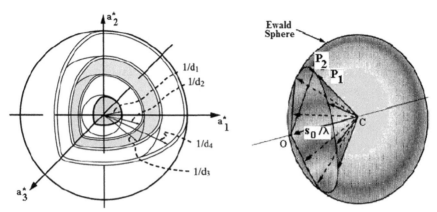

FIGURE 5.19 Crystalline powder diffraction: in left, the scattering spheres in the reciprocal space; in right, the associated Ewald's sphere of diffraction for any of them; after (Matter Diffraction, 2003; X-Rays, 2003; Putz & Lacrămă, 2005; HyperPhysics, 2010).

The crystallographic planes and their representation through Miller indices were rationalized in this section at the level of Ewald sphere of diffraction in the reciprocal space. This complements the structure factors, and implicitly the electronic densities, equally expressed in the previous section through the Miller indexing. However, the electronic localization is made around their attraction centers and the diffraction model should contain also the atomic representation, as basic structural unit. Accordingly, the atomic information should be contained in the diffraction description, and then into the determination of the crystalline structure type, towards the molecular and bio-molecular structural determination by crystallization, as will be exposed in the next section.

5.2.4   X-RAY AND STRUCTURE DETERMINATION

The key in determining the molecular structure (atoms distribution, electronic localization) through the methods of X-ray diffraction consists in the structure factor determination and in the associated electronic density, the relations (5.17) and (5.18), respectively; the present discussion follows Putz and Lacrămă (2005).

In any case, due to the fact that the electrons that are scattered by the X-ray are located around the atoms or groups of constituent atoms of the crystal, the separation in the diffraction space naturally occurs for the structure factors in the components of atomic scattering, called *factors of atomic scattering*, Figure 5.9.

The first step consists in the separation of the total density in the components ρ_j associated to the "*j*" atoms:

$$F\left(\vec{S}\right) = \int\limits_{\substack{CRYSTAL \\ SPACE}} dV\, \rho\left(\vec{r}\right) \exp\left(2\pi i \vec{S} \cdot \vec{r}\right)$$

$$\rightarrow \sum_j \int\limits_{\substack{CRYSTAL \\ SPACE}} dV\, \rho_j\left(\vec{r}\right) \exp\left(2\pi i \vec{S} \cdot \vec{r}\right) \qquad (5.46)$$

after which, the densities ρ_j in relation with the centers r_j of the atoms "*j*" are successively rewritten such as:

$$\int\limits_{\substack{CRYSTAL \\ SPACE}} dV\, \rho_j\left(\vec{r}\right) \exp\left(2\pi i \vec{S} \cdot \vec{r}\right)$$

$$= \int\limits_{\substack{CRYSTAL \\ SPACE}} dV\, \rho_j\left(\vec{r} - \vec{r_j}\right) \exp\left[2\pi i \vec{S} \cdot \left(\vec{r} - \vec{r_j}\right)\right] \exp\left(2\pi i \vec{S} \cdot \vec{r_j}\right)$$

$$= \exp\left(2\pi i \vec{S} \cdot \vec{r_j}\right) \int\limits_{\substack{CRYSTAL \\ SPACE}} dV_{\vec{r'} = \vec{r} - \vec{r_j}}\, \rho_j\left(\vec{r'}\right) \exp\left[2\pi i \vec{S} \cdot \vec{r'}\right]$$

$$\equiv f_j(S) \exp\left(2\pi i \vec{S} \cdot \vec{r_j}\right) \qquad (5.47)$$

wherefrom the new Fourier's expansion of the structure factors based on the factors of atomic scattering f_j is obtained with the alternative forms:

$$F\left(\vec{S}\right) = \sum_j f_j(S) \exp\left(2\pi i \vec{S} \cdot \vec{r_j}\right) \rightarrow F\left(\vec{h}\right) = \sum_j f_j(S) \exp\left(2\pi i \vec{h} \cdot \vec{x_j}\right) \quad (5.48)$$

This way, the diffraction is specialized on the type of the components as atoms, through the associated factors of atomic scattering. In this context, as revealed by the definition of the atomic scattering factors in relation (5.47), the diffraction depends on the scattering vector $\vec{S}$ (by its modulus), and not on the direction of the diffracted rays. Moreover, for small diffraction angles (corresponding at interplanar d distances much larger than the atomic sizes) all the electrons of an atom will be scattering in phase, being the electronic contribution at the total scattering equal to their number from the targeted atom. Instead, while the diffraction angle increases the various zones of the electronic density centered in the origin of an atom will generate the scattered X-ray, partially dephased, so that, with the resolution increasing ($1/d$) will be recorded a decrease of the atomic scattering factors, Figure 5.20, and, respectively, a decrease of their contribution to the total scattering, through the influence of the scattering factor (5.48).

Thus, the power of scattered X-ray decreases with the increasing of the diffraction angle and increases with the increasing of the number of electrons in an atom. Note that for the zero diffraction angle (corresponding to the scattering in transmission) the factor of atomic scattering corresponds to the total number of electrons in an atom.

For practical reasons of localization at the level of crystalline lattice the "continuous" definition of the atomic scattering factor from Eq. (5.47), with the form

$$f\left(S \overset{(5.8)}{=} \frac{2\sin\theta}{\lambda} \right) = \int\limits_{\substack{CRYSTAL \\ SPACE}} dV\, \rho\left(\vec{r} \right) \exp\left[2\pi i \vec{S}\cdot \vec{r} \right] \qquad (5.49)$$

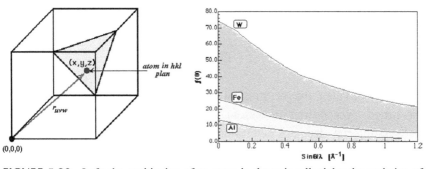

FIGURE 5.20 Left: the positioning of an atom in the unit cell; right: the variation of the atomic scattering factors with the diffraction angle, for some metals; after (Matter Diffraction, 2003; X-Rays, 2003; Putz & Lacrămă, 2005; HyperPhysics, 2010).

is parameterized under a discrete shape, consecrated by the finite form (Cromer & Mann, 1968):

$$f^0_{atom}\left(\frac{\sin\theta}{\lambda}\right) = \sum_{i=1}^{4} a_i \exp\left[-b_i\left(\frac{\sin\theta}{\lambda}\right)^2\right] + c \tag{5.50}$$

wherefrom, knowing the nine coefficients a_i, b_i and c, together with the incident wavelength, the atomic scattering factor can be evaluated for each angle of diffraction.

As an additional effect, firstly, the anomalous dispersion (or the so-called resonance absorption) is noted as becoming important around the absorption edges (representing the characteristic limit of diffraction, review the Figure 5.2) for the targeted atom, so introducing an imaginary component (i.e., dephased with the angle $\pi/2$) along the real one:

$$f = f^0 + f' + if'' \tag{5.51}$$

Moreover, although placed inside of a lattice, the atoms are not immobile, yet having a vibration motion around their equilibrium positions due to the absolute temperature to which the crystal is considered. Accordingly, the X-ray scattering will be considered at the averaged position $<u>$ (in Å) around the center of vibration of each atom through the so-called *Debye-Waller factor*:

$$B = 8\pi^2 \langle u \rangle^2 \tag{5.52}$$

with the result in a decaying for the atomic scattering factor, for the large reflection angles, according to the law:

$$f_B = f \exp\left(-B\frac{\sin^2\theta}{\lambda^2}\right) \tag{5.53}$$

which motivates the crystals' investigation at low temperatures (the so-called cryo-cooling processes).

Aiming the structure determination for the majority of bio-molecular systems the consideration of the average of the isotropic vibrations is enough; still, in order to increase the accuracy of atomic localization, in any case, the temperature of anisotropic effects can be considered at the level of the individual atoms by replacing the average $<u>$ with the tensor u_{ij}.

Early in XXI the electronic density (5.18) was Fourier's mapped as (Putz, 2003),

$$\rho\left(\vec{x}\right) = \sum_{\vec{h}} \left|F\left(\vec{h}\right)\right| \exp\left(i\alpha_{\vec{h}}\right) \exp\left(-2\pi i \vec{h}\cdot\vec{x}_j\right) \qquad (5.54)$$

while modeling the knowledge of the atomic scattering factors driving the modulus and the phase of the structure factor with the general forms, respectively:

$$\left|F\left(\vec{h}\right)\right| \cong \exp[F(\varepsilon)] \left|\sum_j f_j \exp\left(2\pi i \vec{h}\cdot\vec{x}_j\right)\right|,$$

$$\alpha_{\vec{h}} = \arccos\left\{ \frac{1+F(\varepsilon)}{\exp[F(\varepsilon)]} \frac{\sum_j f_j \cos\left(2\pi \vec{h}\cdot\vec{x}_j\right)}{\left|\sum_j f_j \exp\left(2\pi i \vec{h}\cdot\vec{x}_j\right)\right|} \right\} \qquad (5.55)$$

including the case of the crystals elastically deformed. The function $F(\varepsilon)$ is responsible for the embedding of the elastic deformation effects of the crystal in relation with the deformation ε-parameter. In Figure 5.21, an illustration of this model is exposed for the unit cell deformation case (of FCC type) of the Si-crystals.

This model, will be detailed in after-next section, yet worth noting from now how it presumes the knowledge of the Si-atoms position in the unit cell, being therefore reliable only after the determination of their location, for example, from the rationalization of the diffraction experiments. Nevertheless, a measure in estimating the theoretical-to-experimental modeling can be provided by the procedure of the *least (minimum) squares fitting,* based on the formula:

$$D = \sum_{\vec{h}} W_{\vec{h}} \left\{ \left|F_{\exp}\left(\vec{h}\right)\right| - \left|\kappa F_{calc}\left(\vec{h}\right)\right| \right\}^2 \qquad (5.56)$$

with $W_{\vec{h}}$ the weight of a "h" estimate in the structure factor, being calculated as based on the experimental (intensity of) diffraction pattern or by

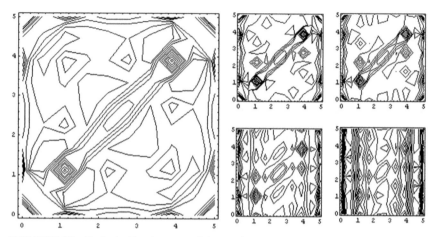

FIGURE 5.21   Left: the Fourier map of electronic density for the perfect unit cell (FCC) of Si; right, from top to bottom, clockwise: various levels of the of elastic deformation degree at the level of the unit cell; (all the units of measurements are in Å); after (Putz, 2003; Putz & Lacrămă, 2005).

the considered theoretical model, and with κ the scale parameter for the allowed model.

However, when the relative positions in the unit cell of the constituent atoms is not yet know, the relation (5.55) can not be anymore directly applied, but another model should be introduced, in direct relation with the experimental diffraction picture. As long as in a diffraction experiment the intensities recorded on the film of diffraction is measured, the idea of considering the Fourier transformations of the measured intensities appears, generating, this time, not the electronic density, but the so-called *Patterson's function*:

$$P\left(\vec{x}\right) = \sum_{\vec{h}} \left| F\left(\vec{h}\right) \right|^2 \exp\left(-2\pi i \vec{h} \cdot \vec{x}\right)$$

(5.57)

The Patterson's function (5.57) does not directly represent the phase information but it contains the measurable ones (such as intensity of scattering X-radiation). Thus, if a protein with unknown structure is investigated, for called sample or parent (P) or the original crystal - the diffraction film is recorded, the resulted intensities are measured and the modulus of the scattering factor F_p is determined; for evaluation there remains the associated phase α_p as the last information required for the structure characterization through the electronic density equation, such as Eq. (5.18) or (5.54).

For this purpose, the additive property of the Fourier transformations and respectively for the scattering factors is used, by rewriting the scattering factor of the sample crystal as the difference between the scattering factor corresponding to the crystal of isomorphic structure (having the same lattice, but with different base – the atoms or the constituent groups of atoms) doped with heavy atoms (strongly targeted by X-ray), i.e., so composing the HP structure ("heavy parent"-heavy sample crystal, in sense of enriched) *and* the one isomorphic structure exclusively formed from the atoms rich in electrons (such as Hg, Pt, Au), the so-called H (heavy) structure:

$$F_P\left(\vec{h}\right) = F_{HP}\left(\vec{h}\right) - F_H\left(\vec{h}\right) \tag{5.58}$$

The phasorial representation (in terms of phase angles and of the scattering amplitude) of the (5.58) relation is shown in Figure 5.22. There is noted how for establishing the right phase (i.e., the angle for calculation) for the structure factor of the sample, F_P, one should consider the diffraction picture of two isomorphic structures, i.e., those derived through the involvement of the "heavy atoms", in order to decide the phase angle of the "vector" F_P consistent with both structures; that position will fix the phase α_p.

Thus, the structure factor, $F_{PH(calc)}$, can be calculated based on the consideration of the isomorphic structures derived with heavy atoms, such as:

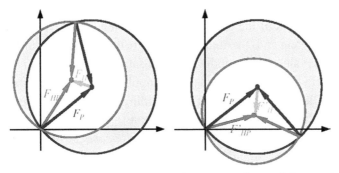

FIGURE 5.22 The phase relations between the structure factor of the sample (parent) crystal (P) and the structure factor of its isomorphic derivatives, with heavy (H) atoms; after (Matter Diffraction, 2003; X-Rays, 2003; Putz & Lacrămă, 2005).

$$F_{HP(calc)} = |F_H| \exp(i\alpha_H) + |F_P| \exp(i\alpha_P) \qquad (5.59)$$

The last equation can be considered in an iterative procedure of the D-factor minimization from (5.56), with the actual form:

$$D = \sum_{\vec{h}} W_{\vec{h}} \left\{ \left| F_{HP\,exp}\left(\vec{h}\right) \right| - \left| \kappa\, F_{HPcalc}\left(\vec{h}\right) \right| \right\}^2 \qquad (5.60)$$

Once "refined" the position of "heavy" atoms, by the refinement of the (5.59) factor, the final set of phases for the sample crystal P is determined (because each set of planes "h"=(hkl) induces a certain phase of scattering of X-rays), so determining the corresponding structure model, i.e., the density, or the electronic localization, such as the relations (5.18) and (5.54):

$$\rho_{Model}(\vec{x}) = \sum_{\vec{h}} \left\{ \left| F_{P(calc)}\left(\vec{h}\right) \right| \exp\left[i\alpha_P\left(\vec{h}\right) \right] \exp[-2\pi i \vec{h} \cdot \vec{x}] \right\} \qquad (5.61)$$

The validation of the model obtained can be made in various ways. A consecrated method consists in the evaluation of the so-called residual index R, a measure of the global relative error of the calculated structure factor respecting the observed one:

$$R = \frac{1}{\sum_{\vec{h}} \left| F_{P(exp)}\left(\vec{h}\right) \right|} \sum_{\vec{h}} \left\| F_{P(exp)}\left(\vec{h}\right) \right| - \left| F_{P(calc)}\left(\vec{h}\right) \right\| \qquad (5.62)$$

Values of the residual index $R \approx 0.25$ indicate that most of atoms were correctly localized upto 0.1Å resolution; the organic structures with small sizes can be refined till $R < 0.05$; for proteins, the index R is usually high in the first stages of the structure determination, and decreases while the effects of solvent and (thermal) vibration are taking into account. However, because the value $R \approx 0.25$ indicates an unsatisfactory degree of refining for the most of structural techniques, the supplementation by diffraction data will be called, yet maintaining the bondings' lengths and the structures' angles, in a reasonable domain of variation.

The procedure is called *hybrid electronic maps* (or the "over-fit") method, Figure 5.23, and can be obtained, for example, by non-refining (i.e., non-minimizing) completely the D factor of (5.60), so generating a structure such as Eq. (5.61), yet unrefined, therefore being called as the *structure by omission*; this is compared (by superimposing) with the *density difference* (or synthesis of the Fourier maps' differences)

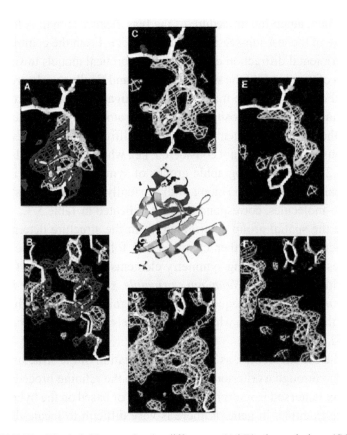

FIGURE 5.23 The hybrid maps for the differences (A&B), the omissions (C&D), and the extrapolations (E&F) of the X-ray determined electronic densities of the chromophore 4-hydroxycinnamyl (up) of the yellow photo-active protein that contains it (down), as extracted from the photo-tropic bacterium Ectothiorhodospira halophila, respectively; after Genick et al. (1997); BioInformatics Protein Models (2013).

$$\Delta\rho\left(\vec{x}\right) = \sum_{\vec{h}}\left\{\left|F_{PH(calc)}\left(\vec{h}\right)\right| - \left|F_{P(calc)}\left(\vec{h}\right)\right|\right\}\exp\left(i\alpha_{P,\vec{h}}\right)\exp\left(-2\pi i\vec{h}\cdot\vec{x}\right)$$

$$(5.63)$$

generating the so-called *extrapolated density*:

$$\rho_{Extra}\left(\vec{x}\right) = \sum_{\vec{h}}\left\{\left|F_{PH(calc)}\left(\vec{h}\right)\right|\exp\left(i\alpha_{P,\vec{h}}\right)\right\}\exp\left(-2\pi i\vec{h}\cdot\vec{x}\right) \quad (5.64)$$

Thus, in fact, again the procedure of the bias electronic map is found, see the before of the last sub-section and Figure 5.11. From the combination of the experimental diffraction data with the theoretical models toward determination of the phase, of the structure factors and finally the electronic density, the Fourier electronic maps of the investigated structures are refined, so resulting in the atoms positioning, or of the atoms groups, at the unit cell level of the crystalline system subject to the diffraction. If the effects of the anomalous diffraction are not considered, i.e., when the Friedel's law is satisfied (5.24), the crystallographic groups of symmetry detected in X-rays (the so-called *Laue groups*) along of the spatially possible groups, specific to the bio-molecules, correspond to those indicated in Table 5.3.

Once the spatial group determined, from the structure determination, this structure can be further characterized following all the correlated properties with the existing symmetry elements and operations (in accordance with the so-called *Neumann's principle, se also the Curie Principle of Section 2.5.8.2*). Consequently, there is again emphasized the importance of the accuracy with which the structure is determined or the method is refined.

For the bio-molecules, as proteins, a validation of the structure is very useful also through a criterion not included in the refining procedure (e.g., bias, or by Patterson isomorphic replacement, or based on the hybrid maps, etc.). For example, in general, there is very difficult to locate the values of the torsion angles of the main polypeptide chain, but their distribution is very restricted in a representation of *Ramachandran* type, Figure 5.24 (Ramachandran, 1952; Ramachandran et al., 1963).

In Figure 5.24, the energy distribution of residues (amino acids of polypeptide chains) is represented in relation to the conformation in the chain to which belong. For example, the most populated regions are those indicated by [A, B, L] – in about 92.9% of cases, the most intensely colored in Figure 5.24; they are followed by the additional permitted regions – in about 6.6% of cases, denominated [a, b, l, p], so being less colored in Figure 5.24; next are the regions allowed with maximum "generosity" relative to the variation of the torsion angles – in about 0.5% of cases – as those indicated as [~a, ~b, ~l, ~p], with very pale coloration in Figure 5.24; finally the forbidden energetic regions (i.e., with the highest conformational energy) – are ideally in about 0.0[...]% of cases – those uncolored.

TABLE 5.3 The 65 Possible Spatial Groups for the Bio-Molecular Crystallization, Satisfying the Friedel's Law (Regarding the Absence of Anomalous Dispersion)*

Syngony	No. of independent parameters	Type	Diffraction symmetry	Spatial groups
Triclinic	6	P	$\bar{1}$	**P1**
Monoclinic	4	P	2/m	**P2, P21, C2**
Orthorhombic	3	P	mmm	**P222, P212121, P2221, P21212**
		C	mmm	**C222, C2221**
		I	mmm	**I222, I212121**
		F	mmm	**F222**
Tetragonal	2	P	4/m	**P4, (P41, P43), P42**
			4/mmm	**P422, (P4122, P4322), P4222, P4212, (P41212, P43212), P42212**
		I	4/m	**I4, I41**
			4/mmm	**I422, I4122**
Trigonal	2	P	$\bar{3}$	**P3, (P31, P32)**
			$\bar{3}$m	**[P321, P312], [(P3121, P3221), (P3112, P3212)]**
Rhombohedral	2	R	$\bar{3}$	**R3**
			$\bar{3}$m	**R32**
Hexagonal	2	P	6/m	**P6, (P61, P65), P63, (P62, P64)**
			6/mmm	**P622, (P6122, P6522), P6322, (P6222, P6422)**
Cubic	1	P	m3	**P23, P213**
			m3m	**P432, (P4132, P4332), P4222**
		I	m3	**[i23, I213]**
			m3m	**I432, I4143**
		F	m3	**F23**
			m3m	**F432, F4132**

*After (Blundell & Johnson, 1976; Sherwood, 1976; Drenth, 1994).

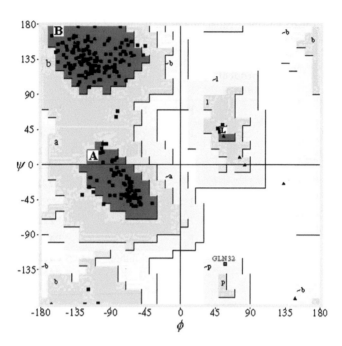

FIGURE 5.24   The Ramachandran map for the energy distribution accounting for the conformation of a significant number (about 120) of proteins; after Laskowski et al. (1993).

Note how, the Gly residues are the most tolerated at the conformation with high energy and are represented by triangles. Excluding them, any location of the residues in areas less allowed, or forbidden, as the case of GLN32 in Figure 5.24, indicates either a refining error of the theoretical model or an unusual conformation, with an important effect.

However, the big problem of the bio-molecular structure determination by the refining of the Fourier synthesis model, with the "crossing" rechecking with the map of Ramachandran type included, consists in the lack of absolute criteria for stopping the procedure, otherwise the refining supports further refining *ad tedium*, in fact until the fatigue-boredom of the investigator (Pauling, 1960; Sasisekharan, 1962; Fraser & Mac Rae, 1969; IUPAC-IUB, 1970; Bernstein et al., 1977, Allen et al., 1979, Kabsch & Sander, 1983; Nishikawa & Ooi, 1986; Branden & Jones, 1990; Engh & Huber, 1991; Morris et al., 1992, Brünger, 1992; Voet & Voet, 1995; Dodson et al., 1996, Vaguine et al., 1999, Hanson & Stevens, 2000; Rupp & Segelke, 2001; Keywegt & Brünger, 1996; Kleywegt & Jones, 1995, 2002).

For example, the IBM researchers have already made the first of the 65,000 types of integrated chips that form the biggest computer projected by then, the so-called *Blue Gene* ready launched in 2006, carrying the mission to decode the configuration "mysteries" in proteins, with a huge role in bio-medicine. As application, has been discovered how the "mad cow" disease is caused not so much by a virus, but by a conformation of the prion protein in a forbidden zone of the areas of Ramachandran map, having the effect of cellular proliferation of a type of tertiary packing that produces, by further quaternary packing, the infection or the cancer. The problem is particularly complex because the number of possible conformations of a single protein overcomes the number of the atoms found in the entire known Universe! Blue Gene is projected in order to be able to solve, or determine, all the possible three-dimensional conformations of an average protein, continuously working for a year!

However, an alternative is about to be developed by adopting an analytical statistical-quantum model, i.e., the Fourier's synthesis combining with the stochastic methods specific to the quantum statistics. The issue remains open!

5.3 FIELDS AND INTENSITIES IN DYNAMICAL THEORY OF X-RAY DIFFRACTION

5.3.1 FUNDAMENTAL EQUATIONS

Within the vacuum-crystal plane separation, the incident field represented by (in vacuum) plane wave has the custom form (James, 1965):

$$\vec{E_0^e} \exp\left(i\omega_0 t - i2\pi \vec{k_0} \cdot \vec{r} \right) \tag{5.65}$$

Inside the crystal, due to the interaction between the incident and the scatted radiation, the internal field will not longer have the form (5.65), but modified such as in eq. (5.66); the present discussion follows (James, 1965; Putz and Lacrămă, 2005);

$$\vec{D_0} \exp\left(i\omega_0 t - i2\pi \vec{K_0} \cdot \vec{r} \right) \tag{5.66}$$

In eq. (5.66), the amplitude $\vec{D}_0$ and the wave vector $\vec{K}_0$ can be established from the following considerations:

   (i) *The continuity condition, involving that the incident wave (5.65) should coincide* with the internal incident wave (5.66). For solving this matter the propagation origin on the vacuum-crystal surface will be considered such as $\vec{n} \cdot \vec{r} = 0$, where n is the normal at the separation surface, and r is the motion vector.

   (ii) *The internal incident wave* and the diffracted wave should form a self-consistent set of equations.

The first condition, when conjugated with the property of the dielectric constant, i.e., to differ from the unit by 10^{-6} parts, in the field's domain of the frequencies specific to the X-ray, it unfolds into the set of equations:

$$\begin{cases} \vec{K}_0 = \vec{k}_0 + \Delta \vec{n} \\ K_0^2 = k_0^2 (1+\delta_0)^2 \approx k_0^2 (1+2\delta_0) \end{cases}$$

(5.67)

being δ_0 a very small and positive quantity.

   The system (5.67) may be directly solved with the respectively resulting expressions:

$$\begin{cases} \Delta = \dfrac{k_0 \delta_0}{\vec{n} \cdot \vec{u}_0^e} = \dfrac{k_0 \delta_0}{\gamma_0} \\ \vec{K}_0 = \vec{k}_0 + \dfrac{k_0 \delta_0}{\gamma_0} \vec{n} \end{cases}$$

(5.68)

where, $\vec{u}_0^e$ is the vector of the incident direction of the external field on the crystal, and γ_0 is the driving cosines of the incident (and transmission) direction with the normal of the crystal surface. The second equation in the Eq. (5.68) set offers a very important result, i.e., the normal external and internal (transmitted) components of the incident wave vectors toward the crystal surface support a jump on this surface, while the tangential components are continuous. Within these considerations one approximates $\vec{E}_0^e \approx \vec{D}_0$ due to the fact that for the X-ray frequencies the dielectric constant is very close to the unity. However, the internal incident wave is not yet completely determined as long as the diffraction angle $1+\delta_0$ remains unknown. In order to eliminate such limitation one passes to the fulfillment of the second condition above exposed, i.e., the construction of a self-consistent set of equations for the propagation in the crystal.

Therefore, the following set of equations related to those phenomenological discussed can be formed:

$$\begin{cases} \nabla \times \vec{H} = \dfrac{\partial \vec{D}}{\partial t} \\[2mm] \nabla \times \vec{E} = -\mu_0 \dfrac{\partial \vec{H}}{\partial t} \\[2mm] \vec{D} = \sum_h \vec{D}_h \exp\left(i\omega_0 t - i2\pi \vec{K}_h \cdot \vec{r} \right) \\[2mm] \vec{K}_h = \vec{K}_0 + \vec{h} \\[2mm] \vec{D} = \kappa \varepsilon_0 \vec{E} \\[2mm] \kappa = 1 + \chi \\[2mm] \chi = \sum_h \chi_h \exp\left(-i2\pi \vec{h} \cdot \vec{r} \right) \\[2mm] \chi_h = -\dfrac{e^2 \lambda_0^2 F_h}{4\pi^2 \varepsilon_0 m_0 c^2 V} \end{cases} \qquad (5.69)$$

where:

- *the first two equations* are the Maxwell's equations (in International System of Units) written for the field of frequencies of the X radiations in the absence of the density of current;
- *the third equation* is the expansion of the diffracted field in the crystal as a superposition of Bloch's waves;
- *the fourth equation* is the equation of Bragg's coupling of the transmitted and diffracted directions;
- *the fifth equation* exposes the connection of the propagated fields through the dielectric κ – constant: an equation of material;
- *the sixth equation* connects the dielectric constant with atomic χ-susceptibility;
- *the seventh equation* shown the Fourier expansion of atomic susceptibility for the crystallographic planes that contributes to diffraction,
- and *the eighth equation* relates the coefficients of atomic susceptibility of generic h-index (order) with the structure factor F_h through the consecrated constants as the frequency of the incident wave and the volume of the unit cell (Batterman & Cole, 1964).

There appears as obvious the interdependence of the Maxwell electromagnetic propagation with the Bragg's coupling condition, along the conditions of material and the associated structure factors for a reflection on crystal and for the given conditions of incidence. This system can be reduced to a resumed one by cumulating all the conditions which should be satisfied for accomplishing the self-consistency. So, by making the required eliminations and the replacements inside of the system (5.69), the condensed Maxwell equations write as the expression:

$$\nabla \times \left[\nabla \times (1-\chi)\vec{D} \right] = -\frac{1}{c^2} \frac{\partial^2 \vec{D}}{\partial^2 t} \tag{5.70}$$

and under the approximation of material's constants

$$1/\kappa \approx 1-\chi \tag{5.71}$$

By using the other equations of the set (5.69) one can evaluate:

$$(1-\chi)\vec{D} = \exp(i\omega_0 t) \left\{ \begin{aligned} &\sum_h \vec{D}_h \exp\left(-i2\pi \vec{K}_h \cdot \vec{r} \right) \\ &-\sum_g \sum_l \chi_g \vec{D}_l \exp\left[-i2\pi \left(\vec{g} + \vec{K}_l \right) \cdot \vec{r} \right] \end{aligned} \right\} \tag{5.72}$$

where, through considering the Ewald's connection

$$\vec{g} + \vec{K}_l = \vec{K}_{g+l} \tag{5.73}$$

along the indices' summation

$$g+l=h \tag{5.74}$$

the last equation becomes:

$$(1-\chi)\vec{D} = \exp(i\omega_0 t) \left[\sum_h \left(\vec{D}_h - \vec{C}_h \right) \exp\left(-i2\pi \vec{K}_h \cdot \vec{r} \right) \right] \tag{5.75}$$

with the introduced notation:

$$\vec{C}_h \equiv \sum_l \chi_{h-l} \vec{D}_l \tag{5.76}$$

With the term (5.75) replaced in Eq. (5.70), there results an expression, similarly to the previous steps, wherefrom, by comparing the corresponding coefficients the new set of equation is obtained:

$$-\vec{K}_h \times \left[\vec{K}_h \times \left(\vec{D}_h - \vec{C}_h \right) \right] = k_0^2 \, \vec{D}_h \qquad (5.77)$$

or, through its expansion and combination with Eq. (5.77) it releases the system:

$$\sum_l \left[\chi_{h-l} \left(\vec{K}_h \cdot \vec{D}_l \right) \vec{K}_h - \chi_{h-l} K_h^2 \, \vec{D}_l \right] = \left(k_0^2 - K_h^2 \right) \vec{D}_h \qquad (5.78)$$

The system (5.78) is fundamental for the dynamical theory of X-ray diffraction and it is at the foreground for studying the interferences which occur during the X-ray propagation in a monocrystal.

However, by this result one can draw the main features regarding the propagation.

Firstly, there are noted the equalities:

$$\vec{D}_h \cdot \vec{K}_h = 0, \; \vec{D}_0 \cdot \vec{K}_0 = 0 \qquad (5.79)$$

which prove that all the propagated waves which are in agreement with the self-consistent set (5.69) are of transversal type.

Secondly, for an incidence where the Bragg's equation is far from being satisfied by any set of crystallographic planes, the diffracted waves will have a negligible amplitude, and, therefore, the special relations can be considered

$$\vec{D}_0 \neq 0, \; \vec{D}_h \approx 0 \; \text{for} \; h \neq 0 \qquad (5.80)$$

into the Eqs. (5.76) and (5.77), wherefrom, the actual system reduces as:

$$K_0^2 = \frac{k_0^2}{1 - \chi_0} \qquad (5.81)$$

or under the approximation:

$$K_0 \approx k_0 \left(1 + \frac{1}{2} \chi_0 \right) \qquad (5.82)$$

showing that the refraction index for the internal (transmitted) incident wave should be equal to the average refraction index for the crystalline system.

In this case, no diffracted wave is generated, so the internal incident wave is completely determined taking into account also the condition of continuity (5.68) with the considered approximations.

Further on, the solution of the case when the incident internal wave generates a single diffracted wave will be exposed, being this case known as "two waves approximation".

5.3.2 DYNAMICAL DIFFRACTION IN TWO WAVES APPROXIMATION

In the case that the incident direction is chosen such as only a single eigen wave vector of the reciprocal lattice can be found on the Ewald's reflection sphere (Figure 5.1), so that the Bragg's equation (5.1) to be satisfied, the system (5.78) is reduced at two equations, specific for the approximations made; the present discussion follows (Azároff et al., 1974; Putz and Lacrămă, 2005).

Imposing the conditions:

$$\vec{D}_0 \neq 0, \vec{D}_h \neq 0, \vec{D}_l \approx 0 \, (\text{for } l \neq h, 0) \tag{5.83}$$

the system characterizing the dynamical diffraction for the two waves approximation (transmitted-diffracted) is obtained as follows

$$\begin{cases} \chi_{\bar{h}} \left(\vec{K}_0 \cdot \vec{D}_h \right) \vec{K}_0 - \chi_{\bar{h}} K_0^2 \vec{D}_h = \left[k_0^2 - K_0^2 (1 - \chi_0) \right] \vec{D}_0 \\ \chi_h \left(\vec{K}_h \cdot \vec{D}_0 \right) \vec{K}_h - \chi_h K_h^2 \vec{D}_0 = \left[k_0^2 - K_h^2 (1 - \chi_0) \right] \vec{D}_h \end{cases} \tag{5.84}$$

where the refraction index of the crystal for the transmitted respectively diffracted wave will be separately considered in agreement with the relation (5.67) to be:

$$K_0^2 = k_0^2 (1 + 2\delta_0), \ K_h^2 = k_0^2 (1 + 2\delta_h) \tag{5.85}$$

The system (5.84) can also be transformed by scalar multiplying its first equation with $\vec{D}_0$, and the second with $\vec{D}_h$, and, then, taking into account of the relations (5.79), and (5.85) there is obtained:

$$\begin{cases} (2\delta_0 - \chi_0)D_0 \; - \; \chi_{\bar{h}} \sin\left(\overset{\wedge}{\overrightarrow{D_0}, \overrightarrow{D_h}} \right) D_h = 0 \\[3mm] -\chi_h \sin\left(\overset{\wedge}{\overrightarrow{D_0}, \overrightarrow{D_h}} \right) D_0 \; + \quad (2\delta_h - \chi_0)D_h = 0 \end{cases} \qquad (5.86)$$

where, based on the function of normal or parallel polarization of the electric induction vector, the polarization factor specializes as:

$$\sin\left(\overset{\wedge}{\overrightarrow{D_0}, \overrightarrow{D_h}} \right) \equiv C = \begin{cases} 1, & \textit{for } \sigma \textit{ polarization} \\ \cos 2\theta_B, & \textit{for } \pi \textit{ polarization} \end{cases} \qquad (5.87)$$

with $2\theta_B$ representing the Bragg's scattering angle.

The linear and homogeneous system (5.86) will have non-trivial solutions if the associated determinant will be identical null, wherefrom results the equation:

$$(2\delta_0 - \chi_0)(2\delta_h - \chi_0) = \chi_h \chi_{\bar{h}} C^2 \qquad (5.88)$$

and it will display the solution:

$$x \equiv \frac{D_h}{D_0} = \frac{2\delta_0 - \chi_0}{C\chi_{\bar{h}}} \qquad (5.89)$$

The solution (5.89) is not yet completed until the possible values for the refraction indices (respectively for δ_0 and δ_h quantities) associated to the transmitted and diffracted waves are determined.

These quantities are connected through the Eq. (5.88) - also called as the *dispersion equation* just for the correlation that it includes, between the refraction indices associated to the transmitted and diffracted directions and along the structure factor of the crystal and the incident wave vector. The last correlation is even better emphasized if the relations (5.1), (5.68), and (5.85) are considered wherefrom another connection between the quantities δ_0 and δ_h can be expressed namely:

$$\delta_h = \frac{1}{b}\delta_0 + \frac{1}{2}\alpha \qquad (5.90)$$

where the notations have been used:

$$\frac{1}{b} = 1 + \frac{\vec{n} \cdot \vec{h}}{\vec{n} \cdot \vec{k}_o} \tag{5.91}$$

$$\alpha = \frac{h^2 + 2\vec{k}_0 \cdot \vec{h}}{k_0^2} \tag{5.92}$$

Remarkable, if the relation (5.91) is inversed, the approximation of the parameter b in relation with the driving cosines of the incident-transmitted and diffracted directions can be considered:

$$b = \frac{\vec{n} \cdot \vec{k}_o}{\vec{n} \cdot \left(\vec{k}_o - \vec{K}_o + \vec{K}_h \right)} \approx \frac{\gamma_o}{\gamma_h} \tag{5.93}$$

From the relation (5.90), with the respective notations, the dependence of the refraction indices associated to the transmitted and diffracted directions of the incident wave vector on the crystal is emphasized, a fact that justifies the given name as dispersion equation for the relation (5.88).

Furthermore, a quadratic equation aiming to determine one of the measures δ_o and δ_h by combining the obtained relations, (5.88) and (5.90), respectively, can be obtained. For instance, if the quantity δ_o associated to the transmitted direction and the normal polarization ($C=1$) is chosen we obtain:

$$(2\delta_o - \chi_0)\left(\frac{2}{b}\delta_0 - \chi_0 + \alpha \right) = \chi_h \chi_{\bar{h}} \tag{5.94}$$

A similar equation is obtained also for the solution (5.89) taking into account the connection between it and the δ_o parameter:

$$x^2 + x\left[(1-b)\frac{\chi_0}{\chi_{\bar{h}}} + \frac{b\alpha}{\chi_{\bar{h}}} \right] - b\frac{\chi_h}{\chi_{\bar{h}}} = 0 \tag{5.95}$$

The analytical solutions of these equations are direct and expressed such as:

$$x_{1,2} = \frac{-z \pm \sqrt{q + z^2}}{\chi_{\bar{h}}} \tag{5.96}$$

$$\left.\begin{array}{r}\delta_0{}' \\ \delta_0{}''\end{array}\right\} = \frac{1}{2}\left(\chi_0 - z \pm \sqrt{q + z^2}\right) \qquad (5.97)$$

where the following notations have been introduced:

$$z = \frac{1-b}{2}\chi_0 + \frac{b}{2}\alpha \;, \quad q = b\chi_h\chi_{\bar{h}} \qquad (5.98)$$

Due to the fact that two solutions have been obtained, the shape of the transmitted and respectively diffracted fields will be, in fact, considered as superposition of the two possible solutions, namely:

(i) for transmitted direction:

$$\exp\left(i\omega_0 t - i2\pi \vec{k_0^e}\cdot\vec{r}\right)\left[D_0{'}\exp\left(-i\varphi_1\tau\right) + D_0{''}\exp\left(-i\varphi_2\tau\right)\right] \qquad (5.99)$$

(ii) for diffracted direction the general solution will be written taking into account the relation $x = D_h/D_0$, such as:

$$\exp\left[i\omega_0 t - i2\pi\left(\vec{k_0^e} + \vec{h}\right)\cdot\vec{r}\right]\left[x_1 D_0{'}\exp\left(-i\varphi_1\tau\right) + x_2 D_0{''}\exp\left(-i\varphi_2\tau\right)\right]$$

$$(5.100)$$

where the second relation of the set (5.68) was considered and the specific notations were introduced:

$$\varphi_1 \equiv 2\pi\frac{k_0\delta_0{'}}{\gamma_0}, \; \varphi_2 \equiv 2\pi\frac{k_0\delta_0{''}}{\gamma_0}, \tau \equiv \vec{n}\cdot\vec{r} \qquad (5.101)$$

Of course, the complete determination of the fields requests also an analytical knowledge of the amplitudes $D_0{'}$, $D_0{''}$. For such determination, the condition of continuity at the crystal surface represents a first connection, where, the amplitude of the incident external wave should coincide with the sum of the amplitudes of the internal waves (solutions) propagation, according to the relation (5.99):

$$D_0{'} + D_0{''} = E_0^e \qquad (5.102)$$

However, the condition of continuity, itself, does not ensure uniqueness for the internal solutions of propagation in the crystal.

Therefore, for completing the analysis to specific cases the implementation of another wave-structure connection should be chosen.

For being able to make a specific evaluation, a plane-parallel geometry of the crystal is chosen, looking for the solutions of the propagation for the two consecrated cases, respectively the Laue and Bragg diffractions, schematically represented in Figure 5.25.

The distinction between the two (Laue vs. Bragg) situations can be essentially made by the sign of the b parameter introduced with the shape from (5.93), of whose analysis the Laue's case for the positive values of this parameter is distinguished, while the Bragg's case will correspond for the negative values. Both cases will be explicated in the sequel.

(i) In Laue's case there are no diffracted waves that go out by the crystal surface characterized by the condition $\vec{n} \cdot \vec{r} = 0$ and therefore, naturally, the condition according which the internal diffracted waves to be cancelled on this surface appears by taking into account the general shape of the diffracted fields of (5.39), and will be written such as:

$$x_1 D_0' + x_2 D_0'' = 0 \tag{5.103}$$

From now on the two constrains, (5.102) and (5.103), give complete solutions for the internal propagation upon the dynamical diffraction:

$$\begin{cases} D_0' = \dfrac{x_2}{x_2 - x_1} E_0^e \\[3mm] D_0'' = -\dfrac{x_1}{x_2 - x_1} E_0^e \end{cases} \tag{5.104}$$

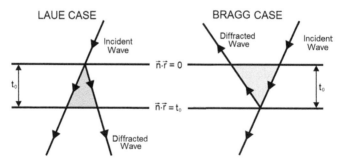

FIGURE 5.25 Representations of Laue and Bragg diffraction cases for a plane-parallel crystal.

Moreover, the ratios of the intensities of diffracted I_h and transmitted I_e^a waves in relation with the intensity of the incident wave I_0^e can be calculated, with the condition $\tau = t_0$ in Eqs. (5.99) and (5.100); by taking into account the Eq. (5.104) they give

$$\frac{I_h}{I_0^e} = \left[\frac{x_1 x_2 (c_1 - c_2)}{x_2 - x_1} \right]^2 \tag{5.105}$$

$$\frac{I_e^0}{I_0^e} = \left[\frac{x_2 c_1 - x_1 c_2}{x_2 - x_1} \right]^2 \tag{5.106}$$

with the notations:

$$c_1 = \exp(-i\varphi_1 t_0), \; c_2 = \exp(-i\varphi_2 t_0) \tag{5.107}$$

The ratio (5.105) is particularly important, so that after the simple algebraic operations it can be exposed also under explicit form:

$$\frac{I_h}{I_0^e} = b^2 |\chi_h|^2 \exp(-\mu_0 t) \frac{\sin^2(av) + \sinh^2(aw)}{|q + z^2|} \tag{5.108}$$

where the next parameters have been introduced:

$$v + iw \equiv \sqrt{q + z^2} \; , a \equiv \pi k_0 \frac{t_0}{\gamma_0} \; , t \equiv \frac{1}{2} \left(\frac{1}{\gamma_0} + \frac{1}{\gamma_h} \right) t_0 \tag{5.109}$$

The μ_0 in Eq. (5.108) is the linear absorption coefficient corresponding to the average path (t) traversed by the diffracted wave in the considered crystal.

(ii) For the Bragg's case, the diffracted wave emerges through the equation surface $\vec{n} \cdot \vec{r} = 0$ and therefore should cancel at the equation surface $\vec{n} \cdot \vec{r} = t_0$, so that, in this case, the second connection will be given by taking into account the Eqs. (5.100) and (5.107), such that:

$$c_1 x_1 D_0' + c_2 x_2 D_0'' = 0 \tag{5.110}$$

In this case, by cumulating the relationships (5.102) and (5.110) the propagation solutions for the Bragg's case will be univocally obtained as:

$$\begin{cases} D_0' = \dfrac{c_2 x_2}{c_2 x_2 - c_1 x_1} E_0^e \\[3mm] D_0'' = -\dfrac{c_1 x_1}{c_2 x_2 - c_1 x_1} E_0^e \end{cases} \qquad (5.111)$$

Using this result along the fields shape given as in Eqs. (5.99) and (5.100) one can evaluate, as in Laue case, the ratio of the *diffracted intensities* related to the incident intensity of an external wave calculated at the surface $\vec{n} \cdot \vec{r} = 0$:

$$\frac{I_h}{I_0^e} = \left[\frac{x_1 x_2 (c_1 - c_2)}{c_2 x_2 - c_1 x_1} \right]^2 \qquad (5.112)$$

and, respectively, the ratio of the *transmitted intensity* related to the incident intensity of the external wave calculated at the surface $\vec{n} \cdot \vec{r} = t_0$:

$$\frac{I_e^0}{I_0^e} = \left[\frac{(x_2 - x_1) c_1 c_2}{c_2 x_2 - c_1 x_1} \right]^2 \qquad (5.113)$$

Again, using the notations (5.109), the ratio (5.112) can be rewritten in an even more elaborate form, namely:

$$\frac{I_h}{I_0^e} = \frac{b^2 |\chi_h|^2 \left[\sin^2(av) + \sinh^2(aw) \right]}{\left\{ \begin{array}{l} |q + z^2| + \left(|q + z^2| + |z|^2 \right) \sinh^2(aw) - \left(|q + z^2| - |z|^2 \right) \sin^2(av) \\[2mm] + \dfrac{1}{2} \left[\left(|q + z^2| + |z|^2 \right)^2 - |q|^2 \right]^{1/2} \sinh|2aw| + \dfrac{1}{2} \left[\left(|q + z^2| - |z|^2 \right)^2 - |q|^2 \right]^{1/2} \sin|2av| \end{array} \right\}}$$

$$(5.114)$$

Obviously, in both cases have been worked for the normal polarization, while the case with parallel polarization may be further considered by replacing everywhere χ_h with $\chi_h \cos 2\theta_B$ and respectively $\chi_{\bar{h}}$ with $\chi_{\bar{h}} \cos 2\theta_B$.

5.3.3 THE CASE OF ZERO ABSORPTION

Furthermore, rather than the effective ratio of diffracted and transmitted intensities related to the incident wave, two measures derived from (5.112) will be used; the present discussion follows(Zachariasen, 1946; Putz and Lacrămă, 2005);

(a) the reflection power:

$$\frac{P_h}{P_0} = \frac{1}{|b|}\frac{I_h}{I_0^e} \tag{5.115}$$

(b) the integrated reflection power:

$$R_h^y = \int \frac{P_h}{P_0} dy \tag{5.116}$$

where, the y-parameter introduced by the expression:

$$y = \frac{z}{\sqrt{|b||C|\chi_h|}} \tag{5.117}$$

$$\overset{(5.98)}{=} \frac{1}{\sqrt{|b||C|\chi_h|}}\left(\frac{1-b}{2}\chi_0 + \frac{b}{2}\alpha\right) \tag{5.118}$$

generates the scale where the integrated reflection power can be represented. Another scale derived from this is that one in relation with the angular deviation from the value of the ideal Bragg angle θ_B; it is obtained by replacing the α-parameter from Eq. (5.118) with its value, approximately calculated from Eq. (5.92) for a rotating crystal, where the wavelength of the incident radiation is constantly kept and the incident direction varies around the Bragg's angle; there is obtained:

$$\alpha \approx 2(\theta_B - \theta)\sin 2\theta_B \tag{5.119}$$

Therefore, on the scale of the angular deviation, the power of integrated reflection will be written such as:

$$R_h^\theta = \int \frac{P_h}{P_0} d(\theta - \theta_B) \tag{5.120}$$

while the relationship between the two forms of the integrated reflection power will be:

$$R_h^\theta = \frac{|C\chi_h|}{\sqrt{b}\sin 2\theta_B} P_h^y \tag{5.121}$$

From now, the useful expressions (5.108) for Laue's case and respectively (5.114) for Bragg's case, for the non-absorbent dynamical propagation, will be next employed.

(i) For the Laue's case ($b > 0$), while considering the atomic susceptibility χ as being eminently real quantity (which corresponds to the fact that the polarizability on the unit volume is real for the zero absorption) has as a consequence the following specializations:

$$\mu_0 = 0 \,,\ w = 0 \,,\ v = \sqrt{q + z^2} = \sqrt{b|\chi_h|^2 + z^2} \qquad (5.122)$$

based on the relations (5.98), (5.108) and (5.109).

If a parameter correlated with the thickness of the crystal is also introduced with the shape:

$$A = a\sqrt{b}|C|\chi_h| \overset{\underset{(5.109),}{(5.93)}}{=} \pi k_0 C|\chi_h| \frac{t_0}{\sqrt{|\gamma_0\gamma_h|}} \qquad (5.123)$$

the reflection power can be calculated as based from the relation (5.108) on the introduced scale parameter y, with simplified shape:

$$\frac{P_h}{P_0} = \frac{\sin^2\left[A\sqrt{1 + y^2}\right]}{1 + y^2} \qquad (5.124)$$

(ii) For the Bragg's case ($b < 0$), and consequently from the analysis of the relations (5.98) and (5.109), another two cases result as based on the sigh of the measure $q + z^2$. If $q + z^2 > 0$ then $w=0$, while $q + z^2 < 0$ will be considered for $v=0$, respectively. For these two considered cases two reduced expressions of the relation (5.114) are obtained, again based on the scale parameter y, and through implementing the trigonometric relationships:

$$\sinh^2 \varphi = -\sin^2(i\varphi) \,,\ \coth^2 \varphi = -\coth^2(i\varphi) \qquad (5.125)$$

so allowing the immediate transformation of one in other. Therefore, there is enough writing just one of them, for example that for $w=0$, with the result:

$$\frac{P_h}{P_0} = \frac{\sin^2\left(A\sqrt{y^2 - 1}\right)}{y^2 - 1 + \sin^2\left(A\sqrt{y^2 - 1}\right)} = \frac{1}{1 + (y^2 - 1)\cot^2\left(A\sqrt{y^2 - 1}\right)} \qquad (5.126)$$

A common analysis of the relations (5.124) and (5.126) shows that they are symmetrical respecting the value reached for $y=0$. Using the relations (5.118) and (5.119) the symmetry condition of the reflection power at diffraction will be rewritten in angular scale, such as:

$$\theta = \theta_B + \frac{1-b}{2b\sin 2\theta_B}\chi_a \qquad (5.127)$$

It can have the significance of *dynamic Bragg angle*, respecting the genuine Bragg angle θ_B specific to the *kinematic theory of diffraction*. Consequently, this expression generates also a deviation from the famous (kinematic) Bragg's law:

$$2d_h \sin\theta_B = \lambda_a \qquad (5.128)$$

where, instead of the Bragg's angle the "dynamical Bragg's angle" is replaced as given by Eq. (5.127). There is also noted that the largest deviation from the kinematic Bragg's law (5.128) is obtained from Eq. (5.127) for small values of the $|b|$ parameter, which corresponds to a grazing incidence at the surface of the crystal.

There will be interesting to explore the specializations of the relations (5.124) and (5.126) for various thicknesses of a crystal with plane-parallel faces; by thickness one means the way in which the relation (5.123) has been introduced, i.e., in relation with the values of the parameter A respecting unity.

5.3.3.1 The Thick Crystal Case

The Laue and Bragg cases will be separately considered, for the approximation $A \gg 1$.

(i) In the Laue's case, from the relation (5.124), more exactly from the shape of the numerator one may deduced the form of interference carried by the reflection power: for a value given by the y-parameter, the reflection power will oscillate between 0 and $1/(1+y^2)$ when A tends to infinite.

Therefore, the term $\sin^2\left(A\sqrt{1+y^2}\right)$ can be replaced with its average, 1/2, and the average of the reflection power becomes:

$$\frac{\overline{P_h}}{P_0} = \frac{1}{2(1+y^2)} \qquad (5.129)$$

with the representation of Figure 5.26.

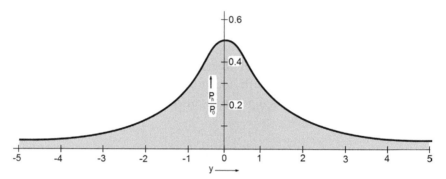

FIGURE 5.26 The power of reflection in the diffracted direction for the Laue's case for the thick crystal without absorption; after Zachariasen (1946).

From the Figure 5.26 there is noted how, in the center of diffraction on the scale of the y parameter, the power of the averaged reflection has a maximum value equal to 1/2, and half of the width of the curve (at half of its maximum) being $W_y^L = 1$. Beside it, also the integrated reflection power can be directly calculated:

$$R_h^y = \int_{-\infty}^{+\infty} \frac{dy}{2(1+y^2)} = \frac{\pi}{2} \tag{5.130}$$

(ii) For Bragg's case, the relation (5.126) can be evaluated in the case of the thick crystal for two domains of values where the scalar y parameter can be found, this way obtaining the so-called Ewald's solution of diffraction, such as:

$$\frac{P_h}{P_0} = \begin{cases} 1, & |y| < 1 \\ 1 - \sqrt{1 - y^{-2}}, & |y| > 1 \end{cases} \tag{5.131}$$

with the representation in the Figure 5.27.

Taking into account the definition of reflection power (5.115) combined with the relation (5.93), the first branch of the expression (5.131) can be described such as:

$$\gamma_0 I_0^e + \gamma_h I_h = 0 \tag{5.132}$$

which shows that for the domain where $|y| < 1$, the energy flow that get into the surface of equation $\vec{n} \cdot \vec{r} = 0$ is identical with the energy flow that

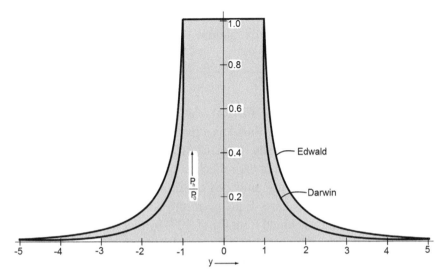

FIGURE 5.27 The reflection power in the diffracted direction for the Bragg's case for the thick crystal without absorption; after Zachariasen (1946).

get out by diffraction through this area. This situation defines the total reflection region, which marks a fundamental difference respecting the Laue case.

Moreover, unlike the Laue's case, also the half of the width of the curve at the half of its maximum is different in this case, i.e., having the actual value $W_y^B = 1.15$.

Regarding the integrated reflection power, on this case it will have the value:

$$R_h^y = 2\left[1 + \int_1^{+\infty}\left(1 - \sqrt{1 - y^{-2}}\right)dy\right] = 2\left[1 + \left(\frac{\pi}{2} - 1\right)\right] = \pi \tag{5.133}$$

i.e., being two folded respecting the recorded value in the Laue's case through the relation (5.130).

5.3.3.2 The Thin Crystal Case

This case actually corresponds to the conditions where the kinematic theory of diffraction is applied, and, in the terms of the parameters here introduced, to the condition $A \ll 1$. For this condition, both relations corresponding to the Laue (5.124) and respectively Bragg (5.126) cases can be rewritten with the approximation:

$$\frac{P_h}{P_0} \approx \frac{\sin^2 Ay}{y^2} \tag{5.134}$$

Worth noting that with the Eq. (5.134) relation an expression of kinematic theory of diffraction may be found as a particular case for the dynamical theory of diffraction. Actually, from Eq. (5.134) the very small values of the parameter A imply that the reflection power in diffraction has very small values respecting the unity, so confirming the neglecting of the extinction effect in this case. Equally, in this kinematical approach, the Bragg's "pure" angle is found from Eq. (5.127) by redoing the calculations considering the crystal refraction index equal with unity ($b \approx 1$).

This way, the integrated power of reflection for the thin crystal will have the form:

$$R_h^y = 2 \int_0^{+\infty} \left(\frac{\sin Ay}{y} \right)^2 dy = \frac{\pi A}{2} \tag{5.135}$$

being the same for the Laue and Bragg cases.

5.3.3.3   The Case of Intermediate Thickness Crystal

From the last two sections analysis there is noted how much different is the manifestation of the diffraction inside the crystal for the extreme cases considered toward the thick and respectively thin crystal. Consequently, also the consideration of the intermediate case appears as natural, i.e., characterized by the $A \approx 1$ condition, and separately imposed to the Laue and Bragg cases. However, reasonable will be that the values obtained in the relations (5.130) and (5.133) coupled with (5.135) to be found at the limit of the general expressions of the integrated reflection power of the cases Laue and Bragg for the crystal of intermediate thickness.

(i) The power of reflection (5.118) for Laue case, is exposed in Figure 5.28 for various values of the A-parameter, respectively for A=0.5, $\pi/2$, and π. From this representation analysis one observes how to the values of A>$\pi/2$ correspond a maximum for the diffractive center on the

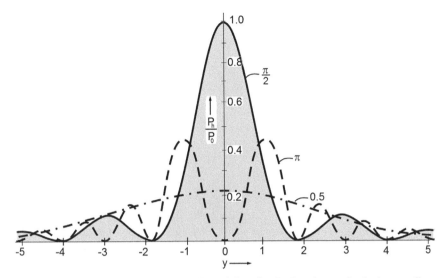

FIGURE 5.28 Reflection power in diffracted direction for Laue's case for the intermediate crystal thickness (A=0.5, π/2, π) without absorption; after Zachariasen (1946).

y-scale, while for the values A>π/2 a narrowing of the central fringe width as well as of the separation of the secondary fringes simultaneously appear, while the reflection power oscillates in the diffraction center between 0 and 1. The integrated reflection power calculation, with the representation in Figure 5.29, represents a particular interest not only for founding values from the extreme cases of the crystal's thickness, but also for the fact that it can be easier obtained from experimental perspective than the ratio of the diffracted intensities relative to that incident.

From Eq. (5.124) the integrated reflection power will be firstly rewritten such as:

$$R_h^y = \int_{-\infty}^{+\infty} \frac{\sin^2\left(A\sqrt{1+y^2}\right)}{1+y^2}dy = \int_0^{\pi/2} \frac{\sin(2A\sin\varphi)}{\sin\varphi}d\varphi \qquad (5.136)$$

Then, by taking into account the zero order Bessel function integral definition:

$$J_0(\rho) \equiv \frac{2}{\pi}\int_0^{\pi/2} \cos(\rho\sin\varphi)d\varphi \qquad (5.137)$$

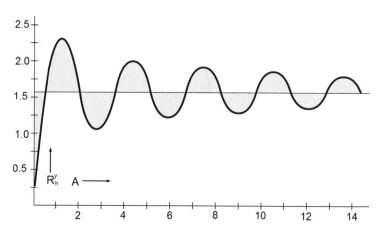

FIGURE 5.29   Integrated reflection power for Laue's case as a function of thickness parameter (A) for the crystal without absorption; after Zachariasen (1946).

while replacing $2A=\rho$ in the Eq. (5.136) and following by its differentiation, one is leaving with the relationship:

$$\frac{dR_h^y}{d\rho} = \frac{\pi}{2} J_0(\rho)$$
(5.138)

Therefore, the integrated reflection power will be successively re-obtained:

$$R_h^y = \frac{\pi}{2}\int_0^{2A} J_0(\rho)d\rho = \pi\sum_{n=0}^{n=\infty} J_{2n+1}(2A) = \pi\begin{cases} A, \text{ for } A <<1 \\ \frac{1}{2}, \text{ for } A >>1 \end{cases}$$
(5.139)

this way recovering the relations (5.130) and (5.135), respectively for the extreme cases of the thick and thin crystal. This fact can be observed also from the Figure 5.29 when the extremes of this representation are under concern.

(ii) The reflection power for Bragg case (5.126) is exposed in Figure 5.30 wherefrom a superior limitation to the value 1 of this measure is noted, in the region of total reflection $|y| < 1$, being this a limiting value as increasing the A-parameter.

The integrated reflection power calculated from the relation (5.126) will be:

$$R_h^y = \int_{-\infty}^{+\infty}\frac{dy}{1+\left(1-y^2\right)\coth^2\left(A\sqrt{1-y^2}\right)} = \pi\tanh A \cong \pi\begin{cases} A, A <<1 \\ 1, A >>1 \end{cases}$$
(5.140)

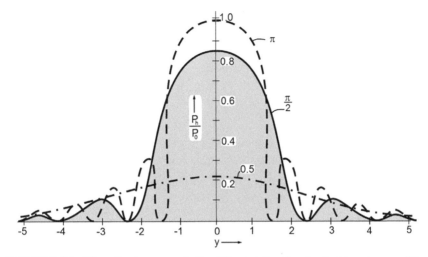

FIGURE 5.30 The reflection power in the diffracted direction for the Bragg's case of the crystal of intermediate thickness (A=0.5, $\pi/2$, π) without absorption; after Zachariasen (1946).

through which, the relations (5.133) and (5.135) are again found as corresponding to the extreme cases of the thick or thin crystal, respectively. Moreover, from the Figure 5.31 associated to the general expressions of the integrated reflection power (5.140), the extreme cases from (5.133) and (5.135) can be rediscovered for an error of 5%, for particular values of the parameter A:

$$R_h^y = \begin{cases} \pi A, \text{ for } A < 0.4 \\ \pi, \text{ for } A > 1.8 \end{cases} \qquad (5.141)$$

Unlike the Laue's case, the crystal's intermediate thickness domain can be restricted according to Eq. (5.141) at the interval (5.142) in the Bragg case.

$$0.4 < A < 1.8 \qquad (5.142)$$

Thus the importance of the thickness parameter A is noted, which, in a more explicit way can take be written by combining the relation (5.123) with the last relation of the set (5.69), regarding $\chi_0 (K_0 = 1/\lambda_0)$, namely as:

$$A = \frac{e^2 \lambda_0 F_h}{4\pi\varepsilon_0 m_0 c^2 V} \frac{t_0}{\sqrt{\gamma_0 \gamma_h}} \qquad (5.143)$$

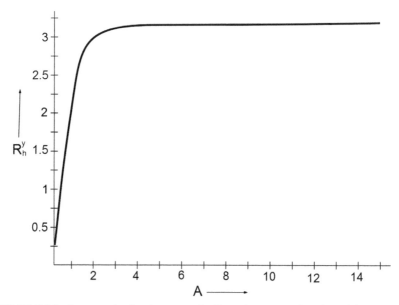

FIGURE 5.31   Integrated reflection power for Bragg's case as a function of the thickness A-parameter for the case of crystal without absorption; after Zachariasen (1946).

It depends on the structure factor and therefore it can be conventionally called the *effective linear dimension* of the crystalline block $t_0 / \sqrt{\gamma_0 \gamma_h}$.

The results obtained in this sub-Section regarding the propagation solutions in the crystal with zero absorption can be applied with good approximation for the crystals of the crystalline powder dimension, respectively for the values of the effective linear dimensions of the crystal of 10^{-4} (cm) order or smaller.

If the effective linear dimensions of the crystal exceed this value, for example are of order 10^{-2} (cm), the formulas for the thin crystal rest valid only for the weak reflections, yet not being anymore valid for the strong reflections, respectively not longer applies to the central fringes of the diffraction.

5.3.4   THE CASE OF THE CRYSTALS WITH ABSORPTION

When the absorption in the crystal is considered, the polarizability in volume unit becomes a complex measure and therefore the atomic susceptibility will be the starting point here; the present discussion follows (Laue, 1960; Putz and Lacrămă, 2005):

$$\chi = \chi' + i\chi'' \qquad (5.144)$$

where the real and imaginary components can be expanded in the Fourier's series:

$$\begin{cases} \chi' = \sum_h \chi_h' \exp\left(-i2\pi \vec{h} \cdot \vec{r}\right) \\ \chi'' = \sum_h \chi_h'' \exp\left(-i2\pi \vec{h} \cdot \vec{r}\right) \end{cases} \qquad (5.145)$$

where, the Fourier's coefficients satisfy the relations

$$\chi_{\bar{h}}' = (\chi_h')^*, \quad \chi_{\bar{h}}'' = (\chi_h'')^* \qquad (5.146)$$

If, moreover, the general case of Fourier's coefficients with complex shape is considered:

$$\begin{cases} \chi_h' = (\chi_h')_r + i(\chi_h')_i \\ \chi_h'' = (\chi_h'')_r + i(\chi_h'')_i \end{cases} \qquad (5.147)$$

then, unlike the case of the crystal without absorption, the next three measures are different in general:

$$\begin{cases} |\chi_h|^2 = |\chi_h'|^2 + |\chi_h''|^2 + 2[(\chi_h')_i (\chi_h'')_r - (\chi_h')_r (\chi_h'')_i] \\ |\chi_{\bar{h}}|^2 = |\chi_h'|^2 + |\chi_h''|^2 - 2[(\chi_h')_i (\chi_h'')_r - (\chi_h')_r (\chi_h'')_i] \qquad (5.148) \\ \chi_h \chi_{\bar{h}} = |\chi_h'|^2 - |\chi_h''|^2 + 2[(\chi_h')_r (\chi_h'')_r + (\chi_h')_i (\chi_h'')_i] \end{cases}$$

These relations represent the base of the theoretical explanation of the *Friedel's law*, that asserts that the diffraction phenomenon is invariant at the crystal inversion respecting the incident radiation. Actually, this invariance is also justified through the existence of a symmetry center, or through the equivalent equality:

$$\chi_{\bar{h}} = \chi_h \qquad (5.149)$$

This equality can be deduced from the relations (5.148) only if the next relation occurs

$$|\chi_h''| << |\chi_h'| \qquad (5.150)$$

unless for the very weak reflections and for the cases where the wavelength for the incident radiation is very close to the critical edge of absorption. For any other combination, the Friedel's law is no longer valid.

Furthermore a crystal with the symmetry center will be considered and for simplicity will be assumed that

$$(\chi_h')_i = (\chi_h'')_i = 0 \tag{5.151}$$

Therefore, the Fourier's coefficients of Eq. (5.147) are considered as real. Moreover, the next parameter is introduced:

$$k \equiv \frac{\chi_h''}{\chi_h'} \ , \ |k| << 1 \ , \ k^2 \approx 0 \tag{5.152}$$

Under these conditions, the relations (5.148) become:

$$\begin{cases} |\chi_h|^2 \cong |\chi_h'|^2 \\ \chi_h \chi_{\bar{h}} \cong |\chi_h'|^2 [1 + i2k] \end{cases} \tag{5.153}$$

If also the relations of the parameters defined in Eqs. (5.118) and (5.123) are adapted as:

$$y = \frac{z_r}{\sqrt{|b||C||\chi_h'|}} = \frac{1}{\sqrt{|b||C||\chi_h'|}} \left(\frac{1-b}{2} \chi_0' + \frac{b}{2} \alpha \right) \tag{5.154}$$

$$g = \frac{z_i}{\sqrt{|b||C||\chi_h'|}} = \frac{1}{\sqrt{|b||C||\chi_h'|}} \frac{1-b}{2} \chi_0'' \tag{5.155}$$

$$A = a\sqrt{|b||C||\chi_h'|} = \pi k_0 C |\chi_h'| \frac{t_0}{\sqrt{|\gamma_0 \gamma_h|}} \tag{5.156}$$

then, the relation (5.109) will be rewritten with the aid of the recalculated terms of Eq. (5.98) by considering the (5.153) expressions for the q-evaluation and respectively, (5.154) and (5.155) for the rewriting of z. Altogether there will be obtained that:

$$\sqrt{q+z^2} = v + iw = C|\chi_h'|\sqrt{|b|}\sqrt{\pm(1+i2k) + (y+ig)^2} \tag{5.157}$$

where, the positive sign under radical corresponds to the Laue's case, while the negative one to the Bragg's case.

Before distinctly treating the two cases, also the form of the absorption linear coefficient, μ_0, will be deduced using the complex form of polarization on the unit volume.

To this aim the classical shape of the linear absorption is considered as the ratio of the intensity of the external transmitted wave relative to the intensity of the external incident wave:

$$\frac{I_e^0}{I_0^e} = \exp\left(-\mu_0 \frac{t_0}{\gamma_0}\right) \qquad (5.158)$$

with t_0/γ_0 being the length of the path through the crystal.

Using the limiting conditions (5.68) for the transmitted wave (5.66), and by taking into account the shape of the refraction index in relation with the crystal dielectric constant for the transmitted wave:

$$n_0 = (1+\delta_0) = \kappa_0^{1/2} = (1+\chi_0)^{1/2} \approx 1 + \frac{1}{2}\chi_0 = 1 + \frac{1}{2}\chi_0' + i\frac{1}{2}\chi_0'' \quad (5.159)$$

the internal transmitted wave becomes:

$$\vec{D}_0 \exp\left\{i\omega_0 t - i2\pi\left[\vec{k}_0 + \frac{k_0}{\gamma_0}\frac{1}{2}(\chi_0' + i\chi_0'')\vec{n}\right]\cdot\vec{r}\right\} \qquad (5.160)$$

Next, one calculates the ratio of the intensities of the incident and transmitted waves through the crystal surfaces, $\vec{n}\cdot\vec{r} = t_0$ and $\vec{n}\cdot\vec{r} = 0$, respectively, with the obtained expression:

$$\frac{I_e^0}{I_0^e} = \exp\left(2\pi k_0 \chi_0'' \frac{t_0}{\gamma_0}\right) \qquad (5.161)$$

By comparing it with the equivalent one of Eq. (5.158) they will generate the formula of absorption linear coefficient as function of imaginary component of the Fourier's coefficient and atomic susceptibility for the transmitted wave, namely:

$$\mu_0 = -\frac{2\pi}{\lambda_0}\chi_0'' \qquad (5.162)$$

(i) For Laue's diffraction case, the symmetrical case will be considered, so that the formula (5.157) can be rewritten by implementing the $b = +1$ value and therefore having also $g = 0$, which will lead to the particular expression:

$$|q + z^2| \approx C^2 |\chi_h|^2 (1 + y^2) \tag{5.163}$$

Moreover, by taking into account the Eq. (5.152) approximation one can also calculate:

$$a(v + iw) \overset{\underset{(5.157)}{(5.163)}}{=} aC |\chi_h| \sqrt{1 + y^2}$$

$$\overset{(5.123)}{=} A \sqrt{1 + y^2}$$

$$\cong A \frac{1 + y^2 + ik}{\sqrt{1 + y^2}}$$

$$= A \sqrt{1 + y^2} + i \frac{Ak}{\sqrt{1 + y^2}} \tag{5.164}$$

$$\Rightarrow \begin{cases} av \cong A \sqrt{1 + y^2} \\ aw \cong \dfrac{Ak}{\sqrt{1 + y^2}} \end{cases} \tag{5.165}$$

With the expressions (5.163) and (5.165), the ration (5.108) becomes:

$$\frac{I_h}{I_0^e} \approx \frac{1}{C^2} \exp\left(-\frac{\mu_0 t}{\gamma_0}\right) \frac{1}{1 + y^2} \left[\sin^2\left(A \sqrt{1 + y^2}\right) + \sinh^2\left(\frac{kA}{\sqrt{1 + y^2}}\right) \right] \tag{5.166}$$

In the relation (2.99), by comparing the exponent $\mu_0 t / \gamma_0$ with the formula of the parameter A from (5.156), in the framework with χ_h'' and χ_0'' much smaller respecting χ_h', there results that unless for the cases for the parallel polarization, $C = |\cos 2\theta_B|$, and the approximation: $\mu_0 t / \gamma_0 \ll A$ occurs, meaning that A should be higher comparing to the unity. Then, one replaces the squared sine from Eq. (5.166) with its average value, $1/2$, so that the intensities ratio becomes:

$$\frac{I_h}{I_0^e} \approx \frac{1}{2C^2(1+y^2)} \exp\left(-\frac{\mu_0 t}{\gamma_0}\right)\left[1 + 2\sinh^2\left(\frac{kA}{\sqrt{1+y^2}}\right)\right] \qquad (5.167)$$

with the representation in the Figure 5.32, for the power of associated reflection – according to (5.115) relation.

There is remarked a steeper attenuation for various values of the parametric product $|k|A$ comparing with the representation of Figure 5.28 for the crystal without absorption.

(ii) For the evaluation of the reflection power (5.114) for the Bragg case, a simplification of this expression will be considered for the case in which the crystal considered as parallel plane is thick enough so that $\sinh^2 aw$ and respectively $\sinh^2 |aw|$ become very large. Under these conditions, the common factor $\sinh^2 aw$ is removed in Eq. (5.114) and the limit $aw \rightarrow \infty$ is considered, so the reflection power shape for Bragg's case in the approximation of the thick crystal will be obtained:

$$\frac{I_h}{I_0^e} = \frac{b^2 C^2 |\chi_h|^2}{\left|q + z^2\right| + |z|^2 + \sqrt{\left(\left|q + z^2\right| + |z|^2\right)^2 - |q|^2}} \qquad (5.168)$$

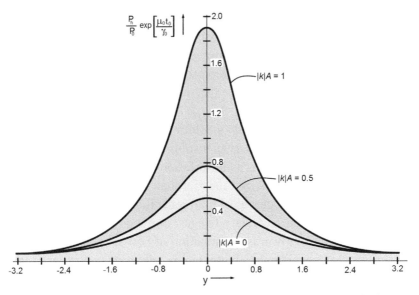

FIGURE 5.32 Reflection power of the Laue's case for the absorbent crystal; after Laue (1960); James (1965).

Using now the Eqs. (5.152)–(5.157), the parameters adapted to the crystal with absorption, when replaced in the last expression, will allow the possibility to write the reflection power such as:

$$\frac{P_h}{P_0} = L - \sqrt{L^2 - (1 + 4k^2)}$$

$$L \equiv \left| \sqrt{(-1 + y^2 - g^2)^2 + 4(gy - k)^2} \right| + y^2 + g^2 \qquad (5.169)$$

with the representation of Figure 5.33.

From the analysis of this representation there is noted that the allure of the curve is unsymmetrical towards the scale center, $y=0$, unless for the $k = 0$ case where the curve is symmetrical. Equally, the maximum of reflection power is reached for $y = k/g$.

If, moreover, the relations (5.169) are considered for the case $k = g = 0$ under the form

$$\frac{P_h}{P_0} = M - \sqrt{M^2 - 1} = \begin{cases} 1, & for \ |y| \le 1 \\ [|y| - \sqrt{y^2 - 1}]^2, & for \ |y| \ge 1 \end{cases}$$

$$L \equiv |y^2 - 1| + y^2 \qquad (5.170)$$

the Darwin's solution associated to the Bragg's absorption case is obtained as represented in Figure 5.27. However, there is noted how, by comparing with the Ewald's solution, both representations give the total reflection in the domain $|y| \le 1$, while, outside this range, the Darwin's solution is more stepper descending.

Moreover, while the Ewald's solution is for the crystals considered with zero absorption, the Darwin's solution corresponds to the crystals with negligible absorption, as an inferior limit for a finite absorption. There is therefore emphasized the fact that, for considering the absorption in the crystals, the curves of total reflection flatten their allure much more "faster" for the secondary fringes, or even losing their scale symmetry (Bragg case) towards the situation of non-absorbent crystals. The first aspect of considerable flattening is tributary to the fact that the crossing path of the waves propagated in the crystal for the secondary (weak) reflections is longer and therefore the absorption increases in its

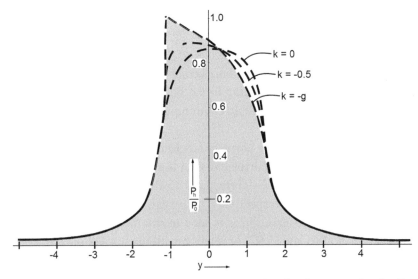

FIGURE 5.33 Reflection power for Bragg's case for absorbent crystal; after Laue (1960); James (1965).

contribution. The phenomenon of asymmetry of the reflection power for the Bragg case with absorption is due to a more complex phenomenon, i.e., the formation of the stationary waves in the crystal during the propagation (Birău & Putz, 2000).

5.4   PENDELLUSUNG PHENOMENA: THE PLANE WAVES APPROXIMATIONS

5.4.1   THE PHENOMENOLOGY OF EWALD'S PENDELLUSUNG

In the previous Section the fundamental equations of the X-ray propagation due to the diffraction on a crystal, considered as perfect, have been formulated and solved, in the two waves approximation: one transmitted and a single one diffracted, and with the unique customizations of Laue's and Bragg's cases for a plane-parallel crystal; the present discussion follows Putz and Lacrămă (2005).

This X-ray diffraction geometrical phenomenology can be nevertheless described by the custom properties of the *dispersion surface* (DS) as a physical dynamic concept of the diffraction as an optical phenomenon inside the crystal.

The dispersion equation is described through a geometrical locus in reciprocal space that, being intersected by the incident $\left(\overrightarrow{K}_0, \overrightarrow{K}_h\right)$ plane, gives the hyperbola with two branches α, β of Figure 5.34. The intersection of the asymptotes is the Lorentz center L_0, which is the equivalent of the Laue center L_a, corresponding to the center of the Ewald's sphere for the case of X-ray diffraction in an environment with the vacuum refraction index.

Thus, the Lorentz center corresponds to the Ewald's sphere when the incident wave vector was correlated with the average refraction index of the crystal, while L_a and L_o correspond to the exact satisfaction of the Bragg's law in vacuum, respectively, in crystal.

This important observation is related to the fact that the solutions of propagations (5.96) and (5.97) in relation with the dispersion surface correspond respectively to the two sheets associated to the hyperbolea, and identified and nominated from now on as solutions corresponding to the (alpha) branch of (1) the dispersion surface, and respectively the solutions corresponding to the (beta) branch (2) of the dispersion surface. Therefore, the equivalent notations will be further considered:

$$\begin{cases} x_{1,2} \leftrightarrow x_{\alpha,\beta} \\ \delta_{0/h}{}' \leftrightarrow \delta_{0/h}^{\alpha} , \quad \delta_{0/h}{}'' \leftrightarrow \delta_{0/h}^{\beta} \end{cases} \tag{5.171}$$

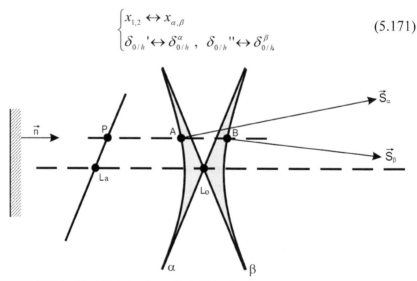

FIGURE 5.34 Modeling the dynamical diffraction by means of the Lorentz and Laue centers along the Poynting vectors $(\vec{S}_{\alpha/\beta})$ associated to the branches of dispersion surface; after Birău and Putz (2000).

The Bragg's coupling relation, Eq. (5.1), will be further fulfilled by each coupled wave vectors corresponding to the transmitted-diffracted waves associated to the branches of dispersion surface for any coupling point of it:

$$\begin{cases} \vec{K}_h^{\alpha} = \vec{K}_0^{\alpha} + \vec{h} \\ \vec{K}_h^{\beta} = \vec{K}_0^{\beta} + \vec{h} \end{cases} \qquad (5.172)$$

In these conditions, for a general writing of the solutions of propagations in the crystal for the transmitted and respectively diffracted waves in the crystal, Eqs. (5.99) and (5.100), there is noted how, on each direction, the contributions brought by the components of the fields associated to the both branches of DS are contained.

Therefore, for a set of coupling (A, B) points associated to the incident P-point (Figure 5.34), the transmitted wave cumulates the contributions in cause on the transmitted direction associated both to the alpha and beta branches of DS in A and B, respectively, and similarly for the diffracted direction.

This observation is essential because it gives the premises of oscillation phenomenon explanation – the so called Pendellösung/the oscillation solution by Ewald. What is in fact this phenomenon? It is just the cumulating of the both energetically contributions associated to the DS branches on the transmitted and diffracted directions and corresponds to the intensities or the powers of effective or integrated reflection as are experimentally calculated and recorded: see Eqs. (5.108), (5.114), (5.124) and (5.126). But not only that! If for various incident beam rays the crossed path through the crystal is different, which is possible by choosing the crystal section of edge shape, for various thickness of the crystal there is noted that the energy is exclusively propagated on the transmitted direction, while for other thickness it propagates also on the diffracted direction.

This phenomenon, apparently weird, leads to the idea that in the intimate dynamical propagation of the waves in the crystal, a coupling between the propagated energies, associated to the DS branches, should occur so that, for certain conditions this coupling interplays the energies from a diffraction direction to another. In this way, the dynamical propagation has a plus of complexity by the assimilation of the specific oscillation of the propagated energies associated to the DS branches, wherefrom the final recorded intensities result; hence, the name of oscillation solution for the dynamical propagation of the energies associated to the conjugate directions. The analytical elucidation of this coupling and the way in which there is made the energetic oscillation of the solutions of dynamical propagation in the crystal for the conjugate (transmitted and diffracted) directions in the approximation of the plane waves, will be in next unfolded.

5.4.2   ANALYTICAL EWALD'S PENDELLÜSUNG

Because there is about the energetic propagation, the form and the behavior of the total Poynting vector associated to the DS branches of the diffracted directions is natural to be analyzed. To this aim, the general form of the solutions of the waves propagated on the diffracted directions associated to the DS branches will be firstly expressed in a form derived from the Laue's (5.104) and Bragg's (5.111) solution in the plane wave approximation; the present discussion follows Birău and Putz (2000) and (Putz & Lacrămă, 2005):

$$\begin{cases} D_{0\alpha} = f(x_\alpha, x_\beta)E_0^e \\ D_{h\alpha} = x_\alpha D_{0\alpha} \\ D_{0\beta} = g(x_\alpha, x_\beta)E_0^e \\ D_{h\beta} = x_\beta D_{0\beta} \end{cases} \tag{5.173}$$

With these, the associated fields can be written, assuming for the wave vectors the generically complex shape:

$$\vec{K}_{0/h(\alpha/\beta)} = \vec{K}'_{0/h(\alpha/\beta)} - i\vec{K}''_{0/h(\alpha/\beta)} \tag{5.174}$$

where, the real component corresponds to the effective propagation in relation to the DS, while the imaginary component corresponds to the absorption in the crystal.

From simple considerations related to the conditions of tangential continuity and Bragg coupling, the complex form (5.174) associated to the wave vectors can be proved that are equal, i.e., the imaginary components corresponding to the different diffracted directions for the same branch of DS, modeling an equal absorption on the diffracted and transmitted directions:

$$\vec{K}''_{0(\alpha/\beta)} = \vec{K}''_{h(\alpha/\beta)}$$

(5.175)

With these considerations, the general solutions take the form:

$$
\begin{cases}
\vec{d}_{0\alpha} = \vec{E}_0^e \, f(x_\alpha, x_\beta) \exp[2\pi i v_o t] \exp\left(-2\pi i \vec{K}'_{0\alpha} \cdot \vec{r} - 2\pi \vec{K}''_{0\alpha} \cdot \vec{r} \right) \\[2mm]
\vec{d}_{h\alpha} = \vec{E}_0^e \, x_\alpha f(x_\alpha, x_\beta) \exp[2\pi i v_o t] \exp\left(-2\pi i \vec{K}'_{h\alpha} \cdot \vec{r} - 2\pi \vec{K}''_{0\alpha} \cdot \vec{r} \right) \\[2mm]
\vec{d}_{0\beta} = \vec{E}_0^e \, g(x_\alpha, x_\beta) \exp[2\pi i v_o t] \exp\left(-2\pi i \vec{K}'_{0\beta} \cdot \vec{r} - 2\pi \vec{K}''_{0\beta} \cdot \vec{r} \right) \\[2mm]
\vec{d}_{h\beta} = \vec{E}_0^e \, x_\beta g(x_\alpha, x_\beta) \exp[2\pi i v_o t] \exp\left(-2\pi i \vec{K}'_{h\beta} \cdot \vec{r} - 2\pi \vec{K}''_{0\beta} \cdot \vec{r} \right)
\end{cases}
$$

(5.176)

where, by recalling the Eqs. (5.104) and (5.111) solutions the general connection can be established (Kato, 1955, 1969)

$$g(x_\alpha, x_\beta) \propto -\frac{x_\alpha}{x_\beta} f(x_\alpha, x_\beta)$$

(5.177)

This way, all the "ingredients" are presented in order to characterize the propagated energy in the crystal under the form of the total Poynting vector,

$$\vec{S} = \vec{d} \times \vec{H}^*$$

(5.178)

However, the punctual expression of this vector is of less use of interest since the temporal and volume averages are under focus to be correlated with the energetic interchange between the two diffracted directions, as a function of penetration depth in the crystal.

5.4.2.1 The Coupling of Poynting Vectors

At the first stage of analysis, the successive expressions of the Poynting vector can be written cumulating the relations (5.176) in a global summation:

$$\vec{S} = \vec{d} \times \vec{H}^*$$

$$= \sum_{u,u',h,h'} \exp\left\{ \begin{array}{c} -2\pi i \left[\left(\vec{K}'_{0u} + \vec{h} \right) - \left(\vec{K}'_{0u'} + \vec{h}' \right) \right] \cdot \vec{r} \\ -2\pi \left(\vec{K}''_{0u} + \vec{K}''_{0u'} \right) \cdot \vec{r} \end{array} \right\} \left(\vec{D}_{hu} \times \vec{H}^*_{h'u'} \right)$$

$$= \sum_{u,u'} \exp\left[-2\pi i \left(\vec{K}'_{0u} - \vec{K}'_{0u'} \right) \cdot \vec{r} \right] \exp\left[-2\pi \left(\vec{K}''_{0u} + \vec{K}''_{0u'} \right) \cdot \vec{r} \right]$$

$$\times \sum_{h,h'} \left(\vec{D}_{hu} \times \vec{H}^*_{h'u'} \right) \exp\left[-2\pi i \left(\vec{h} - \vec{h}' \right) \cdot \vec{r} \right] \qquad (5.179)$$

where the Bragg's relation, $\vec{K}_{hu} = \vec{K}_{0u} + \vec{h}$, has been used and where the indices u can be associated to any (α, β) branch of DS with any polarization.

The expression above can be simplified if its average over the unit cell is considered, a case in which $\vec{r} \rightarrow \vec{R}$, i.e., meaning the neglecting of the variation of the first exponentials inside the cell and their replacement with a constant when $\vec{h} = \vec{h}'$, so that the averaged contribution of the last exponential to be canceled:

$$\left\langle \vec{S} \right\rangle = \frac{1}{2} \mathrm{Re}\left(\vec{d} \times \vec{H}^* \right)$$

$$= \frac{1}{2} \mathrm{Re}\left\{ \sum_{u,u'} \exp\left[\begin{array}{c} -2\pi i \left(\vec{K}'_{0u} - \vec{K}'_{0u'} \right) \cdot \vec{R} \\ -2\pi \left(\vec{K}''_{0u} + \vec{K}''_{0u'} \right) \cdot \vec{R} \end{array} \right] \sum_h \left(\vec{D}_{hu} \times \vec{H}^*_{hu'} \right) \right\} \qquad (5.180)$$

In the last expression, the vectorial product can be rewritten such as:

$$\vec{D}_{hu} \times \vec{H}^*_{hu'} = c \left| \vec{D}_{hu} \right| \left| \vec{D}^*_{hu'} \right| \vec{s}_h \qquad (5.181)$$

with $\vec{s}_h$ the versor of the wave vector K_h associated to the diffracted direction.

Next, by assuming the fact that the differences in the modules for the eigen-wave vectors of the reciprocal lattice are negligible, the expression (5.180) will be rewritten such as:

$$\left\langle \vec{S} \right\rangle = \frac{c}{2} \mathrm{Re}\left(\vec{d} \times \vec{H^*} \right)$$

$$= \frac{c}{2} \mathrm{Re}\left\{ \sum_{u,u'} \exp\left[\begin{array}{c} -2\pi i\left(\vec{K'}_{0u} - \vec{K'}_{0u'} \right) \cdot \vec{R} \\ -2\pi\left(\vec{K''}_{0u} + \vec{K''}_{0u'} \right) \cdot \vec{R} \end{array} \right] \sum_h \left| \vec{D}_{hu} \right| \left| \vec{D}^*_{hu'} \right| \vec{s}_h \right\} \qquad (5.182)$$

If a single active node of the reciprocal lattice, respectively the diffraction on a single set of crystallographic planes (a single couple of diffracted-transmitted H-O waves) is considered, while a pair of coupling points corresponding to the same incidence is placed on the alpha and beta branches of DS, we will further obtain from (5.182):

$$\frac{1}{c}\left\langle \vec{S} \right\rangle = \frac{1}{c}\left(\left\langle \vec{S} \right\rangle_{\alpha\alpha} + \left\langle \vec{S} \right\rangle_{\beta\beta} + 2\left\langle \vec{S} \right\rangle_{\alpha\beta} \right)$$

$$= \frac{1}{2}\exp\left(-4\pi \vec{K''}_{0\alpha} \cdot \vec{R} \right)\left[D^2_{0\alpha}\,\vec{s}_0 + D^2_{h\alpha}\,\vec{s}_h \right] + \frac{1}{2}\exp\left(-4\pi \vec{K''}_{0\beta} \cdot \vec{R} \right)\left[D^2_{0\beta}\,\vec{s}_0 + D^2_{h\beta}\,\vec{s}_h \right]$$

$$+ \exp\left[-2\pi\left(\vec{K''}_{0\alpha} + \vec{K''}_{0\beta} \right) \cdot \vec{R} \right]\left[\begin{array}{c} D_{0\alpha}D_{0\beta}\,\vec{s}_0 \\ + D_{h\alpha}D_{h\beta}\,\vec{s}_h \end{array} \right]\cos\left[2\pi\left(\vec{K'}_{0\alpha} - \vec{K'}_{0\beta} \right) \cdot \vec{R} \right]$$

$$(5.183)$$

Besides the components of total Poynting vector associated to the DS branches the apparition of a coupling component that has an harmonic dependence is noticed. Therefore, the "reduced-R" Poynting vector can be generally written such as:

$$\vec{S}^R = \vec{S}^R_\alpha + \vec{S}^R_\beta + \vec{S}^R_{\alpha\beta} \qquad (5.184)$$

The coupling component divides the propagated energy around the average direction of propagation along the depth of penetration in the crystal.

The general form (5.184) represents the general solution of propagation of the oscillation energy, i.e., the coupling between the distribution of the energy between the DS associated branches, so describing the oscillation propagation solution, or the Ewald's oscillation solution, or, even simpler the *Pendellösung*.

For the coupling term analysis, one notes that if the conditions of continuity (5.68) are written for the real components of the transmitted wave vectors associated to the DS branches, the following relation yields:

$$\vec{K'}_{0\alpha} - \vec{K'}_{0\beta} = \frac{k_0}{\gamma_0}(\delta_0^\alpha - \delta_0^\beta)\vec{n} \qquad (5.185)$$

wherefrom, there is concluded that, for the propagation along the parallel planes with the crystal surface, the coupling term is a measure that harmonically varies with the depth, with the period:

$$P = \frac{1}{\left|\vec{K'}_{0\alpha} - \vec{K'}_{0\beta}\right|} \qquad (5.186)$$

In order to simplify the situation in which the energy propagated on the diffracted direction is "pendulant" and transferred to the transmitted direction through the presence of the coupling term from Eqs. (5.183) and (5.184), a non-absorbent crystal is considered (for simplification), so that, in Eq. (5.183), the condition

$$\vec{K''}_{0\alpha} = \vec{K''}_{0\beta} = 0 \qquad (5.187)$$

will be implemented so resulting the expression:

$$\vec{S}^R = \frac{1}{2}\left(D_{0\alpha}^2\,\vec{s}_0 + D_{h\alpha}^2\,\vec{s}_h\right) + \frac{1}{2}\left(D_{0\beta}^2\,\vec{s}_0 + D_{h\beta}^2\,\vec{s}_h\right)$$

$$+ \left(D_{0\alpha}D_{0\beta}\,\vec{s}_0 + D_{h\alpha}D_{h\beta}\,\vec{s}_h\right)\cos\left(2\pi\frac{Z}{P}\right) \qquad (5.188)$$

with Z the depth of penetration.

Moreover, by taking into account the relations (5.173) and (5.177) the expression (5.188) becomes:

$$\vec{S}^R \cong f(x_\alpha, x_\beta)^2 E_0^{e2} \left[\begin{array}{c} \frac{1}{2}\left(\vec{s_0} + x_\alpha^2 \vec{s_h} \right) + \frac{1}{2}\left(\frac{x_\alpha^2}{x_\beta^2} \vec{s_0} + x_\alpha^2 \vec{s_h} \right) \\ -\left(\frac{x_\alpha}{x_\beta} \vec{s_0} + x_\alpha^2 \vec{s_h} \right) \cos\left(2\pi \frac{Z}{P} \right) \end{array} \right]$$

$$= \begin{cases} f(x_\alpha, x_\beta)^2 E_0^{e2} \frac{1}{2}\left(1 - \frac{x_\alpha}{x_\beta} \right)^2 \vec{s_0} & , for \ \cos\left(2\pi \frac{Z}{P} \right) = 1 \\ f(x_\alpha, x_\beta)^2 E_0^{e2} \left[\frac{1}{2}\left(1 + \frac{x_\alpha}{x_\beta} \right)^2 \vec{s_0} + 2x_\alpha^2 \vec{s_h} \right] & , for \ \cos\left(2\pi \frac{Z}{P} \right) = -1 \end{cases}$$

$$(5.189)$$

Worth noting that for a certain value, say +1, for the trigonometric factor, which corresponds to a certain depth of penetration, the total reduced Poynting vectors carry the entire energy only on the transmitted direction, $\vec{s_0}$, while, at a modified depth of the crystal corresponding to a plus/minus half in the oscillation period (5.186), the total reduced Poynting vectors carry its energy on both the directions of diffraction.

This observation corresponds to the theoretical explanation for the experimental observations for which the thickness of the crystal can influence the diffraction direction where the energy corresponding to the diffraction solution is carried. Moreover, even inside the crystal, based on its penetration, the total energy is pendulant between the directions of diffraction as based on the value for the trigonometric factor of coupling.

This fact reinforces the denomination of Pendellösung as the oscillation solution for the X-ray dynamical propagation in crystals and analytically explains it. Moreover, if in the expressions above the relations (5.96) are replaced with the correspondence (5.171), through taking into account also of the parametrical expressions (5.98) one can evaluate the ratio:

$$\frac{\left(1 - \frac{x_\alpha}{x_\beta} \right)^2}{\left(1 + \frac{x_\alpha}{x_\beta} \right)^2} = \begin{cases} 1 + \frac{4\chi_h \chi_{\bar{h}}}{\alpha^2} & , b = +1 \ for \ Laue \ case \\ 1 - \frac{4\chi_h \chi_{\bar{h}}}{(2\chi_0 - \alpha)^2} & , b = -1 \ for \ Bragg \ case \end{cases}$$

$$(5.190)$$

From this evaluation there appears that, for instance, the energy transported on the transmitted direction is smaller that the Laue's case and respectively bigger than the Bragg's case, for when the total reduced Poynting vectors transport the energy on both diffraction directions – comparing to the case corresponding to the energetic transport exclusively on the transmitted direction.

These theoretical results represent the base and the criteria for the observation and explanation of experimental results.

5.4.2.2   Direction of Propagation of Poynting Vectors

As shown in relation (5.189), although for specific conditions the total reduced Poynting vector carries the energy resulted due to the diffraction distributed between its diffracted and transmitted directions, the contributions coming from the propagations associated to the DS branches are present in both cases. Therefore, the preoccupation for the study of the individual directions with the Poynting vectors associated to the DS branches appears as natural (Birău & Putz, 2000).

In order to keep a general discussion a treatment as free as possible of restrictions and specificities will be approached.

For instance, one will start form reconsidering of the equation (5.70) under the form:

$$\sum_l \chi_{m-l} \left[\vec{K}_m \times \left(\vec{K}_m \times \vec{D}_l \right) \right] = (k_0^{e2} - K_m^2) \vec{D}_m \quad , \quad m = -\infty ... + \infty \quad (5.191)$$

where the wave vectors involved follow the given significances, however with the diffracted wave vector satisfying the Bragg's relation $\left(\vec{K}_m = \vec{K}_0 + \vec{h}_m \right)$.

From the analysis of Eq. (5.191) relation there noted that all the components of the electric induction are transversal respecting the wave vectors corresponding to the diffracted directions of propagation.

This property leads to the simplification of the expression (5.191) if it is developed by the double vectorial product. Moreover, if the decomposition of the field components in the directions corresponding to the σ-normal and π-parallel directions of polarization is considered, we obtain:

$$\frac{k_0^2 - K_m^2}{k_0^2}\left(\vec{D}_{m\sigma} + \vec{D}_{m\pi}\right) = -\sum_l \chi_{m-l}\kappa_m^2\left(\vec{D}_{l\sigma'} + \vec{D}_{l\pi'}\right) \tag{5.192}$$

where the relation

$$K_m^2 = k_0^2\kappa_m^2 \tag{5.193}$$

was used by which the wave vector of diffracted direction is connected to the incident external ray on the crystal through the specific polarization (κ), and implicitly through the refraction index associated with this direction.

If the Eq. (5.192) is respectively multiplied with the unit vectors corresponding to the normal and parallel directions of polarization for the field components of the left side of the relation (5.192), the system of equation formed on the components of polarization associated to the electrical induction results as:

$$\begin{cases} \dfrac{k_0^2 - K_m^2}{k_0^2}D_{m\sigma} = -\sum_l \chi_{m-l}\kappa_m^2 \dfrac{1}{D_{m\sigma}}\left(\vec{D}_{l\sigma'}\cdot\vec{D}_{m\sigma} + \vec{D}_{l\pi'}\cdot\vec{D}_{m\sigma}\right) \\[3mm] \dfrac{k_0^2 - K_m^2}{k_0^2}D_{m\pi} = -\sum_l \chi_{m-l}\kappa_m^2 \dfrac{1}{D_{m\pi}}\left(\vec{D}_{l\sigma'}\cdot\vec{D}_{m\pi} + \vec{D}_{l\pi'}\cdot\vec{D}_{m\pi}\right) \end{cases} \tag{5.194}$$

This system can be equivalently rewritten, through the corresponding relations:

$$\begin{cases} \dfrac{k_0^2 - K_m^2}{k_0^2}D_{m\sigma} = -\sum_l \left[\chi\!\begin{pmatrix} l & m \\ \sigma' & \sigma \end{pmatrix}D_{l\sigma'} + \chi\!\begin{pmatrix} l & m \\ \pi' & \sigma \end{pmatrix}D_{l\pi'}\right] \\[3mm] \dfrac{k_0^2 - K_m^2}{k_0^2}D_{m\pi} = -\sum_l \left[\chi\!\begin{pmatrix} l & m \\ \sigma' & \pi \end{pmatrix}D_{l\sigma'} + \chi\!\begin{pmatrix} l & m \\ \pi' & \pi \end{pmatrix}D_{l\pi'}\right] \end{cases} \tag{5.195}$$

and allows the non-trivial solutions if and only if the associated determinant is vanishing.

For the writing of the determinant associated to the system (5.195) the indices m and l will be considered as taking the values from minus to plus infinite, and, therefore, towards the complete determinant expansion also the inverse values will be reached, so that, highlighting the two groups of coefficients associated to the concerned couple, i.e., (m,l) and (l,m), the determinant will take the form:

$$
\Delta \equiv
\begin{vmatrix}
\dfrac{k_0^2 - K_m^2}{k_0^2} & 0 & \cdots & \chi\!\begin{pmatrix} l & m \\ \sigma' & \sigma \end{pmatrix} & \chi\!\begin{pmatrix} l & m \\ \pi' & \sigma \end{pmatrix} \\[2ex]
0 & \dfrac{k_0^2 - K_m^2}{k_0^2} & \cdots & \chi\!\begin{pmatrix} l & m \\ \sigma' & \pi \end{pmatrix} & \chi\!\begin{pmatrix} l & m \\ \pi' & \pi \end{pmatrix} \\[2ex]
\vdots & \vdots & \ddots & \vdots & \vdots \\[2ex]
\chi\!\begin{pmatrix} m & l \\ \sigma & \sigma' \end{pmatrix} & \chi\!\begin{pmatrix} m & l \\ \pi & \sigma' \end{pmatrix} & \cdots & \dfrac{k_0^2 - K_l^2}{k_0^2} & 0 \\[2ex]
\chi\!\begin{pmatrix} m & l \\ \sigma & \pi' \end{pmatrix} & \chi\!\begin{pmatrix} m & l \\ \pi & \pi' \end{pmatrix} & \cdots & 0 & \dfrac{k_0^2 - K_l^2}{k_0^2}
\end{vmatrix}
= 0 \quad (5.196)
$$

However, in order to focus the attention to a DS associated branch, the alpha branch with the not-reduced Poynting vector corresponding for the Pendellösung period is considered, for example, thus generally rewritten from Eq. (5.183) as:

$$
\vec{S}_\alpha \cong \sum_{m\sigma} \vec{K}_{m\alpha} (D_{m\sigma}^\alpha)^2
\tag{5.197}
$$

If the case of the non-absorbent crystal is considered one can consecutively write that:

$$
\chi_m = \chi_{-m}^* \Rightarrow \chi\!\begin{pmatrix} m & l \\ \sigma & \sigma' \end{pmatrix} = \chi^*\!\begin{pmatrix} l & m \\ \sigma' & \sigma \end{pmatrix} \Rightarrow |l\sigma'; m\sigma| = |m\sigma; l\sigma'|^* \tag{5.198}
$$

where through $|l\sigma'; m\sigma|$ the lower determinant is obtained from the higher (5.196) by removing the column $l\sigma'$ and the row $m\sigma$, accordingly symbolized. Moreover, for a given incident vector satisfying the general dispersion equation (5.196), the amplitudes ratio for an arbitrary set of a two field components propagated in the crystal due to the diffraction will be generally written with the equivalent forms:

$$
\frac{D_{m\sigma}^\alpha}{D_{l\sigma'}^\alpha} = \pm \frac{|m\sigma; m\sigma|^\alpha}{|l\sigma'; m\sigma|^\alpha}, \quad \frac{D_{l\sigma'}^\alpha}{D_{m\sigma}^\alpha} = \pm \frac{|l\sigma'; l\sigma'|^\alpha}{|m\sigma; l\sigma'|^\alpha}
\tag{5.199}
$$

From the relations (5.198) and (5.199) the series of identities cames out:

$$
\frac{(D_{m\sigma}^\alpha)^2}{|m\sigma; m\sigma|^\alpha} = \frac{(D_{l\sigma'}^\alpha)^2}{|l\sigma'; l\sigma'|^\alpha} = \ldots = ct.
\tag{5.200}
$$

with which the expression (5.197) can be rewritten such as:

$$\vec{S}_{\alpha} \cong \sum_{m\sigma} \left[\vec{K}_{m\alpha} \big| m\sigma; m\sigma \big|^{\alpha} \right] \tag{5.201}$$

Aiming to analyze the direction of this vector in relation with the dispersion described by the Eq. (5.196), a tangential plane to DS is considered, so that the modified wave vector of transmitted direction

$$\vec{K}_0 \rightarrow \vec{K}_0 + \vec{\delta\tau} \tag{5.202}$$

satisfies the dispersion equation

$$\delta\Delta = 0 \tag{5.203}$$

with the operator δ acting toward all the wave vectors of diffracted directions.

Expanding the determinant from Eq. (5.196) upon the elements of the $m\sigma$ column and applying the above δ operator, one successively obtains:

$$\delta\Delta = -\frac{2}{k_0^2} \vec{K}_m \cdot \vec{\delta\tau} \big| m\sigma; m\sigma \big|^{\alpha} + \delta_{m\sigma}\Delta$$

$$= -\frac{2}{k_0^2} \vec{K}_m \cdot \vec{\delta\tau} \big| m\sigma; m\sigma \big|^{\alpha} - \frac{2}{k_0^2} \vec{K}_m \cdot \vec{\delta\tau} \big| m\pi; m\pi \big|^{\alpha} + \delta_{m\sigma,m\pi}\Delta$$

$$= \dots = -\frac{2}{k_0^2} \sum_{m\sigma} \vec{K}_m \cdot \vec{\delta\tau} \big| m\sigma; m\sigma \big|^{\alpha} = 0 \tag{5.204}$$

where $\delta_{m\sigma}\Delta$ is the determinant that remains to be evaluated after the elimination of the $m\sigma$ line and column – and so on.

By comparing the relation (5.201) with (5.204) the vectorial identity results:

$$\vec{S}_{\alpha} \cdot \vec{\delta\tau} = 0 \tag{5.205}$$

showing that the Poynting vector associated to the alpha branch of DS (and identical for the beta branch of DS) has the direction of propagation of the normal energy at DS in the coupling point, Figure 5.34.

With this conclusion, the phenomenological picture of the propagation of the energy with oscillation between the Poynting vectors associated to the DS branches is completely described and generalized, allowing an even deeper understanding of the interpretation and observation of the intensities associated to the diffraction directions.

5.4.3 INTEGRATED INTENSITIES OVER THE PENDELLLISUNG PERIOD: THE LAUE'S CASE

From now on, all the relations regarding the intensities, the ratio of intensities, the powers of reflection, and the powers of integrated reflections, calculated in previous Section can be understood in the light of the interference of the energies associated to the DS' branches for the diffraction directions over the Pendellösung period.

Therefore, there becomes instructive reloading both phenomenological and analytical the formulas vindicated in the previous Section for the Laue's case; the used notations and conventions remain valid; the present discussion follows (Azároff et al., 1974; Putz and Lacrămă, 2005).

Accordingly, the relations (5.105) and (5.106) will have the actual forms:

$$\frac{I_h}{I_0^e} = |b| \frac{|q|}{|q+z^2|} \exp(-\mu_0 t)\left[\sin^2(av) + \sinh^2(aw)\right] \tag{5.206}$$

$$\frac{I_e^0}{I_0^e} = \frac{\exp(-\mu_0 t)}{|q+z^2|}\left\{ \begin{array}{l} |q+z^2| + \left(|q+z^2| + |z|^2\right)\sinh^2(aw) - \\ \left(|q+z^2| - |z|^2\right)\sin^2(av) \pm \\ \frac{1}{2}\left[\left(|q+z^2| + |z|^2\right)^2 - |q|^2\right]^{1/2}\sinh|2aw| \pm \\ \frac{1}{2}\left[\left(|q+z^2| - |z|^2\right)^2 - |q|^2\right]^{1/2}\sin|2av| \end{array} \right\} \tag{5.207}$$

where, the plus-and-minus sign is properly associated if the arguments of the modulus is higher than one or zero trigonometric (for the "plus" sign), and respectively smaller than one or zero trigonometric (for the "minus" sign).

For the approximations (5.163) and (5.165) the above expressions are reduced to:

$$\frac{I_h}{I_0^e} \cong \frac{|b|}{1+y^2} \exp(-\mu_0 t) \left[\sin^2(av) + \sinh^2(aw)\right] \tag{5.208}$$

$$\frac{I_e^0}{I_0^e} = \exp(-\mu_0 t) \left[1 + \frac{1+2y^2}{1+y^2}\sinh^2(aw) - \frac{\sin^2(av)}{1+y^2} + \frac{y\sqrt{1+y^2}}{1+y^2}\sinh|2aw|\right] \tag{5.209}$$

with which the integrated reflection powers are properly calculated:

$$\begin{cases} R_h^y = \dfrac{1}{|b|} \int \dfrac{I_h}{I_0^e} dy \\[2mm] R_e^{0y} = \int \left[\dfrac{I_e^0}{I_0^e} - \exp\left(-\dfrac{\mu_0 t_0}{\gamma_0}\right)\right] dy \end{cases} \tag{5.210}$$

Further, the general lines of calculation and the results of these integrations will be exposed, meaning in fact the trigonometric functions averaging over the Pendellösung period.

This way, the integrated reflection power associated to the diffracted direction can take the form:

$$R_h^y = \exp[-\mu_0 t] J_h \tag{5.211}$$

where:

$$J_h = W + V \tag{5.212}$$

$$W = \int_{-\infty}^{+\infty} \frac{\sin^2 A\sqrt{1+y^2}}{1+y^2} dy = \pi \sum_{n=0}^{\infty} J_{2n+1}(2A) \tag{5.213}$$

$$V = \int_{-\infty}^{+\infty} \frac{1}{1+y^2} \sinh^2 \frac{A(k+yg)}{\sqrt{1+y^2}} dy = \frac{1}{2}\int_0^{\pi} \{\cosh[h\cos(\theta+\beta)]-1\}d\theta$$

$$= \frac{\pi}{2}[I_0(h)-1] \tag{5.214}$$

and where the last integrals have been transformed by using the notations (working also as the changing of variables):

$$\begin{cases} h = 2A\sqrt{k^2 + g^2} \\ \beta = \tan^{-1}\dfrac{k}{g} \\ y = -\cot\theta \end{cases} \tag{5.215}$$

Cumulating these results for the integrated reflection power on the diffracted direction, the general expression is obtained:

$$R_h^y = \exp\left(-\mu_0 t\right)\frac{\pi}{2}\left[2\sum_{n=0}^{\infty} J_{2n+1}(2A) + I_0(h) - 1\right] \tag{5.216}$$

where J_m is the first kind m-order Bessel function, and I_m is the modified Bessel function of m-order.

Also, we can comment on the term that contains the Bessel function of the first kind: this term was also previously found as the *Waller term* (5.139) (Waller, 1926), here corresponding to the integrated reflection power in the case of the non-absorbent crystal. Moreover, this term is almost equal to that one for the moderate reflections (characterized by the diffraction fringes not far from the central fringe), so that the relation (5.216) can be also approximated such as:

$$R_h^y = \exp\left(-\mu_0 t\right)\frac{\pi}{2}I_0(h) \tag{5.217}$$

Interesting, from the evaluation of the power of integrated reflection on the transmitted direction turns out that, regardless the energetic coupling associated to the DS branches, the energy propagated on the transmitted directions "survives". In fact, it is for this reason the present analytics complement the calculated intensities in last Section by those relating the transmitted direction, as will be further exposed.

One starts from the evaluation of the total integrated reflection power:

$$R_h^y + R_e^{0y} \equiv \exp\left(-\mu_0 t\right)J_{h+0} \tag{5.218}$$

where:

$$J_{h+0} = J_{h+0}^{(1)} + J_{h+0}^{(2)} \tag{5.219}$$

$$J_{h+0}^{(1)} = \int_{-\infty}^{+\infty} \left[\cosh \frac{2A(k+yg)}{\sqrt{1+y^2}} - \cosh 2Ag \right] dy$$

$$= \int_0^\pi \{\cosh[h\cos(\theta+\beta)] - \cosh[h\cos\beta]\} \frac{d\theta}{\sin^2\theta}$$

$$= -2\pi \sum_{n=0}^\infty (2n)I_{2n}(h)\cos(2n\beta) \tag{5.220}$$

$$J_{h+0}^{(2)} = \int_{-\infty}^{+\infty} \left[\frac{y\sqrt{1+y^2}}{1+y^2} \sinh \frac{2A(k+yg)}{\sqrt{1+y^2}} - \sinh 2Ag \right] dy$$

$$= \int_0^\pi \{\cos\theta\sinh[h\cos(\theta+\beta)] - \sinh[h\cos\beta]\} \frac{d\theta}{\sin^2\theta}$$

$$= -2\pi \sum_{n=0}^\infty (2n+1)I_{2n+1}(h)\cos[(2n+1)\beta] \tag{5.221}$$

In these expressions, the relations (5.215) have been used as the changing of variables, along the connection between the hyperbolic functions and the series of the modified Bessel functions; while for the integrals' calculations the main Cauchy value ($\lim_{\varepsilon\to0} \int_\varepsilon^{\pi-\varepsilon} d\theta$) has been used in order to avoid the divergences in their evaluation.

Again, by cumulating the results, the power of integrated reflection on the transmitted direction can be written under the general final form:

$$R_e^{0y} = \exp(-\mu_0 t)(-2\pi)\left[\sum_{m=1}^\infty mI_m(h)\cos(m\beta)\right] - R_h^y \tag{5.222}$$

Here one should noted that, because of the convergence problems appearing in the Eqs. (5.220) and (5.221) as integrals evaluations, precautions should be taken very careful when analyzing the experimental values of the integrated reflection power on the transmitted direction, especially for the asymmetric curves, see again Figure 5.33.

However, the relation (5.222) can be even simpler rewritten in some particular cases. For the symmetrical Laue's case, there appear the conditions: $\gamma_0 = \gamma_h, g = 0 \Rightarrow \beta = \pi/2$

by which expression (5.222) becomes:

$$R_e^{0y}(g=0) = \exp\left(-\mu_0 t\right)\left(\pi h\right)\left[-I_1(h) + 2\sum_{n=0}^{\infty}(-1)^n I_{2n+1}(h)\right] - R_h^y$$

$$= \exp\left(-\mu_0 t\right)\left(\pi h\right)\left[-I_1(h) + 2\int_0^h dt I_0(t)\right] - R_h^y \qquad (5.223)$$

where, for example, the reduced expression (5.217) for the integrated reflection power on the diffracted direction can be used.

For the case in which $k = 0$, two sub-cases appear: as based on the values of the $\beta = 0, \pi$ parameter the situations in which $g > 0$, $g < 0$ are corresponding and equation (5.222) and can be respectively rewritten:

$$R_e^{0y}(k=0) = \begin{cases} \exp\left(-\mu_0 t\right)\left(\pi h\right)[-I_1(h) - I_0(h)] - R_h^y \;, \; for \; g > 0 \\ \exp\left(-\mu_0 t\right)\left(\pi h\right)[-I_1(h) + I_0(h)] - R_h^y \;, \; for \; g < 0 \end{cases} \qquad (5.224)$$

Moreover, also the approximate formulas derived from Eqs. (5.216) and (5.222) can be established as corresponding to the approximations of the thin and respectively thick crystal. For the thin crystal the reflections powers integrated around the condition $h \cong 0$ are expanded and the approximate expressions result as:

$$R_h^y \cong \exp\left(-\mu_0 t\right)\frac{\pi}{2}\left(1 + \frac{1}{4}h^2 + \frac{1}{64}h^4 + ...\right) \qquad (5.225)$$

$$R_e^{0y} \cong \exp\left(-\mu_0 t\right)\pi\left(-\cos\beta \cdot h - \frac{1}{2}\cos2\beta \cdot h^2 + ...\right) - R_h^y \qquad (5.226)$$

with the representations given in Figure 5.35; in the region corresponding to the $h \cong 0$ condition, for the parameter's value $\beta = \pi/2$, the dynamical X-ray scattering is in the symmetrical Laue's case.

From the graphical and analytical analysis there is noted that for $h \cong 0$ the integrated reflection powers on the diffracted and transmitted directions are equalized (in modulus) for the symmetrical Laue's case.

These results explain the experimental observations (starting with the Bragg, James and Bosanquet 1921' experiment), this time for the crystal thin enough, so that the thin crystal approximation is retained valid.

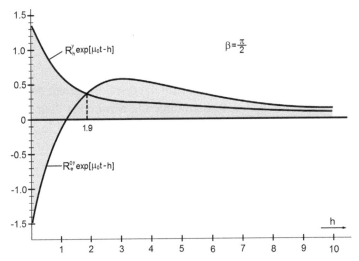

FIGURE 5.35 Integrated reflection powers for the diffracted and transmitted powers in the symmetrical Laue's case; after James (1965); Azároff et al. (1974).

On the other extreme, respectively on the asymptotic region with representation in the Figure 5.35, the thick crystal approximation is governed by the complementary analytical ($h\rightarrow\infty$) expansion:

$$R_h^y \cong \exp(-\mu_0 t)\frac{\pi}{2}\frac{\exp(h)}{\sqrt{2\pi h}}\left(1+\frac{1}{8h}+\frac{9}{2(8h)^2}+...\right) \qquad (5.227)$$

$$R_e^{0y} \cong \exp(-\mu_0 t)\pi\frac{\exp(h)}{\sqrt{2\pi h}}\left(B_0+\frac{B_1}{h}+\frac{B_2}{h^2}+...\right)-R_h^y \qquad (5.228)$$

where, the parameters introduced in the expression (5.228) have in turn the general forms:

$$\begin{cases} B_0 = \frac{1}{2}\frac{1+\cos\beta}{1-\cos\beta}+\frac{1}{2} \\[2mm] B_1 = \frac{3}{4}\left[1+\frac{1}{4}(1-\cos\beta)\right]\frac{1+\cos\beta}{(1-\cos\beta)^2}+\frac{3}{4}\cdot\frac{1}{4} \\[2mm] B_2 = \frac{3\cdot5}{8}\left[1+\frac{1}{4}(1-\cos\beta)+\frac{3}{32}(1-\cos\beta)^2\right]\frac{1+\cos\beta}{(1-\cos\beta)^3}+\frac{3\cdot5}{8}\cdot\frac{3}{32} \\[2mm] ... \end{cases} \qquad (5.229)$$

Some representations for the various values of the β-parameter are exposed in Figure 5.36 for the integrated reflection power on the transmitted direction.

Unlike the symmetrical Laue's case exposed in Figure 5.35, where for the regions corresponding to the thick crystal the integrated reflection power associated to the transmitted direction was higher than corresponding to the diffracted direction, for the curves represented in Figure 5.36, and that corresponding to other cases but the Laue's symmetrical one, the two integrated reflection powers are equalized for relatively small thicknesses of the crystal subject to the diffraction.

With these observations, various conclusions can be interpreted and formulated on the experimental observations in relation with the specific conditions by which they were produced. Some of them were just exposed in this Section.

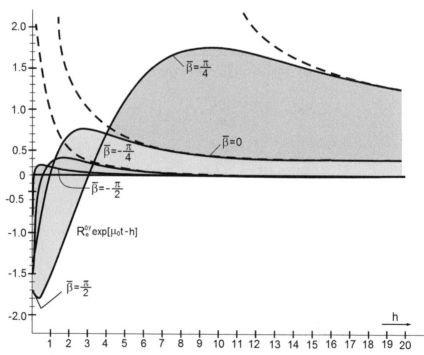

FIGURE 5.36 Integrated reflection power for the transmitted direction for various values of the parameter $\overline{\beta} = \beta - \pi/2$; after James (1965); Azároff et al. (1974).

This way, the phenomenological and analytical bases through which an experiment can be interpreted – or, more, it can be adapted so that the results to correspond to some "expectations" analytically predicted, by the adjustment of the specific parameters, e.g. the crystal thickness, the capacity of absorption, etc. were revealed and discussed.

Until now, this sub-Section aimed a completion or a reformulation of the previous one for X-ray crystallography study.

These because one has defined the qualitative premises of the calculations developed in the previous section by elucidating the solutions of propagation with the oscillation of the energies associated to the DS branches, and between them so consecrating the Pendellösung effect, but also because the previous Section calculations have been quantitatively prolonged so that also the study of the integrated reflection power on the transmitted direction, due to the diffraction on an absorbent crystal, to be included, in condition in which the transmitted direction being the only one "surviving" to any energetic coupling between the two DS branches, during the propagation.

5.4.4 ELECTRONIC DENSITY BY STRUCTURE FACTOR FOR DEFORMED CRYSTALS

Starting from an expression of the integrated reflection/reflectivity power (5.116) or (5.120), the analytical expressions for a Pendellösung period and of the associated structure factor will be further determined, during the symmetrical no-absorption Laue's diffraction. Accordingly, the Fourier's electronic map of the electronic distribution in the considered crystal will be established. As an application, the determination of the electronic distribution in the crystal of Si (FCC) will be considered for both the perfect monocrystal as well for the "film" of the electronic distribution during the slightly deformation of the crystalline lattice, under the elastic action; the present discussion follows (Putz, 2003; Putz and Lacrămă, 2005).

5.4.4.1 Pendellusung and the Electronic Ordered Structure

As a first interest in the solid state chemistry, as well in the biophysical chemistry, there is the evaluation of the electronic distribution in the investigated structure through the X-ray diffraction, under the form of the Fourier's transformation, i.e., the combined relations (5.18) and (5.25):

$$\rho(r) = \sum_{g=0}^{\infty}\left\{\left|F_g\right|\exp(i\phi_g)\exp[-2\pi i\,g\,r]\right\} \qquad (5.230)$$

For generality, one represents by g – the crystalline plane of reflection (diffraction) which allows the planar and spatial nominations, for example, $g \equiv (hk);(hk)$, respectively, being h, k, and l the Miller indices associated to the directions x, y and z of the reference system attached to the unit cell. Thus, the product gr of (5.230) can mean both the abbreviation of the products sum $(hx+ky)$ in plan, and respectively $(hx+ky+lz)$ in space.

The connection between the amplitude of the structure factor of (5.230) and the Pendellösung period (5.186), here denoted with Λ_g for the reflection plane g, can be established as based on the geometrical constants: V – the unit cell volume, λ – the wavelength of incident X-ray on the crystalline plane g, γ_g – the Bragg's angle ($\cos\theta_g$) of diffraction on the crystalline plane g, and on the universal constants: e the elementary electronic charge, m_0 the electronic rest mass and c the light speed propagated in vacuum, with the general expression (Takagi, 1969; Katagawa & Kato, 1974):

$$\left|F_g\right| = \frac{\pi V}{\lambda\gamma_g}\frac{m_0 c^2}{e^2}\frac{1}{\Lambda_g} \qquad (5.231)$$

This way, the problem of structure factor determination was reduced to the problem of expression of Pendellösung period.

5.4.4.2 Modeling of Elastic Deformation of the Crystalline Structure

Aiming to formulate a relation between the structure factor and a experimentally determined measure, one employs the integrated reflectivity (R_g) which, being directly proportional with the diffracted intensity (I_g), at its turn abstracted from a relation such as Eq. (5.115), and integrated like in Eq. (5.116) – for example, it will depend on the square of the amplitude of the structure factor, as based on the Eq. (5.34) prescription; this discussion follows (Putz, 2003).

"The agent" of this relationship between the integrated reflectivity and the amplitude of the structure factor of Eq. (5.231) will be the atomic susceptibility χ_g, defined through the same constants as those of Eq. (5.231), as introduced in Eq. (5.69), but while fixing $4\pi\varepsilon_0 = 1$:

$$\chi_g = \frac{\lambda^2}{\pi V} \frac{e^2}{m_0 c^2} \left| F_g \right| \tag{5.232}$$

In this context, the analytically reflectivity integrated can be introduced (Janáček et al., 1978):

$$R_g^{\text{analitic}} = \frac{2\pi^2 \tau C_g^2 \gamma_g \left|\chi_g\right|^2}{\lambda \sin^2 2\theta_g} \left| {}_1F_1\left(\varepsilon, 1, -\frac{1}{\varepsilon} \frac{\pi^2 \tau^2 C_g^2 \left|\chi_g\right|^2}{4\lambda^2 \cos^2 \theta_g}\right) \right|^2 \tag{5.233}$$

where, the parameters τ, C_g, ε also intervene while respectively representing: the depth of the crystal, the polarization factor (5.87), and the parameter of elastic deformation – inside the confluent hyper-geometric function ${}_1F_1$; they so provide a relation that (quite) generally characterizes the symmetrical Laue's diffraction on a mono-crystal, without absorption.

A simplification of the expression (5.233) can be obtained by introducing the notation:

$$u \equiv -\frac{\pi^2 \tau^2 C_g^2 \left|\chi_g\right|^2}{4\lambda^2 \cos^2 \theta_g} = -\frac{\pi^2 \tau^2 C_g^2}{4\gamma_g^4 \Lambda_g^2} \tag{5.234}$$

with which also the hyper-geometric confluent function ${}_1F_1$ can be simply written by truncating its expansion until the second degree of the deformation ε so that:

$${}_1F_1\left(\varepsilon, 1, \frac{1}{\varepsilon} u\right) \cong F_0(u) + \varepsilon F_1(u) + \varepsilon^2 F_2(u) \tag{5.235}$$

The components in Eq. (5.235) are expressed in relations with the Bessel functions J_0 and J_1, such as (Janáček & Kuběna, 1978; Janáček et al., 1978):

$$
\begin{aligned}
F_0(u) &= J_0(\sqrt{-4u}), \\
F_1(u) &= \frac{u^2}{2} \frac{d^2}{du^2} J_0(\sqrt{-4u}), \\
F_2(u) &= \frac{u^3}{8} \frac{d^3}{du^3} \left[-\frac{1}{3} J_0(\sqrt{-4u}) + \frac{u}{\sqrt{-u}} J_1(\sqrt{-4u}) \right]
\end{aligned}
\tag{5.236}
$$

For practical manipulations, the asymptotic shapes of the Bessel function $J_n(\bullet)$ of Eq. (5.236) will be considered (Mathews & Walker, 1969):

$$J_n(\bullet) \approx \sqrt{\frac{2}{\pi \bullet}} \cos\left(\bullet - \frac{n\pi}{2} - \frac{\pi}{4} \right) \tag{5.237}$$

The Eq. (5.237) limit has also a clear physical relevance, since corresponding to the case of the monocrystal in the approximation of the thick crystal, see the previous Section(s).

5.4.4.3   Pendellusung and the Amplitude of the Structure Factor

Of course, the expression (5.233) is not enough for expressing the Pendellösung period; it should be equated with a "measured" counter-value, for example, R_g^{measured}, which corresponds to experimentally values or by an alternative method deduced. For the direct formulation of such structure factor equation there is convenient to formally written that (Putz, 2003):

$$\sqrt{R_g^{\text{measured}}} = \sqrt{R_g^{\text{analytic}}} \tag{5.238}$$

Thus, the Pendellösung equation can be written such as:

$$\sqrt{R_g^{\text{measured}}} = \sqrt{\frac{\lambda \gamma_g}{2\tau(1-\gamma_g^2)}} \left(\frac{\pi C_g}{\gamma_g^2} \frac{\tau}{\Lambda_g} \right) {}_1F_1\left(\varepsilon, 1, -\frac{1}{\varepsilon} \frac{\pi^2 C_g^2}{4\gamma_g^4} \frac{\tau^2}{\Lambda_g^2} \right) \tag{5.239}$$

If, since the Eq. (5.234), one introduces the notation:

$$\varphi \equiv \sqrt{-4u} = \frac{\pi C_g}{\gamma_g^2} \frac{\tau}{\Lambda_g} \tag{5.240}$$

then, based on it, the expansion of the hyper-geometric confluent function from Eq. (5.235) along the Eq. (5.239) can be both rewritten as (Putz, 2003):

$$\sqrt{R_g^{\text{analytic}}} \cong \sqrt{\frac{\lambda \gamma_g}{2\tau (1-\gamma_g^2)}} \, \varphi \left[F_0(\varphi) + \frac{\varepsilon}{8} F_1(\varphi) + \left(\frac{\varepsilon}{8} \right)^2 F_2(\varphi) \right] \tag{5.241}$$

where, the Eq. (5.236) components, are taking the asymptotically shape, in the approximation of the thick crystal, respectively:

$$F_0(\varphi) \approx \sqrt{\frac{2}{\pi\varphi}} \cos\left(\varphi - \frac{\pi}{4}\right),$$

$$F_1(\varphi) \approx \sqrt{\frac{2}{\pi\varphi}} \left[\left(\frac{3}{4} - \varphi^2\right)\cos\left(\varphi - \frac{\pi}{4}\right) + \varphi\sin\left(\varphi - \frac{\pi}{4}\right)\right],$$

$$F_2(\varphi) \approx \sqrt{\frac{2}{\pi\varphi}} \left[\begin{array}{l}\left(\frac{5}{8} - \frac{1}{8}\varphi^2 + \frac{1}{2}\varphi^4\right)\cos\left(\varphi - \frac{\pi}{4}\right) \\ + \left(\frac{9}{16}\varphi + \frac{5}{12}\varphi^3\right)\sin\left(\varphi - \frac{\pi}{4}\right)\end{array}\right] \qquad (5.242)$$

With the terms (5.242) and (5.241) the Pendellösung equation for the variable φ is obtained (Putz, 2003):

$$\sqrt{R_g^{\text{measured}}} = \sqrt{\frac{\lambda\varphi\gamma_g}{\pi\tau(1-\gamma_g^2)}} \left\{\begin{array}{l}\left[1 + \frac{\varepsilon}{8}\left(\frac{3}{4} - \varphi^2\right) + \left(\frac{\varepsilon}{8}\right)^2\left(\frac{5}{8} - \frac{1}{8}\varphi^2 + \frac{1}{2}\varphi^4\right)\right]\cos\left(\varphi - \frac{\pi}{4}\right) \\ + \left[\frac{\varepsilon}{8}\varphi + \left(\frac{\varepsilon}{8}\right)^2\left(\frac{9}{16}\varphi + \frac{5}{12}\varphi^3\right)\right]\sin\left(\varphi - \frac{\pi}{4}\right)\end{array}\right\} \qquad (5.243)$$

The Eq. (5.243) appears under a non-algebraic form. Yet, fortunately, it can be brought to an algebraic form, firstly by its rearrangement under the equivalent form:

$$1 = \sqrt{\frac{\lambda\varphi\gamma_g}{\pi\tau R_g^{\text{measured}}(1-\gamma_g^2)}} \left[1 + \frac{\varepsilon}{8}\left(\frac{3}{4} - \varphi^2\right) + \left(\frac{\varepsilon}{8}\right)^2\left(\frac{5}{8} - \frac{1}{8}\varphi^2 + \frac{1}{2}\varphi^4\right)\right]\cos\left(\varphi - \frac{\pi}{4}\right)$$

$$+ \sqrt{\frac{\lambda\varphi\gamma_g}{\pi\tau R_g^{\text{measured}}(1-\gamma_g^2)}} \left[\frac{\varepsilon}{8}\varphi + \left(\frac{\varepsilon}{8}\right)^2\left(\frac{9}{16}\varphi + \frac{5}{12}\varphi^3\right)\right]\sin\left(\varphi - \frac{\pi}{4}\right)$$

$$(5.244)$$

and then, by identifying it with a known trigonometric identity:

$$1 = \cos^2\left(\varphi - \frac{\pi}{4}\right) + \sin^2\left(\varphi - \frac{\pi}{4}\right) \qquad (5.245)$$

wherefrom, the identifications of the correspondent terms immediately follow:

$$\cos\left(\varphi-\frac{\pi}{4}\right)=\sqrt{\frac{\lambda\varphi\gamma_g}{\pi\tau R_g^{\text{measured}}(1-\gamma_g^2)}}\left[1+\frac{\varepsilon}{8}\left(\frac{3}{4}-\varphi^2\right)+\left(\frac{\varepsilon}{8}\right)^2\left(\frac{5}{8}-\frac{1}{8}\varphi^2+\frac{1}{2}\varphi^4\right)\right],$$

$$\sin\left(\varphi-\frac{\pi}{4}\right)=\sqrt{\frac{\lambda\varphi\gamma_g}{\pi\tau R_g^{\text{measured}}(1-\gamma_g^2)}}\left[\frac{\varepsilon}{8}\varphi+\left(\frac{\varepsilon}{8}\right)^2\left(\frac{9}{16}\varphi+\frac{5}{12}\varphi^3\right)\right]$$

$$(5.246)$$

with which help, the initial equation (5.243) will be rewritten under the algebraic form:

$$\frac{\pi\tau R_g^{\text{measured}}(1-\gamma_g^2)}{\lambda\gamma_g\varphi}$$

$$=\left[1+\frac{\varepsilon}{8}\left(\frac{3}{4}-\varphi^2\right)+\left(\frac{\varepsilon}{8}\right)^2\left(\frac{5}{8}-\frac{1}{8}\varphi^2+\frac{1}{2}\varphi^4\right)\right]^2+\left[\frac{\varepsilon}{8}\varphi+\left(\frac{\varepsilon}{8}\right)^2\left(\frac{9}{16}\varphi+\frac{5}{12}\varphi^3\right)\right]^2$$

$$(5.247)$$

To find the solution of the Eq. (5.247), a new simplifying notation can be considered, namely:

$$\omega=\frac{\pi\tau R_g^{\text{measured}}(1-\gamma_g^2)}{\lambda\gamma_g}$$

$$(5.248)$$

which, together with the dissolution of the parenthesis in Eq. (5.247) and by neglecting all of the terms over the quadratic order for the elastic deformation ($\varepsilon^n,>2$), the simplified equation is produced:

$$\frac{\omega}{\varphi}=1+\varepsilon\left(\frac{3}{16}-\frac{1}{4}\varphi^2\right)+\varepsilon^2\left(\frac{29}{1024}-\frac{3}{256}\varphi^2\right)$$

$$(5.249)$$

The Eq. (5.249) corresponds from now on to an equation of third order in the φ variable:

$$\omega=\varphi\left(1+\frac{3}{16}\varepsilon+\frac{29}{1024}\varepsilon^2\right)-\varphi^3\left(\frac{1}{4}\varepsilon+\frac{3}{256}\varepsilon^2\right)$$

$$(5.250)$$

The direct solution of the Eq. (5.250) looks like (Putz, 2003):

$$\varphi[\alpha(\omega,\varepsilon),\varepsilon] = \frac{2^{16}\,3\varepsilon + 2^{10}\,45\,\varepsilon^2 + 2^6\,115\,\varepsilon^2 + 261\,\varepsilon^4 + (3\alpha)^{2/3}}{6\varepsilon(64+3\varepsilon)(3\alpha)^{1/3}} \quad (5.251)$$

where:

$$\alpha(\omega,\varepsilon)$$

$$= \left(3\varepsilon^3\right)^{1/2}\left[\frac{2^4\,176947\,\varepsilon\,(64+3\varepsilon)^4\,\omega^2}{-\left(2^{16}+2^{10}\,15\varepsilon + 2^7\,19\varepsilon^2 + 87\varepsilon^3\right)^3}\right]^{1/2} - \omega\varepsilon^2\left(2^{22}\,9 + 2^{17}\,27\varepsilon + 2^{10}\,81\varepsilon^2\right)$$

$$(5.252)$$

Still, the Eq. (5.251) with Eq. (5.252) presumes in fact terms with orders superior to the quadratic approximation in the deformation ε parameter. These can be directly "reduced" by simply neglecting them or, more elegantly (and more consistent in physical terms) by considering the Maclaurin expansion of the solution (5.251) with (5.252) until the second order of the parameter of elastic deformation (Putz, 2003):

$$\varphi(\omega,\varepsilon)$$

$$\cong \lim_{\varepsilon\to0}\varphi[\alpha(\omega,\varepsilon),\varepsilon] + \varepsilon\lim_{\varepsilon\to0}\left\{\frac{\partial}{\partial\varepsilon}\varphi[\alpha(\omega,\varepsilon),\varepsilon]\right\} + \frac{1}{2}\varepsilon^2\lim_{\varepsilon\to0}\left\{\frac{\partial^2}{\partial\varepsilon^2}\varphi[\alpha(\omega,\varepsilon),\varepsilon]\right\}$$

$$(5.253)$$

where the coefficients of the expansion, are found with the expressions:

$$\lim_{\varepsilon\to0}\varphi[\alpha(\omega,\varepsilon),\varepsilon] = \omega,$$

$$\lim_{\varepsilon\to0}\left\{\frac{\partial}{\partial\varepsilon}\varphi[\alpha(\omega,\varepsilon),\varepsilon]\right\} = \frac{1}{16}\omega(4\omega^2-3),$$

$$\lim_{\varepsilon\to0}\left\{\frac{\partial^2}{\partial\varepsilon^2}\varphi[\alpha(\omega,\varepsilon),\varepsilon]\right\} = \frac{1}{512}\omega(192\,\omega^4 - 180\,\omega^2 + 7) \quad (5.254)$$

From now on, with the help of relations (5.253) and (5.254), the real and unique solution of the Eq. (5.250) may be put in an analytical form:

$$\varphi(\omega,\varepsilon) \cong \omega\left[1 + \frac{1}{16}\varepsilon(4\omega^2-3) + \frac{1}{1024}\varepsilon^2(192\,\omega^4 - 180\,\omega^2 + 7)\right] \quad (5.255)$$

However, the solution (5.255) should be conducted under the form dependent to the physical and geometrical parameters specific to the X-ray dynamical diffraction of monocrystals.

To this end, the definition (5.240) will be called in order the Pendellösung period, and then the definition (5.248) be recovered for the involvement of the integrated reflectivity (for the moment, only inside the brackets' parentheses from the (5.255) solution) in order to emphasize and analyze the ideal case of the elastic non-deformation. Thus, we will obtain (Putz, 2003):

$$\frac{1}{\Lambda_g} \cong \frac{\gamma_g (1-\gamma_g^2) R_g^{\text{measured}}}{\lambda C_g} \left[1 + \frac{1}{16} \varepsilon \left(4\omega^2 - 3 \right) + \frac{1}{1024} \varepsilon^2 \left(192\omega^4 - 180\omega^2 + 7 \right) \right]$$

(5.256)

The limitation of the expression (5.256) only to the first term will generate the relation between the measured integrated reflectivity and the Pendellösung period for the non-deformed crystal:

$$R_g^{\text{measured}} \approx \frac{\lambda C_g}{\gamma_g (1-\gamma_g^2) \Lambda_g^{\text{ideal}}}$$

(5.257)

The relation (5.257) combined with the definition (5.231) will also provide the measured integrated reflectivity dependency by the structure factor amplitude of the non-deformed ideal crystal case:

$$R_g^{\text{measured}} \approx \frac{\lambda^2 C_g}{\pi V (1-\gamma_g^2) m_0 c^2} \frac{e^2}{\left| F_g^{\text{ideal}} \right|}$$

(5.258)

Next, by considering Eq. (5.258) in Eq. (5.256) in the rest of the terms the relations (5.231) and (5.248), the amplitude of the structure factor of the slight deformed crystal is finally obtained with the expression (Putz, 2003):

$$\left| F_g \right| \cong \left| F_g^{\text{ideal}} \right| \left\{ \begin{array}{l} 1 + \dfrac{1}{16} \varepsilon \left[4 \dfrac{(1-\gamma_g^2)^2 \tau^2 \pi^2 \left(R_g^{\text{measured}} \right)^2}{\lambda^2 \gamma_g^2} - 3 \right] \\[3mm] + \dfrac{1}{1024} \varepsilon^2 \left[\begin{array}{l} 192 \dfrac{(1-\gamma_g^2)^4 \tau^4 \pi^4 \left(R_g^{\text{measured}} \right)^4}{\lambda^4 \gamma_g^4} \\[3mm] -180 \dfrac{(1-\gamma_g^2)^2 \tau^2 \pi^2 \left(R_g^{\text{measured}} \right)^2}{\lambda^2 \gamma_g^2} + 7 \end{array} \right] \end{array} \right\}$$

(5.259)

From Eq. (5.259) there is clearly how, once with the increase of the contributions of the corrections of the elastically deformation, also increases the amplitude of the structure factor, and, respectively, based on the relation (5.231), also the Pendellösung period properly decreases. This phenomenon corresponds to the experimental observations, as recorded through the contractions effect of the diffraction fringes for the diffraction on the slight bending crystals (Janáček & Kuběna, 1978; Janáček et al., 1978; Authier & Simon, 1968).

From practical reasons, the relation (5.259) is also conveniently written through the resuming of all the corrections due to the slight elastically deformations into a so-called deformation function $F(\varepsilon)$, with which the deformation solution of the amplitude of the structure factor in the equivalent solutions can be written:

$$\left|F_g\right| \equiv \left|F_g^{\text{ideal}}\right|\left[1 + F(\varepsilon)\right] \tag{5.260}$$

$$\cong \left|F_g^{\text{ideal}}\right|\exp\left[F(\varepsilon)\right] \tag{5.261}$$

The last expression, by its shape, will additionally allow the evaluation of the model for the phase associated to the structure factor, relating with the deformation function, so solving in an original way the "phase matter" mentioned at the beginning of this Chapter.

5.4.4.4 Application to the Electronic Maps of Contour

As a first implementation of the previous results, the Bragg's law for the central diffraction fringe can be considered, i.e., adapting the relation (5.5) to the actual needs:

$$\sin\theta_g = \frac{\lambda}{2d_g} \tag{5.262}$$

which can be further included in the expression of the direction cosine of the scattering Bragg's angle specific to the diffracting g-plane:

$$\gamma_g = \sqrt{1 - \sin^2\theta_g} = \sqrt{1 - \frac{\lambda^2}{4d_g^2}} \tag{5.263}$$

In any case, beyond the direct involvement of the Bragg's law in the above results, the problem of the phase determination appears, for the structure factor amplitude of Eqs. (5.259)–(5.261).

Such a determination of the phase may use the real and imaginary components of the structure factor expression, since identified at exponential level with a generalized form of the relation (5.261) (Putz, 2003):

$$F_g = F_g^{\text{Re}} + iF_g^{\text{Im}} = |F_g| \exp(i\phi_g) = F_g^{\text{ideal}} \exp[F(\varepsilon)] \qquad (5.264)$$

Moreover, by Eq. (5.264) the ideal structure factor can be rewritten as a function of atomic scattering factors f_j of the (j) atoms involved, i.e., contained on the diffraction planes and placed at the r_j distance towards the referential centered at the level of the unit cell, under the form prescribed by the transformation (5.48), as adapted to this analysis:

$$F_g^{\text{ideal}} = \sum_j f_j \exp[2\pi i g r_j] \qquad (5.265)$$

where the summation is made upon all the atoms existing in the unit cell.

On the other side, the definition relation for the phase associated to the structure factor in Eq. (5.264) is given, within the complex plane representation, through the consecrate formula:

$$\phi_g = \arccos\left(\frac{F_g^{\text{Re}}}{|F_g|}\right) \qquad (5.266)$$

Nevertheless, in this approach the influence of the deformation function $F(\varepsilon)$ is accounted as follows. For the real component of the structure factor of (5.266) both the expansion of F_g^{ideal} with the expression (5.265) and of $\exp[F(\varepsilon)]$ as in (5.260) will be considered (Putz, 2003):

$$F_g^{\text{Re}} \cong [1 + F(\varepsilon)] \sum_j f_j \cos(2\pi g r_j) \qquad (5.267)$$

while, for the structure factor amplitude, i.e., the denominator from (5.266), the same measures but in a always positive form will be considered, such as:

$$|F_g| \cong \exp[F(\varepsilon)] \left| \sum_j f_j \exp(2\pi i g r_j) \right| \qquad (5.268)$$

as the forms (5.261) and (5.265) would be directly combined.

With the relations (5.267) and (5.266) the present expression of the structure factor phase includes also the effect of the elastically deformation, identically with that anticipated in Eq. (5.55) (Putz, 2003):

$$\phi_g = \arccos\left(\frac{1+F(\varepsilon)}{\exp[F(\varepsilon)]} \frac{\sum_j f_j \cos(2\pi g r_j)}{\left|\sum_j f_j \exp(2\pi i g r_j)\right|}\right) \qquad (5.269)$$

This way, we arrive to the analytical expression of the researched electronic density, the relation (5.230), where the forms (5.261), (5.265) and (5.269), are respectively considered for the amplitude of the deformed structure factor, the amplitude of the non-deformed structure factor, and for the phase associated to the deformed monocrystal, with the general model (Putz, 2003):

$$\rho(r) \cong \exp[F(\varepsilon)]$$

$$\times \sum_{g=0}^{\infty}\left|\sum_j f_j \exp[2\pi i(g r_j)]\right| \exp\left\{i\left[\arccos\left(\frac{1+F(\varepsilon)}{\exp[F(\varepsilon)]} \frac{\sum_j f_j \cos(2\pi g r_j)}{\left|\sum_j f_j \exp(2\pi i g r_j)\right|}\right) - 2\pi g r\right]\right\}$$

$$(5.270)$$

presented, *in anticipo*, by the earlier relations (5.54) and (5.55).

For the practical applications, the only unsolved problem remains the impossibility of infinite summation upon all the possible diffraction's planes. Therefore, the infinite sum of Eq. (5.270) can be limited by using the inverse of the parameters' lengths of the unit cell through the so-called resolution condition (Cantor & Schimmel, 1980):

$$\left|\sum_g \left(g\frac{1}{a}\right)\right| \leq \frac{1}{\text{Resolution }(\overset{0}{A})} \qquad (5.271)$$

wherefrom, the set of diffraction planes follow as:

$$(g) \in \text{Res} \equiv \{(g)\} \qquad (5.272)$$

very relevant for the electronic maps' determination, within the current approach.

The present scheme will be further applied for the case of the (FCC) crystal of Si, for which the planar projection of the electronic density, in a plane-base of the elementary cell, (x, y) will be employed.

As long as a structure of diamond type is about, the unit cell being a cube with the periodic side of the length $a = 5.43 \overset{0}{\text{A}}$, see (The elements, 2003), the projection of the interplanar distance of the shape is imposed as (Verma & Srivastava, 1982):

$$d_g = \frac{5.43}{\sqrt{h^2 + k^2}} (\overset{0}{\text{A}})$$
(5.273)

When the conditions of X-ray diffraction are specified, i.e., the wavelength of the incident radiation, being that determined by the line $K_{\alpha 1}$ of a target of Ar, see again the beginning of the present Chapter, $\lambda = 0.5594075 \overset{0}{\text{A}}$ (Lide & Frederikse, 1995), then, with Eq. (5.273) in Eq. (5.263), the dependence of the direction cosines of the Bragg's diffraction angle as a function of Miller indices associated to the directions (x, y) is advanced:

$$\gamma_g = \sqrt{1 - 0.00265 \left(h^2 + k^2 \right)}$$
(5.274)

The carried energy by this X-ray beam is determined as based on the notorious relation of Planck quantification

$$E = 2\pi\hbar c / \lambda$$
(5.275)

with the value 2.152[KeV], being $\hbar$ the reduced Planck constant. For the atoms of Si-unit cell of the diamond type structure, for such an interaction energy, the atomic scattering factors are about the same at the level of the unit cell and with the numerical value $f_{si} \cong 14.0401$, see (X-Ray Interactions With Matter, 2003).

When an electronic density moderate resolution is chosen around of $1 \overset{0}{\text{A}}$, by the employment of the condition (5.271) adapted to the present restrictions one is leaved with the parametric inequality:

$$|ha* + ka*| \le 1$$
(5.276)

wherefrom, for Si (FCC) the experimental parameter

$$a^* = 1/a = 0.184 \overset{0}{A}{}^{-1} \qquad (5.277)$$

is replaced and the solution of the inequality will prescribe the resolution limit:

$$|(h+k)0.184| \leq 1 \Rightarrow \max(h+k) = 5 \qquad (5.278)$$

which allows the determination of the current set of diffraction planes associated (Putz, 2003):

$$(h,k) \in \text{Res} \equiv \begin{cases} (0,0),(0,1),(0,2),(0,3),(0,4),(0,5), \\ (1,0),(1,1),(1,2),(1,3),(1,4), \\ (2,0),(2,1),(2,2),(2,3), \\ (3,0),(3,1),(3,2), \\ (4,0),(4,1), \\ (5,0) \end{cases} \qquad (5.279)$$

Due to the projection of the electronic density investigated in the plane (x, y) of the unit cell, the coordinates of the atoms of Si, r_j, that contribute to the electronic density will be also limited to the set:

$$(x_j, y_j) \in \{(0,0),(0,1),(1,0),(1,1),(1/2,1/2)\} \qquad (5.280)$$

Here, there should be specified that, although not explicitly taken into account, the presence of the other spatial (including those interstitial) atoms will be recorded in the calculated electronic density; this, because their "trace" is contained in the experimental length of the parameters of the unit cell that, in turn, consider the compactness of the type of the analyzed lattice.

The next measured input that should be specified refers to the measured integrated reflectivity, R_g^{measured}. As previously mentioned, at this point, a direct connection with a measured experimental or with one alternatively calculated value can be made. In this case, the commonly used value deduced in Eq. (5.130) is chosen, i.e., for the Laue's symmetrical diffraction, on a thick crystal, for the so-called Zachariasen dynamical approach, so imposing the $R_g^{\text{measured}} \cong \pi/2$ value.

The last necessary parameter to specify refers to the crystal's thickness: accordingly, in the approximation of the thick crystal, a suitable choice prescribes a value about $\tau \cong 1(mm)$.

By collecting all these information in the expression (5.270), and taking into account of the (5.259)-(5.261), (5.265), (5.273), (5.274), (5.279) and (5.280) connections, the computational expression of the planar projection of the electronic density for the crystal of Si (FCC) analytically results.

At the level of the unit cell, the representations of the Figure 5.37 (a-e) present the "deformation film" of the electronic density for the crystal of Si(FCC), in the plane (x, y) in relation with the values of the elastically deformation ε-parameter.

In Figure 5.37 (a) the ideal case of the planar projection of the electronic density of contour at the ideal unit cell level ($\varepsilon = 0$) is exposed. A good resolution of the electronic distribution is noted, so illustrating the electronic density around the atomic scattering centers.

There is also noted a more persistent electronic presence on a diagonal, along a blurred one on the other one; this effect is due to the interstitial atoms of the unit cell, which, for the diamond structure are alternatively placed respecting the planes of the bases of the unit cell.

Therefore, one records the "traces" of the atomic presences not directly included in the effective set (5.280) of calculation. The appearance of those traces in the electronic distribution, by the dynamical X-ray diffraction, confirms the reliability of the theoretical treatment applied, in agreement with the experimental records.

In Figures 5.37 (b, c), the contributions of the deformation parameter ε towards the electronic density are taken into account, having $\varepsilon = 1,2$ values, respectively. In any case, the effects of the deformation are exposed until the first order of the deformation (left) and respectively for its second order (right).

However, one notes that only around the value of ε having values as units, essential modifications in the electronic density are recorded as the effect of the successive deformation of the crystalline structure. At this point there seems that the condition of "slightly deformation" ($\varepsilon \to 0$) is to be violated or, equivalently, one can conclude that the real slight deformations $\varepsilon \in (10^{-5}, 10^{-1})$ do not produce notable modifications in the (FCC) structure of Si!

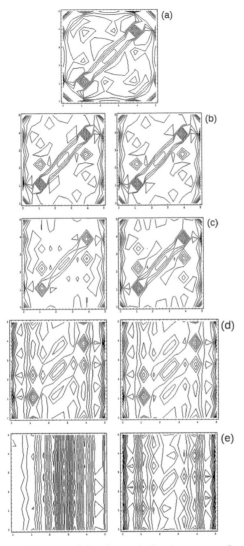

FIGURE 5.37 (a): The map of the electronic density contour for the (001) plane of the Si (FCC) monocrystal for the ideal case of the non-deformed the unit cell ($\varepsilon = 0$); (b,c): the successive deformations of the contour electronic density for the cases of the elastically deformation, taking into account only the deformations of first order (on the left) and respectively until the second order of the deformations (on the right), with ε parameter being fixed at the values 1 (in b) and 2 (in c), respectively (Putz, 2003). (d,e): The successive deformation of the contour electronic density contour for the (001) plane of the Si (FCC) monocrystal for the cases of the elastically deformations as: taking into account only the deformations of the first order (left) and respectively until the second order of the deformations (right), with the ε parameter being fixed at the values 4 (in d) and 8 (in e), respectively (Putz, 2003).

Yet, since noting that the other extreme limit ($\varepsilon \to \infty$) is applied for the treatment of the *kinematic diffraction*, there appears that the values of the ε parameter in the field of units, although in extrapolated cases, do not contravene to the *dynamical diffraction* context where the presented scheme is placed.

Further on, in Figure 5.37 (d, e) the deformations of the contour electronic density at the level of the unit cell are exposed for the $\varepsilon = 4,8$ values of the deformation parameter ε, respectively. This time a consistent disturbance of the structure of the unit cell is noted, relatively to the (FCC) arrangement of Si of the ideal non-deformed case of Figure 5.37 (a).

Anyway, for each structure each interval of deformation around and departing from the ideal case should be explored, so that the significant deformation of the unit cell identification, and respectively to be accordingly emphasized, aiming the associated deformation degree prediction (Putz, 2003).

All these results and application can be extended both in the theoretical and experimental directions. For example, on the theoretical side, similar results by considering the case of Bragg's symmetrical reflection, or of the asymmetrical diffraction, and/or by considering also the absorption effect (e.g., through implementing the Laue's above case) can be formulated; yet, also inside the actual model improvements can be brought: the expression of the phase of (5.269) can be reconsidered by expanding the deformation function $F(\varepsilon)$ in even higher orders of the deformation ε -parameter for the real component of the structure factor; the set of (5.279) planes involved in diffraction can be also extended by choosing a "smoother" resolution in electronic density representation. Finally, the connection with the experiment can be considered by replacing a real-measured value of the integrated reflectivity in the structure factor, starting with the relation (5.259), thus producing the electronic distribution map in direct relation with the scattering one.

5.5   TOTAL SELF-CONSISTENCY IN X-RAY DIFFRACTION

5.5.1   BORRMANN PHENOMENOLOGY

The key of the dynamical propagation is the Bragg's law correlates the dynamical diffraction with the two fields of the diffraction corresponding to the two branches of dispersion surface (DS), so keeping the propagation coupled between the two directions allowed by the condition of continuity of the phases, both at the crystalline surface and for the atomic planes beneath; the present discussion follows (Birău & Putz, 2000).

Therefore, going on the causal phenomenological level, the key of the entire dynamical propagation appears as the continuity of the phases. Thus, there is natural the effects of this propagation to be analyzed in terms of phase's inter-play between the diffraction branches, i.e., between the primary and diffracted branches, as likely for any combination of them.

A note is now required: for obtaining the diffraction's hyperbola as an intersection of the DS with the incident plane, the *Reciprocity Theorem* in propagation is also taken into account: no matter which wave from crystal will be considered the primary wave, the same set of diffracted waves will be generated. This observation should be revealed, since the interpretation of the hyperbolic geometrical locus from the Bragg's law would required that the O and H points of the reciprocal space have the foci role, while, as based on the propagation in the reciprocal space constituted as in Figure 5.38, for the two reflection spheres of which intersection generates the hyperbole, the propagation was considered *from* the hyperbola points to those Bragg coupled points (Birău & Putz, 2000).

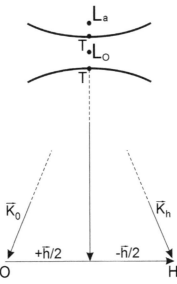

FIGURE 5.38   The Ewald's branching decomposition for the Bragg dynamical coupling of diffraction.

In this context, there will be proved how simply the anomalous absorption on the beta branch can be perceived, Figure 5.38, by the so-called Borrmann effect.

To this aim, only the decomposition of the directions involved in Bragg couplings is necessary in the wave vectors that propagate along the normal direction on OH, i.e., on the directions along the diffraction planes – because both O and H on the Ewald's reflection sphere are attached to a diffraction plane – with the equal wave vectors $+\vec{h}/2$ and $-\vec{h}/2$ for the symmetrical case, formed between the diffraction planes under concern. The last category of vectors formed between the diffraction planes are responsible for the standing waves (SW)' appearance; they are placed with their nodes right on the atomic planes for the propagation associated with the alpha branch of the DS, and with the anti-nodes on the atomic planes for the propagations associated to the beta branch of DS.

Why this way and not otherwise? The answer is simply taken right from the Ewald's approach: for the coupling points (T) from the alpha branch of DS, the one on top in Figure 5.38, since very close to the Laue's diffraction center, the asymptotic behavior for the resonance factor is imposed

$$\left(K_h^2 - k_0^2\right)^{-1} \overset{T \approx l_a}{\longrightarrow} \infty \qquad (5.281)$$

which corresponds to a maximum conversion of the atomic dipoles' amplitudes, so "attacked" by the incident field $\left(\vec{k}_0\right)$, into the amplitudes of diffracted field $\left(\vec{K}_h\right)$, which obviously correspond to a minimum absorption. The opposite case is available for T planes of the beta branch of DS, "away" from the Laue's incidence condition – and therefore the propagations coupled from this branch will suffer an abnormally high absorption respecting a linear absorption.

Such decomposition, which has as subsidiary the "decomposition" of the absorption in a normal component, away from the Bragg angle, and one anomalous, associated to each branch of DS, is directly transposed in the *Borrmann phenomenology*. But the anomalous effect of absorption, i.e., the Borrmann effect, appears only under the condition of the thick crystal, or more accurate for the $\mu t \gg 1$ case (Bormann, 1940, 1950, 1954, 1955, 1959).

The decomposition from Figure 5.38 does not contain any specification regarding this condition, appearing that it can be realized in any condition; then, how does the Borrmann's effect manifests? And moreover, how the condition $\mu t \gg 1$ can be correlated to the appearance of the standing waves?

Certainly, the phase phenomenology gives a qualitative evaluation, simply and unitary for the dynamical diffraction with the formation of the standing waves through the interference of the transmitted-diffracted coherent waves; it is analytically considered through the shape of the Bloch waves, propagated in accordance with the Maxwell equations, and coupled through the Bragg's condition – yet, nowhere in the semi-classical approach the formation of the standing waves can be correlated with the physical $\mu t \gg 1$ condition, when the anomalous effects appear.

From now on the necessity of the quantum approach of the dynamical effects in the crystal fields' propagation is required, within the approximation of the two transmitted-diffracted waves, through which, the Borrmann effect should appear naturally integrated and correlated with the anomalous energetic propagation for the DS.

5.5.2 QUANTUM MODELING OF DYNAMICAL BORMANN EFFECT

5.5.2.1 Formulation of Quantum State Vectors

In terms of phase's continuity at propagation, both between the DS branches but also between the transmitted-diffracted directions, the quantum formalism of the Borrmann effect will be developed for a deeper understanding of the phase transfer, as coming from the photons' transfer phenomenology; the present discussion follows (Biagini, 1990; Birău & Putz, 2000).

Accordingly, the following forms of the state vectors on the two diffraction branches is considered, each having as components the photons number on the transmitted-diffracted directions, written upon the expression proposed by Loudon:

$$|\alpha\rangle \equiv \frac{1}{\sqrt{2l+1}} \sum_{s=-l}^{+l} |n-s, n+s\rangle \qquad (5.282)$$

$$|\beta\rangle \equiv \frac{1}{\sqrt{2l+1}} \sum_{s=-l}^{+l} (-1)^{s+1} |n-s, n+s\rangle \qquad (5.283)$$

where the first component of the state vectors is the number of photons propagated in the transmitted direction with the wave vectors $\vec{K}_0$ and the polarization vectors $\vec{n}_0$, while the second component of the state vectors is the number of the diffracted photons (toward the primary direction, a fact reflected also by the way of writing the sum and the difference of the number of photons, respecting the number of photons for the transmitted direction) characterized by the wave vectors $\vec{K}_h = \vec{K}_0 + \vec{h}$ and the polarization vectors $\vec{n}_h$.

Moreover, also the sign between the state vectors associated to the two diffraction branches is remarked, a sign correlated with the phases' difference between them that essentially depend by the number of photons changed between the transmitted-diffracted directions on the branch in cause.

The number l is an integer that fixes the maximum number of photons that can be changed between the directions of diffraction associated to a branch of DS; *n is an integer number representing the initial number of photons on each of the diffraction directions of DS – and which is modified during the propagation through the exchange of photons that "jump" from a direction to another (for the moment on the same branch), due to the scattering on the atomic planes, defining through this "jump" the essence itself of the intimate quantum description of the diffraction.*

Commonly, n is chosen higher than l but they can be also equalized, which would correspond to an "*emptying of direction*" on the diffraction during the propagation, especially if the continuity for the phases between the coupled directions means coherence in fact. This equality will be used also along the asymptotic limit, just as a measure of the coherence in the phases propagation on the two directions, coherence that will be further analytically argued.

Thus, the specifically introduction to the quantum formalism is imposed, i.e., the description of the phases through the associated operators, by considering the photonic "creation" or "annihilation" definition (with the necessity of the inclusion of a photons number operator on one diffraction direction).

To this aim the definitions of the phase operators associated to the transmitted direction will be assumed:

$$\exp\left[i\hat{\phi}_0\right] = \frac{1}{\sqrt{\hat{n}_0 + 1}}\,\hat{a}_0 \tag{5.284}$$

$$\Rightarrow \exp\left[-i\hat{\phi}_0\right] = \hat{a}_0^+\,\frac{1}{\sqrt{\hat{n}_0 + 1}} \tag{5.285}$$

$$\hat{n}_0 = \hat{a}_0\hat{a}_0^+ - 1 = \hat{a}_0^+\hat{a}_0 \tag{5.286}$$

and similarly for diffracted direction:

$$\hat{n}_h = \hat{a}_h\hat{a}_h^+ - 1 = \hat{a}_h\hat{a}_h^+ \tag{5.287}$$

where the creation-annihilation operators of the Dirac quantum formalism have been obtained, with the following properties:

$$\hat{a}_0\left|n-s, n+s\right\rangle = \sqrt{n-s}\left|n-s-1, n+s\right\rangle \tag{5.288}$$

$$\hat{a}_0^+\left|n-s, n+s\right\rangle = \sqrt{n-s+1}\left|n-s+1, n+s\right\rangle \tag{5.289}$$

$$\hat{a}_h\left|n-s, n+s\right\rangle = \sqrt{n+s}\left|n-s, n+s-1\right\rangle \tag{5.290}$$

$$\hat{a}_h^+\left|n-s, n+s\right\rangle = \sqrt{n+s+1}\left|n-s, n+s+1\right\rangle \tag{5.291}$$

$$\left\langle m\middle|m'\right\rangle = \delta_{m..} \tag{5.292}$$

$$\sum_m \left|m\right\rangle\left\langle m\right| = \hat{1} \tag{5.293}$$

$$\hat{a}_0^+\hat{a}_0\left|m, m'\right\rangle = m\left|m, m'\right\rangle \tag{5.294}$$

$$\hat{a}_h^+\hat{a}_h\left|m, m'\right\rangle = m'\left|m, m'\right\rangle \tag{5.295}$$

$$\left[\hat{a}_0\hat{a}_0^+\right] = \hat{1}, \ \left[\hat{a}_h\hat{a}_h^+\right] = \hat{1} \tag{5.296}$$

With these expressions, the calculation of the measures of interest can be unfolded, towards the proofing of the phases' coherence between the dynamical propagation directions, and, leading with the highlighting of the Borrmann effect as a consequence of this fact.

5.5.2.2 Quantum Dynamic Localization

One will proceed through employing the limiting equality:

$$\lim_{l \to \infty} f(n,l,s) = \lim_{l \to n \to \infty} f(n,l,s) \tag{5.297}$$

towards calculations for the operators' dispersions which quantify the photons number on each direction, while neglecting the exchange photons number (s) in the linear combinations respecting the initial number (n) on each direction,

$$\langle\alpha|\hat{n}_0|\alpha\rangle = \langle\alpha|\hat{a}_0^+ \hat{a}_0|\alpha\rangle \overset{\substack{(5.282),\\(5.294)}}{=} \frac{1}{(2l+1)}\sum_{s=-l}^{+l}(n-s) \overset{n \gg s}{=} n \tag{5.298}$$

$$\langle\alpha|\hat{n}_0^{\,2}|\alpha\rangle = \langle\alpha|\hat{a}_0^+ \hat{a}_0 \hat{a}_0^+ \hat{a}_0|\alpha\rangle \overset{\substack{(5.282),\\(5.294)}}{=} \frac{1}{(2l+1)}\sum_{s=-l}^{+l}(n-s)^2$$

$$\overset{(5.297)}{=} n^2 + \frac{2}{2l+1}\sum_{s=1}^{+l}s^2 \overset{\substack{2l+1 \geq 2l\\ l+1 \geq l}}{=} n^2 + \frac{l^2}{3} \tag{5.299}$$

and similarly for the diffracted direction associated to the alpha and beta branches of the DS; the present discussion follows (Biagini, 1990; Birău & Putz, 2000).

The photons number dispersion is calculated on a propagation direction:

$$\Delta\hat{n}_0 = \sqrt{\langle\alpha|\hat{n}_0^{\,2}|\alpha\rangle - (\langle\alpha|\hat{n}_0|\alpha\rangle)^2} = \frac{l}{\sqrt{3}} \tag{5.300}$$

which confirms the fact that there is no such a determinate number of photons on a direction during the propagation.

Analytically, the action of the photons number operator on the state vector associated to the alpha branch can also be obtained from (5.298):

$$\hat{n}_0|\alpha\rangle = n|\alpha\rangle \tag{5.301}$$

$$\hat{n}_h |\alpha\rangle = n|\alpha\rangle \tag{5.302}$$

which will be used in calculation for the phases dispersions and for the degree of localization for both a singular direction as also for the two correlated directions.

For a single direction there will be successively written:

$$\langle\alpha|\cos\hat{\phi}_0|\alpha\rangle = \langle\alpha|\frac{1}{2}[\exp(i\hat{\phi}_0) + \exp(-i\hat{\phi}_0)]|\alpha\rangle$$

$$= \frac{1}{2}\frac{\langle\alpha|\hat{a}_0|\alpha\rangle}{\sqrt{\hat{n}_0+1}} + \frac{1}{2}\frac{\langle\alpha|\hat{a}_0^+|\alpha\rangle}{\sqrt{\hat{n}_0+1}} \overset{\substack{(5.297)\\(5.298)\\(5.301)}}{=} 0 \tag{5.303}$$

$$\langle\alpha|\sin\hat{\phi}_0|\alpha\rangle = 0 \tag{5.304}$$

$$\langle\alpha|\cos^2\hat{\phi}_0|\alpha\rangle = \langle\alpha|\frac{1}{4}[\exp(i\hat{\phi}_0) + \exp(-i\hat{\phi}_0)][\exp(i\hat{\phi}_0) + \exp(-i\hat{\phi}_0)]|\alpha\rangle$$

$$= \frac{1}{4}\langle\alpha|[\frac{\hat{a}_0}{\sqrt{\hat{n}_0+1}} + \frac{\hat{a}_0^+}{\sqrt{\hat{n}_0+1}}][\frac{\hat{a}_0}{\sqrt{\hat{n}_0+1}} + \frac{\hat{a}_0^+}{\sqrt{\hat{n}_0+1}}]|\alpha\rangle$$

$$\overset{(5.282)}{=} \frac{1}{4}\langle\alpha|[\frac{\hat{a}_0\hat{a}_0^+}{\hat{n}_0+1} + \frac{\hat{a}_0^+\hat{a}_0}{\hat{n}_0+1}]|\alpha\rangle \overset{(5.294)}{=} \frac{1}{2}\sum_{s=1}^{+l}\frac{(n-s+1)+(n-s)}{(n+1)(2l+1)}$$

$$\overset{n>>s}{=} \frac{1}{2}\sum_{s=1}^{+l}\frac{2n}{(n+1)(2l+1)} \overset{\substack{2l+1\cong 2l\\n+1\cong n}}{=} \frac{nl}{n(2l)} = \frac{1}{2} \tag{5.305}$$

wherefrom automatically results also the average measure

$$\langle\alpha|\sin^2\hat{\phi}_0|\alpha\rangle = \frac{1}{2} \tag{5.306}$$

These relations, when coupled with Eqs. (5.303) and (5.304), will give the dispersions of the phases on the individual directions of the alpha and beta branches, respectively:

$$\Delta \cos \hat{\phi}_0 = \frac{1}{\sqrt{2}}, \ \Delta \sin \hat{\phi}_0 = \frac{1}{\sqrt{2}},$$

$$\Delta \cos \hat{\phi}_h = \frac{1}{\sqrt{2}}, \ \Delta \sin \hat{\phi}_h = \frac{1}{\sqrt{2}} \qquad (5.307)$$

which show that, when separately taken, the transmitted and diffracted directions present a dispersion in the photons phases localization in the propagation associated to the branches of DS.

This means that each individual branch does not produce stationary detectable effects.

The next step is the calculation of the phases' dispersions for the transmitted waves when quantum coupled (through the photonic transfer) with those diffracted, as follows (Biagini, 1990; Birău & Putz, 2000):

$$\langle \alpha | \cos(\hat{\phi}_0 - \hat{\phi}_h) | \alpha \rangle = \langle \alpha | \frac{1}{2} \{ \exp[i(\hat{\phi}_0 - \hat{\phi}_h)] + \exp[-i(\hat{\phi}_0 - \hat{\phi}_h)] \} | \alpha \rangle$$

$$= \frac{1}{2} \langle \alpha | \{ \exp[i\hat{\phi}_0] \exp[-i\hat{\phi}_h] + \exp[-i\hat{\phi}_0] \exp[i\hat{\phi}_h] \} | \alpha \rangle$$

$$= \frac{1}{2} \langle \alpha | \{ \frac{\hat{a}_0 \hat{a}_h^+}{\sqrt{(\hat{n}_0+1)(\hat{n}_h+1)}} + \frac{\hat{a}_0^+ \hat{a}_h}{\sqrt{(\hat{n}_0+1)(\hat{n}_h+1)}} \} | \alpha \rangle$$

$$= \frac{1}{(2l+1)(n+1)} \frac{1}{2} \sum_{s=-l}^{+l} \left[\begin{array}{l} \sqrt{(n-s)(n+s+1)} \langle n-s, n+s | n-s-1, n+s+1 \rangle \\ + \sqrt{(n-s+1)(n+s)} \langle n-s, n+s | n-s+1, n+s-1 \rangle \end{array} \right]$$

QUANTUM
TRANSFER

$$n+s+1 \rightarrow n+s$$
$$n-s-1 \rightarrow n-s$$
$$\underline{n-s+1 \rightarrow n-s} \qquad = \frac{1}{(2l+1)(n+1)} \sum_{s=1}^{l} [\sqrt{(n+s)(n-s+1)} + \sqrt{(n-s)(n+s+1)}] =$$
$$\underline{n+s-1 \rightarrow n+s}$$

$$\underline{\underline{n \gg 1}} = \frac{2}{(2l+1)(n+1)} \sum_{s=1}^{l} \sqrt{(n-s)(n+s)}$$

$$\underline{\underline{n \gg s}} = \frac{2l}{2l+1}$$

$$(5.308)$$

and yielding the observable average:

$$\langle\alpha|\sin(\hat{\phi}_0-\hat{\phi}_h)|\alpha\rangle = 0 \qquad (5.309)$$

Regarding the squared average in quantum dispersion one also successively obtains:

$$\langle\alpha|\cos^2(\hat{\phi}_0-\hat{\phi}_h)|\alpha\rangle = \frac{1}{4}\langle\alpha|\left\{\frac{\hat{a}_0\hat{a}_h^+}{\sqrt{(\hat{n}_0+1)(\hat{n}_h+1)}}+\frac{\hat{a}_0^+\hat{a}_h}{\sqrt{(\hat{n}_0+1)(\hat{n}_h+1)}}\right\}$$

$$\times\left\{\frac{\hat{a}_0\hat{a}_h^+}{\sqrt{(\hat{n}_0+1)(\hat{n}_h+1)}}+\frac{\hat{a}_0^+\hat{a}_h}{\sqrt{(\hat{n}_0+1)(\hat{n}_h+1)}}\right\}|\alpha\rangle$$

$$\overset{(5.288)-(5.292)}{=}\frac{1}{4}\langle\alpha|\frac{\hat{a}_0\hat{a}_h^+\hat{a}_0\hat{a}_h^++\hat{a}_0^+\hat{a}_h\hat{a}_0\hat{a}_h^+}{(\hat{n}_0+1)(\hat{n}_h+1)}|\alpha\rangle$$

$$=\frac{1}{4(2l+1)}2\sum_{s=1}^{+l}\frac{(n-s+1)(n+s)+(n-s)(n+s+1)}{(n+1)^2(2l+1)}$$

$$\begin{array}{c}\text{QUANTUM}\\\text{TRANSFER}\\n-s\to n\\n+s\to 2n\\\to\\n\pm l\approx n\end{array}\qquad \frac{1}{2(2l+1)}\sum_{s=1}^{+l}\frac{4n^2}{(n+1)^2}\overset{(5.297)}{=}\frac{2l}{2l+1}\overset{(5.297)}{\to}1 \quad (5.310)$$

wherefrom immediately also results:

$$\langle\alpha|\sin^2(\hat{\phi}_0-\hat{\phi}_h)|\alpha\rangle = \frac{1}{2l+1}\overset{(5.297)}{\to}0 \qquad (5.311)$$

These relations, which give the dispersions between the propagated directions associated to the alpha branch are exactly valid also for the beta branch of DS, namely:

$$\Delta\cos(\hat{\phi}_0-\hat{\phi}_h) = \frac{\sqrt{2l}}{2l+1}\overset{(5.297)}{\to}0 \qquad (5.312)$$

$$\Delta \sin(\hat{\phi}_0 - \hat{\phi}_h) = \frac{1}{\sqrt{2l+1}} \overset{(5.297)}{\to} 0 \qquad (5.313)$$

which show the *localization of the photonic phases* between the directions of propagation associated to the branches of DS (here from *Dynamical Localization*): In this case a quantum transfer between them necessary appears. This is a fundamental observation because it gives the key itself for the understanding the Borrmann effect in terms of quantum causes; as such, the quantum Bormann effect, being constructed on the modeling of the diffraction as a effect of photonic exchange between DS directions, it will be able to explain even the quantitative picture, which eventually will lift any mystery on the anomalous absorption – now understanding it, not necessary through an excessive absorption (which would "warm" the crystal) but through a jump of the number of photons from a branch to another, being this quantum jump caused by the process of scattering itself on the atomic planes. The quantum jump is not random, but appears "immediately after" the diffraction (the scattering on the atomic planes), certifying this way quantum effect of the diffraction.

More on the effect of this quantum jump: it provides "in operatorial phase" the propagations on the two directions obtained after the "impact" with the diffraction planes, after which they remained with a *correlation* (through the Bragg's law) between the individual phases, which will be enforced by the *Dynamical Localization*.

In fact, the quantum jump is a conceptual alternative for the classical absorption; one can equivalently say that two fields associated to the DS correlated branches suffer asymmetric absorption or, equivalently, that the photons that characterize them "jump" from a field to another with the same asymmetrical effect. The last alternative is phenomenologically more appealing, because unequivocally includes the keeping of the energy conservation principle, here quantified through the *number* of photons. In the first version, there can *not* be clearly said what happens with the field abnormally largely absorbed, because the finality is also a re-emission, which it should also be anomalously – and, then would equalize in intensity the field already anomalously transmitted, finally obtaining a doubling of the anomalous transmission, i.e., an amplification of the incident beam! But a common crystal does not show by itself the Laser effect!

Therefore, the classical treatment of the Borrmann effect has the quality of the immediate interpretation of the recorded fluorescence spectra (see Putz, 2014) but does not fully respond to what is happening with this energy, intimate-dynamically manifested in the anomalous absorption. Here's why, the quantum view can causally present the diffraction phenomenology in dynamic evolution, where the asymmetrical propagation naturally derives from the quantum definition itself of diffraction: *the coherent (dynamic) photonic transfer derived from the dynamic localization*, (5.312) and (5.313).

5.5.2.3   The Quantum Diffracted Energies

One may proceed with the effective calculation of the energies asymmetrically distributed between the diffraction branches, which results from the SW (standing waves) composition on each branch by the coherent directions of transmission-diffraction; the present discussion follows (Biagini, 1990; Birău & Putz, 2000).

The general form of the fields on the transmitted-diffracted directions is considered, in the operatorial writing (Biagini, 1990; Birău & Putz, 2000):

$$\hat{\vec{E}}_{0,h} = \sum_{\vec{K}_{0,h},n_{0,h}} i\left(\frac{2\pi\hbar\omega}{V}\right)^{1/2} \vec{n}_{0,h} \left\{ \begin{array}{l} \hat{a}_{0,h}\exp[i(2\pi\vec{K}_{0,h}\,\vec{r}-\omega t)] \\ +\hat{a}_{0,h}^{+}\exp[-i(2\pi\vec{K}_{0,h}\,\vec{r}-\omega t)] \end{array} \right\} \qquad (5.314)$$

immediately resulting in:

$$\langle\alpha|\hat{\vec{E}}_{0,h}|\alpha\rangle = 0, \langle\beta|\hat{\vec{E}}_{0,h}|\beta\rangle = 0 \qquad (5.315)$$

which is in accordance with the phases' indeterminacy on the singular directions, but which for the square average give the SW terms, in agreement with the coherence of the phases of the involved directions, as shown in the following.

From the Eq. (5.314) expression, only the approximation of the two coupled waves in propagation is retained, so that the sums upon the wave vectors will be reduced to a single term with the corresponding wave vector (Birău & Putz, 2000):

$$\langle \alpha | \hat{E}_0^{\ 2} | \alpha \rangle = \frac{2\pi \hbar \omega}{V} \langle \alpha | \left\{ \hat{a}_0^{\ +} \exp[-i(2\pi \vec{K}_0 \cdot \vec{r} - \omega t)] - \hat{a}_0 \exp[i(2\pi \vec{K}_0 \cdot \vec{r} - i\omega t)] \right\}$$

$$\times \left\{ \hat{a}_0 \exp[i(2\pi \vec{K}_0 \cdot \vec{r} - i\omega t)] - \hat{a}_0^{\ +} \exp[-i(2\pi \vec{K}_0 \cdot \vec{r} - i\omega t)] \right\} | \alpha \rangle$$

$$= \frac{2\pi \hbar \omega}{V} \langle \alpha | \left[\hat{a}_0^{\ +} \hat{a}_0 + \hat{a}_0 \hat{a}_0^{\ +} \right] | \alpha \rangle$$

$$= \frac{2\pi \hbar \omega}{V} 2 \sum_{s=1}^{+l} \frac{(n-s) + (n-s+1)}{2l+1} \overset{n \gg s}{=} \frac{2\pi \hbar \omega}{V} [4n \frac{l}{2l+1} + \frac{2l}{2l+1}]$$

$$\overset{(5.297)}{=} \frac{2\pi \hbar \omega}{V} (2n+1)$$

$$(5.316)$$

and similarly for the diffracted direction; while for the combined term one will have:

$$\langle \alpha | \hat{\vec{E}}_0 \hat{\vec{E}}_h | \alpha \rangle = \left(\vec{n}_0 \cdot \vec{n}_h \right) \frac{2\pi \hbar \omega}{V} \langle \alpha | \left\{ \hat{a}_0^{\ +} \exp[-i(2\pi \vec{K}_0 \cdot \vec{r} - i\omega t)] - \hat{a}_0 \exp[i(2\pi \vec{K}_0 \cdot \vec{r} - i\omega t)] \right\}$$

$$\times \left\{ \hat{a}_h \exp[i(2\pi \vec{K}_h \cdot \vec{r} - i\omega t)] - \hat{a}_h^{\ +} \exp[-i(2\pi \vec{K}_h \cdot \vec{r} - i\omega t)] \right\} | \alpha \rangle$$

$$= \left(\vec{n}_0 \cdot \vec{n}_h \right) \frac{2\pi \hbar \omega}{V} \langle \alpha | \left\{ \hat{a}_0^{\ +} \hat{a}_h \exp[i2\pi(\vec{K}_h - \vec{K}_0) \cdot \vec{r}] + \hat{a}_0 \hat{a}_h^{\ +} \exp[-i2\pi(\vec{K}_h - \vec{K}_0) \cdot \vec{r}] \right\} | \alpha \rangle$$

$$= \left(\vec{n}_0 \cdot \vec{n}_h \right) \frac{2\pi \hbar \omega}{V} \sum_{s=-l}^{+l} \left[\begin{array}{l} \frac{\sqrt{(n+s)(n-s+1)} \langle n-s, n+s | n-s+1, n+s-1 \rangle}{2l+1} \exp[i2\pi(\vec{K}_h - \vec{K}_0) \cdot \vec{r}] \\ + \frac{\sqrt{(n-s)(n+s+1)} \langle n-s, n+s | n-s-1, n+s+1 \rangle}{2l+1} \exp[-i2\pi(\vec{K}_h - \vec{K}_0) \cdot \vec{r}] \end{array} \right]$$

$$\overset{\substack{+1 \to -1 \\ -1 \to +1 \\ \text{COHERENT} \\ \text{QUANTUM} \\ \text{TRANSFER}}}{=} \left(\vec{n}_0 \cdot \vec{n}_h \right) \frac{2\pi \hbar \omega}{V} \sum_{s=-l}^{+l} \frac{\sqrt{(n+s)(n-s+1)}}{2l+1} \left\{ \exp[i2\pi(\vec{K}_h - \vec{K}_0) \cdot \vec{r}] + \exp[-i2\pi(\vec{K}_h - \vec{K}_0) \cdot \vec{r}] \right\}$$

$$(5.317)$$

Gathering the relation of Eq. (5.316) type on the two directions with the terms such as, Eq. (5.317), the total energy of the standing waves propagated by quantum jump associated to the alpha branch will have the final form (Biagini, 1990; Birău & Putz, 2000):

$$\langle \alpha | (\hat{\vec{E}_0} + \hat{\vec{E}_h})^2 | \alpha \rangle$$

$$= \frac{2\pi\hbar\omega}{V} \left[4\left(n + \frac{1}{2}\right) + 2(\vec{n}_0 \cdot \vec{n}_h) \sum_{s=-l}^{+l} \frac{\sqrt{(n+s)(n-s+1)}}{2l+1} \left\{ \begin{array}{l} \exp[\ i2\pi(\vec{K}_h - \vec{K}_0) \cdot \vec{r}] \\ + \exp[\ -i2\pi(\vec{K}_h - \vec{K}_{01}) \cdot \vec{r}] \end{array} \right\} \right]$$

$$\overset{\substack{2l+1 \approx 2l \\ n+\frac{1}{2} \approx n \\ n \gg s}}{=} \frac{8\pi\hbar\omega n}{V} \left\{ 1 + \vec{n}_0 \cdot \vec{n}_h \cos[\ 2\pi(\vec{K}_h - \vec{K}_0) \cdot \vec{r}] \right\}$$

$$(5.318)$$

A similar procedure for the beta branch will give:

$$\langle \beta | (\hat{\vec{E}_0} + \hat{\vec{E}_h})^2 | \beta \rangle = \frac{8\pi\hbar\omega n}{V} \left\{ 1 - \vec{n}_0 \cdot \vec{n}_h \cos[2\pi(\vec{K}_h - \vec{K}_0) \cdot \vec{r}] \right\} \qquad (5.319)$$

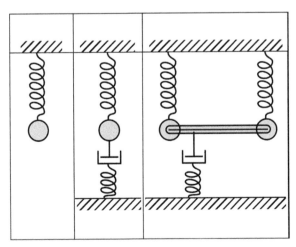

FIGURE 5.39 The Borrmann mechanical oscillatory analogy with dynamical diffraction on dispersion surface branches; after Biagini (1990).

which brings the quantum approach, in position to retrieve the anomalous propagation, with a photonic-plus on the alpha branch and a photonic-minus on the beta branch, in terms of quantum jump, so assimilated as effects of the absorption in the scattering on the diffraction planes.

In fact, the equivalence of the anomalous absorption with the movement of the symmetry of the photonic number propagated on the direction – on branches, from a branch to another is as such confirmed, as can be very well intuited even by an analogy with a system mechanical modeled (Figure 5.39), where: the position (a) corresponds to the free beam, the position (b), to the beam propagated in the crystal in a single way, but forced to change its frequency, and (c) being in fact the Bragg's coupling, which due to the scattering, the diffraction itself, moves the coupling symmetry (propagation) to one of the branches, thus corresponding with the quantum jump (Birău & Putz, 2000).

There is important to specify that the relations (5.318) and (5.319) can also describe the *inverse anomaly*, analytically discussed elsewhere in classical manner (Putz, 2014), in the situation when the centers of coupling with the transmitted and diffracted directions, O and H, are modified so that a π dephased is induced between the vectors of the $\vec{n}_0, \vec{n}_h$ polarizations, through inversing the sign of the $\vec{n}_0 \cdot \vec{n}_h$ products.

Thus, by the present quantum approach a completely generalized diffraction panel is obtained, seen as quantum self-consistency, through the reciprocal (dynamic) photonic transfer between the diffraction directions, i.e., between the allowed branches by the effect of the *dynamic localization* derived in the previous section.

5.5.2.4 The Quantum Dynamic Jump

The energetic calculations associated with the alpha and respectively beta branches propagation, calculated as quantum averages (5.318) and (5.319), allow now the evaluation of the experimentally measurements by calling their classical limit, through which the energetic average for each branch will be given by the average of the energy of the fields in an unitary volume of a dispersive medium; the present discussion follows (Biagini, 1990; Birău & Putz, 2000):

$$\overline{E_\alpha} = \frac{1}{8\pi} \left\{ \frac{d(\omega\varepsilon)}{d\omega} \langle\alpha|\overset{\wedge}{\vec{E}}|\alpha\rangle^2 + \frac{d(\omega\mu)}{d\omega} \langle\alpha|\overset{\wedge}{\vec{H}}|\alpha\rangle^2 \right\}$$

$$\overset{X-RAY}{=} \frac{1}{8\pi} \frac{d(\omega\kappa\varepsilon_0)}{d\omega} \langle\alpha|\overset{\wedge}{\vec{E}}|\alpha\rangle^2 \tag{5.320}$$

In the last equation, by taking into account for the expression of the dielectric constant for the X-ray, elsewhere deduced, (Putz, 2014), and also by the expression (5.318) for the energy distribution, through a quantum jump for the propagation coupled with the alpha branch, we can obtain for the total energy of the propagation of the fields associated to the alpha branch of the DS in the dispersive medium the formulation (Biagini, 1990; Birău & Putz, 2000):

$$\overline{E_\alpha} := E_\alpha = \frac{1}{8\pi} \left\{ \frac{d}{d\omega} \left[\varepsilon_0\omega - \frac{e^2\rho(\vec{r})}{m\omega} \right] \right\} \frac{8\pi nh\omega}{V} \left[1 + \overset{\rightarrow}{n_0}\cdot\overset{\rightarrow}{n_h}\cos(2\pi\vec{h}\cdot\vec{r}) \right]$$

$$= \left[\varepsilon_0 + \frac{e^2\rho(\vec{r})}{m\omega^2} \right] \frac{nh\omega}{V} \left[1 + \overset{\rightarrow}{n_0}\cdot\overset{\rightarrow}{n_h}\cos(2\pi\vec{h}\cdot\vec{r}) \right]$$

$$= \left(\varepsilon_0 \frac{nh\omega}{V} \right)_{\substack{IN\ VACUUM \\ \vec{r}=0}} + \frac{e^2nhNZ}{m\omega} + \frac{e^2nhN}{m\omega}\left(\overset{\rightarrow}{n_0}\cdot\overset{\rightarrow}{n_h} \right)\sum_i \int |\Psi_0|^2 \cos(2\pi\vec{h}\cdot\overset{\rightarrow}{r_i})dv$$

$$= E_0 + E_1 - E_2 \tag{5.321}$$

Similarly, the average energy allocated in diffraction through the quantum dynamic transfer is obtained for the propagated fields associated with the beta branch of DS in the dispersive medium (Biagini, 1990; Birău & Putz, 2000):

$$\overline{E_\beta} := E_\beta = E_0 + E_1 + E_2 \tag{5.322}$$

where, the first two terms are representing the energy "inherited" from the propagation in vacuum, and the uniform energy of propagation in the crystal with an uniform distribution of Z electrons in an unitary cell with a repartition of N unitary cells on volume unit, respectively; while the tertiary term emphasizes on the movement of the energetic symmetry based on the distribution of the unielectronic function Ψ_0 in the unit cell containing "the event" of the standing waves for the transmitted and diffracted directions on the DS branches, i.e., containing "the vent" of the photonic jump or the effect of the diffraction itself and the cause of the energetic distribution asymmetry in the further propagation.

However, by the relations (5.321) and (5.322) the hidden effect of the anomalous absorption was emphasized in the third term by the different signs attached, meaning a lack of the absorption associated to the alpha branch – and, in fact, a plus of free photons transferred, while recording for the beta branch a plus of absorption and a minus of free photons transferred, both effects happened by the *dynamic localization* of the fields between the propagation directions, i.e., the associated branches of DS.

Further on, this transfer between the branches will be emphasized by the mediation of an intermediate absorption followed by an intermediate emission, this way quantum describing the photonic transfer itself between the propagations associated to the branches DS, i.e., the Borrmann quantum jump.

Consider a static distribution of the photons of the fields associated to the two branches of the DS, at equilibrium, so being described by the consecrated expressions (Biagini, 1990; Birău & Putz, 2000):

$$n_\alpha = \frac{1}{\exp\left(\dfrac{E_0 + E_1}{kT}\right) - 1} \tag{5.323}$$

$$n_\beta = \frac{1}{\exp\left(\dfrac{E_0 + E_1}{kT}\right) - 1} \tag{5.324}$$

Then, one considers that the diffraction process takes place. This way, the *dynamic localization* (see Eqs. (5.312) and (5.313)) of the phases between the directions of propagation and consequently between the fields associated to the DS branches through a direct photonic transfer "arises"– and the relations above become asymmetric:

$$n_\alpha = \frac{1}{\exp\left(\dfrac{E_\alpha}{kT}\right) - 1} \tag{5.325}$$

$$n_\beta = \frac{1}{\exp\left(\dfrac{E_\beta}{kT}\right) - 1} \tag{5.326}$$

case in which the scattering amplitude writes as:

$$P_{\alpha \to \beta} = K_{\alpha \to \beta} n_\alpha (n_\beta + 1) \tag{5.327}$$

$$P_{\beta \to \alpha} = K_{\beta \to \alpha} n_\beta (n_\alpha + 1) \tag{5.328}$$

wherefrom, by the coherence condition for the photonic transfer between the fields associated to the DS branches, the transfer rate will turn out with the expression:

$$P_{\alpha \to \beta} = P_{\beta \to \alpha} \Rightarrow \frac{K_{\alpha \to \beta}}{K_{\beta \to \alpha}} = \exp\left(-\frac{2E_2}{kT}\right) \tag{5.329}$$

From the last relationship the energetic symmetry displacement is noted, to the alpha branch, as an exponential function of the term which represents the effect of dynamic localization manifested by the standing waves; this way, the quantum dynamic localization of the phases between the directions (on DS branches) is realized such that the photons pass from the field associated with the beta branch of the DS to the one associated with the alpha branch of DS, in order to keep the statistic equilibrium.

The same ratio satisfies also the photonic jump process mediated by a photonic absorption between the DS branches in the crystal, followed by a photonic emission with the transfer on the other branch, namely:

$$P_{\alpha \to \beta} = K n_\alpha (n_\beta + 1) n_\nu \tag{5.330}$$

$$P_{\beta \to \alpha} = K n_\beta (n_\alpha + 1)(n_\nu + 1) \tag{5.331}$$

which, in the same conditions of dynamic localization and taking into account of the Eqs. (5.327) and (5.328) one will obtain for the transfer constants the relations:

$$K_{\alpha \to \beta} = \frac{K}{\exp\left(\dfrac{2E_2}{kT}\right) - 1} \tag{5.332}$$

$$K_{\beta \to \alpha} = \frac{K}{1 - \exp\left(-\dfrac{2E_2}{kT}\right)} \tag{5.333}$$

which satisfy the ratio (5.329). Therefore, there is proved that by the consideration of the absorption in a quantum view, the explicit view of the anomalous absorption effect consists in quantum jumps, i.e., the intermediation of the photonic transfer (exchanges) allowed by the dynamic localization.

Here's how in this approach, the absorption is naturally integrated in the quantum transfer phenomenology, so that what was abnormal in the semi-quantum approach becomes dynamic localization, with the effect of the quantum jump caused by the scattering itself on the diffraction planes (by the "rise" of the propagation directions associated to the DS branches along the dispersion relations between them).

As long as the quantum transfer takes place reciprocally (dynamic, self-consistently), i.e., between the two fields associated to the two branches of DS, the Borrmann effect appears only as an effect of the photonic statistic asymmetry caused by diffraction, while canceling the phasic dispersion effect.

The phases' coherence which accompanies any wave scattering, doubled by the correlation of the dynamic localization, leads in corpuscular understanding of the photonic asymmetric transfer. Therefore, the *dynamic asymmetry* corresponds to the dynamic anomaly in the absorption language, having its bases in the dynamic localization.

The fact that the Borrmann's effect is in fact an effect that appears only in the condition of the thick crystal, $\mu t \gg 1$, calculating the photonic flows associated to the alpha and beta branches of DS can be highlighted, allowing the coherent (dynamic) reciprocal exchange between them:

$$\frac{df_\alpha}{dt} = N'(P_{\beta \to \alpha} - P_{\alpha \to \beta}) \qquad (5.334)$$

where t is the traversed distance in the crystal, and N' is the number of atoms in the unit volume, while

$$f_\alpha = \frac{n_\alpha c}{V} \qquad (5.335)$$

is the photonic flow associated to the alpha branch of DS.

Finally, through imposing the constancy of the total photonic flow,

$$f = f_\alpha + f_\beta \qquad (5.336)$$

the next approximations are allowed

$$n_\alpha + 1 \cong n_\alpha, n_\beta + 1 \cong n_\beta \tag{5.337}$$

leading to the rewriting of Eq. (5.334) by using of Eqs. (5.330) and (5.331), such as:

$$\frac{df_\alpha}{dt} = af_\alpha(f - f_\alpha), a = \frac{KV^2N'}{c^2} \tag{5.338}$$

whose solution leads to the photonic flows associated to the DS branches (Biagini, 1990; Birău & Putz, 2000):

$$f_\alpha(t) = \frac{f}{1 + \exp(-aft)} \rightarrow \begin{cases} \dfrac{f}{2}, at \rightarrow 0 \\ f, at \rightarrow \infty \end{cases} \tag{5.339}$$

$$f_\beta(t) = \frac{f}{1 + \exp(aft)} \rightarrow \begin{cases} \dfrac{f}{2}, at \rightarrow 0 \\ 0, at \rightarrow \infty \end{cases} \tag{5.340}$$

Therefore, these results are confirmed here, i.e. by the Borrmann effect based on the previous discussions, and, consecutively, the appearance of the dynamic standing waves, as a directly correlated effect with the thick crystal condition, here $at \gg 1$.

5.6 CONCLUSION

This chapter approached the X-ray diffraction on perfect crystals in terms of semi-classical dynamical theory (Maxwell equations + the shape of the wave Bloch functions associated to the fields propagated in the crystal), with the accent on the calculation of the intensities associated to the transmitted and diffracted waves in the two waves approximation: the diffraction associated to the production of a single diffracted wave respecting the transmitted one.

We did not insisted on the properties of dispersion equation, but on the shape and concrete conditions for which the propagation solutions can be univocally and explicitly written (phases and amplitudes). These, together

TABLE 5.4 The Synopsis Table of Total Self-Consistencies in Dynamical X-Ray Diffraction; after (Biagini, 1990; Birău & Putz, 2000).

$1{\rightarrow}(\alpha\backslash\beta){\rightarrow}1$	$\beta{\rightarrow}(1\backslash2){\rightarrow}\beta$	$2{\rightarrow}(\beta\backslash\alpha){\rightarrow}2$	$\alpha{\rightarrow}(2\backslash1){\rightarrow}\alpha$	$\langle = \rangle$	Self consistency effect
0	0	0	$\pi.$	π	$SW_{\alpha+}^{\beta-}$
0	π	0	π	2π	Quantum Jump
0	0	0	0	0	Pendellösung
0	π	0	0	π	$SW_{\alpha-}^{\beta+}$
π	0	π	π	3π	$SW_{\alpha+}^{\beta-}$
π	π	π	π	4π	Quantum Jump
π	0	π	0	2π	Pendellösung
π	π	π	0	3π	$SW_{\alpha-}^{\beta+}$

with the intensities, the reflection powers and the integrated reflection powers form the base for the analysis of the way in which the energy is dissipated, respectively propagated in the crystal, wherefrom the Ewald (Pendellösung) oscillation solution arises, as being the most natural way in which the waves that follow the diffraction are dynamically coupled in the crystal, wherefrom also the possibility of the multiple experimental characterization regarding the structure of analyzed crystals.

At the end of a long series of conclusions, pointed on the way, the analytical speech exposed can be resumed, in a systematic panel, Table 5.4 (Birău & Putz, 2000).

Worth specifying that after the presented detailed study centered on the matter of the propagation with the formation of Standing Waves –SW, we have reached the conclusion that the Borrmann effect is not a random phenomenon, but it is always the companion of the propagation in the crystal that "suffered" a dynamic scattering with X-ray on the thick crystal $\mu t \gg 1$.

The effect is experimentally highlighted by the recording of the fluorescence curves and can be both revealed, in a *direct way* ($SW_{\alpha-}^{\beta+}$), as in the most cases (with the abnormal absorption on the beta branch and the abnormal transmission on the alpha branch), but also under the form of the *inverse abnormal* $(SW_{\alpha+}^{\beta-})$ with the abnormal absorption on the alpha branch and the abnormal transmission on the beta branch in special conditions of dispersion, see (Putz, 2014).

Moreover, the quantum modeling of the Borrmann effect gives a general view of the self-consistency, seen as the mutual (dynamic) free photonic transfer under the conditions of the *dynamic localization.*

The generalized quantum transfer is a good win in the understanding of the X-ray propagation in crystal, being present both between the directions (in the table being simply called *quantum jump*), but also between the fields associated to the DS branches when become the full (direct and inverse) Borrmann effect, both being correlated by the asymmetric form of the photonic distribution.

The total self-consistency is exposed within a unified dynamic table, leaded by the Bragg's law along the phases' coherence, when the asymmetric quantum transfer is favored (with the "roots" in the *dynamic localization*).

Within the table have been noted also the cases that appear when the tops of the alpha and beta branches are unified in the Lorentz point of DS, for a complete generalization, since these cases correspond to the high energy X-ray diffraction (very short wavelengths), respectively reducing the interaction (e.g., for the neutrons instead of the X-ray).

Therefore, the premises of a generalized treatment of the diffraction with bosons (X-ray, photons), respectively with fermions (neutrons, electrons) have been clarified on quantum bases, while further theoretical and experimental quest on bosonic interaction (X-rays) on the bosonized electrons in chemical bonding, according with the theory exposed in Volume III of the present five-volume set (Putz, 2016b), is expected to be explored in the years to come.

KEYWORDS

- **Bormann phenomenology and triangle**
- **elastic deformation of solids**
- **electronic density**
- **electronic maps of contours**
- **Ewald sphere of diffraction**
- **fields and intensities sin crystals**
- **Laue and Bragg diffraction**
- **Pendellusung phenomenology**
- **plane waves approximation**
- **Poynting vectors**
- **quantum diffracted energies**
- **self-consistency in dynamic X-ray diffraction**
- **structure determination**
- **X-ray diffraction**
- **zero absorption**

REFERENCES

AUTHOR'S MAIN REFERENCES

Putz, M. V. (2016a). *Quantum Nanochemistry. A Fully Integrated Approach: Vol. I. Quantum Theory and Observability*. Apple Academic Press & CRC Press, Toronto-New Jersey, Canada-USA.

Putz, M. V. (2016b). *Quantum Nanochemistry. A Fully Integrated Approach: Vol. III. Quantum Molecules and Reactivity*. Apple Academic Press & CRC Press, Toronto-New Jersey, Canada-USA.

Putz, M. V. (2014). *Quantum and Optical Dynamics of Matter for Nanotechnology*. IGI Global, Hershey Passadena, USA (DOI: 10.4018/978–1–4666–4687–2).

Putz, M. V., Lacrămă, A. M. (2005). *Exploring the Complex Natural Systems* (in Romanian), Mirton Publishing House, Timişoara, Chapters 1–3.

Putz, M. V. (2003). Electronic Density from Structure Factor Determination in Small Deformed Crystals. *Int. J. Quantum Chem.* 94(4), 222–231 (DOI: 10.1002/qua.10475).

Birău, O., Putz, M. V. (2000). The "Standing Waves" Concept in Dynamical Theory of X-Ray Diffraction (in Romanian), Mirton Publishing House, Timişoara, Chapter 5.

SPECIFIC REFERENCES

Allen, F. H., Bellard, S., Brice, M. D., Cartwright, B. A., Doubleday, A., Higgs, H., Hummelink, T., Hummelink-Peters, B. G., Kennard, O., Motherwell, W. D. S., Rodgers, J. R., Watson, D. G. (1979). The Cambridge Crystallographic Data Centre: computer-based search, retrieval, analysis and display of information. *Acta Cryst.* B35, 2331–2339.

Authier, A., Simon, D. (1968). Application de La théorie dynamique de, S. Tagaki au contraste d'un défaut plan en topographie par rayons, X. I. Faute d'empilement. *Acta Cryst. A* 24, 517–526; Simon, D., Authier, A. ibidem, Application de La théorie dynamique de, S. Takagi au contraste d'un défaut plan en topographie par rayons, X. II. Franges de moiré, 527–534.

Azároff, L. V., Kaplow, R., Kato, N., Weiss, R. J., Wilson, A. J. C., Young, R. A., (Eds.) (1974), *X-Ray Diffraction*, McGraw-Hill, New York.

Batterman, B. W., Cole, H. (1964). Dynamical diffraction of X-rays by perfect crystals. *Rev. Mod. Phys.* 36, 681–717.

Bernstein, F. C., Koetzle, T. F., Williams, G. J. B., Meyer, E. F. Jr., Brice, M. D., Rogers, J. R., Kennard, O., Shimanouchi, T., Tasumi, M. (1977). The Protein Data Bank: a computer-based archival file for macromolecular structures. *J. Mol. Biol.* 112(3), 535–542.

Biagini, M. (1990). Quantum theory of the Borrmann effect. *Phys. Rev. A* 42, 3695–3702.

BioInformatics Protein Models (2013), https://bioweb.uwlax.edu/GenWeb/Molecular/Bioinformatics/Temp/Unit_4_temp/Proteins_3/proteins_3.htm

Blundell, T. L., Johnson, L. N. (1976). *Protein Crystallography*, Academic Press, London.

Borrmann, G. (1941). Über extinktionsdiagramme der röntgenstrahlen von quarz. *Phys. Z.* 42, 157–162.

Borrmann, G. (1950). Die absorption von röntgenstrahlen in fall der interferenz. *Z. Phys.* 127, 297–323.

Borrmann, G. (1954). Der kleinste absorbtion koeffizint interfierender röntgenstrahlung. *Z. Kristallogr.* 106, 109–121.

Borrmann, G. (1959). Röntgenwellenfelder. *Beitr. Phys. Chem. 20Jahrhunderts,* Braunschweig: Vieweg und Sohn, 262–282.

Borrmann, G. (1955). Vierfachbrechung der Rontgenstrahlen durch das ideale Kistallgitter. *Naturwissenschaften* 42, 67–68.

Branden, C. I., Jones, T. A. (1990). Between objectivity and subjectivity *Nature* 343, 687–689.

Brünger, A. T. (1992). Free R value: a novel statistical quantity for assessing the accuracy of crystal structures. *Nature* 355(6359), 472–475.

Cantor, C. R., Schimmel, P. R. (1980), *Biophysical Chemistry. Part II Techniques for the Study of Biological Structure and Function,* W. H. Freeman & Co., New York, Chapter 13, X-Ray Crystallography.

Cromer, D. T., Mann, J. B. (1968). X-ray scattering factors computed from numerical Hartree-Fock wave functions *Acta Cryst. A* 24, 321–324.

Dodson, E. J., Kleywegt, G. J., Wilson, K. S. (1996). Report of a workshop on the use of statistical validators in protein X-ray crystallography. *Acta Cryst.* D52, 228–234.

Drenth, J. (1994). *Principles of Protein X-ray Crystallography*, Springer-Verlag, New York.

Engh, R. A., Huber, R. (1991). Accurate bond and angle parameters for X-ray protein structure refinement. *Acta Cryst.* A47, 392–400.

Ewald, P. P. (1965). Crystal optics for visible light and X-rays. *Rev. Mod. Phys.* 37, 46–56.

Ewald, P. P. (1969). Introduction to the dynamical theory of X-ray diffraction *Acta Cryst. A* 25, 103–105.

Fanchon, E., Hendrickson, W. A. (1991). *Crystallographic Computing*, IUCr/Oxford University Press Vol. 5.

Fraser, R. D. B., Mac Rae, T. P. (1969). *X-ray Methods;* In: Leach, S. J. (Ed.) *Physical Principles and Techniques of Protein Chemistry* part A, Academic Press, New York.

Genick, U. K., Borgstahl, G. E. O., Ng, K., Ren, Z., Pradervand, C., Burke, P. M., Šrajer, V., Teng, T.-Y., Schildkamp, W., McRee, D. E., Moffat, K., Getzoff, E. D. (1997). Structure of a protein photocycle intermediate by millisecond time-resolved crystallography. Science. *Science* 275, 1471–1475.

Hanson, M. A. Stevens, R. C. (2000). Cocrystal structure of synaptobrevin-II bound to botulinum neurotoxin type B at 2.0 |[Aring]| resolution. *Nature Struct. Biol.* 7, 687–692.

IUPAC-IUB, Commission on Biochemical Nomenclature (1970). *J. Mol. Biol.*, 52, 1.

James, R. W. (1965), *The Optical Principles of The Diffraction of X-Rays*, Cornell University Press, Ithaca, New York.

Janáček, Z., Kuběna, J. (1978). Integrated intensity of the Laue diffraction from an elastically bent crystal. *Phys. Stat. Sol. (a)* 45(1), K55-K57.

Janáček, Z., Kuběna, J., Holý, V. (1978). X-ray Laue diffraction from a bent crystal integrated intensity. *Phys. Stat. Sol. (a)* 50(1), 285–291.

Kabsch, W., Sander, C. (1983). Dictionary of protein secondary structure: Pattern recognition of hydrogen-bonded and geometrical features. Biopolymers 22(12), 2577–2637.

Katagawa, T., Kato, N. (1974). The exact dynamical wave fields for a crystal with a constant strain gradient on the basis of the Takagi-Taupin equations. *Acta Cryst. A* 30, 830–836.

Kato, N. (1955). Integrated intensities of the diffracted and transmitted X-rays due to ideally perfect crystal (Laue Case). *J. Phys. Soc. Jap.* 10, 46–55.

Kato, N. (1969). The determination of structure factors by means of Pendell"osung fringes *Acta Cryst. A* 25, 119.

Kleywegt, G. J., Brünger, A. T. (1996). Checking your imagination: applications of the free R value. *Structure* 4(8), 897–904.

Kleywegt, G. J., Jones, T. A. (1995). Where freedom is given liberties are taken. *Structure* 3, 535–540.

Kleywegt, G. J., Jones, T. A. (2002). Homo crystallographicus—quo vadis? *Structure* 10(4), 465–472.

Laskowski, R. A., Mac Arthur, M. W., Moss, D. S., Thornton, J. M. (1993). PROCHECK: a program to check the stereochemical quality of protein structures. *J. Appl. Cryst.* 26, 283–291.

Laue, M. Von (1952). Die energiströmung bei röntgenstrahl-interferenzen in kristallen. *Acta Cryst.* 5, 619–625.

Laue, M.von (1953). Der teilchenstrom bei raumgitterinterferenzen von materi-ewellen. *Acta Cryst.* 6, 217–218.

Laue, M.von (1960), *Röntgenstrahlinterferenzen*, Akademische Verlagsgesellshaft Frankfurt/Main.

Lide, D. R., Frederikse, H. P. R., Eds. (1995), *Handbook of Chemistry and Physics*, CRC Press Inc., New York, Sections 10 (Atomic, Molecular, and Optical Physics) and 1 (Basic Constants, Units, and Conversion Factors).

Mathews, J., Walker, R. L. (1969). *Mathematical Methods of Physics*, Second Edition, Addison-Wesley Publishing Company Inc., Chapter 7, Special Functions.

Matter Diffraction (2003, 2013), University of Liverpool: http://www.matter.org.uk/diffraction/

Morris, A. L., McArthur, M. W., Hutchinson, E. G., Thornton, J. M. (1992). Stereochemical quality of protein structure coordinates. *Proteins* 12(4), 345–364.

Nishikawa, K., Ooi, T. (1986). Radial locations of amino acid residues in a globular protein: Correlation with the sequence. *J. Biochem.* 100, 1043–1047.

Pauling, L. (1960), *The Nature of the Chemical Bond*, Third Edition, Cornell University Press, New York.

Ramachandran, G. N. (1952). The transmission of X-rays at settings near the Bragg angle. *J. Appl. Phys.* 23, 500.

Ramachandran, G. N., Ramakrishnan, C., Sasisekharan, V. (1963). Stereochemistry of polypeptide chain configurations. *J. Mol. Biol.* 7, 95–99.

Rupp, B., Segelke, B. W. (2001). Questions about the structure of the botulinum neurotoxin B light chain in complex with a target peptide. *Nature Struct. Biol.* 8, 643–664.

Sasisekharan, V. (1962). *Stereochemical criteria for polypeptide and protein structures*, Wiley and Sons, Madras.

Sherwood, D. (1976). *Crystals, X-ray and Proteins*, Longman, London.

Takagi, S. (1969). A Dynamical Theory of Diffraction for a Distorted Crystal. *J. Phys. Soc. Japan* 26, 1239–1253.

Vaguine, A. A., Richelle, J., Wodak, S. J. (1999). SFCHECK: a unified set of procedures for evaluating the quality of macromolecular structure-factor data and their agreement with the atomic model. *Acta Cryst.* D55, 191–205.

Verma, A. R., Srivastava, O. N. (1982). *Crystallography for Solid State Physics*, Wiley Eastern Limited, New Delhi, Chapter 13, Morphology and Angular Relationships.

Voet, D., Voet, J. G. (1995), *Biochemistry*, Second Edition, John Willey & Sons, New York, Chapter 7, The Three-Dimensional Structure of Proteins.

Waller, I. (1926). *Ann. Phys.* 79, 261–273.

X-Ray Interactions With Matter (2003, 2013), http://henke.lbl.gov/optical_constants/

Zachariasen, W. S. (1946). *X-Ray Diffraction in Cristals*, Wiley & Sons, New York.

FURTHER READINGS

Authier, S., Lagomarsino, S., Tanner, B. K. (Ed.) (1996). *X – Ray and Neutron Dynamical Diffraction: Theory and Applications*, NATO ASI Ser., Ser. B: Physics.

Baym, G. (1969). *Lectures on Quantum Mechanics*, Benjamin – Cummings, NewYork.

Loudon, R. (1973). *The Quantum Theory of Light*, Oxford University Press, Oxford.

Mandl, F. (1992). *Quantum Mechanics*, John Wiley & Sons Ltd.

Schiff, L. I. (1986). *Quantum Mechanics*, McGraw-Hill, New York.

X-Rays (2003): [http://ceaspub.eas.asu.edu/concrete/xray/index.htm], [http://isites.bio.rpi.edu/bystrc/pub/Xtal/xtal17/sld004.htm], [http://bioinformatics.weizmann.ac.il/iucr-top/comm/cteach/pamphlets/2/2.html (IUCR)], [http://www-cxro.lbl.gov/optical_constants/pert_form.html], [http://www-structure.llnl.gov/xray/101index.html].

INDEX